Urban Planning Law under EU-Influence

Städtebaurecht unter EU-Einfluss

BERLINER SCHRIFTEN ZUR STADT- UND REGIONALPLANUNG

Herausgegeben von Stephan Mitschang

Band 16

PETER LANG
Internationaler Verlag der Wissenschaften

Stephan Mitschang (ed. / Hrsg.)

Urban Planning Law under EU-Influence
Städtebaurecht unter EU-Einfluss

PETER LANG
Frankfurt am Main · Berlin · Bern · Bruxelles · New York · Oxford · Wien

Bibliografische Information der Deutschen Nationalbibliothek
Die Deutsche Nationalbibliothek verzeichnet diese Publikation in der Deutschen Nationalbibliografie; detaillierte bibliografische Daten sind im Internet über http://dnb.d-nb.de abrufbar.

Gedruckt auf alterungsbeständigem, säurefreiem Papier.

ISSN 1861-762X
ISBN 978-3-631-61878-3

www.peterlang.de

Preface

Over the last two decades, Community law has played a major role in shaping national urban building law in general and particularly with regard to the use of ground space. This includes primarily European habitat and species protection law based on the Fauna-Flora-Habitat Directive (96/42/EC) and the Bird Protection Directive (79/409/EEC). Development planning has also been included by the SEA Directive (2001/42/EC) and the Seveso-II Directive (96/82/EC). Similarly, the EC Services Directive (2006/123/EC) and the Renewable Energy Directive (2009/28/EC) can also have implications on local development planning.

Given this background, the current research project produced on behalf of the Federal Institute for Research on Building, Urban Affairs and Spatial Development (BBSR) within the Federal Office for Building and Regional Planning (BBR) as part of general departmental research in 2010 entitled "Town planning law under the influence of the EU Sharing experience among EU member states on inner-city development, particularly with regard to European species and bird protection together with the town planning law aspects of the EC Services Directive (2006/123/EC)" presents the results of a survey conducted as a means of sharing experience with the EU Member States on this issue. It offers suggestions for both the legislator and scientific disciplines with regard to the possible further development of town planning law.

Many thanks to Dr.-Ing. Tim Schwarz for his extensive support.

Berlin, February 2011

Ulrich Battis
Stephan Mitschang
Christian-W. Otto

Vorwort

In den letzten zwei Jahrzehnten wurde das nationale Städtebaurecht, insbesondere soweit es auf die Inanspruchnahme von Bodenflächen ausgerichtet ist, maßgeblich von gemeinschaftsrechtlichen Vorgaben mitgeprägt. Anzuführen sind insoweit zunächst das europäische Habitat- und Artenschutzrecht auf der Grundlage der Fauna-Flora-Habitat-Richtlinie (RL 96/42/EG) und der Vogelschutz-Richtlinie (RL 79/409/EWG). Daneben haben aber auch die Plan-UP-Richtlinie (RL 2001/42/EG) und die Seveso-II-Richtlinie (RL 96/82/EG) Einfluss auf die Bauleitplanung. Gegebenenfalls können sich auch noch aus der EG-Dienstleistungsrichtlinie (RL 2006/123/EG) sowie aus der Erneuerbare-Energien-Richtlinie (RL 2009/28/EG) Auswirkungen auf die örtliche Bauleitplanung ergeben.

Vor diesem Hintergrund enthält die vorliegende Forschungsarbeit, die wir im Auftrag des Bundesinstituts für Bau-, Stadt- und Raumforschung (BBSR) im Bundesamt für Bauwesen und Raumordnung (BBR) im Rahmen der Allgemeinen Ressortforschung im Jahr 2010 erstellt haben, unter dem Titel „Städtebaurecht unter EU-Einfluss – Erfahrungsaustausch mit EU-Mitgliedstaaten zur Innenstadtentwicklung, insbesondere europäischer Arten- und Vogelschutz sowie bauplanungsrechtliche Aspekte der EG-Dienstleistungsrichtlinie (Richtlinie 2006/123/EG)“ die Ergebnisse eines mit EU-Mitgliedsstaaten zu diesen Themenbereichen geführten Erfahrungsaustausches. Sie geben für den Gesetzgeber und die fachwissenschaftlichen Disziplinen Anregungen für eine mögliche Weiterentwicklung des Städtebaurechts.

Wir danken Herrn Dr.-Ing. Tim Schwarz für die vielfältige Unterstützung.

Berlin, im Februar 2011

Ulrich Battis
Stephan Mitschang
Christian-W. Otto

Table of Contents/ Inhaltsverzeichnis

I. English Summary

1. Problem

In recent decades, national planning law and particularly area-related urban planning law has been shaped extensively by Community regulations. These include in particular European habitat and species protection law, the SEA Strategic Environmental Assessment Directive (2001/42/EC) and the SEVESO-II Directive (96/82/EC). Although improbable, it cannot be ruled out from the outset that urban planning law may also be affected by the implementation of the EC Services Directive (2006/123/EC) and the Renewable Energies Directive (2009/28/EC).

Against this specific background, the research project "Sharing experience on European urban planning law" aims to obtain findings as to how the Member States of the European Union react to certain urban planning requirements while taking account of the abovementioned community law guidelines that are valid for all. The project aims to make the various strategies of the Member States comparable by analyzing the requirements made of implementation and application in the Member States. In this way, leeway can be identified while at the same time ascertaining how requirements can be met, particularly when it comes to boosting inner-city development with its manifold social, ecological, economical and legal reciprocal effects, also in line with the species protection requirements, European habitat protection and the SEA Directive. The results can provide important impetus for future amendments of the building code and the land use act.

The research project makes reference to previous conferences of European experts in urban planning law and is based among others on a written survey among the EU Member States.

2. Study methods

The research project is based on a written survey in the EU Member States together with a subsequent expert conference.

The written survey was prepared by developing a questionnaire based on current findings and requirements in the above topics and compiling a corresponding answer sheet for the German legal situation. After translating into English and French, the questionnaires and answer sheets were sent out to previously researched contacts in the corresponding ministries of the EU Member States.

3. Results of the research project

Both the feedback from the written survey in the Member States and the conference in Berlin produced extensive findings with regard to the addressed topics.

Habitat and species protection law

In all Member States, area protection is part of the local planning procedures, where the requirements for area protection are heeded and planning adapted accordingly. The exemption ruling in Article 6 (4) of the FFH Directive has been implemented in all Member States.

Clear differences emerge in the actual organisation of species protection in urban land use planning of the Member States. Various Member States take account of species protection already in the planning stage, some in both the planning and the approval stage and others only in the approval stage. Together with the existing environmental assessment procedures as part of urban land use planning in the Member States, various Member States also proceed with a separate species-protection assessment. In some cases, species protection is integrated in the environmental assessment or in the assessment of implications for the site under the Habitats Directive. The species protection requirements to be considered in the planning or approval stage can pose an obstacle particularly to inner-city development.

SEA Directive (2001/42/EC)

The Strategic Environment Assessment has been integrated as an environment assessment in the urban land use planning of the Member States. Various Member States make exemptions from environment assessment to simplify the planning procedures. A uniform system of exemptions is not apparent, even after considering the written opinions. In spite of differing issues, the Habitats Directive assessment is usually integrated in the (strategic) environment assessment.

SEVESO-II Directive (96/82/EC)

The requirements of the SEVESO-II Directive (96/82/EC) are implemented in the Member States. Conflicts are possible particularly in densely populated countries or regions when it comes to the allocation of use both from advancing building development and also from the settlement of new uses in the vicinity of protectable uses or existing industrial facilities. However, application of the national regulations that are used to implement the SEVESO-II Directive would appear to be fairly uncomplicated.

EC Services Directive (2006/123/EC)

Both the written answers and also the statements of those attending the conference confirm the opinion of the Federal Ministry of Transport, Building and Urban Affairs that the EC Services Directive (2006/123/EC) does not apply to the plans and programmes involved in regional planning and especially to urban land use planning on a local level.

Renewable Energies Directive (2009/28/EC)

Most Member States have still not yet implemented the Renewable Energies Directive (2009/28/EC). However, no direct need for implementation is seen from Article 13 of the Directive.

I. Kurzfassung

1. Problemstellung

Das nationale Planungsrecht, insbesondere das flächenbezogene Städtebaurecht, ist in den vergangenen Jahrzehnten maßgeblich von gemeinschaftsrechtlichen Vorgaben geprägt worden. Zu nennen sind hierbei das europäische Habitat- und Artenschutzrecht auf der Grundlage der Vogelschutz-Richtline (79/409/EWG) und der Fauna-Flora-Habitat-Richtlinie (96/42/EG). Weiter haben die Plan-UP-Richtlinie (2001/42/EG) und die SEVESO-II-Richtlinie (96/82/EG) Einfluss auf die Bauleitplanung. Unwahrscheinlich – aber nicht von vornherein auszuschließen – sind die möglichen Auswirkungen auf das Städtebaurecht, die sich aus der Umsetzung der EG-Dienstleistungsrichtlinie (2006/123/EG) sowie der Erneuerbare-Energien-Richtlinie (2009/28/EG) ergeben können.

Vor diesem thematischen Hintergrund soll das Forschungsprojekt „Erfahrungsaustausch Europäisches Städtebaurecht" Erkenntnisse darüber verschaffen, wie die Mitgliedstaaten der Europäischen Union unter Beachtung der für alle gleichen o. g. gemeinschaftsrechtlichen Vorgaben auf bestimmte städtebauliche Erfordernisse reagieren. Durch die Betrachtung der Anforderungen an die Umsetzung und die Anwendung in den Mitgliedstaaten soll es möglich werden, die unterschiedlichen Strategien der Mitgliedstaaten vergleichbar zu machen. So können Spielräume erkannt und ermittelt werden, wie sich Anforderungen, gerade im Hinblick auf eine Stärkung der Innenstadtentwicklung mit ihren vielfältigen sozialen, ökologischen, ökonomischen und rechtlichen Wechselwirkungen auch unter Beachtung artenschutzrechtlicher Vorgaben, dem Habitatschutz sowie der SUP-Richtlinie bewältigen lassen.

Die Ergebnisse können für zukünftige Novellierungen des Baugesetzbuchs und der Baunutzungsverordnung wichtige Impulse vermitteln.

2. Untersuchungsmethoden

Das Forschungsprojekt basiert auf einer schriftlichen Umfrage in den Mitgliedstaaten der EU sowie einer darauf aufbauenden Experten-Fachtagung.

Zur Vorbereitung der schriftlichen Umfrage wurde auf der Grundlage aktuell bestehender Erkenntnisse und Anforderungen in den oben genannten Themenbereichen ein Fragebogen entwickelt und für die deutsche Rechtslage ein entsprechender Antwortbogen erstellt. Frage- und Antwortbogen wurden nach Anfertigung einer englischen sowie französischen Übersetzung an vorab recherchierte Ansprechpartner in den jeweils zuständigen Ministerien der EU-Mitgliedstaaten versandt.

Auf der Basis der eingehenden Antworten wurden Vertreter einzelner Mitgliedstaaten zu einer zweitägigen Tagung in Berlin am 29. und 30.7.2009 eingeladen. Hierbei wurden wiederum Aspekte der oben genannten Themenbereiche vertiefend erörtert.

3. Ergebnisse des Forschungsprojektes

Sowohl der Rücklauf aus der schriftlichen Befragung in den Mitgliedstaaten als auch die Konferenz in Berlin führten zu umfangreichen Erkenntnissen im Hinblick auf die behandelten Themenfelder.

Habitat- und Artenschutzrecht

Der Gebietsschutz ist in allen Mitgliedstaaten Bestandteil der örtlichen Planung. Die Anforderungen des Gebietsschutzes werden dort beachtet, die Planung entsprechend angepasst. Das Genehmigungsverfahren ist dementsprechend von den vorgelagerten Planungsentscheidungen geprägt. Sofern eine Planung nicht zuvor erfolgt ist, werden die Anforderungen der FFH-RL im Genehmigungsverfahren beachtet. Die Ausnahmeregelung des Art. 6 Abs. 4 FFH-Richtlinie wurde in allen Mitgliedstaaten umgesetzt. Für den Bereich der städtebaulichen Planung spielt die Ausnahmeregelung jedoch nur eine untergeordnete Rolle.

Deutliche Unterschiede ergeben sich im Hinblick auf die Ausgestaltung des Artenschutzes in der Bauleitplanung der Mitgliedstaaten. So wird der Artenschutz in verschiedenen Mitgliedstaaten bereits im Rahmen der Planung berücksichtigt (Frankreich, Irland, Lettland, Österreich, Spanien, Ungarn und Zypern). Gerade in den neueren Mitgliedstaaten findet sich dahingegen der Artenschutz sowohl auf der Planungs- als auch der Zulassungsebene, ohne dass eine klare Akzentuierung oder Abschichtung erkennbar ist. Selten ist dahingegen die Berücksichtigung des Artenschutzes ausschließlich im Rahmen des Genehmigungsverfahrens (Niederlande). Neben den bestehenden Umweltprüfverfahren im Rahmen der Bauleitplanung in den Mitgliedstaaten erfolgt eine eigenständige artenschutzrechtliche Prüfung in Großbritannien, Lettland, Österreich und den Niederlanden. Zum Teil wird der Artenschutz aber auch in andere Umweltprüfverfahren integriert und ist Bestandteil der Umweltprüfung oder der FFH-Verträglichkeitsprüfung.

Insbesondere im Rahmen der Konferenz in Berlin wurde deutlich, dass die Anforderungen des Artenschutzes die innerstädtische Entwicklung in den Mitgliedstaaten hemmen kann. Übereinstimmend waren die Teilnehmer der Konferenz daher der Auffassung, dass für die Bebauung innerhalb der Städte Erleichterungen in Bezug auf den Artenschutz möglich sein sollten.

Plan-UP-Richtlinie (2001/42/EG)

Die Strategische Umweltprüfung ist als Umweltprüfverfahren in die Bauleitplanung der Mitgliedstaaten integriert worden. Verschiedene Mitgliedstaaten kennen Ausnahmen von der Umweltprüfung zur Vereinfachung der Planverfahren. Ein einheitliches System von Ausnahmen ist auch unter Berücksichtigung der schriftlichen Stellungnahmen nicht erkennbar. Teilweise wird mit Ausnahmekatalogen (Frankreich), Grenzwerten (Niederlande – dort steht die Überarbeitung der Kriterien an) oder mit Vorprüfungen (Irland) operiert. Das spanische System ähnelt sehr stark dem Deutschen, in dem bei Vorliegen bereits erfolgter Umweltprüfungen auf die Ergebnisse dieser Prüfungen zurückgegriffen werden kann (Abschichtung). In den neuen Mitgliedstaaten wurden noch keine Ausnahmeregelungen entwickelt. Die Verträglichkeitsprüfung ist, trotz unterschiedlicher Fragestellungen, zumeist in die (Strategische) Umweltprüfung integriert.

EG-Dienstleistungsrichtlinie (2006/123/EG)

Sowohl die schriftlichen Antworten als auch die Aussagen der Konferenzteilnehmer bestätigen die Auffassung des BMVBS, dass die EG-Dienstleistungsrichtlinie (2006/123/EG) keine Anwendung auf die Pläne und Programme der Raumplanung und speziell der örtlichen Bauleitplanung findet.

Erneuerbare-Energien-Richtlinie (2009/28/EG)

Die Umsetzung der Erneuerbare-Energien-Richtlinie (2009/28/EG) steht in den meisten Mitgliedstaaten bislang noch aus. Grundsätzlich kann die räumliche Planung über Standortausweisungen einen Beitrag zur Erhöhung des Anteils Erneuerbarer Energien und damit der Zielerreichung der Richtlinie leisten. Es bestand jedoch Einvernehmen bei den Vertretern der Mitgliedstaaten, dass sich aus Artikel 13 der Richtlinie kein direkter Umsetzungsbedarf für den Bereich des Städtebaus und der Bauleitplanung ergibt. Auch die schriftlich vorliegenden Antworten der anderen Mitgliedstaaten bestätigen dies.

SEVESO-II-Richtlinie (96/82/EG)

Die Anforderungen der SEVESO-II-Richtlinie (96/82/EG) sind in den Mitgliedstaaten umgesetzt. Insbesondere in der Diskussion mit den Vertretern der Mitgliedstaaten im Rahmen der Konferenz wurde deutlich, dass sich gerade in dicht besiedelten Ländern oder Regionen Konflikte im Hinblick auf die Zuordnung von Nutzungen sowohl durch heranrückende Bebauung als auch durch die Ansiedlung von neuen Nutzungen in der Umgebung schutzwürdiger Nutzungen oder bestehender Industrieanlagen ergeben. Die Anwendung der nationalen Regelungen, mit denen die SEVESO-II-Richtlinie umgesetzt wurde, erscheint jedoch weitgehend unproblematisch.

II. Introduction

1. Initial situation and objective of the research project

In recent decades, national planning law and particularly area-related town planning law has been shaped extensively by Community regulations. This is indicated not only by the recent extensive amendments to the Building Code in reaction to transposition of the SEA Directive[1], but also by numerous court decisions putting Community species protection law into effect. The need for national adjustments has been revealed among others by the verdict reached by the European Court of Justice on 10 January 2006 (C-98/03). Transposition of the SEVESO-II Directive[2] is also relevant with regard to the objective of inner development. Nor can it be ruled out from the outset that town planning law and in particular development planning may also be affected by the Services Directive[3] and the Renewable Energies Directive[4] of the European Union when transposed into national law.

Given these procedural and substantive requirements made by Community law, it is questionable which national scope remains for adequately addressing the demands of town planning. On the one hand, there is a need in particular to support inner development, contributing to this objective with further simplifications to accelerate corresponding implementation, particularly by low-cost citation of inner-city land resources. On the other hand, town planning law has to meet environmental and nature protection requirements, particularly in terms of climate and species protection, so that by sounding out the available legal and functional scope, account has to be taken of the requirements for inner development as well as the environment-related requirements of the European directives.

This research project aims to obtain findings as to how the Member States of the European Union react to certain town planning requirements while taking account

1 Directive 2001/42/EC of the European Parliament and of the Council of 27 June 2001 on the assessment of the effects of certain plans and programmes on the environment (Official Journal of the EC No. L 197, p. 30–37).

2 Council Directive 96/82/EC of 9 December 1996 on the control of major-accident hazards involving dangerous substances (Official Journal of the EC No. L 010 p. 13-33), as amended by Directive 2003/105/EC of 16 December 2003 of the European Parliament and of the Council amending Directive 96/82/EC (Official Journal of the EC No. L 345 p. 97-105).

3 Directive 2006/123/EC of the European Parliament and of the Council of 12 December 2006 on services in the internal market (Official Journal of the EC No. L 376, p. 36–68).

4 Directive 2009/28/EC of the European Parliament and of the Council of 23 April 2009 on the promotion of the use of energy from renewable sources and amending and subsequently repealing Directives 2001/77/EC and 2003/30/EC (Official Journal of the EC, No. L 140, p. 16–62).

of the abovementioned Community law guidelines that are valid for all. It should thus be possible to make the various strategies of the Member States comparable. In this way, leeway can be detected for coping in particular with inner-city development with its manifold social, ecological, economical and legal reciprocal effects, also in line with the species protection requirements, habitat protection and the SEA Directive.

When it comes to the requirements of the Services Directive, the findings obtained in terms of how the individual Member States handle the Services Directive provide a basis for discussing whether and with which justification this Directive applies to town planning law or not. Discussion of the possible scope of application to development planning is also the focus of Article 13 of the Renewable Energies Directive.

As a result of the research project, it should be possible to detect and discuss which leeway can be accorded in the regulations.

2. Method

Against the background of the stipulated objective for the research project, a two-stage approach was chosen to deal with the issue.

Firstly, a written questionnaire was developed to look into the implementation and application of certain directives in the Member States (cf. Chapter III). At the same time, research was carried out to ascertain appropriate contacts in the ministries responsible for legislation with regard to construction, planning and environmental issues for the other 26 Member States of the European Union. The questionnaire together with a master response for Germany was sent out to the ascertained contacts in the ministry administrations of the Member States (cf. Chapter III 3.1). Furthermore, research was also carried out to ascertain appropriate contacts in the European Commission who were informed about the objective, course and results of the research project.

The feedback from the contacts provided the basis for the **second part** of the research project. This consisted in holding and evaluating a two-day conference with the chosen contacts and representatives of the European Union, where the already presented written survey answers provided the basis for in-depth discussion of the various individual aspects of the issues (cf. Chapter IV).

II. Einleitung

1. Ausgangslage und Ziel des Forschungsprojektes

Das nationale Planungsrecht, insbesondere das flächenbezogene Städtebaurecht, ist in den vergangenen Jahrzehnten maßgeblich von gemeinschaftsrechtlichen Vorgaben geprägt worden. Davon zeugen nicht nur die letzten umfangreichen Novellen des Baugesetzbuchs in Reaktion auf die Umsetzung der SUP-Richtlinie[5], sondern auch zahlreiche gerichtliche Entscheidungen, in denen dem artenschutzrechtlichen Gemeinschaftsrecht zur Geltung verholfen wurde. Nicht zuletzt das Urteil des EuGH vom 10.1.2006 (C-98/03) hat den nationalen Anpassungsbedarf belegt. Im Hinblick auf das Ziel der Innenentwicklung ist auch die Umsetzung der SEVESO-II-Richtlinie[6] von Bedeutung. Weitere Vorgaben, bei denen ein Einfluss auf das Städtebaurecht und speziell die Bauleitplanung nicht von vornherein auszuschließen sind, stellen die Dienstleistungsrichtlinie[7] sowie die Erneuerbare-Energien-Richtlinie[8] der Europäischen Union dar, die in nationales Recht umzusetzen sind.

Angesichts dieser verfahrens- und materiellrechtlichen Vorgaben des Gemeinschaftsrechts ist fraglich, welche nationalen Spielräume bestehen bleiben, um den Herausforderungen an die Städtebauplanung sachgerecht zu begegnen. Einerseits gilt es, insbesondere die Innenentwicklung zu unterstützen und durch weitere Vereinfachungen zu einer beschleunigten Umsetzung dieser Zielsetzung, insbesondere durch eine kostengünstige Heranziehung der innerstädtischen Flächenressourcen, beizutragen. Andererseits werden an das Städtebaurecht umwelt- und naturschutzbezogene Anforderungen gestellt, insbesondere in Bezug auf den Klima- und den Artenschutz, sodass es unter Auslotung der zur Verfügung stehenden rechtlichen und fachlichen Spielräume darauf ankommt, sowohl den Anforderungen an die Innenentwicklung als auch den umweltbezogenen Anforderungen der europäischen Richtlinien Rechnung zu tragen.

---

5 Richtlinie 2001/42/EG des Europäischen Parlaments und des Rates vom 27.6.2001 über die Prüfung der Umweltauswirkungen bestimmter Pläne und Programme (ABl. EG Nr. L 197, S. 30–37).

6 Richtlinie 96/82/EG des Rates vom 9.12.1996 zur Beherrschung der Gefahren bei schweren Unfällen mit gefährlichen Stoffen (ABl. EG Nr. L 010 S. 13-33), zuletzt geändert durch die Richtlinie 2003/105/EG vom 16.12.2003 des Europäischen Parlaments und des Rates zur Änderung der Richtlinie 96/82/EG (ABl. EG Nr. L 345 S. 97-105).

7 Richtlinie 2006/123/EG des Europäischen Parlaments und des Rates vom 12.12.2006 über Dienstleistungen im Binnenmarkt (ABl. EG Nr. L 376, S. 36–68).

8 Richtlinie 2009/28/EG des Europäischen Parlaments und des Rates vom 23.4.2009 zur Förderung der Nutzung von Energie aus erneuerbaren Quellen und zur Änderung und anschließenden Aufhebung der Richtlinien 2001/77/EG und 2003/30/EG (ABl. EG, Nr. L 140, S. 16–62).

Dieses Forschungsprojekt soll Erkenntnisse darüber verschaffen, wie die Mitgliedstaaten der Europäischen Union unter Beachtung der für alle gleichen o. g. gemeinschaftsrechtlichen Vorgaben auf bestimmte städtebauliche Erfordernisse reagieren. Dadurch soll es möglich werden, die unterschiedlichen Strategien der Mitgliedstaaten vergleichbar zu machen. So können Spielräume erkannt und ermittelt werden, die sich gerade in der Innenstadtentwicklung mit ihren vielfältigen sozialen, ökologischen, ökonomischen und rechtlichen Wechselwirkungen auch unter Beachtung artenschutzrechtlicher Vorgaben, dem Habitatschutz sowie der SUP-Richtlinie bewältigen lassen.

In Bezug auf die Anforderungen der Dienstleistungsrichtlinie ist auf den bisher vorliegenden Erkenntnissen über den Umgang einzelner Mitgliedstaaten mit der Dienstleistungsrichtlinie zu diskutieren, ob und mit welcher Begründung diese Richtlinie im Städtebaurecht Anwendung findet oder nicht. Die Diskussion des möglichen Anwendungsbereichs auf die Bauleitplanung steht auch bei Art. 13 der Erneuerbare-Energien-Richtlinie der EU im Mittelpunkt.

Im Ergebnis des Forschungsprojektes sollen mögliche Regelungsspielräume erkannt und diskutiert werden.

2. Vorgehensweise

Vor dem Hintergrund der oben formulierten Zielsetzung des Forschungsprojektes erfolgte eine zweistufige Vorgehensweise in der Bearbeitung.

Im **ersten Teil** des Forschungsprojektes erfolgte die Entwicklung eines schriftlichen Fragebogens, der die Umsetzung und Anwendung bestimmter Richtlinien in den Mitgliedstaaten abfragte (vgl. Kapitel III). Parallel hierzu wurden Ansprechpartner in den für die Gesetzgebung im Bereich des Bau-, Planungs- und Umweltrecht zuständigen Ministerien der anderen 26 Mitgliedstaaten der Europäischen Union recherchiert. Der Fragenkatalog wurde mit einer Musterantwort für Deutschland an die ermittelten Ansprechpartner in den Ministerialverwaltungen in den Mitgliedstaaten versendet (vgl. Kapitel III 3.1). Darüber hinaus wurden auch Ansprechpartner in der Europäischen Kommission recherchiert, die über die Zielsetzung, den Verlauf und die Ergebnisse des Forschungsprojektes informiert wurden.

Die Rückantworten der Ansprechpartner bilden die Grundlage für den **zweiten Teil** des Forschungsprojektes. Dies bestand aus der Durchführung und Auswertung einer zweitägigen Konferenz, zu der ausgewählte Ansprechpartner sowie Vertreter der Europäischen Kommission eingeladen wurden. Vor dem Hintergrund der bereits vorgelegten schriftlichen Antworten wurden die verschiedenen Einzelaspekte der Themenkomplexe vertiefend diskutiert (vgl. Kapitel IV).

III. Written survey in the EU Member States

1. Development of the survey

The survey covered the following aspects:

- European habitat and species protection law
- SEA Directive (2001/42/EC)
- Services Directive (2006/123/EC)
- Renewable Energies Directive (2009/28/EC)
- SEVESO-II Directive (96/82/EC)

The survey focuses on the issues covered by European habitat and species protection law together with the SEA Directive, each featuring several questions. The individual questions for the individual topical areas are discussed in greater detail below.

1.1 European habitat and species protection law

Community species protection law, in particular the FFH Directive[9] and the Bird Protection Directive[10], also makes demands with regard to the planning and approval procedures for building projects. Here the Directives make no explicit distinction between inner areas (cohesively developed settlements for example in the city centres) and the outer areas (that have no cohesive development). But it is obvious that the presence of a beaver (Castor fiber)[11] at Alexanderplatz in Berlin or a dormouse (Glis glis)[12] at London's Piccadilly Circus has to be evaluated differently than in virtually uninhabited areas, such as the Masurian Lake District. Similarly, in principle it can be presumed that greater development of the inner areas helps to protect the outer areas. Furthermore, the EC Directives feature legal definitions and exemptions that are open to interpretation and which are relevant for use in

---

9 Council Directive 92/43/EC of 21 May 1992 on the conservation of natural habitats and of wild fauna and flora (Official Journal of the EC No. L 206, p. 7), as amended by Council Directive 2006/105/EC of 20 November 2006 (Official Journal of the EC No. L 363, p. 368).

10 Council Directive 79/409/EEC of 2 April 1979 on the conservation of wild birds (Official Journal of the EC No. L 103, p. 1), as amended by Directive 2008/102/EC of the European Parliament and of the Council of 19 November 2008 (Official Journal of the EC No. L 323, p. 31).

11 cf. Annex IV lit. a) FFH Directive.

12 cf. Annex IV lit. a) FFH Directive.

development planning. After all, it is possible for a species to have its natural habitat in outer areas and even occur in larger quantities here while still being deemed worthy of protection when found in inner areas. And so the question arises as to how individual countries evaluate the relationship between conservation and development planning, which need for reform is seen and how the legal situation is assessed with regard to the following questions:

What is the relationship between national nature conservation law and the building planning law? Are nature conservation law requirements taken into account in development planning or not dealt with until a subsequent approval procedure takes place?

Within the scope of the assessment of the implications of a plan or project for a site[13]*, how has the exception provided by Article 6 Paragraph 4 of the FFH Directive been transposed into national law in the field of building planning and how is it applied? Is there any distinction regarding whether a project is built within or outside an area which is already continuously built up?*

Apart from the conservation area related assessment of implications for the site pursuant to Article 6 Paragraph 3 of the FFH Directive, is an assessment under species protection law also always performed according to Article 16 of the FFH Directive?

Is there a need to exempt such construction projects from prohibitions under species protection law which are erected in areas which have already been developed? What role could it play if the protected species had their natural habitat outside the development area and also occurred in greater number there?

How is the element of "intention" in Article 12 Paragraph 1 Letter a, b and c of the FFH Directive interpreted for construction projects in the light of the legal decisions taken by the European Court of Justice?

1.2 SEA Directive (2001/42/EC)

In Germany, the SEA Directive has been implemented by integrating a general environment assessment in the procedure for preparing land use and construction plans. Limitations of the general environment assessment obligation are derived from Articles 3 Paragraph 3 to 5 of the SEA Directive and refer to amending and supplementing development plans, as well as preparing status-oriented construction plans and plans for inner development.

This led to the following questions with regard to implementation and simplification of use:

Which levels and instruments of development planning exist and with respect to this, how was the SEA Directive transposed into national law?

---

13 Article 6 Paragraph 3 FFH Directive.

How was the SEA Directive transposed into national law?

Are options, especially the option of insignificance with the context of preparing, amending and supplementing the development plans, used to accelerate and simplify procedures for particular cases?

What is the relationship between the environmental assessment and the European species protection in implementation and practical application?

1.3 EC Services Directive (2006/123/EC)

Directive 2006/123/EC of the European Parliament and the Council of 12 December 2006 on services in the internal market (Services Directive) aims to boost the economic reform process in the EU. The Directive aims to eliminate a large number of internal market barriers and create a legal area that guarantees the freedom of establishment and the freedom to provide services between the Member States. To this end, the Member States have to review their legal and administration regulations and ascertain whether these fulfil the stipulations made in the Directive, subsequently amending the regulations accordingly. Germany has examined whether and to which extent the planning regulations in the Building Code fall within the scope of the Services Directive.

The following question arose with regard to possible applicability of the Services Directive in the sphere of building and planning law:

Is it deemed necessary to transpose the Services Directive (Directive 2006/123/EC) for the area of national urban planning law and development planning?

1.4 Renewable Energy Directive (2009/28/EC)

For the first time, the Renewable Energies Directive contains binding targets for all Member States regarding the share of renewable energies and which measures should be used to achieve the targets. Accordingly, Article 13 Paragraph 4 Renewable Energies Directive states that by 31 December 2014, Member States must, in their building regulations and codes or by other means with equivalent effect, require the use of minimum levels of energy from renewable sources in new buildings and existing buildings that are subject to major renovation.

This led to the following question with regard to possible impacts on building and planning law:

Is it considered necessary to transpose Article 13 of the Renewable Energy Directive (2009/28/EC) for the area of national urban planning law and development planning?

1.5 SEVESO-II Directive (96/82/EC)

The SEVESO-II Directive obliges the Member States of the European Union to prevent major accidents with dangerous substances and to limit their consequences for man and the environment with a view to ensuring high levels of protection throughout the Community in a consistent and effective manner. Accordingly, Article 12 of the Directive expects that Member States shall ensure that preventive consideration already has to be given to appropriate distances in regional and land-use planning. In Germany, Article 12 Paragraph 1 Subparagraph 2 of the SEVESO-II Directive has been transposed into national law by the provision of Section 50 of the Federal Immissions Control Act[14].

Given the national transposition into the Federal Immissions Control Act, the following question was developed regarding the transposition of Article 12 Paragraph 1 Subparagraph 2 of the SEVESO-II Directive:

> *How has Article 12 Paragraph 1 Subparagraph 2 of Directive 96/82/EC of 9 December 1996 the control of major-accident hazards involving dangerous substances (Official Journal of the EC No. L 010 p. 13; "SEVESO-II Directive) been transposed into national law?*

2. Research to ascertain the contacts in the Member States

While the questionnaire was being developed, at the same time research was carried out to ascertain appropriate contacts in the ministries responsible for legislation with regard to construction, planning and environmental issues for the Member States of the European Union in order to send out the questionnaires and also with regard to the subsequent conference. The target group consisted of experts working in the construction planning law departments of the corresponding ministries involved in drawing up the legislation.

The questionnaire was translated into English and French and a master answer sheet was produced for Germany (cf. Chapter 3.1). These documents were sent out to the ascertained contacts on 15 February 2010, asking for the individual Member States to complete and return the questionnaires by 12 April 2010.

Furthermore, research was carried out to ascertain appropriate contacts for the European Commission in the Directorate General for the Environment and in the Directorate General for Regional Policy, to inform them about the contents and course of the research project.

---

14 Act to prevent harmful effects on the environment from air pollution, noise levels, vibrations and similar (Federal Immissions Control Act – BImSchG) as amended in the proclamation of 26 September 2002 (German Federal Gazette I p. 3830), as amended by Article 2 Legal Repeal Act Environment of 11 August 2009 (German Federal Gazette I p. 2723).

3. Feedback from the Member States

In the framework of the written survey, responses were received from altogether 13 Member States of the European Union: Estonia, France, the United Kingdom, Ireland, Latvia, the Netherlands, Austria (Vienna), Poland, Portugal, Slovakia, Spain, Hungary and Cyprus. The responses are provided below, together with the master answer for Germany.

III. Schriftliche Umfrage in den EU-Mitgliedstaaten

1. Entwicklung des Fragebogens

Folgende Themenbereiche wurden durch den Fragebogen erfasst:

- Europäisches Habitat- und Artenschutzrecht
- Plan-UP-Richtlinie (2001/42/EG)
- Dienstleistungsrichtlinie (2006/123/EG)
- Erneuerbare-Energien-Richtlinie (2009/28/EG)
- SEVESO-II-Richtlinie (96/82/EG)

Die Schwerpunkte des Fragebogens bilden die Themenkomplexe zum Europäischen Habitat- und Artenschutzrecht sowie der Plan-UP-Richtlinie, die jeweils mehrere Fragen umfassten. Im Folgenden werden die einzelnen Fragestellungen näher erläutert, die zu den genannten Themenbereichen erarbeitet wurden.

1.1 Europäisches Habitat- und Artenschutzrecht

Das gemeinschaftliche Artenschutzrecht, insbesondere die FFH-Richtlinie[15] und die Vogelschutzrichtlinie[16], stellt auch an die Planung und Genehmigung von baulichen Vorhaben Anforderungen. Die Richtlinien unterscheiden dabei nicht ausdrücklich zwischen Innenbereich (dem im Zusammenhang bebauten, etwa innerstädtischen, Siedlungsbereich) und dem Außenbereich (dem nicht im Zusammenhang bebauten Bereich). Dass aber die Existenz eines Bibers (Castor fiber)[17] am Berliner Alexanderplatz oder eines Siebenschläfers (Glis glis)[18] am Londoner Piccadilly Circus anders zu bewerten ist, als auf unbesiedelten Bereichen, wie z. B. der masurischen Seenplatte, liegt nahe. Ebenso ist grundsätzlich davon auszugehen, dass die Stärkung der Innenentwicklung zur Schonung des Außenbereichs beiträgt. Zudem sehen die EG-Richtlinien selbst auslegungsfähige Rechtsbegriffe und Aus-

---

15 Richtlinie 92/43/EWG des Rates vom 21.5.1992 zur Erhaltung der natürlichen Lebensräume sowie der wildlebenden Tiere und Pflanzen (Abl. EG Nr. L 206, S. 7), zuletzt geändert durch Richtlinie 2006/105/EG des Rates vom 20.11.2006 (ABl. EG Nr. L 363, S. 368).

16 Richtlinie 79/409/EWG des Rates vom 2.4.1979 über die Erhaltung der wildlebenden Vogelarten (Abl. EG Nr. L 103, S. 1), zuletzt geändert durch Richtlinie 2008/102/EG des Europäischen Parlaments und des Rates vom 19.11.2008 (ABl. EG Nr. L 323, S. 31).

17 Vgl. Anhang IV lit. a) FFH-Richtlinie.

18 Vgl. Anhang IV lit. a) FFH-Richtlinie.

nahmeregelungen vor, deren Nutzung in der Bauleitplanung von Interesse ist. Schließlich kann eine Population ihren natürlichen Lebensraum im Außenbereich haben und dort ggf. auch in größerer Zahl auftreten, gleichwohl diese Arten aber - bei Auftreten im Innenbereich - als schützenswert gelten. Von daher stellt sich die Frage, wie das Verhältnis von Naturschutz und Bauleitplanung in den einzelnen Ländern beurteilt wird, welcher Reformbedarf gesehen wird und wie die Rechtslage bei folgenden Fragen bewertet wird:

In welchem Verhältnis stehen das nationale Naturschutzrecht und das Bauplanungsrecht? Werden die Anforderungen des Naturschutzrechts bereits in der Bauleitplanung bewältigt oder erst im nachfolgenden Genehmigungsverfahren?

Wie ist im Rahmen der Verträglichkeitsprüfung[19] *die Ausnahmeregelung des Art. 6 Abs. 4 FFH-RL im Bereich des Bauplanungsrechts in nationales Recht umgesetzt worden und wie wird diese angewandt? Gibt es eine Differenzierung in Bezug darauf, dass ein Vorhaben innerhalb oder außerhalb eines im Zusammenhang bebauten Gebietes errichtet wird?*

Wird neben der schutzgebietsbezogenen Verträglichkeitsprüfung des Art. 6 Abs. 3 FFH-Richtlinie bei baulichen Vorhaben auch stets eine artenschutzrechtliche Prüfung nach Art. 16 FFH-Richtlinie durchgeführt?

Wird ein Bedarf dafür gesehen, solche Bauvorhaben von artenschutzrechtlichen Verboten freizustellen, die in Gebieten errichtet werden, die bereits im Zusammenhang bebaut sind? Welche Rolle könnte es hierbei haben, wenn die geschützte Art ihre natürliche Verbreitung außerhalb des Siedlungsbereichs hat und dort auch in größerer Zahl vorkommt?

Wie wird das Tatbestandsmerkmal der „Absicht" in Artikel 12 Absatz 1 Buchstabe a, b und c FFH-Richtlinie bei Bauvorhaben im Lichte der Rechtsprechung des Europäischen Gerichtshofs verstanden?

1.2 Plan-UP-Richtlinie (2001/42/EG)

Die Umsetzung der Plan-UP-Richtlinie erfolgte in der Bundesrepublik Deutschland durch Integration einer generellen Umweltprüfung in das Verfahren zur Aufstellung von Flächennutzungs- und Bebauungsplänen. Einschränkungen der generellen Umweltprüfungspflicht gehen auf Art. 3 Abs. 3 bis 5 der Plan-UP-Richtlinie zurück und betreffen die Änderung und Ergänzung von Bauleitplänen sowie die Aufstellung von bestandsorientierten Bebauungsplänen und solchen zur Innenentwicklung.

Folgende Fragestellungen im Hinblick auf die Umsetzung sowie die Vereinfachung der Anwendung wurden hieraus entwickelt:

---

19 Art. 6 Abs. 3 FFH-Richtlinie.

Welche Ebenen und Instrumente städtebaulicher Planungen bestehen und wie erfolgte diesbezüglich die Umsetzung der Plan-UP-Richtlinie in nationales Recht?

Wie wurde die Plan-UP-Richtlinie in das nationale Recht umgesetzt?

Werden für bestimmte Fälle, insbesondere der Geringfügigkeit im Rahmen der Aufstellung, Änderung und Ergänzung von Bauleitplänen, Möglichkeiten zur Beschleunigung sowie Verfahrenserleichterungen genutzt?

In welchem Verhältnis stehen Umweltprüfung und Europäischer Artenschutz in der Umsetzung und der praktischen Anwendung zueinander?

1.3 EG-Dienstleistungsrichtlinie (2006/123/EG)

Mit der Richtlinie 2006/123/EG des Europäischen Parlaments und des Rates vom 12. Dezember 2006 über Dienstleistungen im Binnenmarkt (Dienstleistungsrichtlinie – DL-RL) soll der wirtschaftliche Reformprozess in der EU vorangetrieben werden. Die Richtlinie zielt darauf ab, eine Vielzahl von Binnenmarktschranken zu beseitigen und einen Rechtsrahmen zu schaffen, der die Niederlassungsfreiheit und den freien Dienstleistungsverkehr zwischen den Mitgliedstaaten garantiert. Die Mitgliedstaaten haben zu diesem Zweck ihre Rechts- und Verwaltungsvorschriften zu prüfen und festzustellen, ob diese die in der Richtlinie genannten Vorgaben erfüllen, und die Vorschriften ggf. anzupassen. Deutschland hat untersucht, ob und in welchem Umfang die planungsrechtlichen Regelungen des Baugesetzbuchs in den Anwendungsbereich der Dienstleistungsrichtlinie fallen.

Im Hinblick auf eine mögliche Anwendbarkeit der Dienstleistungsrichtlinie auf den Bereich des Bau- und Planungsrechts ergab sich folgende Fragestellung:

Wird ein Umsetzungsbedarf der Dienstleistungsrichtlinie (Richtlinie 2006/123/EG) für den Bereich des nationalen Städtebaurechts und der Bauleitplanung gesehen?

1.4 Erneuerbare-Energien-Richtlinie (2009/28/EG)

Die Erneuerbare-Energien-Richtlinie enthält erstmals verbindliche Ziele für alle Mitgliedstaaten zum Anteil Erneuerbarer Energien und durch welche Maßnahmen diese Ziele erreicht werden sollen. In diesem Sinne regelt Art. 13 Abs. 4 EE-RL, dass die Mitgliedstaaten verpflichtet sind, in Bauvorschriften oder auf andere Weise mit vergleichbarem Ergebnis – soweit angemessen – zu verlangen, dass bis spätestens zum 31.12.2014 in neuen Gebäuden und in bestehenden Gebäuden, an denen erhebliche Renovierungsarbeiten vorgenommen werden, ein Mindestmaß an Erneuerbaren Energien genutzt wird.

Im Hinblick auf mögliche Auswirkungen auf das Bau- und Planungsrecht wurde hieraus folgende Fragestellung formuliert:

Wird ein Umsetzungsbedarf hinsichtlich des Artikels 13 Erneuerbare-Energien-Richtlinie (2009/28/EG) für den Bereich des nationalen Städtebaurechts und der Bauleitplanung gesehen?

1.5 SEVESO-II-Richtlinie (96/82/EG)

Die SEVESO-II-Richtlinie verpflichtet die Mitgliedstaaten der Europäischen Union zur Verhütung schwerer Unfälle mit gefährlichen Stoffen und zur Begrenzung der Unfallfolgen für Mensch und Umwelt, um auf abgestimmte und wirksame Weise in der ganzen Gemeinschaft ein hohes Schutzniveau zu gewährleisten. Dementsprechend fordert Art. 12 der Richtlinie von den Mitgliedstaaten, dass bereits in der Raum- und Flächenplanung angemessene Abstände präventiv beachtet werden müssen. In Deutschland ist Artikel 12 Absatz 1 Unterabsatz 2 der Seveso-II-Richtlinie durch die Regelung des § 50 BImSchG[20] umgesetzt worden.

Vor dem Hintergrund der nationalen Umsetzung im Bundes-Immissionsschutzgesetz wurde folgende Fragestellung zur Umsetzung des Artikels 12 Abs. 1 Unterabs. 2 der SEVESO-II-Richtlinie in den Mitgliedstaaten entwickelt:

Wie ist Artikel 12 Absatz 1 Unterabsatz 2 der Richtlinie 96/82/EG vom 09.12.1996 zur Beherrschung der Gefahren bei Unfällen mit gefährlichen Stoffen (ABl. L 10 vom 14.01.1997, S. 13; „Seveso-II-Richtlinie“) in nationales Recht umgesetzt worden?

2. Recherche der Ansprechpartner in den Mitgliedstaaten

Zum Versand des Fragebogens sowie im Hinblick auf die Konferenz im zweiten Teil des Forschungsprojektes erfolgte parallel zu der Entwicklung des Fragebogens eine Recherche von Ansprechpartnern in den für die Bereiche Bau-, Planungs- und Umweltrecht zuständigen Ministerien in den Mitgliedstaaten der Europäischen Union. Die Zielgruppe waren hierbei Experten, die im Bereich Bauplanungsrecht in den jeweiligen Ministerien tätig sind und im Rahmen der Gesetzgebung inhaltlich mitwirken.

Von dem Fragebogen wurde zusammen mit den Musterantworten für Deutschland (vgl. Kap. 0) eine englische sowie eine französische Übersetzung angefertigt. Diese

20 Gesetz zum Schutz vor schädlichen Umwelteinwirkungen durch Luftverunreinigungen, Geräusche, Erschütterungen und ähnliche Vorgänge (Bundes-Immissionsschutzgesetz – BImSchG), in der Fassung der Bekanntmachung vom 26.9.2002 (BGBl. I S. 3830), zuletzt geändert durch Art. 2 RechtsbereinigungsG Umwelt v. 11.8.2009 (BGBl. I S. 2723).

wurden am 15. Februar 2010 an die ermittelten Ansprechpartner gesendet, mit der Bitte, den Fragebogen für den jeweiligen Mitgliedstaat bis zum 12. April 2010 zu beantworten und zurückzusenden.

Darüber hinaus wurden auch Ansprechpartner der Europäischen Kommission in der Generaldirektion Umwelt sowie der Generaldirektion Regionalpolitik recherchiert, um diese über die Inhalte und den Verlauf des Forschungsprojektes zu informieren.

3. Rücklauf aus den Mitgliedstaaten

Im Rahmen der schriftlichen Befragung erfolgte eine Rückantwort aus insgesamt 13 Mitgliedstaaten der Europäischen Union: Estland, Frankreich, Großbritannien, Irland, Lettland, Niederlande, Österreich (Wien), Polen, Portugal, der Slowakei, Spanien, Ungarn und Zypern. Zusammen mit der Musterantwort für Deutschland werden diese im Folgenden wiedergegeben.

3.1 Germany

3.1.1 Question 1: Systematic preliminary question

What is the relationship between binding land use plans and the approval of individual construction projects?

Under German building law, the binding land use plan which defines the use of land plots for building and other usages, and the approval of individual construction projects have different functions. The planning permission/ building permit is granted if the construction project fulfils the legal requirements. If a binding land use plan exists, the construction project must also be measured against the requirements of the binding land use plan.

3.1.2 Question 2: European Habitat and Protection of Species Law

a) *What is the relationship between national nature conservation law and the building planning law? Are nature conservation law requirements taken into account in development planning or not dealt with until a subsequent approval procedure takes place?*

Under German nature conservation law, which is primarily defined in the Federal Nature Conservation Act (Bundesnaturschutzgesetz), a distinction is made between

- provisions with national significance only concerning interventions in nature and the landscape,
- the provisions on the "Natura 2000" network and
- the provisions concerning the protection of wild animal and plant species, their habitats and biotopes under national and European law.

As a rule, nature conservation related requirements are checked when individual construction projects are approved. This principle is broken in the following cases:

- If binding land use plans have been prepared, assessments regarding the expected interventions in nature and the landscape take place at the binding land use planning level for the construction projects to be approved within the scope of development planning. With the exception of the circumstances of the intervention, the provisions for this are not laid down in the Federal Nature Conservation Act but in the Federal Building Code. No assessments take place at approval level.
- Equally, with respect to the protection and conservation objectives of the FFH and bird conservation areas, the compatibility of the planned projects is already assessed during the binding land use plan

procedure and is then no longer assessed at approval level. Here the Federal Building Code refers to the requirements stipulated in the Federal Nature Conservation Act.

The following applies to species protection: The prohibitions on intervention and impairment for especially protected animal and plant species (cf. Article 12 and 16 FFH Directive, Article 5 Bird Conservation Directive) prohibit specific acts. Not the provisions of the binding land use plan but their realisation and implementation therefore represent the prohibited intervention. However, if the realisation of the binding land use plan is faced with prohibitions under species protection law, this would also have repercussions on the binding land use plan, because a plan which cannot be enforced is void and invalid due to lack of necessity. Therefore, during the binding land use plan procedure it is necessary to check whether the species protection law exceptions and exemptions which may be required at the subsequent level of project approval can be granted in principle or not. However, the binding land use plan itself does not determine the nature conservation permissibility nor are required exceptions and exemptions bindingly promised; this is left to the project approval level.

b) Within the scope of the assessment of the implications of a plan or project for a site, how has the exception provided by Article 6 Paragraph 4 of the FFH Directive been transposed into national law in the field of building planning and how is it applied?

Is there any distinction regarding whether a project is built within or outside an area which is already continuously built up?

For the FFH assessment of implications for the site, the Federal Building Code refers to the Federal Nature Conservation Act, where the requirements of the exception provided under Article 6 Paragraph 4 of the FFH Directive have been transposed into national law. The corresponding provisions must be taken into account in the preparation of development plans.

There is no distinction regarding whether a project is built within or outside an area which is already continuously built up.

c) Apart from the conservation area related assessment of implications for the site pursuant to Article 6 Paragraph 3 of the FFH Directive, is an assessment under species protection law also performed according to Article 12 of the FFH Directive?

Species protection law assessments are performed in addition to the assessment of implications for the site if there are indications that the prohibitions on interventions for protected species could be violated. In particular, reason for this could exist if, in the context of a construction project, a binding land use plan is implemented which includes breeding sites or resting places of protected species.

Thesis:

To avoid duplicate assessments, it would be desirable, to grant the assessment of implications for the site priority over prohibitions on interventions under species protection law in those cases in which the prohibitions under species protection law target the same species as the conservation objectives or the protective purpose of the relevant FFH area.

d) Is there a need to exempt such construction projects from prohibitions under species protection law which are erected in areas which have already been developed? What role could it play if the protected species had their natural habitat outside the development area and also occurred in greater number there?

In Germany, there are some considerations to limit the scope of prohibitions under species protection law for construction projects in already continuously built-up areas provided the protected species concerned have their natural habitat outside the already continuously built-up areas and also actually occur in greater numbers there and the construction projects are not subject to an environmental impact assessment (EIA) obligation or an assessment of implications for the site. These considerations are based on the expectation that if it were made easier to build within existing continuously built-up areas, the inner development of towns and cities would be facilitated and the use of new areas in previously undeveloped areas would be reduced. The prohibitions under species protection law of Article 12 of the FFH Directive require measures to introduce a strict protection system for the protected species in their **natural habitats**. There is also a reference to natural habitats in Article 1 letter i and Article 16 Paragraph 1 of the FFH Directive. If the strict protection system with the prohibitions listed in Article 12 Paragraph 1 FFH Directive was only effective in the natural habitats, it would be generally appropriate to waive the strict prohibitions of the FFH Directive relating to species protection - and not only by way of exceptional decisions, if the populations of the species concerned remain in a favourable conservation condition, which can be assumed if the populations as such are retained as a viable element within their natural range. In this way accesses and interventions within already continuously built-up areas can be approved provided this area is not the natural habitat of the species concerned without any need to fulfill the prerequisites of Article 16 FFH Directive. Due to the population-related approach which provides the basis for the conservation condition, it is permissible for individual settlement areas of a species to be lost or for individuals to be killed by a project. A worsening of the conservation condition of the populations is not assumed until the number of individuals forming the populations is substantially reduced. On the other hand, the conservation condition of a species is unaffected by an intervention if the populations per se are retained as a viable element in their natural range which can extend beyond the plan area.

e) How is the element of "intention" in Article 12 Paragraph 1 Letter a, b and c of the FFH Directive interpreted for construction projects in the light of the legal decisions taken by the European Court of Justice?

On the basis of the Caretta judgment of the European Court of Justice dated 30.01.2002 – C 103/00 - and the judgment on trap hunting dated 20.10.2005 – C 6/04 - intention is understood to be not only the deliberate but also the accepted killing or destruction, i. e. the result of the action is not wanted but is known and nevertheless the action resulting in disruption or death is undertaken. According to the judgements of the Federal Administrative Court an "intentional" realisation of facts is not deemed not to apply even if a deterioration as defined by Article 12 Paragraph 1 Letter a, b and c of the FFH Directive proves to be an inevitable consequence of a legal action under German law (e.g. making use of a legally valid building permit/ planning permission).

3.1.3 Question 3: SEA Directive (2001/42/EC)

a) Which levels and instruments of development planning exist and with respect to this, how was the SEA Directive transposed into national law?

The Building Code provides local authorities the instrument of two-stage development planning to control urban development. The planning instruments are the preparatory land use plan and the binding land use plan. The preparatory land use plan is the preparatory development plan in which the future urban development of the entire municipality's territory is shown. The preparatory land use plan is not a legal standard.

The binding land use plan defines the planning law for certain areas within the local authority in a legally binding way. The binding land use plan is a statute and is also approved and issued by the local authority.

b) How was the SEA Directive transposed into national law?

The SEA Directive has been transposed into national building planning law (Building Code) in that an environmental assessment must be performed in all procedures for the preparation of preparatory land use or binding land use plans. The procedural requirements under EU legislation (such as scoping, environmental report, participation of the public and authorities, consideration in the decision, announcement of the decision and the summary explanation, monitoring) have been integrated in the existing procedural steps for the preparation of a development plan.

The environmental assessment in the binding land use plan procedure fulfils both all requirements of the SEA Directive and the requirements of an environmental impact assessment (EIA). This applies regardless of whether the binding land use plan is subject to an actual EIA obligation or not.

At national level the SEA Directive has also been transposed by corresponding amendments to the EIA law, which contains basic provisions for the environmental assessment. Further provisions exist in environmental law, for example for regional development plans. The federal states have also created provisions for transposing the SEA Directive.

c) Are options, especially the option of insignificance within the context of preparing, amending and supplementing the development plans, used to accelerate and simplify procedures for particular cases?

It is possible to make use of procedural simplifications and accelerations (e. g. dispensing with two-stage public participation, participation of the affected public only, omission of the environmental assessment) in the preparation of development plans, which only slightly change existing development plans, in the preparation of binding land use plans which only cover continuously built-up areas and do not significantly change the criterion of permissibility or in the preparation of binding land use plans which define the use of small areas with the aim of making areas reusable, of subsequently increasing density or other inner development measures.

In all cases, it is a prerequisite that the procedure only justifies the permissibility of projects which are not subject to any EIA obligation and which do not put at risk areas of European significance (FFH and bird conservation areas).

d) What is the relationship between the environmental audit and the FFH assessment of implications for the site?

The FFH assessment of implications for the site as defined by Article 6 Paragraph 3 of the FFH Directive, like the environmental audit and the EIA, is integrated as a dependent element in the development planning and approval procedure. The environmental audit can adopt the function of a supporting procedure, especially for the definition of scope and its evaluation to be undertaken within the context of the FFH assessment of implications. Implementation of the FFH assessment of implications for the site is documented in an independent chapter of the environmental report.

3.1.4 Question 4: EU Services Directive (2006/123/EC)

Is it deemed necessary to transpose the EC Services Directive (Directive 2006/123/EC) for the area of national urban planning law and development planning?

The Services Directive coordinates national legal and administrative regulations which specifically define the freedom to provide services and the freedom of establishment. Legal and administrative regulations which do not define or concern the service activity per se but the behaviour of persons in general, i. e. including persons who do not work as service providers, do not specifically refer to services. The primary law enabling principles and the content of the Services Directive itself suggest that the said directive is not to be applied to such general provisions. This results from consideration 9 of the Directive:

> *"This Directive applies only to requirements which affect the access to, or the exercise of, a service activity. Therefore, it does not apply to requirements, such as road traffic rules, rules concerning the development or use of land, town and country planning, building standards as well as administrative penalties imposed for non-compliance with such rules which do not specifically regulate or specifically affect the service activity but have to be respected by providers in the course of carrying out their economic activity in the same way as by individuals acting in their private capacity."*

Urban planning provisions concerning the planning law permissibility of construction projects do not define a service activity per se but instead, as abstract general provisions, are aimed at any person if and provided they use a plot of land for building or other purposes. These provisions lack a special service relationship. They are general provisions in the sense of Consideration 9 of the Services Directive. Therefore there is no need for transposition as regards urban planning provisions which concern the planning law permissibility of projects.

3.1.5 Question 5: Renewable Energy Directive (2009/28/EC)

Is it considered necessary to transpose Article 13 of the Renewable Energy Directive (Directive 2009/28/EC) for the area of national urban planning law and development planning?

With Directive 2009/28/EC of the European Parliament and the Council on the promotion of the use of energy from renewable sources and amending and subsequently repealing Directives 2001/77/EC and 2003/30/EC (RE Directive), the law of renewable energy in Europe is redefined and especially the promotion of all energy sectors is combined in a single legal act. In addition, the Directive contains binding targets for a further development of renewable energy. In this sense, according to Article 3 of the RE Directive, by the target year 2020, energy from renewable sources is to be developed to account for a share of at least 20 % of the Community's gross final consumption of energy. To achieve this target, each member state is obliged to take measures to encourage and promote energy from renewable sources. To this end, each state has been assigned a development target. For Germany, this provides for the development of energy from renewable sources until it accounts for 18 % of Germany's gross final consumption of energy by this target date (compared to 5.8 % in 2005). In addition, each member state must guarantee that their share of renewable energy in the transport sector increases to at least 10 % by the target year 2020.

Article 13 Paragraph 1 of the Directive concerns the administrative procedures, legal regulations and codes of the member states. It should be considered whether this results in requirements for urban planning law too. At the moment no further transposition need is considered to exist for development planning law; deadline provisions and national approvals should only be introduced for specific urban planning authorisations. It would be interesting to know the status of the considerations in other member states.

The Directive provision concerned states:

"Article 13

Administrative procedures, legal regulations and standards

(1) Member States shall ensure that any national rules concerning the authorisation, certification and licensing procedures that are applied to plants and associated transmission and distribution network infrastructures for the production of electricity, heating or cooling from renewable energy sources, and to the process of transformation of biomass into biofuels or other energy products, are proportionate and necessary. Member States shall, in particular, take the appropriate steps to ensure that:

a) subject to differences between Member States in their administrative structures and organisation, the respective responsibilities of national, regional and local administrative bodies for authorisation, certification and licensing procedures including spatial planning are clearly coordinated and defined, with transparent timetables for determining planning and building applications;

b) comprehensive information on the processing of authorisation, certification and licensing applications for renewable energy installations and on available assistance to applicants are made available at the appropriate level;

c) administrative procedures are streamlined and expedited at the appropriate administrative level;

d) rules governing authorisation, certification and licensing are objective, transparent, proportionate, do not discriminate between applicants and take fully into account the particularities of individual renewable energy technologies;

e) administrative charges paid by consumers, planners, architects, builders and equipment and system installers and suppliers are transparent and cost-related; and

f) simplified and less burdensome authorisation procedures, including through simple notification if allowed by the applicable regulatory framework, are established for smaller projects and for decentralised devices for producing energy from renewable sources, where appropriate."

3.1 Deutschland

3.1.1 Frage 1: Systematische Vorfrage

In welchem Verhältnis stehen Bebauungspläne und die Genehmigung einzelner Bauvorhaben?

Im deutschen Baurecht haben der Bebauungsplan, der die bauliche und sonstige Nutzung von Grundstücken regelt, und die Genehmigung einzelner Bauvorhaben unterschiedliche Funktionen. Die Baugenehmigung wird erteilt, wenn das Bauvorhaben den rechtlichen Vorgaben entspricht. Wenn ein Bebauungsplan besteht, ist das Bauvorhaben auch an den Vorgaben des Bebauungsplans zu messen.

3.1.2 Frage 2: Europäisches Habitat- und Artenschutzrecht

a) In welchem Verhältnis stehen das nationale Naturschutzrecht und das Bauplanungsrecht? Werden die Anforderungen des Naturschutzrechts bereits in der Bauleitplanung oder erst in einem nachfolgenden Genehmigungsverfahren bewältigt?

Im deutschen Naturschutzrecht, das primär im Bundesnaturschutzgesetz (BNatSchG) geregelt ist, wird zwischen

- den nur national bedeutsamen Bestimmungen über Eingriffe in Natur und Landschaft,
- den Bestimmungen über das Netz „Natura 2000" und
- den Regelungen über den Schutz der national und gemeinschaftsrechtlich geschützten wild lebenden Tier- und Pflanzenarten, ihrer Lebensstätten und Biotopen

unterschieden.

Grundsätzlich werden bei der Zulassung einzelner Bauvorhaben naturschutzbezogene Anforderungen geprüft. Dieser Grundsatz wird in folgenden Fällen durchbrochen:

- Soweit Bebauungspläne aufgestellt werden, findet bereits auf der Ebene der Bebauungsplanung die Prüfung der zu erwartenden Eingriffe in Natur und Landschaft für die im Geltungsbereich des Bebauungsplans zuzulassenden Bauvorhaben statt. Die Regelungen hierfür finden sich mit Ausnahme des Eingriffstatbestandes nicht im BNatSchG, sondern im Baugesetzbuch (BauGB). Auf Zulassungsebene findet keine Prüfung statt.
- Ebenso wird in Bezug auf die Schutz- und Erhaltungsziele der FFH- und Vogelschutzgebiete die Verträglichkeit der geplanten Vorhaben bereits im Bebauungsplanverfahren und dann grundsätz-

lich nicht mehr auf Genehmigungsebene geprüft. Das BauGB verweist insoweit auf die Vorgaben des BNatSchG.

Beim Artenschutz gilt Folgendes: Die Zugriffs- und Beeinträchtigungsverbote für besonders geschützte Tier- und Pflanzenarten (vgl. Artikel 12 und 16 FFH-Richtlinie, Artikel 5 Vogelschutz-Richtlinie) untersagen bestimmte Tathandlungen. Nicht die Festsetzungen des Bebauungsplans, sondern erst deren Verwirklichung und Umsetzung stellen somit den untersagten Eingriff dar. Wenn allerdings der Verwirklichung des Bebauungsplans artenschutzrechtliche Verbote entgegenstehen, hätte dies auch Rückwirkungen auf den Bebauungsplan, denn ein vollzugsunfähiger Plan ist wegen mangelnder Erforderlichkeit nichtig. Daher ist im Bebauungsplanverfahren zu prüfen, ob auf der nachfolgenden Ebene der Vorhabenzulassung die artenschutzrechtlich ggf. erforderlichen Ausnahmen und Befreiungen prinzipiell erteilt werden können. Durch den Bebauungsplan selbst jedoch wird weder die naturschutzrechtliche Zulässigkeit festgestellt noch werden die ggf. erforderlichen Ausnahmen und Befreiungen verbindlich in Aussicht gestellt; dies bleibt der Vorhabenzulassung vorbehalten.

b) Wie ist im Rahmen der Verträglichkeitsprüfung die Ausnahmeregelung des Artikels 6 Absatz 4 FFH-Richtlinie im Bereich des Bauplanungsrechts in nationales Recht umgesetzt worden und wie wird diese angewandt?

Gibt es eine Differenzierung in Bezug darauf, dass ein Vorhaben innerhalb oder außerhalb eines bereits im Zusammenhang bebauten Gebietes errichtet wird?

Für die FFH-Verträglichkeitsprüfung verweist das BauGB auf das BNatSchG, in dem auch die Vorgaben der Ausnahmeregelung des Artikels 6 Absatz 4 FFH-Richtlinie umgesetzt worden sind. Die entsprechenden Regelungen sind bei der Aufstellung von Bauleitplänen zu beachten.

Es gibt keine Differenzierung danach, ob ein Vorhaben innerhalb oder außerhalb eines bereits im Zusammenhang bebauten Gebiets errichtet wird.

c) Wird neben der schutzgebietsbezogenen Verträglichkeitsprüfung des Artikels 6 Absatz 3 FFH-Richtlinie bei baulichen Vorhaben auch eine artenschutzrechtliche Prüfung nach Artikel 12 FFH-Richtlinie durchgeführt?

Artenschutzrechtliche Prüfungen erfolgen neben der Verträglichkeitsprüfung, wenn Hinweise darauf vorliegen, dass die Zugriffsverbote auf geschützte Arten verletzt werden könnten. Anlass dazu kann insbesondere dann bestehen, wenn mit dem Bauvorhaben ein Bebauungsplan umgesetzt wird, der Fortpflanzungs- oder Ruhestätten geschützter Arten überplant.

These:

Zur Vermeidung von Doppelprüfungen wäre erstrebenswert, der Verträglichkeitsprüfung in den Fällen Vorrang vor den artenschutzrechtlichen Zugriffsverboten einzuräumen, in denen die artenschutzrechtlichen Verbote auf den Schutz derselben Arten zielen wie die Erhaltungsziele oder der Schutzzweck des betreffenden FFH-Gebiets.

d) Wird ein Bedarf dafür gesehen, solche Bauvorhaben von artenschutzrechtlichen Verboten freizustellen, die in Gebieten errichtet werden, die bereits im Zusammenhang bebaut sind? Welche Rolle könnte es hierbei haben, wenn die geschützte Art ihre natürliche Verbreitung außerhalb des Siedlungsbereichs hat und dort auch in größerer Zahl vorkommt?

In Deutschland gibt es teilweise Überlegungen, den Geltungsbereich artenschutzrechtlicher Verbote bei Bauvorhaben innerhalb der bereits im Zusammenhang bebauten Gebiete einzuschränken, sofern die betroffenen geschützten Arten ihren natürlichen Lebensraum außerhalb der im Zusammenhang bebauten Gebiete haben, dort auch tatsächlich in größerer Zahl vorkommen und die Bauvorhaben weder einer UVP-Pflicht noch einer Verträglichkeitsprüfung unterliegen. Diesen Überlegungen liegt die Erwartung zugrunde, dass bei einer Erleichterung des Bauens innerhalb bestehender Siedlungsbereiche die Innenentwicklung der Städte gestärkt und die Flächenneuinanspruchnahme in bislang unbebauten Bereichen verringert würde. Die artenschutzrechtlichen Verbote des Artikels 12 FFH-Richtlinie verlangen Maßnahmen, um ein strenges Schutzsystem für die geschützten Arten in deren natürlichen Lebensräumen einzuführen. Eine Bezugnahme auf die natürlichen Lebensräume findet sich auch in Artikel 1 Buchstabe i und Artikel 16 Absatz 1 FFH-Richtlinie. Soll das strenge Schutzsystem mit den in Artikel 12 Absatz 1 FFH-Richtlinie aufgeführten Verboten nur in den natürlichen Lebensräumen greifen, so wäre es außerhalb dieser Lebensräume sachgerecht, generell - und nicht nur im Wege der Ausnahmeentscheidung - von den strengen artenschutzrechtlichen Verboten der FFH-Richtlinie abzusehen, wenn die Populationen der betroffenen Art in einem günstigen Erhaltungszustand bleiben, wovon ausgegangen werden kann, wenn die Populationen als solche in ihrem natürlichen Verbreitungsgebiet als lebensfähiges Element erhalten bleiben. Damit können Zugriffe und Eingriffe innerhalb der bereits im Zusammenhang bebauten Gebiete, soweit dieser Bereich für die betroffene Art nicht der natürliche Lebensraum ist, zugelassen werden, ohne dass die Voraussetzungen nach Artikel 16 FFH-Richtlinie vorliegen müssen. Aufgrund des dem Erhaltungszustand zugrunde liegenden populationsbezogenen Ansatzes ist es zulässig, dass durch ein Vorhaben einzelne Siedlungsräume einer Art verloren gehen oder Individuen getötet werden. Eine Verschlechterung des Erhaltungszustands der Populationen wird erst angenommen, wenn die Zahl der die Populationen bildenden Individuen wesentlich verkleinert wird. Hingegen bleibt der Erhaltungszustand einer Art von einem Eingriff unberührt, wenn die Populationen als solche in ihrem natürlichen Verbreitungsgebiet, das über das Plangebiet hinausreichen kann, als lebensfähiges Element erhalten bleiben.

e) Wie wird das Tatbestandsmerkmal der „Absicht" in Artikel 12 Absatz 1 Buchstabe a, b und c FFH-Richtlinie bei Bauvorhaben im Lichte der Rechtsprechung des Europäischen Gerichtshofs verstanden?

Auf der Grundlage des Caretta-Urteils des EuGH vom 30.01.2002 – C 103/00 und des Fallenjagd-Urteils vom 20.10.2005 – C 6/04 wird unter Absicht nicht nur die zielgerichtete, sondern auch die in Kauf genommene Tötung oder Störung verstanden, wenn also der Handlungserfolg zwar nicht gewollt, aber erkannt und gleichwohl die zur Störung oder Tötung führende Handlung vorgenommen wird. Nach der Rechtsprechung des Bundesverwaltungsgerichts entfällt eine „absichtliche" Tatbestandsverwirklichung auch dann nicht, wenn sich eine Beeinträchtigung i. S. d. Artikel 12 Absatz 1 Buchstabe a, b und c FFH-Richtlinie als unausweichliche Konsequenz eines nach deutschem Recht rechtmäßigen Handelns (z. B. Gebrauchmachen von einer rechtmäßigen Baugenehmigung) erweist.

3.1.3 Frage 3: Plan-UP-Richtlinie (2001/42/EG)

a) Welche Ebenen und Instrumente städtebaulicher Planungen bestehen und wie erfolgte diesbezüglich die Umsetzung der Plan-UP-Richtlinie in nationales Recht?

Zur Lenkung der städtebaulichen Entwicklung stellt das Baugesetzbuch den Gemeinden eine zweistufige Bauleitplanung zur Verfügung. Planungsinstrumente sind der Flächennutzungsplan und der Bebauungsplan. Der Flächennutzungsplan ist der vorbereitende Bauleitplan, in dem die künftige städtebauliche Entwicklung des gesamten Gemeindegebiets dargestellt ist. Der Flächennutzungsplan ist keine Rechtsnorm.

Der Bebauungsplan setzt das Baurecht für bestimmte Gebiete in der Gemeinde rechtsverbindlich fest. Der Bebauungsplan ist eine Satzung und wird ebenfalls von der Gemeinde erlassen.

b) Wie wurde die Plan-UP-Richtlinie in das nationale Recht umgesetzt?

Die Plan-UP-Richtlinie ist in das nationale Bauplanungsrecht (Baugesetzbuch) in der Weise umgesetzt worden, dass bei grundsätzlich allen Verfahren zur Aufstellung von Flächennutzungs- oder Bebauungsplänen eine Umweltprüfung zu erfolgen hat. Die gemeinschaftsrechtlichen Verfahrensvorgaben (wie Scoping, Umweltbericht, Öffentlichkeits- und Behördenbeteiligung, Berücksichtigung bei der Entscheidung, Bekanntgabe der Entscheidung und der zusammenfassenden Erklärung, Monitoring) sind in die bestehenden Verfahrensschritte zur Aufstellung eines Bauleitplans integriert worden.

Die Umweltprüfung im Bebauungsplanverfahren erfüllt sowohl sämtliche Anforderungen der Plan-UP-Richtlinie als auch die Anforderungen an eine Umweltverträglichkeitsprüfung (UVP). Dies gilt unabhängig davon, ob der Bebauungsplan tatsächlich einer UVP-Pflicht unterliegt.

Auf Bundesebene wurde die Plan-UP-Richtlinie daneben durch entsprechende Ergänzungen des UVP-Gesetzes, das grundsätzliche Regelungen zur Umweltprüfung enthält, umgesetzt. Weitere Regelungen finden sich im Fachrecht, zum Beispiel für Raumordnungspläne. Die Bundesländer haben zur Umsetzung der Plan-UP-Richtlinie ebenfalls Regelungen geschaffen.

c) Werden für bestimmte Fälle, insbesondere der Geringfügigkeit im Rahmen der Aufstellung, Änderung und Ergänzung von Bauleitplänen, Möglichkeiten zur Beschleunigung sowie Verfahrenserleichterungen genutzt?

Verfahrenserleichterungen und -beschleunigungen (z. B. Verzicht auf zweistufige Öffentlichkeitsbeteiligung, Beteiligung nur der betroffenen Öffentlichkeit, Absehen von der Umweltprüfung) sind möglich bei Aufstellung von Bauleitplänen, die bestehende Bauleitpläne nur geringfügig ändern, bei Aufstellung von Bebauungsplänen, die nur im Zusammenhang bebaute Flächen erfassen und den Zulässigkeitsmaßstab nicht wesentlich ändern, oder bei Aufstellung von Bebauungsplänen, die die Nutzung kleiner Gebiete mit dem Ziel der Wiedernutzbarmachung von Flächen, der Nachverdichtung oder anderer Maßnahmen der Innenentwicklung regeln.

Voraussetzung ist in allen Fällen, dass durch die Verfahren die Zulässigkeit nur solcher Vorhaben begründet wird, die keiner UVP-Pflicht unterliegen und die Gebiete von gemeinschaftlicher Bedeutung (FFH- und Vogelschutzgebiete) nicht beeinträchtigen können.

d) In welchem Verhältnis stehen Umweltprüfung und FFH-Verträglichkeitsprüfung zueinander?

Die FFH-Verträglichkeitsprüfung im Sinne von Artikel 6 Absatz 3 FFH-RICHTLINIE ist ebenso wie die Umweltprüfung bzw. die UVP als unselbstständiges Element in die Bauleitplanung bzw. in Genehmigungsverfahren integriert. Die Umweltprüfung kann dabei die Funktion eines Trägerverfahrens übernehmen, insbesondere für die im Rahmen der FFH-Verträglichkeitsprüfung vorzunehmende Grundlagenermittlung und ihre Bewertung. Dokumentiert wird die Durchführung der FFH-Verträglichkeitsprüfung in einem eigenständigen Kapitel des Umweltberichts.

3.1.4 Frage 4: Dienstleistungsrichtlinie (2006/123/EG)

Wird ein Umsetzungsbedarf der Dienstleistungsrichtlinie (Richtlinie 2006/123/EG) für den Bereich des nationalen Städtebaurechts und der Bauleitplanung gesehen?

Durch die Dienstleistungsrichtlinie werden nationale Rechts- und Verwaltungsvorschriften koordiniert, die speziell die Dienstleistungs- und Niederlassungsfreiheit reglementieren. Ohne einen speziellen Dienstleistungsbezug sind solche Rechts- und Verwaltungsvorschriften, die nicht die Dienstleistungstätigkeit als solche, sondern generell das Verhalten von Personen, also auch solcher Personen, die nicht als Dienstleister tätig sind, regeln oder betreffen. Aus den primärrechtlichen Ermächtigungsgrundlagen und dem Inhalt der Dienstleistungsrichtlinie ergibt sich, dass diese nicht auf derartige Jedermann-Regelungen anzuwenden ist. Dies ergibt sich aus Erwägungsgrund 9 der Richtlinie:

> *„Diese Richtlinie findet nur auf die Anforderungen für die Aufnahme oder Ausübung einer Dienstleistungstätigkeit Anwendung. Sie findet somit keine Anwendung auf Anforderungen wie Straßenverkehrsvorschriften, Vorschriften bezüglich der Stadtentwicklung oder Bodennutzung, der Stadtplanung und der Raumordnung, Baunormen sowie verwaltungsrechtliche Sanktionen, die wegen der Nichteinhaltung solcher Vorschriften verhängt werden, die nicht die Dienstleistungstätigkeit als solche regeln oder betreffen, sondern von Dienstleistungserbringern im Zuge der Ausübung ihrer Wirtschaftstätigkeit genauso beachtet werden müssen wie von Privatpersonen."*

Städtebauliche Regelungen über die planungsrechtliche Zulässigkeit von baulichen Vorhaben regeln nicht eine Dienstleistungstätigkeit als solche, sondern richten sich als abstrakt-generelle Regelungen gegen jede Person, wenn und soweit sie ein Grundstück in baulicher oder sonstiger Weise nutzt. Diesen Regelungen fehlt ein spezieller Dienstleistungsbezug. Sie sind Jedermann-Regelungen im Sinne des Erwägungsgrundes 9 der Dienstleistungsrichtlinie. Für städtebauliche Regelungen, die die planungsrechtliche Zulässigkeit von Vorhaben betreffen, besteht deshalb kein Umsetzungsbedarf.

3.1.5 Frage 5: Erneuerbare-Energien-Richtlinie (2009/28/EG)

Wird ein Umsetzungsbedarf hinsichtlich des Artikels 13 Erneuerbare-Energien-Richtlinie (2009/28/EG) für den Bereich des nationalen Städtebaurechts und der Bauleitplanung gesehen?

Mit der Richtlinie 2009/28/EG des Europäischen Parlaments und des Rates zur Förderung der Nutzung von Energie aus erneuerbaren Quellen und zur Änderung und anschließenden Aufhebung der Richtlinien 2001/77/EG und 2003/30/EG (EE-Richtlinie) wird das Recht der erneuerbaren Energien in Europa neu geordnet und insbesondere die Förderung aller Energiesektoren, im Rahmen eines einheitlichen Rechtsaktes vereint. Außerdem enthält die Richtlinie verbindliche Zielwerte für den weiteren Ausbau der erneuerbaren Energien. In diesem Sinne soll nach Artikel 3 EE-RL bis zum Zieljahr 2020 der Anteil erneuerbarer Energien am gesamteuropäischen Endenergieverbrauch auf mindestens 20 % ausgebaut werden. Zur Erreichung dieses Ziels wird jeder Mitgliedstaat verpflichtet, Maßnahmen zur Förderung von erneuerbaren Energien zu ergreifen. Dazu wird dem jeweiligen Mitgliedstaat ein Ausbauziel zugeordnet. Dieses sieht für die Bundesrepublik Deutschland vor, den Anteil erneuerbarer Energien am Bruttoendenergieverbrauch bis zum zu diesem Zeitpunkt auf 18 % auszubauen (gegenüber 5,8 % im Jahr 2005). Außerdem muss jeder Mitgliedstaat gewährleisten, dass sein Anteil an erneuerbaren Energien im Verkehrssektor bis zum Zieljahr 2020 auf mindestens 10 % ansteigt.

Artikel 13 Absatz 1 der Richtlinie betrifft das Verwaltungsverfahren, die Rechtsvorschriften und die Regelwerke der Mitgliedstaaten. Zu überlegen ist, ob sich hieraus Anforderungen auch für das Städtebaurecht ergeben. Für das Recht der Bauleitplanung wird derzeit kein weiterer Umsetzungsbedarf gesehen; nur für bestimmte städtebauliche Genehmigungen sollen Fristenregelungen und Genehmigungsfiktionen eingeführt werden. Von Interesse ist, welchen Stand die Überlegungen in weiteren Mitgliedstaaten haben.

Die betreffende Richtlinienvorschrift lautet:

„Artikel 13

Verwaltungsverfahren, Rechtsvorschriften und Regelwerke

(1) Die Mitgliedstaaten stellen sicher, dass einzelstaatliche Vorschriften für die Genehmigungs-, Zertifizierungs- und Zulassungsverfahren, die auf Anlagen zur Erzeugung von Elektrizität, Wärme oder Kälte aus erneuerbaren Energiequellen und die angegliederten Infrastrukturen der Übertragungs- und Verteilernetze sowie auf den Vorgang der Umwandlung von Biomasse in Biokraftstoffe oder sonstige Energieprodukte angewandt werden, verhältnismäßig und notwendig sind. Die Mitgliedstaaten ergreifen insbesondere angemessene Maßnahmen, um sicherzustellen, dass

a) vorbehaltlich der Unterschiede zwischen den Mitgliedstaaten hinsichtlich ihrer Verwaltungsstruktur und -organisation die entsprechenden Zuständigkeiten der nationalen, regionalen und lokalen Verwaltungsstellen für die Genehmigungs-, Zertifizierungs- und Zulassungsverfahren – auch im Hinblick auf die Raumplanung – eindeutig koordiniert und festgelegt sind und transparente Zeitpläne für Entscheidungen über Planungs- und Bauanträge genau bestimmt sind;

b) auf der geeigneten Ebene umfassende Informationen über die Bearbeitung von Genehmigungs-, Zertifizierungs- und Zulassungsanträgen für Anlagen zur Nutzung von erneuerbarer Energie und über die den Antragstellern zur Verfügung stehende Unterstützung angeboten werden;

c) die Verwaltungsverfahren auf der geeigneten Verwaltungsebene gestrafft und beschleunigt werden;

d) die Vorschriften für Genehmigung, Zertifizierung und Zulassung objektiv, transparent und verhältnismäßig sind, nicht zwischen Antragstellern diskriminieren und den Besonderheiten der einzelnen Technologien für erneuerbare Energie vollständig Rechnung tragen;

e) Verwaltungsgebühren, die die Verbraucher, Planungsbüros, Architekten, Bauunternehmen sowie die Geräte- und Systeminstallateure und -lieferanten entrichten müssen, transparent und kostenbezogen sind; und

f) gegebenenfalls vereinfachte und weniger aufwendige Genehmigungsverfahren, unter anderem der Ersatz des Genehmigungsverfahrens durch eine einfache Mitteilung, falls dies im Rahmen des einschlägigen Rechtsrahmens zulässig ist, für kleinere Projekte und gegebenenfalls für dezentrale Anlagen zur Produktion von Energie aus erneuerbaren Quellen eingeführt werden."

3.1.6 Frage 6: SEVESO-II-Richtlinie

Wie ist Artikel 12 Absatz 1 Unterabsatz 2 der Richtlinie 96/82/EG vom 09.12.1996 zur Beherrschung der Gefahren bei Unfällen mit gefährlichen Stoffen (ABl. L 10 vom 14.01.1997, S. 13; „Seveso-II-Richtlinie") in nationales Recht umgesetzt worden?

In Deutschland ist Artikel 12 Absatz 1 Unterabsatz 2 der Seveso-II-Richtlinie durch folgende Regelung im Bundes-Immissionsschutzgesetz umgesetzt worden:

„§ 50 Planung

Bei raumbedeutsamen Planungen und Maßnahmen sind die für eine bestimmte Nutzung vorgesehenen Flächen einander so zuzuordnen, dass schädliche Umwelteinwirkungen und von schweren Unfällen im Sinne des Artikels 3 Nr. 5 der Richtlinie 96/82/EG in Betriebsbereichen hervorgerufene Auswirkungen auf die ausschließlich oder überwiegend dem Wohnen dienenden Gebiete sowie auf sonstige schutzbedürftige Gebiete, insbesondere öffentlich genutzte Gebiete, wichtige Verkehrswege, Freizeitgebiete und unter dem Gesichtspunkt des Naturschutzes besonders wertvolle oder besonders empfindliche Gebiete und öffentlich genutzte Gebäude, so weit wie möglich vermieden werden. Bei raumbedeutsamen Planungen und Maßnahmen in Gebieten, in denen die in Rechtsverordnungen nach § 48a Abs. 1 festgelegten Immissionsgrenzwerte nicht überschritten werden, ist bei der Abwägung der betroffenen Belange die Erhaltung der bestmöglichen Luftqualität als Belang zu berücksichtigen."

Die betreffende Richtlinienvorschrift lautet:

„Artikel 12

Überwachung der Ansiedlung

(1) […]

Die Mitgliedstaaten sorgen dafür, dass in ihrer Politik der Flächenausweisung oder Flächennutzung und/oder anderen einschlägigen Politiken sowie den Verfahren für die Durchführung dieser Politiken langfristig dem Erfordernis Rechnung getragen wird, dass zwischen den unter diese Richtlinie fallenden Betrieben einerseits und Wohngebieten, öffentlich genutzten Gebäuden und Gebieten, wichtigen Verkehrswegen (so weit wie möglich), Freizeitgebieten und unter dem Gesichtspunkt des Naturschutzes besonders wertvollen bzw. besonders empfindlichen Gebieten andererseits ein angemessener Abstand gewahrt bleibt und dass bei bestehenden Betrieben zu-

sätzliche technische Maßnahmen nach Artikel 5 ergriffen werden, damit es zu keiner Zunahme der Gefährdung der Bevölkerung kommt.“

3.2 Estonia

3.2.1 Question 1: Systematic preliminary question

What is the relationship between binding land use plans and the approval of individual construction projects?

Conditions for construction of individual projects are issued based on binding land use plans, i. e. on approved detailed plans in areas where detailed plans are required (e. g. towns) and on approved comprehensive plans where detailed plans are not required.

3.2.2 Question 2: European Habitat and Protection of Species Law

a) *What is the relationship between national nature conservation law and the building planning law? Are nature conservation law requirements taken into account in development planning or not dealt with until a subsequent approval procedure takes place?*

Nature conservation law requirements (and all other legal requirements) are taken into account during the entire planning process. Nature conservation experts are normally invited to participate in the preparation of spatial plans.

b) *Within the scope of the assessment of the implications of a plan or project for a site, how has the exception provided by Article 6 Paragraph 4 of the FFH Directive been transposed into national law in the field of building planning and how is it applied?*

The Habitats Directive art 6.3 and 6.4 have been transposed into Environmental Impact Assessment and Environmental Management System Act. According to this act environmental impact shall be assessed if activities (e. g. building activities) are proposed which alone or in conjunction with other activities may potentially significantly affect a Natura 2000 site and a strategic environmental assessment shall be carried out during the preparation of a strategic planning document before its adoption by a legal act, if the document is the basis for activities which are likely to significantly affect a Natura 2000 site.

Is there any distinction regarding whether a project is built within or outside an area which is already continuously built up?

No.

c) *Apart from the conservation area related assessment of implications for the site pursuant to Article 6 Paragraph 3 of the FFH Directive, is an assessment under species protection law also performed according to Article 12 of the FFH Directive?*

According to the Environmental Impact Assessment and Environmental Management System Act if the assessments should be taken into account also protected species and for example if proposed activities may potentially affect species protection sites or habitats of protected species the decision-maker shall obtain approval for the draft decision to refuse initiation of environmental impact assessment of the proposed activities with the administrator of the specified protected natural objects.

d) Is there a need to exempt such construction projects from prohibitions under species protection law which are erected in areas which have already been developed? What role could it play if the protected species had their natural habitat outside the development area and also occurred in greater number there?

If the species is in favourable conservation status and the population number is increasing, in this case there might be possibility for exception.

e) How is the element of "intention" in Article 12 Paragraph 1 Letter a, b and c of the FFH Directive interpreted for construction projects in the light of the legal decisions taken by the European Court of Justice?

If the presence of Art 12 species is recorded, then it must be taken into consideration also for the development of construction projects.

3.2.3 Question 3: SEA Directive (2001/42/EC)

a) Which levels and instruments of development planning exist and with respect to this, how was the SEA Directive transposed into national law?

Concerning plans SEA is mandatory in the case of national, county and comprehensive plan. SEA is also mandatory in the case of certain detailed plans. The necessity of SEA of the planning documents must also be considered for example if amendments are made to the mentioned documents (through so-called screening).

b) How was the SEA Directive transposed into national law?

The SEA directive was transposed into national law in the Environmental Impact Assessment and Environmental Management System Act which came into force 3 April 2005.

c) Are options, especially the option of insignificance within the context of preparing, amending and supplementing the development plans, used to accelerate and simplify procedures for particular cases?

The overall SEA procedure is the same for all strategic planning documents. Nevertheless there are some provisions which stipulate the minimum requirements and so in practice the processes might differ in the case of different planning documents. For example the public display of the SEA report must be at least 21 days but of course it can also be longer (depending of the complexity of the specific planning document etc).

d) What is the relationship between the environmental audit and the FFH assessment of implications for the site?

No answer

3.2.4 Question 4: EU Services Directive (2006/123/EC)

Is it deemed necessary to transpose the EC Services Directive (Directive 2006/123/EC) for the area of national urban planing law and development planning?

No.

3.2.5 Question 5: Renewable Energy Directive (2009/28/EC)

Is it considered necessary to transpose Article 13 of the Renewable Energy Directive (Directive 2009/28/EC) for the area of national urban planning law and development planning?

The Article 13 requirements will be transposed into the building legislation.

3.2.6 Question 6: SEVESO-II-Directive

How has Article 12 Paragraph 1 Subsection 2 Council Directive 96/82/EC of 9 December 1996 on the control of major-accident hazards involving dangerous substances (OJ L 10, 14.1.1997, p. 13; "Seveso-II-Directive") been transposed into national law?

Land-use planning

(1) [...]

Member States shall ensure that their land-use and/ or other relevant policies and the procedures for implementing those policies take account of the need, in the long term, to maintain appropriate distances between establishments covered by this Directive and residential areas, areas of public use and areas of particular natural sensitivity or interest, and, in the case of existing establishments, of the need for additional technical measures in accordance with Article 5 so as not to increase the risks to people.

In Estonia, areas around establishments are zoned as result of risk analysis. The decision making of land-use planning is done by local government. According to law (Chemicals Act, Planning Act) local government must consider in decision making also risks of major accidents. For ensuring that requirement, planning and building documents must be approved by competent authority (Rescue Centre).

In case of an existing situation, information to the public must be provided and also early warning system must be in place. Regular supervision is carried out by Rescue Centres and Technical Surveillance Board to ensure the safety. In case of an existing situation, where distances between establishments and residential areas are too small, further development of establishment or residential areas is limited (distance is not allowed to diminish even smaller).

In addition to major accidents risks also environmental impact assessment and air quality requirements must be considered.

3.2 Estland

3.2.1 Frage 1: Systematische Vorfrage

In welchem Verhältnis stehen Bebauungspläne und die Genehmigung einzelner Bauvorhaben?

Die Anforderungen an den Bau von Einzelprojekten ergeben sich aus Bebauungsplänen. In Gebieten (z. B. in Städten), in denen ein Detailplan erforderlich ist, ergeben sich die Anforderungen aus dem genehmigten Plan. In Gebieten, in denen Detailpläne nicht erforderlich sind, werden die Vorgaben aus einem genehmigten Flächennutzungsplan entnommen.

3.2.2 Frage 2: Europäisches Habitat- und Artenschutzrecht

a) *In welchem Verhältnis stehen das nationale Naturschutzrecht und das Bauplanungsrecht? Werden die Anforderungen des Naturschutzrechts bereits in der Bauleitplanung oder erst in einem nachfolgenden Genehmigungsverfahren bewältigt?*

Die Anforderungen des nationalen Naturschutzrechts (und alle weiteren rechtlichen Vorgaben) werden während des gesamten Planungsprozesses berücksichtigt. Experten des Naturschutzes werden bei der Erarbeitung von Bauleitplänen beteiligt.

b) *Wie ist im Rahmen der Verträglichkeitsprüfung die Ausnahmeregelung des Artikels 6 Absatz 4 FFH-Richtlinie im Bereich des Bauplanungsrechts in nationales Recht umgesetzt worden und wie wird diese angewandt?*

Die Absätze 3 und 4 des Artikels 6 der FFH-Richtlinie wurden in die Umweltverträglichkeitsprüfung und das Gesetz über Umweltmanagementsysteme übernommen. Laut diesem Gesetz sollen Umweltauswirkungen abgeschätzt werden, wenn Vorhaben (z. B. Bauvorhaben) beantragt werden, die alleine oder in Verbindung mit anderen Vorhaben mögliche erhebliche Auswirkungen auf Natura 2000-Gebiete hervorrufen können. Eine strategische Umweltprüfung soll während der Erstellung strategischer Planungen vorgenommen werden, bevor diese rechtskräftig werden, wenn die Planung den Rahmen für eine erhebliche Beeinträchtigung von Natura 2000-Gebieten bildet.

Gibt es eine Differenzierung in Bezug darauf, dass ein Vorhaben innerhalb oder außerhalb eines bereits im Zusammenhang bebauten Gebietes errichtet wird?

Nein.

c) Wird neben der schutzgebietsbezogenen Verträglichkeitsprüfung des Artikels 6 Absatz 3 FFH-Richtlinie bei baulichen Vorhaben auch eine artenschutzrechtliche Prüfung nach Artikel 12 FFH-Richtlinie durchgeführt?

Entsprechend den Vorgaben des Gesetzes über die Umweltverträglichkeitsprüfung und des Gesetzes über Umweltmanagementsysteme werden im Rahmen der Verträglichkeitsprüfung auch die Auswirkungen auf geschützte Tier- und Pflanzenarten erfasst. Zum Beispiel sind bei beabsichtigten Maßnahmen, die mögliche Auswirkungen auf geschützte Arten bzw. Habitate geschützter Arten haben, die Entscheidungsträger dazu angehalten, den Entwurf mit der für den Schutz der Gebiete zuständigen Behörde abzustimmen, um die Einleitung einer Umweltverträglichkeitsprüfung abzuwenden.

d) Wird ein Bedarf dafür gesehen, solche Bauvorhaben von artenschutzrechtlichen Verboten freizustellen, die in Gebieten errichtet werden, die bereits im Zusammenhang bebaut sind? Welche Rolle könnte es hierbei haben, wenn die geschützte Art ihre natürliche Verbreitung außerhalb des Siedlungsbereichs hat und dort auch in größerer Zahl vorkommt?

Wenn der Erhalt einer Art gesichert ist und die Population zunimmt, besteht die Möglichkeit für eine Ausnahme.

e) Wie wird das Tatbestandsmerkmal der „Absicht" in Artikel 12 Absatz 1 Buchstabe a, b und c FFH-Richtlinie bei Bauvorhaben im Lichte der Rechtsprechung des Europäischen Gerichtshofs verstanden?

Soweit Arten nach Artikel 12 der FFH-Richtlinie vorgefunden werden, muss dies bei der Entwicklung eines Bauvorhabens berücksichtigt werden.

3.2.3 Frage 3: Plan-UP-Richtlinie (2001/42/EG)

a) Welche Ebenen und Instrumente städtebaulicher Planungen bestehen und wie erfolgte diesbezüglich die Umsetzung der Plan-UP-Richtlinie in nationales Recht?

Im Hinblick auf die Planung ist die Strategische Umweltprüfung verpflichtend anzuwenden bei Plänen auf Bundes-, Kreis- oder kommunaler Ebene. Eine SUP ist ebenfalls in bestimmten Fällen der Detailplanung verpflichtend. Die Durchführung einer SUP muss auch bei der Änderung der oben genannten Planungsdokumente berücksichtigt werden (durch ein Screening).

b) Wie wurde die Plan-UP-Richtlinie in das nationale Recht umgesetzt?

Die Strategische Umweltprüfung wurde durch das Gesetz über die Umweltverträglichkeitsprüfung und Umweltmanagementsysteme in nationales Recht umgesetzt, das am 03. April 2005 in Kraft getreten ist.

c) Werden für bestimmte Fälle, insbesondere der Geringfügigkeit im Rahmen der Aufstellung, Änderung und Ergänzung von Bauleitplänen, Möglichkeiten zur Beschleunigung sowie Verfahrenserleichterungen genutzt?

Das Verfahren der Strategischen Umweltprüfung ist bei allen strategischen Planungen gleich. Bestimmte Vorgaben stellen die Mindestanforderungen dar, wodurch sich die Verfahren in der Planungspraxis unterscheiden können. So muss zum Beispiel der Umweltbericht für 21 Tage öffentlich ausgelegt werden, wobei diese Frist auch verlängert werden kann, in Abhängigkeit von dem Umfang des jeweiligen Plandokuments etc.

d) In welchem Verhältnis stehen Umweltprüfung und FFH-Verträglichkeitsprüfung zueinander?

Keine Antwort

3.2.4 Frage 4: Dienstleistungsrichtlinie (2006/123/EG)

Wird ein Umsetzungsbedarf der Dienstleistungsrichtlinie (Richtlinie 2006/123/EG) für den Bereich des nationalen Städtebaurechts und der Bauleitplanung gesehen?

Nein.

3.2.5 Frage 5: Erneuerbare-Energien-Richtlinie (2009/28/EG)

Wird ein Umsetzungsbedarf hinsichtlich des Artikels 13 Erneuerbare-Energien-Richtlinie (2009/28/EG) für den Bereich des nationalen Städtebaurechts und der Bauleitplanung gesehen?

Die Anforderungen des Artikels 13 werden in das Baurecht übernommen.

3.2.6 Frage 6: SEVESO-II-Richtlinie

Wie ist Artikel 12 Absatz 1 Unterabsatz 2 der Richtlinie 96/82/EG vom 09.12.1996 zur Beherrschung der Gefahren bei Unfällen mit gefährlichen Stoffen (ABl. L 10 vom 14.01.1997, S. 13; „Seveso-II-Richtlinie") in nationales Recht umgesetzt worden?

Flächennutzungsplanung

(1) [...]

Damit Wohngebiete, öffentlich genutzte Gebiete und unter dem Gesichtspunkt des Naturschutzes besonders wertvolle bzw. besonders empfindliche Gebiete besser vor den Gefahren schwerer Unfälle geschützt werden können, müssen die Mitgliedstaaten in ihren Politiken hinsichtlich der Zuweisung oder Nutzung von Flächen und/ oder anderen einschlägigen Politiken berücksichtigen, dass langfristig zwischen diesen Gebieten und gefährlichen Industrieansiedlungen ein angemessener Abstand gewahrt bleiben muss und dass bei bestehenden Betrieben ergänzende technische Maßnahmen vorgesehen werden, damit es zu keiner stärkeren Gefährdung der Bevölkerung kommt.

In Estland werden Gebiete im Umfeld von Siedlungen infolge einer Risikoanalyse in einem Flächennutzungsplan erfasst. Die Entscheidung über die Flächennutzungsplanung liegt bei der lokalen Verwaltung. Laut Gesetz (Chemikaliengesetz, Planungsgesetz) muss die lokale Verwaltung bei ihrer Entscheidungsfindung auch das Risiko schwerer Unfälle beachten. Um dies sicherzustellen, müssen Antrags-, Planungs- und Bauunterlagen von der zuständigen Behörde (Rettungszentrum) genehmigt werden.

Im Falle einer bestehenden Bebauung ist die Öffentlichkeit zu unterrichten und es muss außerdem ein Frühwarnsystem eingerichtet werden. Die Rettungszentren und der technische Überwachungsdienst führen regelmäßige Überprüfungen durch, um die Sicherheit zu gewährleisten. Bei einer vorliegenden Bebauung, wo die Abstände zwischen Anlagen und Wohngebieten zu gering sind, wird der Bau weiterer Anlagen oder die weitere Wohnbebauung eingeschränkt (der Abstand darf nicht noch geringer werden).

Neben den Risiken schwerwiegender Unfälle sind auch Umweltverträglichkeitsprüfungen und Anforderungen an die Luftqualität zu beachten.

3.3 France

3.3.1 Question 1: Systematic preliminary question

What is the relationship between binding land use plans and the approval of individual construction projects?

The currently valid French urban development law in article L 121-1 contains above all the full set of basic principles to be heeded by all urban development documents: a master plan for compliance with territorial coherence (SCOT), local urban development plans (PLU) and municipal charts (CC). These principles are highly schematic: principles of balance, of diversity in urban functions and social mixture together with environmental protection.

Building permits or preliminary clarification for minor work as well as planning permits are granted if the project complies with the PLU or CC regulations, while a SCOT permit is not used directly for building permits, apart from very large projects. If there is neither a local development plan nor a municipal chart, then building permits are subject to the National Urban Development Ordinance (RNU) (articles L.111-1 f. and R 111-1 f. of the Urban Development Law).

3.3.2 Question 2: European Habitat and Protection of Species Law

a) *What is the relationship between national nature conservation law and the building planning law? Are nature conservation law requirements taken into account in development planning or not dealt with until a subsequent approval procedure takes place?*

The Nature Conservation Law 76-629 dated 10 July 1976 stipulates the principle of this protection and ascertains that compliance with biological balance is a matter of general interest. Accordingly, the urban development texts have to comply with the environmental regulations if they want to remain within the law.

The law 2000-1208 "Urban solidarity and renewal" dated 13 December 2000 as amended by the law 2003-590 "Urbanism and habitat" dated 2 July 2003 has increased the statutory requirements made of the local authorities in compiling their urban development documents.

New requirements have been stipulated with regard to

- the initial status of the environment,
- estimating the probable impacts on the environment of the scheme or plan, and
- the way in which the document is taken into account in order to sustain and improve the environment.

The procedures have also been rendered more democratic by generalising the concerted approach and the objections procedures.

Ordinance 2004-489 dated 3 June 2004 transposes European Directive 2001/42 dated 28 June 2001 on the assessment of the effects of certain plans and programmes on the environment and formalises these new requirements for urban development documents, with possibly extensive effects. The urban development documents also have to take account of France's international and EU obligations (summit of Rio, the Natura 2000 network, the Kyoto protocol and the environmental standards) (law on mountains, law on the coastal regions, law on air, law on water…)

Decree 2005-608 dated 27 May 2005 on assessing the environmental effects of the urban development documents that replaced the Urban Development Law stipulates the documents that have to undergo a strategic environmental assessment, i. e. the territorial land development guidelines (DTA), the preparatory land use plan for the region Ile-de-France (SDRIF), the regional binding land use plans (SAR) for the overseas regions, the binding land use and sustainable development plan for Corsica, the SCOT and some PLU, particularly those whose work and buildings could be detrimental to the creation of a Natura 2000 site and those whose areas are not covered by an environmentally assessed SCOT, with the exception of a few highly specific conditions.

The Charter adopted on 28 February 2005 formalises new environmental obligations and requirements: the precautionary principle, the obligation to prevent environmental hazards respectively to limit the consequences, repairing damage that has been caused to the environment …

b) Within the scope of the assessment of the implications of a plan or project for a site, how has the exception provided by Article 6 Paragraph 4 of the FFH Directive been transposed into national law in the field of building planning and how is it applied?

Excerpt from Article R 414-23 of the Environment Law

III. *If the analysis named under point II reveals that the planning document or programme, the project, the measures or the intervention during or after realisation or during the validity of the planning document can have significant harmful effects on the state of preservation of the natural habitats and the species that had formed the basis for designating one or several protection areas, the evaluation document shall state the measures to be taken to eliminate or reduce the harmful effects.*

IV. *If despite the measures named under point II, significant harmful effects remain on the state of preservation of the natural habitats and the species that had formed the basis for designating one or several protection areas, the evaluation document shall also contain the following:*

1. *Description of possible alternative solutions, the reason why there is no other than the chosen solution, and the elements justifying approval of the planning document, or execution of the programme, the project, the measure or the intervention under the conditions stated in point VII and point VIII of article L. 414-4;*

2. *Description of the measures planned as compensation for the harmful effects that cannot be eliminated by the measures stated under point III above. The compensation measures permit appropriate, efficient compensation in view of the damage caused to the objective of sustaining the Natura 2000 protection area or protection areas. These compensation measures shall be taken according to a timeframe that safeguards continuity of the capabilities of the Natura 2000 network to preserve the natural habitats and the species. Even if these compensation measures are spread out in time and space, they are still the result of a holistic approach to guarantee this continuity.*

3. *For the planning documents, the corresponding expenditure and modalities for taking on the compensation measures lie in the hands of the authority responsible for granting approval; for the programmes, projects and intervention: in the hands of the building developer or party applying for approval, and for the measures: in the hands of the party responsible for corresponding organisation.*

Is there any distinction regarding whether a project is built within or outside an area which is already continuously built up?

No.

c) Apart from the conservation area related assessment of implications for the site pursuant to Article 6 Paragraph 3 of the FFH Directive, is an assessment under species protection law also always performed according to Article 12 of the FFH Directive?

There was unfortunately no time to research this question.

d) Is there a need to exempt such construction projects from prohibitions under species protection law which are erected in areas which have already been developed? What role could it play if the protected species had their natural habitat outside the development area and also occurred in greater number there?

No answer

e) How is the element of "intention" in Article 12 Paragraph 1 Letter a, b and c of the FFH DIRECTIVE interpreted for construction projects in the light of the legal decisions taken by the European Court of Justice?

No answer

3.3.3 Question 3: SEA Directive (2001/42/EC)

a) Which levels and instruments of development planning exist and with respect to this, how was the SEA Directive transposed into national law?

No answer

b) How was the SEA Directive transposed into national law?

Already answered, see above.

c) Are options, especially the option of insignificance within the context of preparing, amending and supplementing the development plans, used to accelerate and simplify procedures for particular cases?

The environmental assessment (SEA) is not necessary for a PLU contained in a SCOT.

d) What is the relationship between the environmental audit and the FFH assessment of implications for the site?

There was unfortunately no time to answer this question

3.3.4 Question 4: EU Services Directive (2006/123/EC)

Is it deemed necessary to transpose the EC Services Directive (Directive 2006/123/EC) for the area of national urban planning law and development planning?

Only along commercial lines: the law to modernise the economy 2008-776 dated 4 August 2008 has precisely phrased the contents of the SCOT and some PLU, stipulating restrictive criteria for commercial settlement possibilities that have to become part of the criteria stipulating the field of construction and the environment. And so this need has been taken into account accordingly in the area of urban planning and development.

3.3.5 Question 5: Renewable Energy Directive (2009/28/EC)

Is it considered necessary to transpose Article 13 of the Renewable Energy Directive (Directive 2009/28/EC) for the area of national urban planning law and development planning?

A priori no need, but there was not enough time to find out in detail.

3.3.6 Question 6: SEVESO-II-Directive

How has Article 12 Paragraph 1 Subsection 2 Council Directive 96/82/EC of 9 December 1996 on the control of major-accident hazards involving dangerous substances (OJ L 10, 14.1.1997, p. 13; "Seveso-II-Directive") been transposed into national law?

No answer

3.3 Frankreich

3.3.1 Frage 1: Systematische Vorfrage

In welchem Verhältnis stehen Bebauungspläne und die Genehmigung einzelner Bauvorhaben?

Das gegenwärtig geltende französische Stadtentwicklungsgesetz sieht einen Leitplan zur Erhaltung der territorialen Kohärenz (SCOT), lokale Stadtentwicklungspläne (PLU) und kommunale Karten (CC) vor. Die Prinzipien, die diesen Plänen zugrunde liegen, sind das Prinzip der Ausgewogenheit, der Diversität der städtischen Funktionen und sozialen Durchmischung sowie des Umweltschutzes (Artikel L 121-1 Stadtentwicklungsgesetz).

Die Baugenehmigung oder die Voraberklärung für geringfügige Arbeiten sowie eine Planungsgenehmigung werden erteilt, wenn das Projekt den Verfügungen des PLU oder des CC entspricht, während eine SCOT-Genehmigung – abgesehen von sehr umfangreichen Vorhaben – nicht direkt für Baugenehmigungen zur Anwendung kommt. Gibt es weder einen lokalen Entwicklungsplan noch eine kommunale Karte, kommt für Baugenehmigungen die Nationale Stadtentwicklungsverordnung (RNU) zur Anwendung (Artikel L.111-1 f. und R 111-1 f. des Stadtentwicklungsgesetzes).

3.3.2 Frage 2: Europäisches Habitat- und Artenschutzrecht

a) *In welchem Verhältnis stehen das nationale Naturschutzrecht und das Bauplanungsrecht? Werden die Anforderungen des Naturschutzrechts bereits in der Bauleitplanung oder erst in einem nachfolgenden Genehmigungsverfahren bewältigt?*

Das Naturschutzgesetz 76-629 vom 10. Juli 1976 legt die Prinzipien des Naturschutzes fest. Danach ist die Erhaltung des biologischen Gleichgewichts von allgemeinem Interesse. Die Stadtentwicklungsdokumente müssen den darin enthaltenen Umweltanforderungen entsprechen.

Das Gesetz 2000-1208 „Städtische Solidarität und Erneuerung" vom 13. Dezember 2000, in seiner durch das Gesetz 2003-590 „Urbanismus und Habitat" vom 2. Juli 2003 abgeänderten Fassung hat die gesetzlichen Anforderungen der Gebietskörperschaften bei der Erarbeitung ihrer Stadtentwicklungsdokumente noch erhöht.

Es stellt neue Anforderungen in Bezug auf:

- den Ausgangszustand der Umwelt,
- die Abschätzung der voraussehbaren Auswirkungen des Schemas oder Plans auf die Umwelt,

- die Art der Berücksichtigung des Dokuments zur Erhaltung und Aufwertung der Umwelt.

Es hat außerdem eine Demokratisierung der Verfahren durch die Verallgemeinerung der Konzertierung und des Einspruchverfahrens bewirkt.

Die Verordnung 2004-489 vom 3. Juni 2004 überträgt die Anforderungen der europäischen Richtlinie 2001/42 vom 27. Juni 2001 über die Prüfung der Umweltauswirkungen bestimmter Pläne und Programme in nationales Recht. Die Stadtentwicklungsdokumente, die beträchtliche Auswirkungen auf die Umwelt haben können, müssen der Umweltprüfung unterzogen werden. Die Stadtentwicklungsdokumente müssen außerdem Frankreichs internationalen sowie europäischen Verpflichtungen (Konferenz von Rio, das Netz Natura 2000, das Kyoto-Protokoll sowie die den Umweltbereich betreffenden Normen) berücksichtigen (Gesetz zum Gebirge, Gesetz zu den Küstenregionen, Gesetz zur Luft, Gesetz zum Wasser…).

Das an die Stelle des Stadtentwicklungsgesetzes getretene Dekret 2005-608 vom 27. Mai 2005 über die Prüfung der Umweltauswirkungen der Stadtentwicklungsdokumente präzisiert die Dokumente, die einer Umweltprüfung unterzogen werden müssen. Hierzu gehören die territorialen Raumentwicklungsrichtlinien (DTA), der Flächennutzungsplan für die Region Ile-de-France (SDRIF), die regionalen Bebauungspläne (SAR) der Überseeregionen, der Plan zur Bebauung und nachhaltigen Entwicklung Korsikas, die SCOT und die PLU, durch die ein Natura 2000-Standort erheblich beeinträchtigt werden kann, sowie die PLU, die nicht aus einem SCOT mit Umweltprüfung entwickelt sind, mit Ausnahme ganz bestimmter Bedingungen.

Die am 28 Februar 2005 angenommene Charta formalisiert neue Umweltpflichten und -anforderungen: das Vorsorgeprinzip, die Pflicht zur Verhütung von Umweltgefahren bzw. zur Eingrenzung der Folgen und das Reparieren von verursachten Umweltschäden.

b) Wie ist im Rahmen der Verträglichkeitsprüfung die Ausnahmeregelung des Artikels 6 Absatz 4 FFH-Richtlinie im Bereich des Bauplanungsrechts in nationales Recht umgesetzt worden und wie wird diese angewandt?

Auszug aus Artikel R 414-23 des Umweltgesetzes:

III. *Ergibt sich aus der unter Pkt. II genannten Analyse, dass das Planungsdokument oder das Programm, das Projekt, die Maßnahmen oder der Eingriff während oder nach der Ausführung oder während der Gültigkeit des Planungsdokuments signifikante schädliche Auswirkungen auf den Erhaltungszustand der natürlichen Habitate und der Arten haben kann, die der Ausweisung eines oder mehrerer Schutzgebiete zugrunde lagen, enthält das Evaluierungsdokument eine Darlegung der Maßnahmen, die zur Beseitigung oder Minderung der schädlichen Auswirkungen ergriffen werden.*

IV. Bestehen trotz der unter Pkt. II genannten Maßnahmen weiterhin signifikante schädliche Auswirkungen auf den Erhaltungszustand der natürlichen Habitate und der Arten, die der Ausweisung eines oder mehrerer Schutzgebiete zugrunde lagen, enthält das Evaluierungsdokument zusätzlich Folgendes:

1. *Die Beschreibung von möglichen Alternativlösungen, die Gründe dafür, dass es keine andere als die gewählte Lösung gibt, und die Elemente, mit denen sich die Billigung des Planungsdokuments begründen lässt, oder die Ausführung des Programms, des Vorhabens, der Maßnahme oder des Eingriffs unter den in Pkt. VII und Pkt. VIII des Artikels L. 414-4 vorgesehenen Bedingungen;*

2. *Die Beschreibung der Maßnahmen, die vorgesehen sind, um die schädlichen Auswirkungen, die von den unter oben stehendem Pkt. III aufgeführten Maßnahmen nicht beseitigt werden können, zu kompensieren. Die Kompensationsmaßnahmen ermöglichen eine angemessene und effiziente Kompensation in Anbetracht der Beeinträchtigung, die sich für die Ziele der Erhaltung des Schutzgebiets oder der Schutzgebiete Natura 2000 ergeben haben. Diese Kompensationsmaßnahmen werden nach einem Zeitplan ergriffen, der eine Kontinuität in den Fähigkeiten des Netzes Natura 2000 zur Erhaltung der natürlichen Habitate und der Arten sichert. Auch wenn diese Kompensationsmaßnahmen zeitlich und räumlich aufgeteilt sind, sind sie doch das Ergebnis eines gesamtheitlichen Herangehens, das diese Kontinuität garantiert.*

3. *Die Einschätzung der entsprechenden Ausgaben und der Modalitäten zur Übernahme der Kompensationsmaßnahmen werden für die Planungsdokumente von der mit der Billigung beauftragten Behörde wahrgenommen, für die Programme, Vorhaben und Eingriffe vom Bauherren bzw. vom antragstellenden Begünstigten und für die Maßnahmen vom organisierenden Begünstigten.*

Gibt es eine Differenzierung in Bezug darauf, dass ein Vorhaben innerhalb oder außerhalb eines bereits im Zusammenhang bebauten Gebietes errichtet wird?

Nein.

c) Wird neben der schutzgebietsbezogenen Verträglichkeitsprüfung des Artikels 6 Absatz 3 FFH-Richtlinie bei baulichen Vorhaben auch eine artenschutzrechtliche Prüfung nach Artikel 12 FFH-Richtlinie durchgeführt?

Keine Antwort

d) Wird ein Bedarf dafür gesehen, solche Bauvorhaben von artenschutzrechtlichen Verboten freizustellen, die in Gebieten errichtet werden, die bereits im Zusammenhang bebaut sind? Welche Rolle könnte es hierbei haben, wenn die geschützte Art ihre natürliche Verbreitung außerhalb des Siedlungsbereichs hat und dort auch in größerer Zahl vorkommt?

Keine Antwort

e) Wie wird das Tatbestandsmerkmal der „Absicht" in Artikel 12 Absatz 1 Buchstabe a, b und c FFH-Richtlinie bei Bauvorhaben im Lichte der Rechtsprechung des Europäischen Gerichtshofs verstanden?

Keine Antwort

3.3.3 Frage 3: Plan-UP-Richtlinie (2001/42/EG)

a) Welche Ebenen und Instrumente städtebaulicher Planungen bestehen und wie erfolgte diesbezüglich die Umsetzung der Plan-UP-Richtlinie in nationales Recht?

Keine Antwort

b) Wie wurde die Plan-UP-Richtlinie in das nationale Recht umgesetzt?

Wurde schon beantwortet, s. o.

c) Werden für bestimmte Fälle, insbesondere der Geringfügigkeit im Rahmen der Aufstellung, Änderung und Ergänzung von Bauleitplänen, Möglichkeiten zur Beschleunigung sowie Verfahrenserleichterungen genutzt?

Die Umweltprüfung (EE) ist für die in einem SCOT enthaltenen PLU nicht erforderlich.

d) In welchem Verhältnis stehen Umweltprüfung und FFH-Verträglichkeits-prüfung zueinander?

Keine Antwort

3.3.4 Frage 4: Dienstleistungsrichtlinie (2006/123/EG)

Wird ein Umsetzungsbedarf der Dienstleistungsrichtlinie (Richtlinie 2006/123/EG) für den Bereich des nationalen Städtebaurechts und der Bauleitplanung gesehen?

Nur im kommerziellen Sinne: Das Gesetz zur Modernisierung der Wirtschaft 2008-776 vom 4. August 2008 hat den Inhalt der SCOT und einiger PLU präzisiert. Aus den planungs- und umweltrechtlichen Anforderungen können sich Vorgaben ergeben, die sich einschränkend auf die kommerzielle Niederlassungsfreiheit auswirken. Insofern wird dem Umsetzungsbedarf im Bereich der Stadtplanung und -entwicklung Rechnung getragen.

3.3.5 Frage 5: Erneuerbare-Energien-Richtlinie (2009/28/EG)

Wird ein Umsetzungsbedarf hinsichtlich des Artikels 13 Erneuerbare-Energien-Richtlinie (2009/28/EG) für den Bereich des nationalen Städtebaurechts und der Bauleitplanung gesehen?

A priori kein Bedarf, ich hatte jedoch keine Zeit, mich kundig zu machen.

3.3.6 Frage 6: SEVESO-II-Richtlinie

Wie ist Artikel 12 Absatz 1 Unterabsatz 2 der Richtlinie 96/82/EG vom 09.12.1996 zur Beherrschung der Gefahren bei Unfällen mit gefährlichen Stoffen (ABl. L 10 vom 14.01.1997, S. 13; „Seveso-II-Richtlinie") in nationales Recht umgesetzt worden?

Keine Antwort

3.4 Great Britain

3.4.1 Question 1: Systematic preliminary question

What is the relationship between binding land use plans and the approval of individual construction projects?

No answer

3.4.2 Question 2: European Habitat and Protection of Species Law

a) *What is the relationship between national nature conservation law and the building planning law? Are nature conservation law requirements taken into account in development planning or not dealt with until a subsequent approval procedure takes place?*

In national nature conservation law a distinction is made between provisions which relate to:

- the protection of European/ Natura 2000 sites;
- the protection of nationally designated sites (SSSIs and ASSSs);
- the protection of habitats; and
- the protection of fauna and flora.

The protection of European/ Natura 2000 sites is achieved by means of provisions which:-

- require an assessment of the implications of any plan or project, which may have a significant effect of such a site; and
- state that a plan or project, which may adversely affect such a site, cannot be given consent unless the decision-maker is satisfied that there are no alternative solutions and the plan or project must be carried out for imperative reasons of overriding importance. The question of compensatory measures also needs to be addressed.

[Case law (Waddenzee) says that if there *may be* (as opposed to *is likely to* be) a significant effect then an appropriate assessment should be carried out.]

As a matter of policy (not law) the above provisions also apply to in respect of Ramsar sites and potential Special Protections Areas.

The protection of other habitats is achieved by other means, one of which is provisions for the notification, protection and management of Sites of Special Scientific Interest ("SSSIs"). A notification includes details of the flora, fauna or geological or physiographical features by reason of which the land is of special interest and any operations that are likely to damage that flora or fauna, or those features. Authorities exercising duties which are likely to affect the features of SSSIs are subject to a general overarching duty to take reasonable steps consistent with the proper exercise of the authority's functions, to further the conservation and enhancement of the features for which site are of special interest. There are also procedural safeguards which apply when an authority in the exercise of its functions proposes to carry out or authorise operations which are likely to damage the special interest features of SSSIs.

The protection of particular species of fauna and flora is achieved by;-

- a duty on local planning authorities, in the exercise of their functions, to have regard to the Directive;
- the creation of a number of specific offences in respect of the harming, destruction, capturing, picking etc of protected species; and
- a licensing system for activities which may infringe protected species protection provisions.

Nature conservation requirements are taken into account by:-

- plan-making authorities when preparing national, regional and local land use plans; and
- decision-makers when considering applications for particular plans and projects.

[Need something on National Parks/ AONBs/ local sites etc?]

There is a section 40 duty under the Natural Environment and Rural Communities Act 2006 whereby "Every public authority must, in exercising its functions, have regard, so far as is consistent with the proper exercise of those functions, to the purpose of conserving biodiversity.'

These formal statutory provisions are supplemented by planning policy guidance which is a material issue for the courts in any legal challenge that may be mounted.

b) Within the scope of the assessment of the implications of a plan or project for a site, how has the exception provided by Article 6 Paragraph 4 of the FFH Directive been transposed into national law in the field of building planning and how is it applied?

In this context, the exception provided by paragraph 4 of article 6 has been transposed by provisions in the Conservation of Habitats and Species Regulations 2010. Under the Regulations, the competent authority may only agree to the plan/ project after ascertaining that it will not adversely affect the integrity of the site. In its considerations, the authority has to have regard to the manner in which the plan/ project is proposed to be carried out, or to any conditions or restrictions that would be imposed on such consent. For further details, see regulations 61-67. A plan or project, which may (see comment re 'is likely to' above) adversely affect a site, cannot be given consent unless the decision-maker is satisfied that there are no alternative solutions and the plan or project must be carried out for imperative reasons of overriding importance. The appropriate authority must secure that any necessary compensatory measures are taken to ensure the overall coherence of Natura 2000 is protected.

Is there any distinction regarding whether a project is built within or outside an area which is already continuously built up?

There is no distinction regarding whether a project is built within or outside an area which is already continuously built up.

c) Apart from the conservation area related assessment of implications for the site pursuant to Article 6 Paragraph 3 of the FFH Directive, is an assessment under species protection law also performed according to Article 12 of the FFH Directive?

Guidance, which planning authorities are required to have regard to, states that "The presence of a protected species is material consideration when a planning authority is considering a development proposal that, if carried out, would be likely to result in hard to the species or its habitat. Local authorities should consult English Nature before granting planning permission."(from paragraph 98 of Circular 06/2005). [the wording differs slightly from the wording that is currently out to consultation] They should also consider attaching appropriate planning conditions or entering into planning obligations under which the developer would take steps to secure the long-term protection of the species concerned. Planning authorities should advise developers that they must comply with any statutory species' protection provisions affecting the site.

The guidance also says [this is taken from the consultation document] that 'it is essential that the presence or otherwise of protected species, and the extent to which they may be affected by the proposed development, is established and taken into account before the planning application is determined, otherwise all relevant material considerations may not have been addressed in making the decision. The need to ensure that ecological surveys are carried out should be left to coverage under planning conditions only in exceptional circumstances. Such surveys should normally be carried out before planning permission is granted. However, bearing in mind the delay and cost that may be involved, developers should not be required to undertake surveys for protected species unless there is a reasonable likelihood of the species being present and affected by the development. Where this is the case, the survey should be completed and any necessary measures to protect the species should be in place, through conditions and/ or planning obligations, before the permission is granted. In appropriate circumstances the permission may also im-pose a condition preventing the development from proceeding without the prior acquisition of a licence.'

Regulation 9(5) of the Habitats Regulations provides that "competent authority, in exercising any of their functions, must have regard to the requirements of the Habitats Directive so far as they may be affected by the exercise of those functions."

[We could also mention the Wildlife and Countryside Act 1981 which also provides protection for certain species?]

d) *Is there a need to exempt such construction projects from prohibitions under species protection law which are erected in areas which have already been developed? What role could it play if the protected species had their natural habitat outside the development area and also occurred in greater number there?*

No. The Directive's requirements include a strict system for the protection of European protected species whether they are present in areas that have already been developed or elsewhere. [In considering derogations under Article 16, there are limited circumstances under which a derogation licence can be issued. Under regulation 53(9) of the Habitats Regulations the relevant licensing body must not grant a licence unless satisfied that (1) there is no satisfactory alternative, and (2) the action authorised will not be 'detrimental to the maintenance of the population of the species concerned at a favourable conservation status in their natural range'. The presence of significant numbers of the species outside the development area might be a material consideration, but this would depend on the specific plan/ project, the species involved, and the potential effects of the plan/ project on the species.]

e) *How is the element of "intention" in Article 12 Paragraph 1 Letter a, b and c of the FFH Directive interpreted for construction projects in the light of the legal decisions taken by the European Court of Justice?*

Commission guidance on Article 12 (2007) discusses two ECJ judgements which considered the term 'deliberate' in the context of the operation of the Habitats Directive (Commission v Greece, Case C-103/00, Commission v Spain, case C-221/04). Further to the guidance, deliberate actions are to be understood as actions by a person who knows, in the light of the relevant legislation that applies to the species involved, and the general information delivered to the public, that his action will most likely lead to an offence against a species, but intends this offence or, if not, consciously accepts the foreseeable results of his action.

Regulation 3(3) of the Habitats Regulations provides that, if terms are used but not defined in the Habitats Regulations, and are used in the Habitats Directive, they have the same meaning as in the Habitats Directive.

3.4.3 Question 3: SEA Directive (2001/42/EC)

a) Which levels and instruments of development planning exist and with respect to this, how was the SEA Directive transposed into national law?

In England, the SEA Directive was transposed in to national law by the Environmental Assessment of Plans and Programmes Regulations 2004 (SI 2004/1633). These regulations apply to all plans and programmes which meet the criteria in the SEA Directive. However, further requirements exist in relation to town and country planning development plans. At the national level, national non-statutory planning policies are not usually subject to strategic environmental assessment. However, regional and local plans which set the framework for development consent in town and country planning both require SEA, not solely by virtue of the Environmental Assessment of Plans and Programmes Regulations 2004 but by virtue of town and country planning legislation encompassing wider considerations. The Planning and Compulsory Purchase Act 2004 introduced a requirement for Sustainability Appraisals of regional and local plans. Sustainability Appraisals are meant to cover the requirements for SEA under the Directive, but go wider than that by also requiring assessment of economic and social impacts. For major infrastructure development projects, development consents are made in accordance with a new and separate statutory framework to other town and country planning decision making. The primary legislation is contained in the Planning Act 2008. There are no plans as such but an Independent Planning Commission makes development consent decisions on the basis of national policy statements. The Planning Act 2008 requires an Appraisal of Sustainability to be undertaken for all national policy statements. Like Sustainability Appraisals for regional and local plans, this covers the requirements of the SEA Directive but goes wider than that to consider other impacts.

b) How was the SEA Directive transposed into national law?

As answer to 3(a)

c) Are options, especially the option of insignificance within the context of preparing, amending and supplementing the development plans, used to accelerate and simplify procedures for particular cases?

No answer

d) What is the relationship between the environmental audit and the FFH assessment of implications for the site?

No answer

3.4.4 Question 4: EU Services Directive (2006/123/EC)

Is it deemed necessary to transpose the EC Services Directive (Directive 2006/123/EC) for the area of national urban planing law and development planning?

Question 4 (Is it deemed necessary to transpose the Services Directive for the area of national planning law and development planning?) - the short answer is no. Looking back at the correspondence I have been given, I think that the reasoning set out in the German model answer reflects the reasoning that CLG took at the time, including with regard to recital 9, so I would be a bit reluctant to set out a lengthy answer that would pretty much just paraphrase what they have said. There is one point in the COLA advice which isn't covered in the German answer, and which might be helpful to highlight as follows:

"It is not deemed necessary to transpose the Services Directive for the area of national planning law and development planning. Consideration was given to recital 9 of the Directive, and it was concluded that this recital is intended to exclude rules of universal application (that is, rules affecting all persons, service providers and private individuals alike, and which form part of the "framework" in which the market in a particular Member State operates). Member States should not rely on general groups of rules falling within a particular category (e. g. planning) in order to exclude a whole category from the scope of the Directive on the basis that in the generality of cases that category does not specifically concern service providers. Rather, Member States should consider whether specific rules within each category specifically affect service providers. Having conducted such an exercise, the HMG view is that there are no rules within its planning legislation which specifically affect service providers and therefore no transposition is required."

3.4.5 Question 5: Renewable Energy Directive (2009/28/EC)

Is it considered necessary to transpose Article 13 of the Renewable Energy Directive (Directive 2009/28/EC) for the area of national urban planning law and development planning?

A UK National Action Plan for the transposition of the Renewable Energy Directive is being prepared and will be published on the Commission website in due course.

3.4.6 Question 6: SEVESO-II-Directive

How has Article 12 Paragraph 1 Subsection 2 Council Directive 96/82/EC of 9 December 1996 on the control of major-accident hazards involving dangerous substances (OJ L 10, 14.1.1997, p. 13; "Seveso-II-Directive") been transposed into national law?

No answer

3.4 Großbritannien

3.4.1 Frage 1: Systematische Vorfrage

In welchem Verhältnis stehen Bebauungspläne und die Genehmigung einzelner Bauvorhaben?

Keine Antwort

3.4.2 Frage 2: Europäisches Habitat- und Artenschutzrecht

a) *In welchem Verhältnis stehen das nationale Naturschutzrecht und das Bauplanungsrecht? Werden die Anforderungen des Naturschutzrechts bereits in der Bauleitplanung oder erst in einem nachfolgenden Genehmigungsverfahren bewältigt?*

Im nationalen Naturschutzgesetz wird unterschieden zwischen Bestimmungen zum:

- Schutz der europäischen Natura 2000-Gebiete,
- Schutz national ausgewiesener Flächen (Sites of Special Scientific Interest – SSSI und ASSS), [A. d. Ü: Ass = Archeologically Sensitive Site]
- Schutz von Lebensräumen,
- Schutz von Fauna und Flora.

Europäische/ Natura 2000-Gebiete werden mithilfe von Bestimmungen geschützt, die:

- Eine Bewertung der Auswirkungen jedes Plans oder Vorhabens verlangen, das ein solches Gebiet in erheblichem Maße beeinträchtigen könnte, und
- festlegen, dass ein Plan oder Vorhaben, das sich in erheblichem Maße auf ein solches Gebiet auswirken könnte, nicht genehmigt werden darf, es sei denn, der Entscheidungsträger wird überzeugt, dass es keine Alternativlösungen gibt und der Plan oder das Vorhaben aus zwingenden Gründen des überwiegenden öffentlichen Interesses ausgeführt werden muss. Die Frage von Ausgleichsmaßnahmen muss ebenfalls behandelt werden.

[In dem Urteil zum Präzedenzfall (Waddenzee) wurde entschieden, dass eine zweckdienliche Bewertung vorzunehmen ist, wenn es erhebliche Auswirkungen *geben könnte* (im Gegensatz zu „Auswirkungen *wahrscheinlich sind*").]

In der politischen Leitlinie (nicht aber dem Gesetz) ist festgelegt, dass die obigen Bestimmungen auch für Gebiete nach der Ramsar-Konvention und mögliche besondere Schutzgebiete (*Special Protection Areas*) gelten.

Der Schutz anderer Lebensräume wird mit anderen Mitteln erreicht, zu denen u. a. die Bestimmungen zur Bekanntmachung, zum Schutz und zur Verwaltung von Gebieten gehören, die von besonderem wissenschaftlichen Interesse sind (*Sites of Special Scientific Interest* – SSSI). Eine Bekanntmachung enthält Angaben über die Flora, Fauna oder die geologischen und physiogeografischen Merkmale, aufgrund derer das Gebiet von besonderem Interesse ist, sowie Handlungen, die diese Pflanzen- und Tierarten oder diese Merkmale wahrscheinlich beeinträchtigen würden. Behörden, die Befugnisse ausüben, die sich wahrscheinlich auf die Merkmale eines SSSI auswirken, unterliegen einer Pflicht, angemessene Schritte zu ergreifen, die mit der ordnungsgemäßen Ausübung ihrer Funktionen vereinbar sind, um den Erhalt und die Stärkung der Merkmale zu fördern, aufgrund derer das Gebiet von besonderem Interesse ist. Es gibt außerdem Sicherheitsmechanismen im Verfahren, die greifen, wenn eine Behörde in Ausübung ihrer Befugnisse Maßnahmen zur Durchführung oder Genehmigung vorschlägt, die wahrscheinlich die Merkmale eines SSSI, die das besondere Interesse begründen, beeinträchtigen.

Der Schutz besonderer Tier- und Pflanzenarten wird erreicht durch:

- Die Pflicht der lokalen Planungsbehörden, in der Ausübung ihrer Funktionen die Richtlinie zu berücksichtigen.
- Die Schaffung einer Reihe spezifischer Straftatbestände mit Bezug auf die Beeinträchtigung, Zerstörung, das Fangen, Pflücken etc. geschützter Arten und.
- Ein Genehmigungssystem für Aktivitäten, die gegen Bestimmungen zum Schutz bestimmter Arten verstoßen könnten.

Naturschutzanforderungen finden Berücksichtigung durch die:

- Planungsbehörden bei der Aufstellung nationaler, regionaler und lokaler Flächennutzungspläne und
- Entscheidungsträger bei der Prüfung von Anträgen für besondere Pläne und Vorhaben.

[Brauchen Sie Angaben zu Nationalparks/ AONB – *Area of outstanding natural beauty* – Gebiete außergewöhnlicher natürlicher Schönheit/ lokal ausgewiesene Gebiete etc.?]

Im nationalen Gesetz über das natürliche Umfeld und ländliche Gemeinschaften (*Natural Environment and Rural Communities Act*) von 2006 wird in § 40 eine Verpflichtung definiert: „Jede öffentliche Behörde muss in Ausübung ihrer Funktionen das Ziel des Erhalts der Artenvielfalt berücksichtigen, sofern dieses mit der ordnungsgemäßen Ausübung dieser Aufgaben vereinbar ist.“

Diese formellen rechtlichen Bestimmungen werden durch Leitlinien der Planungspolitik ergänzt, die für die Gerichte im Falle der Erhebung von Einsprüchen relevant sind.

b) Wie ist im Rahmen der Verträglichkeitsprüfung die Ausnahmeregelung des Artikels 6 Absatz 4 FFH-Richtlinie im Bereich des Bauplanungsrechts in nationales Recht umgesetzt worden und wie wird diese angewandt?

In diesem Kontext wurde die in Artikel 6 Absatz 4 der Richtlinie zur Erhaltung der natürlichen Lebensräume sowie der wildlebenden Tiere und Pflanzen festgelegte Ausnahme durch die Rechtsnormen von 2010 über den Erhalt von Lebensräumen und Arten umgesetzt. Nach diesen Rechtsnormen darf die zuständige Behörde nur dann einem Plan/ Vorhaben zustimmen, wenn sie sich vergewissert hat, dass dieses das Gebiet nicht beeinträchtigen wird. In der Abwägung muss die Behörde auch die Art und Weise berücksichtigen, wie der Plan/ das Vorhaben ausgeführt werden soll, oder andere Bedingungen oder Voraussetzungen, die im Fall der Genehmigung einzuhalten sind. Zu weiteren Einzelheiten siehe Rechtsnormen 61-67. Ein Plan oder Vorhaben, das ein Gebiet beeinträchtigen könnte (im Gegensatz zu „es wahrscheinlich beeinträchtigen wird“, siehe oben) darf nicht genehmigt werden, es sei denn, der Entscheidungsträger gelangt zu dem Schluss, dass es keine Alternativlösung gibt und der Plan oder das Vorhaben aus zwingenden Gründen des überwiegenden öffentlichen Interesses ausgeführt werden muss. Die zuständige Behörde muss sicherstellen, dass alle notwendigen Ausgleichsmaßnahmen durchgeführt werden, um die Gesamtkohärenz von Natura 2000 zu schützen.

Gibt es eine Differenzierung in Bezug darauf, dass ein Vorhaben innerhalb oder außerhalb eines bereits im Zusammenhang bebauten Gebietes errichtet wird?

Es wird keine Unterscheidung dahin gehend getroffen, ob ein Vorhaben innerhalb oder außerhalb eines bereits im Zusammenhang bebauten Gebietes durchgeführt wird.

c) Wird neben der schutzgebietsbezogenen Verträglichkeitsprüfung des Artikels 6 Absatz 3 FFH-Richtlinie bei baulichen Vorhaben auch eine artenschutzrechtliche Prüfung nach Artikel 12 FFH-Richtlinie durchgeführt?

Der Leitfaden, den die Planungsbehörden berücksichtigen müssen, besagt, dass „das Vorkommen einer geschützten Art eine wesentliche Erwägung ist, wenn eine Planungsbehörde einen Bauantrag prüft, der im Falle seiner Ausführung diese Art oder ihren Lebensraum beeinträchtigen würde. Lokale Behörden sollten sich mit English Nature (Naturschutzverband) beraten, bevor sie die Planungsgenehmigung erteilen.” (aus Absatz 98 des Rundschreibens 06/2005) [der Wortlaut unterscheidet sich geringfügig von dem Wortlaut, der sich derzeit in der Konsultation befindet].

Sie sollten weiterhin erwägen, zweckdienliche Planungsbedingungen oder –verpflichtungen festzulegen, die den Bauträger verpflichten, Schritte zu unternehmen, um den langfristigen Schutz der betroffenen Art zu gewährleisten. Die Planungsbehörden sollten Bauträger darauf hinzuweisen, dass sie die gesetzlichen Artenschutzbestimmungen, die für das Gebiet gelten, einhalten müssen.

Die Leitlinien besagen weiterhin [Auszug aus dem Konsultationsdokument], dass es wichtig ist, dass das Vorkommen oder ein anderes Vorliegen einer geschützten Art und das Maß, in dem diese durch das beabsichtigte Vorhaben beeinträchtigt werden könnte, ermittelt und vor der Entscheidung über den Planungsantrag berücksichtigt werden müssen; ansonsten werden eventuell nicht alle rechtsrelevanten Aspekte bei der Entscheidung beachtet. Die Notwendigkeit ökologischer Erhebungen sollte nur unter außergewöhnlichen Umständen erst unter Planungsbedingungen sichergestellt werden. Solche Erhebungen sollten in der Regel vorgenommen werden, bevor die Planungsgenehmigung erteilt wird. Angesichts der Verzögerung und dafür anfallenden Kosten sollten Bauträger jedoch nur dann verpflichtet werden, Erhebungen für geschützte Arten durchzuführen, wenn eine begründete Wahrscheinlichkeit vorliegt, dass diese Art vorkommt und durch das Vorhaben beeinträchtigt wird. Ist dies der Fall, müssen die Erhebungen durchgeführt werden und notwendige Maßnahmen zum Schutz der Art in Form von Bedingungen und/ oder Planungsverpflichtungen vorhanden sein, bevor die Genehmigung erteilt wird. Wo die Umstände dies zulassen, kann in der Genehmigung auch die Bedingung auferlegt werden, dass das Vorhaben nur vorbehaltlich des vorherigen Einholens einer Lizenz ausgeführt werden kann.'

Absatz 9(5) der Habitat-Rechtsnormen sieht vor, ‚dass die zuständige Behörde in der Ausübung ihrer Funktionen die Anforderungen der FFH-Richtlinie berücksichtigen muss, sofern diese durch die Ausübung dieser Funktionen berührt werden könnten."

[Wir könnten auch das Gesetz zu wild lebenden Tierarten und zur Landschaft - *Wildlife and Countryside Act* – von 1981 erwähnen, das auch bestimmte Arten schützt?]

d) Wird ein Bedarf dafür gesehen, solche Bauvorhaben von artenschutzrechtlichen Verboten freizustellen, die in Gebieten errichtet werden, die bereits im Zusammenhang bebaut sind? Welche Rolle könnte es hierbei haben, wenn die geschützte Art ihre natürliche Verbreitung außerhalb des Siedlungsbereichs hat und dort auch in größerer Zahl vorkommt?

Nein. Die Anforderungen in der Richtlinie umfassen ein strenges System zum Erhalt geschützter Arten, unabhängig davon, ob sie in bereits bebauten Gebieten oder woanders vorkommen. [In Erwägung der Ausnahmeregelungen in Artikel 16 gibt es eingeschränkte Umstände, in denen eine Ausnahmelizenz erteilt werden kann. Nach Absatz 53(9) der Habitat-Rechtsnormen darf die zuständige Genehmigungsbehörde nur dann eine Genehmigung erteilen, wenn sie sichergestellt hat, dass es (1) keine zufriedenstellende Alternative gibt und (2) die genehmigte Maßnahme für den Fortbestand der Population der betroffenen Art in ihrem natürlichen Verbreitungsgebiet in einem günstigen Erhaltungszustand nicht nachteilig ist. Ein zahlenmäßig beträchtliches Vorkommen der Art außerhalb des Erschließungsgebiets kann eine rechtsrelevante Überlegung sein, aber dies hängt vom jeweiligen Plan/ Vorhaben und den möglichen Auswirkungen des Plans/ Vorhabens auf die Art ab.]

e) Wie wird das Tatbestandsmerkmal der „Absicht" in Artikel 12 Absatz 1 Buchstabe a, b und c FFH-Richtlinie bei Bauvorhaben im Lichte der Rechtsprechung des Europäischen Gerichtshofs verstanden?

Der Leitfaden der Kommission zu Artikel 12 (2007) erläutert zwei EuGH-Urteile, in denen der Begriff „absichtlich" im Kontext der FFH-Richtlinie erörtert wird (Kommission gegen Griechenland, Rechtssache C-103/00; Kommission gegen Spanien, Rechtssache C-221/04). Nach dem Leitfaden sind als absichtliche Handlungen Schritte einer Person zu verstehen, die angesichts der für die betreffenden Art geltenden einschlägigen Rechtsvorschriften und aufgrund der allgemeinen Unterrichtung der Öffentlichkeit weiß, dass ihre Handlungen höchstwahrscheinlich zu einem Verstoß gegen die Artenschutzbestimmungen führen, diesen Verstoß jedoch gewollt begeht oder die vorhersehbaren Folgen ihrer Handlung bewusst in Kauf nimmt.

Absatz 3(3) der Habitat-Rechtsvorschriften legt fest, dass Begriffe, die zwar in den Habitat-Rechtsvorschriften verwendet und dort nicht definiert werden, jedoch in der FFH-Richtlinie verwendet werden, die gleiche Bedeutung wie in der FFH-Richtlinie haben.

3.4.3 Frage 3: Plan-UP-Richtlinie (2001/42/EG)

a) Welche Ebenen und Instrumente städtebaulicher Planungen bestehen und wie erfolgte diesbezüglich die Umsetzung der Plan-UP-Richtlinie in nationales Recht?

In England wurde die Plan-UP-Richtlinie mittels der Rechtsvorschriften zur Umweltprüfung von Plänen und Programmen von 2004 (*Environmental Assessment of Plans and Programmes Regulations* – SI 2004/1633) in nationales Recht umgesetzt. Diese Rechtsvorschriften gelten für alle Pläne und Programme, die die Kriterien der Plan-UP-Richtlinie erfüllen. Es gelten jedoch noch weitere Anforderungen an Stadt- und Landbauleitpläne. Auf nationaler Ebene unterliegen die nationalen, gesetzlich nicht vorgeschriebenen Planungspolitiken in der Regel keiner strategischen Umweltprüfung. Regionale und lokale Pläne hingegen, die den Rahmen für die Genehmigung von Vorhaben in der Stadt- und Landplanung vorgeben, erfordern beide die SUP, nicht nur Kraft der Rechtsvorschriften zur Umweltprüfung von Plänen und Programmen von 2004, sondern auch nach dem Stadt- und Landplanungsrecht, das darüber hinausgehende Überlegungen umfasst. Das Planungs- und Zwangsenteignungsgesetz von 2004 (*Planning and Compulsory Purchase Act*) machte Nachhaltigkeitsbewertungen für regionale und lokale Pläne verpflichtend. Diese Nachhaltigkeitsbewertungen sollen die Anforderungen der SUP-Richtlinie abdecken, gehen aber darüber hinaus, da sie auch die Bewertung wirtschaftlicher und sozialer Auswirkungen verlangen. Bei großen Infrastrukturbauvorhaben werden die Genehmigungen nach einem neuen, eigenen Rechtsrahmen erteilt, der getrennt vom restlichen Stadt- und Landplanungsentscheidungsverfahren ist. Das Primärrecht ist im Planungsgesetz von 2008 geregelt. Es gibt keine Pläne als solches, sondern eine unabhängige Planungskommission (*Independent Planning Commission*) trifft die Entscheidungen über die Planungsgenehmigung anhand nationaler Politikleitlinien. Das Planungsgesetz von 2008 verlangt, dass für alle nationalen Politiken eine Nachhaltigkeitsbewertung durchzuführen ist. Wie bei den Nachhaltigkeitsbewertungen für regionale und lokale Pläne deckt diese die Anforderungen der Plan-UP-Richtlinie ab, geht aber darüber hinaus, da auch andere Auswirkungen erfasst werden.

b) Wie wurde die Plan-UP-Richtlinie in das nationale Recht umgesetzt?

Siehe Antwort 3(a)

c) Werden für bestimmte Fälle, insbesondere der Geringfügigkeit im Rahmen der Aufstellung, Änderung und Ergänzung von Bauleitplänen, Möglichkeiten zur Beschleunigung sowie Verfahrenserleichterungen genutzt?

Keine Antwort

d) In welchem Verhältnis stehen Umweltprüfung und FFH-Verträglichkeitsprüfung zueinander?

Keine Antwort

3.4.4 Frage 4: Dienstleistungsrichtlinie (2006/123/EG)

Wird ein Umsetzungsbedarf der Dienstleistungsrichtlinie (Richtlinie 2006/123/EG) für den Bereich des nationalen Städtebaurechts und der Bauleitplanung gesehen?

Frage 4 (Wird ein Umsetzungsbedarf der Dienstleistungsrichtlinie für den Bereich des nationalen Städtebaurechts und der Bauleitplanung gesehen?) – die kurze Antwort darauf lautet „nein“. Nach Durchsicht des mir vorliegenden Schriftverkehrs denke ich, dass die in der deutschen Beispielantwort geschilderte Begründung der Argumentation entspricht, die CLG seinerzeit vertrat, einschließlich der zu Erwägungsgrund 9. Daher würde ich ungern eine lange Antwort formulieren, die mehr oder weniger den Inhalt der deutschen Antwort aufgreift. Es gibt jedoch einen Punkt in der Stellungnahme von COLA, der in der deutschen Antwort nicht enthalten ist und hilfreich sein könnte:

„Es wird nicht als erforderlich angesehen, die Dienstleistungsrichtlinie für den Bereich des nationalen Planungsrechts und der Bauleitplanung umzusetzen. Erwägungsgrund 9 der Richtlinie wurde betrachtet und es wurde die Schlussfolgerung gezogen, dass der Erwägungsgrund Allgemeinverbindlichkeitsregeln ausschließen soll (d. h. Regeln, die für alle Personen, Dienstleistungsanbieter und Privatpersonen gleichermaßen gelten und Teil des „Rahmens" sind, innerhalb dessen der Markt in einem bestimmten Mitgliedstaat arbeitet). Die Mitgliedstaaten sollten sich nicht darauf verlassen, dass allgemeine Gruppen von Rechtsvorschriften in eine bestimmte Kategorie fallen (z. B. Planung), um so eine gesamte Kategorie vom Geltungsbereich der Richtlinie auf der Grundlage auszunehmen, dass in der Mehrheit der Fälle diese Kategorie die Dienstleistungsanbieter nicht spezifisch betrifft. Mitgliedstaaten sollten vielmehr erörtern, ob spezifische Rechtsvorschriften innerhalb jeder Kategorie sich spezifisch auf Dienstleistungsanbieter auswirken. Nach Durchführung einer solchen Bewertung ist die Meinung von HMG, dass es keine Rechtsvorschriften im Planungsgesetz gibt, die sich spezifisch auf Dienstleistungsanbieter auswirken, und daher ist keine Umsetzung in nationales Recht erforderlich."

3.4.5 Frage 5: Erneuerbare-Energien-Richtlinie (2009/28/EG)

Wird ein Umsetzungsbedarf hinsichtlich des Artikels 13 Erneuerbare-Energien-Richtlinie (2009/28/EG) für den Bereich des nationalen Städtebaurechts und der Bauleitplanung gesehen?

Das Vereinigte Königreich erarbeitet derzeit einen nationalen Aktionsplan zur Umsetzung der Erneuerbare-Energien-Richtlinie, der zum gegebenen Zeitpunkt auf der Webseite der Kommission veröffentlicht wird.

3.4.6 Frage 6: SEVESO-II-Richtlinie

Wie ist Artikel 12 Absatz 1 Unterabsatz 2 der Richtlinie 96/82/EG vom 09.12.1996 zur Beherrschung der Gefahren bei Unfällen mit gefährlichen Stoffen (ABl. L 10 vom 14.01.1997, S. 13; „Seveso-II-Richtlinie") in nationales Recht umgesetzt worden?

Keine Antwort

3.5 Ireland

3.5.1 Question 1: Systematic preliminary question

What is the relationship between binding land use plans and the approval of individual construction projects?

In Ireland, while there is a spatial strategy at national level (National Spatial strategy 2002) and regional planning guidelines at regional level (eight regions) land use planning is generally carried out at the city or county administrative area level. Each City or County administrative area, or Town Council area (where appropriate) must prepare a development plan every six years.

The development plan sets out the overall strategy for the proper planning and sustainable development of the area of the development plan. The (private) individual construction projects must seek planning permission from the planning authority and the planning authority then makes the decision to grant, grant subject to conditions, or to refuse permission for stated reasons. These decisions can then be appealed to an appeals board. (An Bord Pleanala) In making the planning decision the planning authority must consider the proper planning and sustainable development of the area, regard being had to the development plan, provisions of any special amenity area order, any European site or other area prescribed, where appropriate, the policy of the Government, the Minister or any other Minister of the Government and any other relevant provision or requirement of the planning act or regulations made under the Act.

Where the individual construction projects are public projects e.g. being carried out by the planning authority and permission is being sought under the Planning and Development acts, a different procedure is followed but such development cannot contravene materially the development plan.

3.5.2 Question 2: European Habitat and Protection of Species Law

a) *What is the relationship between national nature conservation law and the building planning law? Are nature conservation law requirements taken into account in development planning or not dealt with until a subsequent approval procedure takes place?*

All development plans (land use plans) either require or are screened for SEA (2001/42/EC) or are now recommended through guidance to be screened for Habitats Directive Assessment (92/43/EEC) (Legislation is being prepared to formally require planning authorities to screen all plans for HDA)

Under the Planning and Development Act 2000 (as amended) development plans shall include objectives for 'the conservation and protection of the environment including, in particular, the archaeological and natural heritage and the conservation and protection of European Sites and any other sites which may be prescribed for the purposes of this paragraph'.

Nature conservation experts from the Department of the Environment, Heritage and Local Government comment on and input into the process of drafting of plans, so that any issues relating to compliance with nature conservation law are brought to notice for consideration.

At Development management stage (project stage) planning applications are assessed/ screened for Environment Impact Assessment (85/337/EEC as amended by 97/11/EC.... Planning and Development Regulations 2001 as amended, European Communities (Environmental Impact Assessment) (Amendment) Regulations 2006) and screened for HDA (92/43/EEC) (European Communities (Natural Habitats) Regulations 1997 as amended).

The existing transposition of the Habitats Directive has been found to be defective and new legislation and regulation is currently being prepared to give a greater alignment between the nature conservation type of legislation and regulation and the land use legislation and regulation).

b) Within the scope of the assessment of the implications of a plan or project for a site, how has the exception provided by Article 6 Paragraph 4 of the FFH Directive been transposed into national law in the field of building planning and how is it applied?

Article 6 Paragraph 4 is transposed in the 1997 Habitats Regulations and to date no Article 6 paragraph 4 development has yet been finalised.

Is there any distinction regarding whether a project is built within or outside an area which is already continuously built up?

With reference to article 6 paragraph 4 above there is no distinction in national legislation between whether a project is built within or outside an area which is already continuously built up. Each project is considered on its own merits.

c) Apart from the conservation area related assessment of implications for the site pursuant to Article 6 Paragraph 3 of the FFH Directive, is an assessment under species protection law also performed according to Article 12 of the FFH Directive?

There is at present no statutory requirement to carry out an assessment under the Habitats Directive. However, it is a criminal offence in Irish law to damage or destroy the breeding site or resting place of an animal listed in Annex IV of the Habitats Directive, or to interfere with protected flora. EIA or SEA should identify the presence of such fauna or flora.

d) Is there a need to exempt such construction projects from prohibitions under species protection law which are erected in areas which have already been developed? What role could it play if the protected species had their natural habitat outside the development area and also occurred in greater number there?

Such cases have not been raised by developers for consideration. They would fall to be considered under the derogation provisions either in accordance with Article 6.4 or Article 16 of the Habitats Directive

e) How is the element of "intention" in Article 12 Paragraph 1 Letter a, b and c of the FFH Directive interpreted for construction projects in the light of the legal decisions taken by the European Court of Justice?

By the requirement that derogation licences be obtained in accordance with Article 16 of the habitats Directive where EIA or SEA identifies a risk in relation to the protected species.

3.5.3 Question 3: SEA Directive (2001/42/EC)

a) Which levels and instruments of development planning exist and with respect to this, how was the SEA Directive transposed into national law?

As indicated in response to question 1 planning authorities must make a development plan for their administrative area every 6 years. These plans are informed by the Regional Planning Guidelines which in turn are informed by the National Spatial Strategy. All city and county development plans, and development plans by town councils with a population over 10.000 require an SEA to be carried out and to inform the making of the plan. All other statutory land use plans under the planning and development acts require to be screened for SEA.

The SEA Directive was transposed into national law by way of regulation. (European Communities (Environmental Assessment of Certain Plans and Programmes) Regulations 2004; Planning and Development (Strategic Environmental Assessment) Regulations 2004)

b) How was the SEA Directive transposed into national law

As indicated above the SEA Directive was transposed into national law by way of regulation. (European Communities (Environmental Assessment of Certain Plans and Programmes) Regulations 2004:SI 435; Planning and Development (Strategic Environmental Assessment) Regulations 2004: SI 436)

SI 436 includes the procedural requirements under the Directive (such as screening, scoping, environmental report, consultation with the designated environmental authorities, the public and if necessary adjacent EU member States, consideration in the making of the plan, monitoring measures and final statement on how the environmental considerations were taken into account in the adopted plan.)

c) Are options, especially the option of insignificance within the context of preparing, amending and supplementing the development plans, used to accelerate and simplify procedures for particular cases?

Where development plans are being varied or amended, the variation or the amendment is screened to ascertain if it is necessary to prepare an SEA. This includes consulting with the Designated Environmental Authorities as appropriate.

d) What is the relationship between the environmental audit and the FFH assessment of implications for the site?

Under the current regulations, planning authorities and An Bord Pleanala on appeal, must ensure that an appropriate assessment (Habitats Directive)is undertaken if necessary. Where an environmental impact assessment accompanies an application for development this is considered to contain the elements of an appropriate assessment.

Recent guidance issued by the Department of Environment, Heritage and Local Government indicated that the Appropriate Assessment may be prepared as part of the EIS but if prepared as part of the EIS it must be a distinct and discrete section of the EIS and clearly marked as such.

Legislation is currently being prepared to incorporate the provisions of the Habitats Directive into planning legislation, rather than, as at present, having separate provisions relating to planning in the regulations transposing the Birds and Habitats Directives (the European communities (Habitats) Regulations 1997-2005).

3.5.4 Question 4: EU Services Directive (2006/123/EC)

Is it deemed necessary to transpose the EC Services Directive (Directive 2006/123/EC) for the area of national urban planning law and development planning?

It is considered that land use plans do not require to be screened for Directive 2006/123/EC, as it is felt that the application of the planning legislation is applied in a uniform manner to all applicants and that the planning system in Ireland does not put undue barriers to service providers from across the EU who wish to operate in Ireland i. e the planning rules do not discriminate or facilitate discrimination against potential entrants to Irish markets.

Having regard to the various Articles and Recitals (particularly Recitals # 9 and 40), it is felt that our planning policies and guidelines for the preparation of development plans (which include, for example, retail planning strategies) seek to strike a balance among economic, social and environmental objectives and such strategies and ensuing policy implementation applies to all (Irish and other EU commercial or business entities).

3.5.5 Question 5: Renewable Energy Directive (2009/28/EC)

Is it considered necessary to transpose Article 13 of the Renewable Energy Directive (Directive 2009 /28/EC) for the area of national urban planning law and development planning?

Currently, no further transposition need in Ireland is considered to exist for development planning law.

Major/ strategic renewable energy infrastructure projects

There have been a number of developments in recent years in Irish Planning legislation to help facilitate the development of renewable energy, in particular in relation to micro-renewable projects and the 'fast-tracking' of projects for obtaining permits for projects deemed to be of a strategic nature.

The 2006 Planning and Development (Strategic Infrastructure) Act came into effect on 1st January 2007 and was designed to ensure that the development of grid infrastructure is underpinned by coordinated local, regional and national approaches to issues, which balance local interests with the national imperative to deliver strategic energy infrastructure projects.

Exemptions for micro-renewable technologies

In order to facilitate the development of certain renewable energy projects and to reduce unnecessary obstacles or non-proportionate requirements, exemptions for certain micro renewable projects were introduced by the Department of the Environment, Heritage and Local Government in 2007 and 2008. These measures reduced the regulatory burden in order to promote the use of certain renewable technologies (ground/ air source heat pump, biomass, and wind turbines for example), and to achieve a better alignment with key national policy objectives including, in particular, addressing climate change through the planning system.

Information is available on the Department on Environment, Heritage & Local Government website: http://www.environ.ie and from the Renewable Energy Information Office http://www.seai.ie/Renewables/REIO/

3.5.6 Question 6: SEVESO-II-Directive

How has Article 12 Paragraph 1 Subsection 2 Council Directive 96/82/EC of 9 December 1996 on the control of major-accident hazards involving dangerous substances (OJ L 10, 14.1.1997, p. 13; "Seveso-II-Directive") been transposed into national law?

No answer

3.5 Irland

3.5.1 Frage 1: Systematische Vorfrage

In welchem Verhältnis stehen Bebauungspläne und die Genehmigung einzelner Bauvorhaben?

In Irland gibt es zwar eine nationale Raumstrategie (*National Spatial Strategy 2002*) und regionale Raumordnungsleitlinien (acht Regionen), aber die Flächennutzungsplanung erfolgt in der Regel auf Ebene des Stadt- oder Verwaltungskreisgebiets (*county*). Jede Stadt, jeder Kreis oder ggf. Stadtrat muss alle sechs Jahre einen Bauleitplan für sein Gebiet aufstellen.

Der Bauleitplan legt die Gesamtstrategie für die angemessene Planung und nachhaltige Entwicklung des im Plan erfassten Gebiets fest. Für (private) Einzel-Bauvor-haben ist die Baugenehmigung der Planungsbehörde einzuholen; diese entscheidet, die Genehmigung zu erteilen, sie vorbehaltlich bestimmter Bedingungen zu erteilen oder unter Angabe der Gründe zu verweigern. Gegen diese Bescheide kann vor einer Beschwerdekammer (An Bord Pleanala) Einspruch eingelegt werden. Bei der Entscheidung muss die Planungsbehörde die angemessene Planung und nachhaltige Entwicklung ihres Zuständigkeitsgebiets beachten, dabei den Bauleitplan, Vorgaben für schutzwürdige Gebiete, ggf. europäische oder andere ausgewiesene Stätten, die Politik der Regierung, des Fachministers oder anderer Minister sowie andere relevante Bestimmungen oder Anforderungen des Planungsgesetzes oder nach Maßgabe des Gesetzes festgelegter Rechtsnormen berücksichtigen.

Handelt es sich bei Einzelbauvorhaben um öffentliche Projekte, d. h. werden sie von der Planungsbehörde ausgeführt und wird die Genehmigung nach dem Planungs- und Bauleitrecht (*Planning Act und Development Act*) beantragt, greift ein anderes Verfahren, jedoch darf ein solches Bauvorhaben nicht maßgeblich gegen den Bauleitplan verstoßen.

3.5.2 Frage 2: Europäisches Habitat- und Artenschutzrecht

a) In welchem Verhältnis stehen das nationale Naturschutzrecht und das Bauplanungsrecht? Werden die Anforderungen des Naturschutzrechts bereits in der Bauleitplanung oder erst in einem nachfolgenden Genehmigungsverfahren bewältigt?

Alle Bauleitpläne (Flächennutzungspläne) erfordern entweder eine Plan-UP (2001/42/EG) oder werden auf die Notwendigkeit einer Plan-UP hin überprüft oder es wird durch die Leitlinien empfohlen, sie auf die Notwendigkeit einer Prüfung nach FFH-Richtlinie (92/43/EG) zu bewerten. (Ein Gesetz ist in Vorbereitung, das von den Planungsbehörden verlangen wird, alle Pläne auf die Notwendigkeit einer Prüfung nach FFH-Richtlinie zu bewerten.)

Nach dem Planungs- und Bauleitgesetz von 2000 (in der geänderten Form) müssen Bauleitpläne Ziele ‚für den Erhalt und Schutz der Umwelt, insbesondere des archäologischen und natürlichen Erbes, und den Erhalt und Schutz europäischer und anderer Stätten, die für die Zwecke dieses Absatzes festgelegt werden, vorsehen'.

Naturschutzexperten vom Ministerium für Umwelt, Erbe und lokale Verwaltung (Department of the Environment, Heritage and Local Government) kommentieren die Aufstellung der Pläne und leisten einen Beitrag, sodass mögliche Probleme bezüglich der Einhaltung von Naturschutzgesetzen festgestellt werden und beachtet werden können.

In der Bebauungsverwaltungsphase (Projektstadium) werden Bauanträge einer UVP unterworfen/ auf UVP-Pflicht geprüft (85/337/EG in der durch 97/11/EG geänderten Form…, auf die Planungs- und Bauleitrechtsnormen von 2001 in der geänderten Form, Rechtsnormen der Europäischen Gemeinschaften (Umweltverträglichkeitsprüfung) 2006 (Änderung)) und auf die Prüfungspflicht nach FFH-Richtlinie (92/43/EG) (Rechtsnormen der Europäischen Gemeinschaften (Natürliche Lebensräume) 1997, in der geänderten Form) bewertet.

Die derzeitige Umsetzung der FFH-Richtlinie wurde für unzureichend befunden und neue Gesetze und Rechtsnormen befinden sich derzeit in der Erarbeitung, um die Gesetzgebung und Rechtsnormen zum Naturschutz besser mit der Gesetzgebung und den Rechtsnormen zur Flächennutzung in Einklang zu bringen.

b) Wie ist im Rahmen der Verträglichkeitsprüfung die Ausnahmeregelung des Artikels 6 Absatz 4 FFH-Richtlinie im Bereich des Bauplanungsrechts in nationales Recht umgesetzt worden und wie wird diese angewandt?

Artikel 6 Absatz 4 wurde in der Habitat-Rechtsnorm von 1997 umgesetzt und bisher wurden keine Bauvorhaben nach Artikel 6 Absatz 4 abgeschlossen.

Gibt es eine Differenzierung in Bezug darauf, dass ein Vorhaben innerhalb oder außerhalb eines bereits im Zusammenhang bebauten Gebietes errichtet wird?

Hinsichtlich Artikel 6 Absatz 4 unterscheidet die nationale Gesetzgebung nicht zwischen Bauvorhaben innerhalb oder außerhalb eines bereits zusammenhängend bebauten Gebiets. Jedes Vorhaben wird eigenständig bewertet.

c) Wird neben der schutzgebietsbezogenen Verträglichkeitsprüfung des Artikels 6 Absatz 3 FFH-Richtlinie bei baulichen Vorhaben auch eine artenschutzrechtliche Prüfung nach Artikel 12 FFH-Richtlinie durchgeführt?

Derzeit ist eine Prüfung nach der FFH-Richtlinie nicht vorgeschrieben. Die Zerstörung von Brut- oder Ruhestätten von Tierarten, die in Anhang IV der FFH-Richtlinie aufgeführt sind, oder der Eingriff in geschützte Flora stellen nach irischem Recht jedoch einen Straftatbestand dar. Um festzustellen, ob solche Tier- oder Pflanzenarten vorliegen, sollte eine UVP oder SUP vorgenommen werden.

d) Wird ein Bedarf dafür gesehen, solche Bauvorhaben von artenschutzrechtlichen Verboten freizustellen, die in Gebieten errichtet werden, die bereits im Zusammenhang bebaut sind? Welche Rolle könnte es hierbei haben, wenn die geschützte Art ihre natürliche Verbreitung außerhalb des Siedlungsbereichs hat und dort auch in größerer Zahl vorkommt?

Diese Fragen wurden von den Bauträgern bisher nicht angesprochen. Sie würden unter die Ausnahmeregelungen nach Artikel 6.4 oder Artikel 16 der FFH-Richtlinie fallen.

e) Wie wird das Tatbestandsmerkmal der „Absicht" in Artikel 12 Absatz 1 Buchstabe a, b und c FFH-Richtlinie bei Bauvorhaben im Lichte der Rechtsprechung des Europäischen Gerichtshofs verstanden?

Durch die Anforderung, dass Ausnahmelizenzen nach Artikel 16 der FFH-Richtlinie eingeholt werden müssen, wenn eine UVP oder SUP ein Risiko für die bedrohte Art feststellt.

3.5.3 Frage 3: Plan-UP-Richtlinie (2001/42/EG)

a) Welche Ebenen und Instrumente städtebaulicher Planungen bestehen und wie erfolgte diesbezüglich die Umsetzung der Plan-UP-Richtlinie in nationales Recht?

Wie bereits zu Frage 1 oben ausgeführt, müssen die Planungsbehörden alle sechs Jahre einen Bauleitplan für ihr Verwaltungsgebiet erstellen. Diese Pläne richten sich nach den regionalen Planungsleitlinien, die sich wiederum an der nationalen Raumstrategie orientieren. Alle Bauleitpläne für Städte und Kreise und Bauleitpläne, die von Stadträten für ein Gebiet mit mehr als 10,000 Einwohnern erstellt werden, erfordern eine SUP, die in die Planerstellung einfließen muss. Alle anderen gesetzlich vorgeschriebenen Flächennutzungspläne nach dem Planungs- und Bauleitrecht müssen auf die Notwendigkeit einer Umweltprüfung hin bewertet werden.

Die Plan-UP-Richtlinie wurde über eine Verordnung in nationales Recht umgesetzt (Rechtsnormen der Europäischen Gemeinschaften (Umweltprüfung bestimmter Pläne und Programme) 2004; Planungs- und Bauleitrecht (Strategische Umweltprüfung) 2004).

b) Wie wurde die Plan-UP-Richtlinie in das nationale Recht umgesetzt?

Wie oben ausgeführt, wurde die Plan-UP-Richtlinie mittels einer Verordnung in nationales Recht umgesetzt. (Rechtsnormen der Europäischen Gemeinschaften (Abschätzung der Umweltfolgen bestimmter Pläne und Programme) 2004 – SI 435; Planungs- und Bauleitrecht (Strategische Umweltprüfung) 2004 – SI 436).

Die Verordnung SI 436 umfasst die Verfahrensvorgaben nach der Richtlinie (wie Überprüfung, Scoping, Umweltbericht, Beteiligung der zuständigen Umweltbehörden, der Öffentlichkeit und, sofern erforderlich, benachbarten EU-Mitgliedstaaten, Berücksichtigung bei der Planaufstellung, Monitoringmaßnahmen und zusammenfassende Erklärung, wie die Umweltaspekte im verabschiedeten Plan berücksichtigt wurden.)

c) Werden für bestimmte Fälle, insbesondere der Geringfügigkeit im Rahmen der Aufstellung, Änderung und Ergänzung von Bauleitplänen, Möglichkeiten zur Beschleunigung sowie Verfahrenserleichterungen genutzt?

Wenn Bauleitpläne geändert oder fortgeschrieben werden, wird die Änderung oder Fortschreibung überprüft, um festzustellen, ob eine SUP erforderlich ist. Dies beinhaltet auch eine Einbeziehung der zuständigen Umweltbehörden, sofern erforderlich.

d) In welchem Verhältnis stehen Umweltprüfung und FFH-Verträglichkeitsprüfung zueinander?

Nach den derzeit geltenden Bestimmungen müssen die Planungsbehörden und im Falle eines Einspruchs (auch An Bord Pleanala) sicherstellen, dass bei Bedarf eine zweckdienliche Bewertung (FFH-Richtlinie) vorgenommen wird. Wird bei einem Bauantrag eine Umweltverträglichkeitsprüfung durchgeführt, gilt diese als ausreichend und es wird davon ausgegangen, dass diese alle Elemente einer zweckdienlichen Bewertung enthält.

Die jüngsten Leitlinien des Ministeriums für Umwelt, Erbe und kommunale Verwaltung besagen, dass eine zweckdienliche Bewertung als Teil einer Umweltverträglichkeitserklärung (UVE) erstellt werden kann, in diesem Fall aber ein eigenständiger Teil der UVP und als solcher gekennzeichnet sein muss.

Es wird derzeit ein Gesetz erarbeitet, um die Bestimmungen der FFH-Richtlinie in irisches Recht zu überführen, statt der bisher separaten Bestimmungen zur Planung in den Rechtsnormen, die die Vogelschutz- und FFH-Richtlinie in irisches Recht umsetzen (Rechtsnormen der Europäischen Gemeinschaften (Habitate) 1997-2005).

3.5.4 Frage 4: Dienstleistungsrichtlinie (2006/123/EG)

Wird ein Umsetzungsbedarf der Dienstleistungsrichtlinie (Richtlinie 2006/123/EG) für den Bereich des nationalen Städtebaurechts und der Bauleitplanung gesehen?

Man ist der Ansicht, dass Flächennutzungspläne nicht auf die Notwendigkeit einer Prüfung nach Richtlinie 2006/123/EG hin bewertet werden müssen, da das Planungsrecht für alle Antragsteller einheitlich angewandt wird und das Planungssystem in Irland Dienstleistern aus anderen EU-Ländern, die ihre Dienste in Irland erbringen möchten, keine ungebührlichen Hemmnisse aufbürdet, d. h. die Planungsvorschriften mögliche Einsteiger auf dem irischen Markt weder diskriminieren noch deren Diskriminierung fördern.

Bezüglich der verschiedenen Artikel und Erwägungsgründe (vor allem Erwägungsgründe # 9 und 40) ist man der Meinung, dass unsere Planungspolitiken und -leitlinien für die Aufstellung der Bauleitpläne (die unter anderem die Einzelhandelsplanungsstrategie umfassen) versuchen, das richtige Gleichgewicht zwischen wirtschaftlichen, sozialen und ökologischen Zielsetzungen herzustellen; diese Strategien und die darauf folgende Umsetzung der Politik gilt für alle (kaufmännische oder unternehmerische Einheiten aus Irland und der restlichen EU).

3.5.5 Frage 5: Erneuerbare-Energien-Richtlinie (2009/28/EG)

Wird ein Umsetzungsbedarf hinsichtlich des Artikels 13 Erneuerbare-Energien-Richtlinie (2009/28/EG) für den Bereich des nationalen Städtebaurechts und der Bauleitplanung gesehen?

Derzeit wird in Irland kein weiterer Umsetzungsbedarf für den Bereich der Bauleitplanung gesehen.

Große/ strategische Infrastrukturvorhaben für erneuerbare Energien

In den letzten Jahren hat es eine Reihe von Änderungen im irischen Planungsrecht gegeben, um die Entwicklung der erneuerbaren Energien zu fördern, insbesondere bei Kleinstprojekten in diesem Bereich und ein ‚Eilverfahren' für die Erteilung von Genehmigungen für Projekte, die als strategisch eingestuft werden.

Das Planungs- und Bauleitrecht von 2006 (Gesetz über strategische Infrastruktur) trat am 1. Januar 2007 in Kraft und wurde konzipiert, um sicherzustellen, dass der Bau der Netzinfrastruktur durch abgestimmte lokale, regionale und nationale Ansätze in Fragen unterstützt wird, die lokale Interessen gegenüber der nationalen Notwendigkeit abwägen, strategische Energieinfrastrukturprojekte durchzuführen.

Ausnahmen für Mikroprojekte im Bereich der erneuerbaren Energie

Um die Durchführung bestimmter Projekte im Bereich der erneuerbaren Energien zu fördern und unnötige Hemmnisse oder unverhältnismäßige Anforderungen zu beseitigen, hat das Ministerium für Umwelt, Erbe und lokale Verwaltung 2007 und 2008 Ausnahmen für bestimmte Kleinstprojekte eingeführt. Diese Maßnahmen verringern den Verwaltungsaufwand, um eine bestimmte Form von erneuerbaren Energien zu fördern (z. B. Wärmepumpen im Boden/ Wasserquelle, Biomasse und Windräder) und eine bessere Ausrichtung an wichtigen nationalen Politikzielen zu erreichen, unter anderem der Bekämpfung des Klimawandels durch das Planungssystem.

Informationen finden Sie auf der Webseite des Ministeriums für Umwelt, Erbe und lokale Verwaltung: http://www.environ.ie und des Informationsamts für Erneuerbare Energien (Renewable Energy Information Office http://www.seai.ie/Renewables/REIO/).

3.5.6 Frage 6: SEVESO-II-Richtlinie

Wie ist Artikel 12 Absatz 1 Unterabsatz 2 der Richtlinie 96/82/EG vom 09.12.1996 zur Beherrschung der Gefahren bei Unfällen mit gefährlichen Stoffen (ABl. L 10 vom 14.01.1997, S. 13; „Seveso-II-Richtlinie") in nationales Recht umgesetzt worden?

Keine Antwort

3.6 Latvia

3.6.1 Question 1: Systematic preliminary question

What is the relationship between binding land use plans and the approval of individual construction projects?

Territorial Planning Law (2002, amended in 2002, 2003, 2004, 2005, 2006, 2007, 2008) of Latvia defines that local self-government (municipality) ensures development and implementation of land use plan, detailed plan and/ or amendments to them and authorises a land use plan as binding regulations.

Building Law (1995, amended in 1997, 2002, 2003, 2004, 2005, 2006, 2007, 2009) prescribes that building on a land plot can be executed, if it is carried out in accordance with a land use plan of local municipality, detailed plan (if requested by laws and regulations) and building regulations that is a part of a land use plan. Individual construction projects are checked to be in line with a land use plan and building permission is issued by the Building Board of the municipality.

3.6.2 Question 2: European Habitat and Protection of Species Law

a) *What is the relationship between national nature conservation law and the building planning law? Are nature conservation law requirements taken into account in development planning or not dealt with until a subsequent approval procedure takes place?*

Law on Specially Protected Nature Territories (1997, amended in 2002, 2003, 2005, 2007, 2009) defines that protected nature areas and their zoning is included/ indicated in a land use plan of municipality. Besides, the Law prescribes types of regulations to ensure conservation of protected nature areas – i. e., general and individual regulations, authorised by the Cabinet of Ministers as well as management plans, authorised by the Minister for Environment.

Law also defines that any development planning must take into consideration regulations of conservation and usage as well as management plan of the protected nature area. Furthermore, general and individual regulations for protected nature areas have hierarchically higher juridical binding power than land use plan. Thus nature conservation requirements are strictly incorporated into overall development planning process and govern over the other sectors regarding protected nature areas.

b) Within the scope of the assessment of the implications of a plan or project for a site, how has the exception provided by Article 6 Paragraph 4 of the FFH Directive been transposed into national law in the field of building planning and how is it applied?

Exception provided by Article 6 Paragraph 4 of the FFH Directive has been transposed by Law on Specially Protected Nature Territories that prescribes a require of compensatory measures and defines that initiator of intended activity or producer of the planning document covers all necessary compensation measures, including monitoring. According to the Law, Environment State Bureau informs the Commission of the compensatory measures adopted.

Furthermore, Law on Environmental Impact Assessment (2001, amended in 2003, 2004, 2005, 2006) specifies for which intended activities (incl. building) in all Natura 2000 sites the impact assessment must be done and delegates the procedure of the impact assessment to the Regulations of the Cabinet of Ministers. Building is permitted according to the conclusions of initial assessment result or, if requested by responsible authorities, - the environmental impact assessment report.

Territorial Planning Law delineates the task to maintain natural and cultural heritage, landscapes and biodiversity. In its turn, relevant Regulations of the Cabinet of Ministers on land use planning require to consider specially protected nature areas, microreserves and landscape areas to be safeguarded when developing land use plans.

Is there any distinction regarding whether a project is built within or outside an area which is already continuously built up?

There is no distinction regarding whether a project is being to be developed in or outside already continuously built up area.

c) Apart from the conservation area related assessment of implications for the site pursuant to Article 6 Paragraph 3 of the FFH Directive, is an assessment under species protection law also performed according to Article 12 of the FFH Directive?

The same regulations/ procedure applies for species protection (see answers above under b)).

In addition, Law on the Conservation of Species and Biotopes (2005, amended in 2006, 2009) deals with issues regarding species and habitats protection, management and monitoring to ensure their favourable conservation status.

d) Is there a need to exempt such construction projects from prohibitions under species protection law which are erected in areas which have already been developed? What role could it play if the protected species had their natural habitat outside the development area and also occurred in greater number there?

In Latvia it depends from case to case, both situations have taken place and relevant decisions are being taken.

e) How is the element of "intention" in Article 12 Paragraph 1 Letter a, b and c of the FFH Directive interpreted for construction projects in the light of the legal decisions taken by the European Court of Justice?

Law on the Conservation of Species and Biotopes prescribes necessary measures to ensure requirements of FFH Directive Article 12 Paragraph1 Letter a,b,c. No court cases for Latvia have been possessed.

3.6.3 Question 3: SEA Directive (2001/42/EC)

a) Which levels and instruments of development planning exist and with respect to this, how was the SEA Directive transposed into national law?

Development Planning System Law (2009, amended in 2010) foresees long-term, mid-term and short-term development planning. Long-term plans (up to 25 years) define long-term development priorities and spatial perspective, while mid-term plans – mid-term priorities and action plan either for national, regional or local level. These are not legally binding acts but identify and show future development of the particular area (state, planning region or local/ municipal level). Territorial Planning Law grants municipalities a legal instrument for development and building preparation of a land use plan. The binding land use plan of municipality is authorised by local self-government.

b) How was the SEA Directive transposed into national law?

SEA Directive is fully transposed in national legislation by Law on Environmental Impact Assessment. It defines criteria, cases when SEA should be applied – for planning documents incl. land-use plan, procedure, responsibilities etc. The competent authority –Environment State Bureau takes a decision whether SEA should be applied for planning documents that covers small areas of local municipalities and technical amendments to land use plans, except cases when there likely could be substantial impact to environment. SEA must be applied for planning documents that might affect Natura 2000 sites.

c) *Are options, especially the option of insignificance within the context of preparing, amending and supplementing the development plans, used to accelerate and simplify procedures for particular cases?*

Not foreseen in the current legislation on development and territorial planning.

Simplifying procedure on (1) arrangement of preparation of detailed plan and of construction design to be carried out at the same planning stage as a single document and (2) identification of areas as 'further planning territories/ zones' in case if not concrete potential land use obvious at the moment is envisaged in a new draft Territorial Planning Law. The draft Law is currently submitted to relevant Ministries for concertation.

d) *What is the relationship between the environmental audit and the FFH assessment of implications for the site?*

No additional environmental audit needed as assessment of implications on Natura 2000 sites and species is being performed during EIA/ SEA process that is ensured by applying Law on Environmental Impact Assessment.

3.6.4 Question 4: EU Services Directive (2006/123/EC)

Is it deemed necessary to transpose the EC Services Directive (Directive 2006/123/EC) for the area of national urban planing law and development planning?

No, the Directive issues do not concern development planning laws. Building services -engineering, designing, building expertise, supervising are regulated by Building Law. Associated Regulations of the Cabinet of Ministers prescribes procedure for certification and register of building engineers and architects.

3.6.5 Question 5: Renewable Energy Directive (2009/28/EC)

Is it considered necessary to transpose Article 13 of the Renewable Energy Directive (Directive 2009/28/EC) for the area of national urban planning law and development planning?

Not at the moment, but might be considered. Guidelines for Renewable Energy Resources in 2006-2013 have been adopted by the Cabinet of Ministers in 2006 and draft Law on Renewable Energy Resources is under preparation.

3.6.6 Question 6: SEVESO-II-Directive

How has Article 12 Paragraph 1 Subsection 2 Council Directive 96/82/EC of 9 December 1996 on the control of major-accident hazards involving dangerous substances (OJ L 10, 14.1.1997, p. 13; "Seveso-II-Directive") been transposed into national law?

Objects and restriction areas defined by SEVESO Directive are accordingly foreseen in land use plans of municipalities. Risk areas and objects, unfavourable sites for building, areas of specific requirements and protection belts/ zones are to be considered when land use plans prepared as Regulations of the Cabinet of Ministers 'Local Governments Territorial Planning Regulations' define. The Regulations are consequential Act of Territorial Planning Law.

Also, Protected Belts Law defines types of protected belts int. al. along roads, rails, gas pipelines and storages, and their size.

3.6 Lettland

3.6.1 Frage 1: Systematische Vorfrage

In welchem Verhältnis stehen Bebauungspläne und die Genehmigung einzelner Bauvorhaben?

Das Raumordnungsgesetz (2002, geändert 2002, 2003, 2004, 2005, 2006, 2007, 2008) Lettlands legt fest, dass die kommunale Selbstverwaltung (Kommune) für die Ausarbeitung und Umsetzung des Bauleitplans, des detaillierten Plans und/ oder deren Änderungen Sorge trägt und einen Bauleitplan als verbindliche Regelung genehmigt.

Das Baurecht (1995, geändert 1997, 2002, 2003, 2004, 2005, 2006, 2007, 2009) legt fest, dass auf einem Grundstück gebaut werden kann, wenn dies entsprechend dem Bauleitplan des örtlichen Gemeinwesens, einem **detaillierten Plan** (sofern von Gesetzen und Verordnungen vorgesehen) und den baurechtlichen Vorschriften, die Teil eines Bauleitplans sind, geschieht. Einzelne Bauvorhaben werden auf die Einhaltung des Bauleitplans und der vom Bauamt der Kommune ausgestellten Baugenehmigung überprüft.

3.6.2 Frage 2: Europäisches Habitat- und Artenschutzrecht

a) In welchem Verhältnis stehen das nationale Naturschutzrecht und das Bauplanungsrecht? Werden die Anforderungen des Naturschutzrechts bereits in der Bauleitplanung oder erst in einem nachfolgenden Genehmigungsverfahren bewältigt?

Das Gesetz über besonders geschützte Naturgebiete (1997, geändert 2002, 2003, 2005, 2007, 2009) legt fest, dass Naturschutzgebiete und ihre Zonierung in einem Bauleitplan der Kommune eingeschlossen/ angegeben sein müssen. Außerdem schreibt das Gesetz verschiedene Arten von Regelungen zur Gewährleistung des Schutzes besonderer Naturschutzgebiete vor, d. h. vom Ministerkabinett genehmigte allgemeine Regelungen und Einzelfallregelungen sowie vom Umweltminister genehmigte Managementpläne.

Das Gesetz legt außerdem fest, dass jede Bauleitplanung die Schutz- und Nutzungsbestimmungen sowie den Managementplan des Naturschutzgebietes berücksichtigen muss. Außerdem haben allgemeine Regelungen und Einzelfallregelungen eine stärkere rechtsverbindliche Wirkung als ein Bauleitplan. Auf diese Weise werden Anforderungen des Naturschutzes in den gesamten Bauleitplanungsprozess einbezogen und regeln die anderen Fachbereiche bezüglich der Naturschutzgebiete.

b) Wie ist im Rahmen der Verträglichkeitsprüfung die Ausnahmeregelung des Artikels 6 Absatz 4 FFH-Richtlinie im Bereich des Bauplanungsrechts in nationales Recht umgesetzt worden und wie wird diese angewandt?

Die in Artikel 6 Absatz 4 FFH-Richtlinie vorgesehene Ausnahmeregelung wurde durch das Gesetz über besonders geschützte Naturräume umgesetzt. Es schreibt Ausgleichsmaßnahmen vor und legt fest, dass der Initiator des beabsichtigten Eingriffs oder der Verfasser des Planungsdokuments alle notwendigen Ausgleichsmaßnahmen, einschließlich des Monitorings, abdecken muss. Entsprechend dem Gesetz informiert die Staatliche Umweltbehörde die Kommission von den ergriffenen Ausgleichsmaßnahmen.

Ferner legt das Gesetz über die Umweltverträglichkeitsprüfung (2001, geändert 2003, 2004, 2005, 2006) fest, für welche beabsichtigten Eingriffe (einschließlich Bauen) in allen Natura 2000-Standorten eine Umweltverträglichkeitsprüfung erfolgen muss und unterwirft das Prüfungsverfahren den Regelungen des Ministerkabinetts. Bautätigkeit ist nach den Schlussfolgerungen aus der ersten Prüfung oder, falls von den zuständigen Behörden gefordert, dem Bericht über die Umweltverträglichkeitsprüfung zulässig.

Das Raumordnungsgesetz beschreibt die Aufgabe, das natürliche und kulturelle Erbe, die Landschaften und Artenvielfalt zu bewahren. Die entsprechenden Regelungen des Ministerkabinetts über die Bauleitplanung sehen ihrerseits vor, dass besondere Naturschutzgebiete, Mikroreservate und Landschaftsschutzgebiete bei der Ausarbeitung von Bauleitplänen zu berücksichtigen sind.

Gibt es eine Differenzierung in Bezug darauf, dass ein Vorhaben innerhalb oder außerhalb eines bereits im Zusammenhang bebauten Gebietes errichtet wird?

Es gibt keine Differenzierung in Bezug darauf, dass ein Vorhaben innerhalb oder außerhalb eines bereits bebauten Gebietes entwickelt wird.

c) Wird neben der schutzgebietsbezogenen Verträglichkeitsprüfung des Artikels 6 Absatz 3 FFH-Richtlinie bei baulichen Vorhaben auch eine artenschutzrechtliche Prüfung nach Artikel 12 FFH-Richtlinie durchgeführt?

Dieselben Bestimmungen gelten/ dasselbe Verfahren gilt auch für den Artenschutz (siehe Antworten unter b).

Außerdem befasst sich das Gesetz über den Artenschutz und die Biotope (2005, geändert 2006, 2009) mit Fragen des Arten- und Habitatschutzes, des Managements und des Monitorings, um ihren günstigen Erhaltungszustand sicherzustellen.

d) Wird ein Bedarf dafür gesehen, solche Bauvorhaben von artenschutzrechtlichen Verboten freizustellen, die in Gebieten errichtet werden, die bereits im Zusammenhang bebaut sind? Welche Rolle könnte es hierbei haben, wenn die geschützte Art ihre natürliche Verbreitung außerhalb des Siedlungsbereichs hat und dort auch in größerer Zahl vorkommt?

In Lettland ist dies vom Einzelfall abhängig; beide Situationen sind eingetreten und entsprechende Entscheidungen wurden getroffen.

e) Wie wird das Tatbestandsmerkmal der „Absicht" in Artikel 12 Absatz 1 Buchstabe a, b und c FFH-Richtlinie bei Bauvorhaben im Lichte der Rechtsprechung des Europäischen Gerichtshofs verstanden?

Das Gesetz über den Artenschutz und die Biotope schreibt notwendige Maßnahmen zur Erfüllung der Anforderungen der FFH-Richtlinie Artikel 12 Absatz 1 Buchstabe a, b und c vor. Gegen Lettland sind keine Gerichtsverfahren anhängig.

3.6.3 Frage 3: Plan-UP-Richtlinie (2001/42/EG)

a) Welche Ebenen und Instrumente städtebaulicher Planungen bestehen und wie erfolgte diesbezüglich die Umsetzung der Plan-UP-Richtlinie in nationales Recht?

Das Gesetz über das Bauplanungssystem (2009, geändert 2010) sieht die langfristige, mittelfristige und kurzfristige Bauleitplanung vor. Langfristige Pläne (bis zu 25 Jahre) legen langfristige Entwicklungsprioritäten und die räumliche Perspektive fest, während mittelfristige Prioritäten und Maßnahmenpläne entweder für die nationale, regionale oder lokale Ebene in mittelfristigen Plänen festgeschrieben werden. Dies sind keine rechtsverbindlichen Handlungen; vielmehr wird die künftige Entwicklung eines bestimmten Gebiets (Staat, Planungsregion, örtliche/ kommunale Ebene) ermittelt und aufgezeigt. Das Raumordnungsgesetz gibt den Kommunen ein Rechtsinstrument zur Erarbeitung und Bauvorbereitung eines Bauleitplans an die Hand. Der bindende Bauleitplan der Kommune wird von den Organen der kommunalen Selbstverwaltung genehmigt.

b) Wie wurde die Plan-UP-Richtlinie in das nationale Recht umgesetzt?

Die Plan-UP-Richtlinie wird in der nationalen Gesetzgebung durch das Gesetz über die Umweltverträglichkeitsprüfung uneingeschränkt umgesetzt. Es legt die Kriterien sowie die Fälle fest, in denen die Strategische Umweltprüfung (SUP) angewendet werden sollte – für Planungsunterlagen einschließlich des Bauleitplans, das Verfahren, die Zuständigkeiten usw. Die zuständige Behörde, das Staatliche Umweltamt, entscheidet über die Anwendung der SUP für Planungsunterlagen, die kleine Gebiete von Kommunen und technische Änderungen der Bauleitpläne betreffen, mit Ausnahme von Fällen erheblicher Umwelteinwirkung. Die SUP ist bei Planungsunterlagen durchzuführen, die sich auf Natura 2000-Standorte auswirken könnten.

c) Werden für bestimmte Fälle, insbesondere der Geringfügigkeit im Rahmen der Aufstellung, Änderung und Ergänzung von Bauleitplänen, Möglichkeiten zur Beschleunigung sowie Verfahrenserleichterungen genutzt?

Dies ist in der gegenwärtigen Gesetzgebung zur Bauleitplanung und Raumordnungsplanung nicht vorgesehen.

Der neue Entwurf des Raumordnungsgesetzes sieht Verfahrenserleichterungen vor für (1) Vorkehrungen für die gleichzeitige Erstellung des detaillierten Plans und des Bauentwurfs in einer Planungsphase als ein einziges Dokument und (2) die Identifizierung von Flächen als „weitere Planungsbereiche/-zonen“, falls keine konkrete potenzielle Bebauung zum Zeitpunkt des neuen Gesetzentwurfs klar ist. Der Gesetzentwurf wird den zuständigen Ministerien gegenwärtig zur Konzertation vorgelegt.

d) In welchem Verhältnis stehen Umweltprüfung und FFH-Verträglichkeitsprüfung zueinander?

Es ist keine zusätzliche Umweltprüfung erforderlich, da die Bewertung der Auswirkungen auf Natura 2000-Standorte während des UVP/ SUP-Prozesses erfolgt, der durch die Anwendung des Gesetzes über die Umweltverträglichkeitsprüfung gewährleistet ist.

3.6.4 Frage 4: Dienstleistungsrichtlinie (2006/123/EG)

Wird ein Umsetzungsbedarf der Dienstleistungsrichtlinie (Richtlinie 2006/123/EG) für den Bereich des nationalen Städtebaurechts und der Bauleitplanung gesehen?

Nein, die in der Richtlinie behandelten Fragen betreffen die Gesetze über die Bauleitplanung nicht. Baudienstleistungen - technische Planung, Projektierungsarbeiten, Bauexpertisen, Bauüberwachung - sind im Baurecht geregelt. Damit im Zusammenhang stehende Regelungen des Ministerkabinetts schreiben das Verfahren für die Zertifizierung und Eintragung von Bauingenieuren und Architekten vor.

3.6.5 Frage 5: Erneuerbare-Energien-Richtlinie (2009/28/EG)

Wird ein Umsetzungsbedarf hinsichtlich des Artikels 13 Erneuerbare-Energien-Richtlinie (2009/28/EG) für den Bereich des nationalen Städtebaurechts und der Bauleitplanung gesehen?

Zum gegenwärtigen Zeitpunkt geschieht dies nicht, es könnte jedoch in Erwägung gezogen werden. Leitlinien zu Quellen erneuerbarer Energien im Zeitraum 2006-2013 wurden vom Kabinett 2006 verabschiedet und ein Gesetzentwurf über Quellen erneuerbarer Energien wird zurzeit erarbeitet.

3.6.6 Frage 6: SEVESO-II-Richtlinie

Wie ist Artikel 12 Absatz 1 Unterabsatz 2 der Richtlinie 96/82/EG vom 09.12.1996 zur Beherrschung der Gefahren bei Unfällen mit gefährlichen Stoffen (ABl. L 10 vom 14.01.1997, S. 13; „Seveso-II-Richtlinie") in nationales Recht umgesetzt worden?

Die in der Seveso-Richtlinie definierten Objekte und beschränkten Bereiche sind in den Bauleitplänen der Kommunen entsprechend vorgesehen. Gefahrengebiete und Objekte, ungünstige Baustandorte, Bereiche mit speziellen Anforderungen und Schutzstreifen/-zonen sind bei der Erstellung der Bauleitpläne gemäß den vom Kabinett erlassenen „Raumordnungsregelungen der örtlichen Behörden" zu berücksichtigen. Die Regelungen sind ein aus dem Raumordnungsrecht folgendes Gesetz.

Das Gesetz über die Schutzstreifen legt außerdem die Schutzstreifenarten, u. a. entlang von Straßen, Schienenwegen, Gasleitungen und Speicheranlagen sowie ihre Größe fest.

3.7 Netherlands

3.7.1 Question 1: Systematic preliminary question

What is the relationship between binding land use plans and the approval of individual construction projects?

A land use plan based on article 3.1 of the Spatial Planning Act defines the usages of land plots for building and other usages.

It is binding for the approval of an application for a building permit, without which it is not allowed to build. If the application for the building permit does not meet the requirements of the land use plan, it is possible for the City Council to grant an exemption from the land use plan (a planning permission) if the application is in their opinion in accordance with good spatial planning. If the exemption is granted, the building permission can follow. Nevertheless, in a Habitat area building is only allowed if one also has permit based on the Nature Conservation Act. If the building disrupts protected flora and fauna, one needs a permit based on the Flora and Fauna Act.

3.7.2 Question 2: European Habitat and Protection of Species Law

a) What is the relationship between national nature conservation law and the building planning law? Are nature conservation law requirements taken into account in development planning or not dealt with until a subsequent approval procedure takes place?

Nature conservation in the Netherlands is aimed at species protection under the Flora and Fauna Act and protection of special areas of conservation under the Nature Conservation Act.

Flora and Fauna Act

This Act regulates protection of plant and animal species living in the wild. It includes those aspects of the Habitats Directive and Wild Birds Directive dealing with species protection, and the Convention on International Trade in Endangered Species of Wild Fauna and Flora. Activities which may be damaging to protected species are essentially prohibited. In addition the Act comprises a permit system, for all activities which could harm the protected species. These activities include building. Permits are issued by the provincial governments or the Ministry of Agriculture, Nature and Food Quality. The Act includes a duty of care, applying to all animals and plants. Finally, the Act also allows landscape elements to be designated protected habitat.

Nature Conservation Act

The Nature Conservation Act 1998 comprises a procedure for designation of nature areas under the Wild Birds and Habitats Directives. For all those nature areas a management plan has to be drawn up.

In addition the Act comprises a permit system (articles 19d – 19kc), for all activities which can harm the protected qualities. These activities include building. Permits are issued by the provincial governments or the Ministry of Agriculture, Nature and Food Quality. This ensures that projects which may affect the Natura 2000 areas will be evaluated most carefully. In addition to nature area protection regulated under the Wild Birds and Habitats Directives, the Nature Conservation Act also regulates the conservation of wetlands and nature monuments.

In the framework of the Wild Birds Directive, 79 areas have been designated in the Netherlands, with a total area of 1,010,910 hectares. The Netherlands has registered 141 Habitats areas with a total area of 750,841 hectares with the European Union. The Wild Birds and Habitats Directives areas together make up the Natura 2000 areas in the Netherlands.

There is no connection between the Housing Act and the Nature Conservation Act or the Flora and Fauna Act. There is no possibility to take these elements into account.

There is also no formal connection between a land use plan and the nature conservation Act or Flora and Fauna Act, but good spatial planning requires that in drafting a land use plan or an exemption from the land use plan, all relevant circumstances should be taken into account, including the effects on nature.

So nature conservation law aims are at a global level taken into account in development planning, but are not dealt with in the building permit until a subsequent approval procedure takes place.

b) Within the scope of the assessment of the implications of a plan or project for a site, how has the exception provided by Article 6 Paragraph 4 of the FFH Directive been transposed into national law in the field of building planning and how is it applied?

See answer to question 2 paragraph a.

Is there any distinction regarding whether a project is built within or outside an area which is already continuously built up?

Neither in the Nature Conservation Act nor in the Buidling Act such distinction has been made. But of course in practice, while establishing if the project causes any harm in the view of the conservation objectives of the Natura 2000-site, this distinction may be important.

c) Apart from the conservation area related assessment of implications for the site pursuant to Article 6 Paragraph 3 of the FFH Directive, is an assessment under species protection law also performed according to Article 12 of the FFH Directive?

Yes (see answer to question 2 paragraph a) species protection law assessments are performed in addition to the assessment, required in article 6, paragraph 3, of the Habitats Directive, if there is any indication that the building project could cause violation of the prohibitions on interventions for protected species (for instance killing of specimens, disturbance during the period of breeding, deterioration of breeding sites and so forth).

d) Is there a need to exempt such construction projects from prohibitions under species protection law which are erected in areas which have already been developed? What role could it play if the protected species had their natural habitat outside the development area and also occurred in greater number there?

Such exception could be helpful for the inner development of towns and would not be a problem from the point of view of biodiversity aims as long as the species are maintained in a favourable conservation status outside the development area. There seems to be no room for a general exception in national law as long as article 16 of the Habitats Directive and article 9 of the Wild Birds Directive are not amended. The possibilities for derogation of the prohibitions in those articles for strictly protected species and birds are very restricted and require an assessment on local level of the effects for the conservation status of the species in the natural range. Building activities are pursuant to article 16 of the Habitats Directive only a reason for derogation if building is in the interest of public health or public safety, or if building is needed for other imperative reasons of overriding public interest. Article 9 of the Wild Birds Directive doesn't even mention 'imperative reasons of overriding public interest' as a justification for derogation.

e) How is the element of "intention" in Article 12 Paragraph 1 Letter a, b and c of the FFH Directive interpreted for construction projects in the light of the legal decisions taken by the European Court of Justice?

On the basis of the Caretta judgement of the European Court of Jusitice (C 103/00) and the judgement on trap hunting (C-6/04) 'intention' is understood to be not only the deliberate killing or destruction but also the accepted killing or destruction, i. e.: the result of the action may be not wanted but the result is known and nevertheless the action resulting in disruption or death is carried out. The legal practice in the Netherlands is in accordance with these rulings.

3.7.3 Question 3: SEA Directive (2001/42/EC)

a) Which levels and instruments of development planning exist and with respect to this, how was the SEA Directive transposed into national law?

On the basis of the Spatial Planning Act municipalities, provinces and the national government are required to set out their spatial policy in one or more structural visions. In some cases an Strategic Impact Assessment is required.

The central instrument of spatial policy is the binding land use plan, issued by the municipal council. The same type of plan can be drafted by the province or the minister of Housing, Spatial planning and the Environment, and is then called an integration plan. All these government bodies can also draft a binding project decision. To the land use plan and project decision the Environmental Impact Assessment is linked.

There are also other Acts which provide the basis for development plans, for example the Water Management Plan based on the Water Act.

The IEA and SEA directives are implemented in chapter 7 of the Dutch Environmental Management Act (Wm) and in the Environmental Impact Assessment Decree 1994 (Besluit M. e. r. 1994).

This Act is a framework act describing the basic principles of environmental policy. The details are provided for in orders in council (AMvB). The Environmental Impact Assessment Decree 1994 is such an order in council. Other important passages in the Environmental Protection Act on the EIA. besides chapter 7 can be found in chapter 2.2 (about the EIA Commission), chapter 14.2 (about the coordination required when drawing up an environmental impact report) and chapter 20 (appeal). The Environmental Impact Assessment Decree 1994 states when an EIA. should be carried out. The decree contains appendices, which include the C and D lists. The C list indicates which activities and decisions require a mandatory environmental impact report. The D list sums up the activities and decisions for which a so-called 'article 7.8a/7.8d procedure' is required. These activities and decisions are evaluated on an individual basis to see whether an EIA. is necessary. (The lists can be found in the Besluit m. e. r. 1994 (in Dutch)).

The EU Strategic Environmental Assessment Directive (Directive 2001/42/EC) was adopted in July 2001. The legislative proposal that transposes this Directive into Dutch legislation mentions an EIA. for plans. To make the difference between the EIA. and the EIA. for plans clear, the latter has been renamed SEA. In practice there is no difference between a SEA and an EIA. for plans as described in the legislative proposal. Article 5 and appendix I of the SEA Directive state precisely what the environmental report should contain. For most activities the Strategic impact Assessment is linked to the structural vision for the activity.

b) How was the SEA Directive transposed into national law?

See before.

c) Are options, especially the option of insignificance within the context of preparing, amending and supplementing the development plans, used to accelerate and simplify procedures for particular cases?

In the Netherlands a choice was made for clear tresholds, below which an environmental impact assessment was not required. For example a plan for the erection of less than 2000 houses did not require such an assessment. However this was not acceptable in the view of the European Commission, because size is not necessarily a sound indication of environmental impact, so now the bill is amended.

d) What is the relationship between the environmental audit and the FFH assessment of implications for the site?

The environmental assessment takes a broad look at the environmental impact including health issues, but the FFH assessment takes a much more specific look at flora and fauna in view of the goals of the directive. EU Services Directive (2006/123/EC)

3.7.4 Question 4: EU Services Directive (2006/123/EC)

Is it deemed necessary to transpose the EC Services Directive (Directive 2006/123/EC) for the area of national urban planing law and development planning?

No, this was not necessary. The EC Services Directive has been implemented by the Services Act. The Services Act has brought no change in the Spatial Planning Act or the Housing Act. Consideration 9 of the directive states: this directive applies only to requirements which affect the access to, or the exercise of, a service activity. Therefore, it does not apply to requirements, such as road traffic, rules concerning the development or use of land, town and country planning, building standards (...). the Spatial Planning Act (land use plans) and the Housing Act only provide general provisions which do not specifically regulate or affect the service activity.

3.7.5 Question 5: Renewable Energy Directive (2009/28/EC)

Is it considered necessary to transpose Article 13 of the Renewable Energy Directive (Directive 2009 /28/EC) for the area of national urban planning law and development planning?

There is no necessity to implement Article 13, Paragraphs 1, 2, 3, 5 and 6, of the Renewable Energy Directive. Article 13, Paragraph 4, does not need implementation, because the Dutch Building Decree already contains adequate energy-perfomance requirements for the erection, alteration and renovation of buildings.

3.7.6 Question 6: SEVESO-II-Directive

How has Article 12 Paragraph 1 Subsection 2 Council Directive 96/82/EC of 9 December 1996 on the control of major-accident hazards involving dangerous substances (OJ L 10, 14.1.1997, p. 13; "Seveso-II-Directive") been transposed into national law?

Article 12 of the EU's Seveso II Directive stipulates, among other things, that Member States must ensure that the objectives of the directive – to prevent major accidents involving dangerous substances and to limit the human and environmental consequences of such accidents – are taken into account in:

- decisions on the siting of new establishments or modifications to existing establishments,
- decisions on new developments in the vicinity of existing establishments.

In the Netherlands, this article has been implemented in national legislation by means of the External Safety (Establishments) Decree. The decree regulates the environmental quality requirements to be set for external safety when land-use planning decisions are taken and when environmental permits are issued for high-risk establishments. It also provides guarantees that, whenever decision-making in these areas impacts on the probability of a disaster occurring and claiming many victims, the procedures will be transparent for the general public.

The decree applies to establishments that fall within the scope of the Hazards of Major Accidents Decree of 1999 (BRZO); stevedoring establishments where dangerous substances are stored as part of the transport process; LPG filling stations; CPR 15-2 and -3 storage depots and establishments containing large-scale ammonia cooling installations. The scope of the decree can be extended by order in council to include other establishments with an external location-based risk exceeding 10-6 per annum (such as propane gas storage depots and certain railway marshalling yards). The decree does not apply to operators handling fireworks or storing explosive substances or objects, unless they are subject to the Hazards of Major Accidents Decree.

Firstly, the decree imposes limit values for the location-based risk to vulnerable objects. The distinction between vulnerable and less vulnerable objects is relevant only to location-based risk, not to societal risk. The term 'vulnerable objects' means residential properties and equivalent premises, such as schools and hospitals. A limit value is a minimum standard to be achieved by a given date and to be maintained thereafter. Limit values must be 'observed' and not exceeded. In setting the limit value, a distinction is drawn between existing and new situations (10-5 per annum for the first and 10-6 per annum for the second). An existing situation is one where an object is present or permissible under the current land use plan. Secondly, the decree imposes a target value for the location based risk to less vulnerable objects.

This term means, for example, workplaces employing relatively few people, such as offices with fewer than 50 members of staff and industrial or commercial premises which themselves pose no risk to the surrounding area and which are generally not occupied by large numbers of people for large parts of the day. A target value is a standard to be achieved so far as possible by a specified date and to be maintained so far as possible thereafter. Authorities must take the target value for location-based risk into account.

The main statutory basis of the External Safety (Establishments) Decree is chapter 5 of the Environmental Management Act. This states that orders in council can be passed establishing environmental quality standards for particular sectors of the environment; the standards may apply either generally or in designated areas, and may be either compulsory limit values or target values to be met so far as possible by a specified date (see further explanation below). The environmental quality standards are directed in the first instance at the public sector. When issuing environmental permits for establishments where dangerous substances are to be stored, produced, processed, etc, and also when making land-use planning decisions relating to the area surrounding such an establishment. Target values must be taken into account and may only be exceeded when there are compelling reasons for doing so.

Location-based risk and societal risk

The decree includes definitions of location-based risk and societal risk. Location-based risk is defined as the risk to individuals in a particular place, expressed as the annual probability of dying in the vicinity of an establishment where dangerous substances are (legally) present as a direct result of an on-site accident involving those substances. The decree establishes environmental quality standards in the form of limit values for location-based risk, e. g. 10-6 per annum for vulnerable objects and for sites in the process of remediation. Where the location-based risk for an existing vulnerable object is between 10-5 and 10-6 per annum, a standstill provision applies: i. e. the establishment concerned may not be modified in any way that will have the effect of increasing that risk. These limit values are primarily binding on the public authorities concerned. Like the duty to explain decisions relating to the level of societal risk, their function is to guarantee a basic level of protection against the risks associated with activities involving dangerous substances in establishments. The duty to explain decisions relating to the level of societal risk in the vicinity of potentially dangerous establishments means that public authorities provide access to information about their choices in this respect. Members of the public can force compliance with the decree via the courts. Societal risk takes account of population density in the vicinity of a high-risk activity and expresses the probability of a given minimum number of fatalities occurring as a result of a single accident involving dangerous substances. The focus is on large scale mortality in case of an accident. When taking the kinds of decisions specified in the decree, provincial and municipal authorities must explain how they have taken account of societal risk in the decision-making process. They must also explain the effect of the environmental license issued or the relevant planning decision on the societal risk. The explanation must include information on what the affected population can do to protect themselves and on the plans for emergency response in the event of an accident. The fire services have powers to advise in this respect.

Land use plan and safety zone

Because the decree also applies to the adoption of land use plans and the granting of exemptions from these, parts of it are based on the Spatial Planning Act. The Act makes municipalities responsible for land-use planning in their areas, and hence for guaranteeing adequate safety in the vicinity of potentially dangerous establishments. Much of this work is done by way of land use plans, which determine how land can be developed and used. A municipality may only designate a piece of land for a high-risk activity in an establishment as referred to in this decree if the associated risks to the vicinity do not exceed the limit values laid down in the decree. In addition, the municipality must judge whether those risks are socially acceptable.

The competent authority can designate a safety zone surrounding the site of an individual high-risk establishment or group of such establishments, for example on an industrial estate. The authority must base the designation of any such zone on the current 10-6 contour calculated when issuing environmental permits to the individual establishments. However, it can take account of any relevant developments that may reasonably be expected to occur in future and of the possible accumulation of location-based risks where more than one establishment is present. This allows the competent authority to reserve space for the industrial estate to expand and to absorb any consequent increase in the risk posed. Once this safety zone has been designated in the development plan, house-building or the construction or establishment of other vulnerable objects or of less vulnerable objects with no functional connection to the relevant establishments or to the area will not longer be permitted within the zone. When designating the safety zone, the competent authority can also choose to reduce the level of risk that establishments are allowed to pose and to modify the environmental permits of the relevant establishments accordingly. Once a safety zone has been designated, the limit values for location-based risk no longer apply within it. However, this is on condition that all vulnerable or less vulnerable objects or premises other than high-risk establishments currently present within the zone or permitted to be established there have a functional connection with one or more of the high-risk establishments in the relevant area or with the relevant area around which the zone has been designated.

3.7 Niederlande

3.7.1 Frage 1: Systematische Vorfrage

In welchem Verhältnis stehen Bebauungspläne und die Genehmigung einzelner Bauvorhaben?

Ein auf Artikel 3 Absatz 1 Raumplanungsgesetz beruhender Bebauungsplan legt die Nutzung von Grundstücken für Gebäude und andere Verwendungen fest.

Er ist verbindlich für die Genehmigung eines Bauantrages, ohne die nicht gebaut werden darf. Entspricht der Bauantrag nicht den Anforderungen des Bebauungsplans, kann der Stadtrat eine Ausnahmeregelung vom Bebauungsplan (Baugenehmigung) erteilen, sofern der Antrag seiner Ansicht nach im Einklang mit (guter) einer nachhaltigen räumlichen Planung ist. Wird die Ausnahmeregelung erteilt, kann danach die Baugenehmigung folgen. Bautätigkeiten in einem Habitat sind nur gestattet, wenn eine Genehmigung auf der Grundlage des Naturschutzgesetzes vorliegt. Stört die Bautätigkeit die unter Schutz gestellte Flora und Fauna, benötigt man eine Genehmigung auf der Grundlage des Flora- und Faunagesetzes.

3.7.2 Frage 2: Europäisches Habitat- und Artenschutzrecht

a) *In welchem Verhältnis stehen das nationale Naturschutzrecht und das Bauplanungsrecht? Werden die Anforderungen des Naturschutzrechts bereits in der Bauleitplanung oder erst in einem nachfolgenden Genehmigungsverfahren bewältigt?*

Der Naturschutz in den Niederlanden hat den Artenschutz nach dem Flora- und Faunagesetz und den Naturschutz nach dem Naturschutzgesetz zum Ziel.

Flora- und Faunagesetz

Dieses Gesetz regelt den Schutz freilebender Pflanzen- und Tierarten. Es schließt die den Artenschutz betreffenden Aspekte der Habitat-Richtlinie, der Vogelschutz-Richtlinie und des Übereinkommens über den internationalen Handel mit gefährdeten Arten freilebender Pflanzen und Tiere ein. Aktivitäten, die geschützte Arten möglicherweise schädigen, sind im Wesentlichen untersagt. Darüber hinaus umfasst das Gesetz ein Genehmigungssystem für alle Aktivitäten, die geschützte Arten gefährden könnten. Bautätigkeit zählt zu diesen Aktivitäten. Genehmigungen werden von den Provinzregierungen oder dem Ministerium für Landwirtschaft, Natur und Lebensmittelqualität erteilt. Das Gesetz schließt eine für alle Tiere und Pflanzen geltende Sorgfaltspflicht ein. Schließlich erlaubt das Gesetz, Landschaftselemente als geschützte Lebensräume auszuweisen.

Naturschutzgesetz

Das Naturschutzgesetz 1998 umfasst ein Verfahren für die Ausweisung von Naturräumen nach der Vogelschutz-Richtlinie und der Habitat-Richtlinie. Für alle diese Naturräume ist ein Managementplan zu erstellen.

Ferner umfasst das Gesetz ein Genehmigungssystem (Artikel 19d – 19k) für alle Aktivitäten, die die geschützten Eigenschaften gefährden können. Bautätigkeit zählt zu diesen Aktivitäten. Genehmigungen werden von den Provinzregierungen und dem Ministerium für Landwirtschaft, Natur und Lebensmittelqualität erteilt. Dies stellt sicher, dass Vorhaben, die Natura 2000-Gebiete möglicherweise beeinträchtigen, besonders sorgfältig untersucht werden. Neben dem in der Vogelschutz-Richtlinie und der Habitat-Richtlinie geregelten Schutz von Naturräumen regelt das Naturschutzgesetz auch den Schutz von Feuchtgebieten und Naturdenkmalen.

Im Rahmen der Vogelschutz-Richtlinie wurden in den Niederlanden 79 Gebiete mit einer Gesamtfläche von 1.010.910 ha ausgewiesen. Die Niederlande haben bei der Europäischen Union 141 Lebensräume mit einer Gesamtfläche von 750.841 ha registriert. Die nach der Vogelschutz-Richtlinie und der Habitat-Richtlinie geschützten Gebiete bilden zusammen die Natura 2000-Gebiete der Niederlande.

Es besteht kein Zusammenhang zwischen dem Wohnungsbaugesetz und dem Naturschutzgesetz oder dem Flora- und Faunagesetz. Eine Möglichkeit, diese Elemente zu berücksichtigen, besteht nicht.

Ferner besteht keinen formaler Zusammenhang zwischen einem Bebauungsplan und dem Naturschutzgesetz oder dem Flora- und Faunagesetz, aber eine nachhaltige Raumplanung verlangt, dass bei der Erarbeitung eines Bebauungsplans oder einer Ausnahmeregelung vom Bebauungsplan alle relevanten Umstände, einschließlich der Auswirkungen auf die Natur, berücksichtigt werden.

So werden die Ziele des Naturschutzgesetzes in der Bauleitplanung global in Betracht gezogen, aber in der Baubewilligung erst behandelt, wenn ein anschließendes Genehmigungsverfahren stattfindet.

b) Wie ist im Rahmen der Verträglichkeitsprüfung die Ausnahmeregelung des Artikels 6 Absatz 4 FFH-Richtlinie im Bereich des Bauplanungsrechts in nationales Recht umgesetzt worden und wie wird diese angewandt?

Siehe Antwort zu 2a).

Gibt es eine Differenzierung in Bezug darauf, dass ein Vorhaben innerhalb oder außerhalb eines bereits im Zusammenhang bebauten Gebietes errichtet wird?

Weder im Naturschutzgesetz noch im Baugesetz ist eine solche Unterscheidung vorgesehen. Eine Differenzierung erscheint für die Praxis relevant, soweit sich durch die Realisierung eines Projekts Auswirkungen auf die Erhaltungsziele eines Natura 2000-Gebiets ergeben.

c) Wird neben der schutzgebietsbezogenen Verträglichkeitsprüfung des Artikels 6 Absatz 3 FFH-Richtlinie bei baulichen Vorhaben auch eine artenschutzrechtliche Prüfung nach Artikel 12 FFH-Richtlinie durchgeführt?

Ja (siehe Antwort zu 2a). Eine artenschutzrechtliche Prüfung ist neben der Verträglichkeitsprüfung nach Artikel 6 Absatz 3 FFH Richtlinie notwendig, wenn sich Hinweise ergeben, dass durch ein Bauvorhaben der Lebensraum einer geschützten Art beeinträchtigt oder zerstört wird (beispielsweise das Töten von Arten, die Störung während der Brutzeit, die Zerstörung von Nestern etc.).

d) Wird ein Bedarf dafür gesehen, solche Bauvorhaben von artenschutzrechtlichen Verboten freizustellen, die in Gebieten errichtet werden, die bereits im Zusammenhang bebaut sind? Welche Rolle könnte es hierbei haben, wenn die geschützte Art ihre natürliche Verbreitung außerhalb des Siedlungsbereichs hat und dort auch in größerer Zahl vorkommt?

Eine entsprechende Ausnahme wäre für die Innenentwicklung der Städte hilfreich und würde im Hinblick auf die Biodiversitätsziele kein Problem darstellen, soweit der Bestand einer Art außerhalb des zu entwickelnden Gebietes ausreichend gesichert ist. Solange Artikel 16 der FFH-Richtlinie und Artikel 9 der Vogelschutzrichtlinie nicht geändert wird, besteht jedoch keine Möglichkeit für eine generelle Ausnahme im nationalen Recht. Die Möglichkeiten einer Änderung der Richtlinien für streng geschützte Arten und Vögel sind sehr begrenzt. Dies erfordert eine Prüfung auf lokaler Ebene, welchen Effekt die Änderung für die geschützten Arten in ihrem natürlichen Lebensraum hat. Ausnahmen von Artikel 16 FFH-Richtlinie können durch Bauvorhaben notwendig werden, wenn die Errichtung des Gebäudes im Interesse der öffentlichen Gesundheit, der öffentlichen Sicherheit oder aus zwingenden Gründen des überwiegenden öffentlichen Interesses notwendig ist. Artikel 9 sieht eine Ausnahme aus zwingenden Gründen des überwiegenden öffentlichen Interesses nicht vor.

e) Wie wird das Tatbestandsmerkmal der "Absicht" in Artikel 12 Absatz 1 Buchstabe a, b und c FFH-Richtlinie bei Bauvorhaben im Lichte der Rechtsprechung des Europäischen Gerichtshofs verstanden?

Auf der Grundlage des Caretta-Urteils des Europäischen Gerichtshofs (C-103/00) und dem Urteil zur Jagd mit Fallen (C-6/04) bezieht sich die „Absicht" nicht nur auf das Töten oder Zerstören, sondern auch darauf, dass die Zerstörung oder Tötung als Resultat einer Aktion zwar nicht gewollt, aber bekannt ist und hingenommen wird.

Die juristische Praxis in den Niederlanden wird entsprechend diesen Entscheidungen gehandhabt.

3.7.3 Frage 3: Plan-UP-Richtlinie (2001/42/EG)

a) Welche Ebenen und Instrumente städtebaulicher Planungen bestehen und wie erfolgte diesbezüglich die Umsetzung der Plan-UP-Richtlinie in nationales Recht?

Entsprechend dem Raumplanungsgesetz müssen Gemeinden, Provinzen und die nationale Regierung ihre räumliche Politik in einem oder mehreren Leitbildern festhalten. In einigen Fällen ist hierfür eine SUP notwendig. Das zentrale Instrument der Raumplanung ist der Bebauungsplan, der durch den Gemeinderat beschlossen wird. Derselbe Plantyp kann von einer Provinz oder dem Ministerium für Wohnen, Raumplanung und Umwelt aufgestellt werden und wird dann integrativer Plan genannt. Diese Körperschaften können ebenso eine verbindliche Projekt-Entscheidung verfassen. Die SUP ist mit dem Bebauungsplan sowie den Projekt-Entscheidungen verbunden. Darüber hinaus bilden andere Gesetze die Grundlage für Entwicklungspläne, wie zum Beispiel den Wasser-Management-Plan, der auf dem Wassergesetz basiert.

Die Richtlinien über die Umweltprüfung und die Strategische Umweltprüfung werden durch Kapitel 7 des Niederländischen Umweltmanagement-Gesetzes und die Verordnung über die Prüfung der Umweltauswirkungen von 1994 umgesetzt. Als Rahmengesetz beschreibt das Umweltmanagement-Gesetz die Grundsätze der Umweltpolitik. Details werden durch Verfügungen des Rates (AMvB) geregelt. Die Verordnung über die Prüfung der Umweltauswirkungen von 1994 stellt eine entsprechende Verfügung des Rates dar. Neben dem Kapitel 7 des Umweltschutzgesetzes finden sich weitere wichtige Abschnitte zur Umweltprüfung in Kapitel 2.2. (über die Umweltprüfung-Behörde), Kapitel 14.2 (über die Koordination im Rahmen der Aufstellung des Umweltberichts) und Kapitel 20 (Einspruch). Die Verordnung über die Prüfung der Umweltauswirkungen von 1994 bestimmt die Anwendung der Umweltprüfung. Die Verordnung beinhaltet Anhänge, darunter auch die Listen C und D. Die Liste C bestimmt, welche Aktivitäten und Entscheidungen einen obligatorischen Bericht über die Umweltauswirkungen (Umweltbericht) erfordern. Die Liste D fasst die Aktivitäten und Entscheidungen zusammen, für die ein sogenanntes Artikel 7.8a/7.8d-Verfahren notwendig ist. Hierbei wird für die Aktivitäten und Entscheidungen geprüft, ob eine Umweltprüfung notwendig ist. (Die Liste befindet sich in dem Beschluss m. e. r. 1994 (in Niederländisch)).

Die Richtlinie über die Strategische Umweltprüfung der EU (Richtlinie 2001/42/EG) wurde im Juli 2001 angenommen. Der gesetzliche Vorschlag zur Umsetzung der Richtlinie in das niederländische Recht betrifft eine Umweltprüfung für Pläne. Zur klaren Unterscheidung zwischen der Umweltprüfung und der Umweltprüfung für Pläne wurde Letztere umbenannt in Strategische Umweltprüfung. In der Praxis gibt es keinen Unterschied zwischen einer Strategischen Umweltprüfung und der Umweltprüfung für Pläne, die in der Gesetzesvorlage beschrieben wird. Artikel 5 und Anhang I der Plan-UP-Richtlinie beschreiben präzise die notwendigen Inhalte des Umweltberichts. Die Strategische Umweltprüfung ist für Leitbilder durchzuführen, mit der bestimmte Aktivitäten vorbereitet werden.

b) Wie wurde die Plan-UP-Richtlinie in das nationale Recht umgesetzt?

Siehe oben.

c) Werden für bestimmte Fälle, insbesondere der Geringfügigkeit im Rahmen der Aufstellung, Änderung und Ergänzung von Bauleitplänen, Möglichkeiten zur Beschleunigung sowie Verfahrenserleichterungen genutzt?

In den Niederlanden wurde eine Entscheidung für klare Grenzwerte getroffen, unterhalb denen eine Prüfung der Umweltauswirkungen nicht erforderlich ist. Zum Beispiel erforderte ein Plan zur Errichtung von weniger als 2000 Häusern keine Prüfung. Dies wurde jedoch nicht von der Europäischen Kommission akzeptiert, da die Größe nicht notwendigerweise ein klarer Hinweis auf die Umweltauswirkungen ist, weshalb das Gesetz geändert wurde.

d) In welchem Verhältnis stehen Umweltprüfung und FFH-Verträglichkeitsprüfung zueinander?

Die Umweltprüfung erfasst ein breites Spektrum von Umweltauswirkungen, darunter auch Gesundheitsaspekte. Dahingegen erfolgt bei der Verträglichkeitsprüfung eine spezifische Betrachtung von Tieren und Pflanzen im Hinblick auf die Ziele der Richtlinie.

3.7.4 Frage 4: Dienstleistungsrichtlinie (2006/123/EG)

Wird ein Umsetzungsbedarf der Dienstleistungsrichtlinie (Richtlinie 2006/123/EG) für den Bereich des nationalen Städtebaurechts und der Bauleitplanung gesehen?

Nein, das war nicht notwendig. Die Dienstleistungsrichtlinie wurde durch das Dienstleistungsgesetz durchgesetzt. Dieses Gesetz brachte keine Veränderungen im Raumordnungsgesetz oder im Wohnungsbaugesetz mit sich. Im Erwägungsgrund 9 der Richtlinie heißt es: Diese Richtlinie findet nur auf Anforderungen für die Aufnahme oder Ausübung einer Dienstleistungstätigkeit Anwendung. Sie findet somit keine Anwendung auf Anforderungen wie Straßenverkehrsvorschriften, Vorschriften bezüglich der Stadtentwicklung oder Bodennutzung, der Stadtplanung und der Raumordnung, Baunormen (...) das Raumordnungsgesetz und (Flächennutzungspläne) und das Wohnungsbaugesetz enthalten nur allgemeine Bestimmungen, die die Dienstleistungstätigkeit nicht speziell regulieren oder beeinflussen.

3.7.5 Frage 5: Erneuerbare-Energien-Richtlinie (2009/28/EG)

Wird ein Umsetzungsbedarf hinsichtlich des Artikels 13 Erneuerbare-Energien-Richtlinie (2009/28/EG) für den Bereich des nationalen Städtebaurechts und der Bauleitplanung gesehen?

Es besteht keine Notwendigkeit, Artikel 13 Absätze 1, 2, 3, 5 und 6 Erneuerbare Energien-Richtlinie umzusetzen. Artikel 13 Absatz 4 muss nicht umgesetzt werden, da die niederländische Bauverordnung bereits entsprechende Anforderungen an die Energieeffizienz bei der Errichtung, dem Umbau und der Sanierung von Gebäuden umfasst.

3.7.6 Frage 6: SEVESO-II-Richtlinie

Wie ist Artikel 12 Absatz 1 Unterabsatz 2 der Richtlinie 96/82/EG vom 09.12.1996 zur Beherrschung der Gefahren bei Unfällen mit gefährlichen Stoffen (ABl. L 10 vom 14.01.1997, S. 13; „Seveso-II-Richtlinie") in nationales Recht umgesetzt worden?

Artikel 12 der Seveso-II-Richtlinie der EU legt unter anderem fest, dass die Mitgliedstaaten dafür Sorge zu tragen haben, dass die Ziele der Richtlinie – Verhinderung schwerer Unfälle mit gefährlichen Stoffen und die Beschränkung ihrer Folgen für Mensch und Umwelt – Berücksichtigung finden in:

- Entscheidungen über die Ansiedlung neuer Betriebe und Änderungen bestehender Betriebe,
- Entscheidungen über neue Entwicklungen in der Nachbarschaft bestehender Betriebe.

In den Niederlanden wurde dieser Artikel mit der Verordnung über die Anlagenumgebungssicherheits-Verordnung in nationales Recht umgesetzt. Die Verordnung regelt die Umweltqualitätsanforderungen an die äußere Sicherheit, wenn Entscheidungen über die Bauleitplanung getroffen werden und wenn Umweltgenehmigungen für Betriebe mit hohem Gefährdungspotenzial erteilt werden. Sie gewährleistet ferner, dass in allen Fällen, in denen die Entscheidungsfindung in diesen Bereichen sich auf die Möglichkeit des Eintritts einer Havarie mit zahlreichen Opfern auswirkt, die Verfahren für die breite Öffentlichkeit transparent sind.

Die Verordnung gilt für Betriebe, die in den Geltungsbereich der Verordnung über schwere Unfälle mit gefährlichen Stoffen von 1999 (BRZO) fallen; Lade- und Löschanlagen, in denen gefährliche Stoffe als Teil des Transportprozesses gelagert werden; Flüssiggas-Tankstellen; Lager nach CPR 15-2 und -3 und Betriebe, die große Ammoniak gekühlte Anlagen umfassen. Der Anwendungsbereich der Verordnung kann auf Anordnung des Rates auf weitere Betriebe mit einem äußeren standortbezogenen Risiko von mehr als 10-6 pro Jahr (wie Propangaslager und bestimmte Eisenbahnrangieranlagen) ausgedehnt werden. Die Verordnung gilt nicht für Betreiber, die mit Feuerwerkskörpern umgehen oder Sprengstoffe oder explosionsgefährdete Gegenstände lagern, es sei denn, sie unterliegen der Verordnung über schwere Unfälle.

Die Verordnung legt erstens Grenzwerte für das standortbezogene Risiko gefährdeter Objekte fest. Die Unterscheidung zwischen gefährdeten und weniger gefährdeten Objekten ist nur für das standortbezogene Risiko, nicht für das gemeinschaftsbezogene Risiko relevant. Der Begriff „gefährdete Objekte" bedeutet Wohngrundstücke und gleichwertige Gebäude wie Schulen und Krankenhäuser. Ein Grenzwert ist ein bis zu einem bestimmten Datum zu erreichender und danach aufrechtzuerhaltender Mindeststandard. Grenzwerte müssen „eingehalten" und dürfen nicht überschritten werden. Bei Festlegung des Grenzwertes ist zwischen bestehenden und neuen Situationen (für Letztere gilt 10-5 pro Jahr) zu unterscheiden. Wenn ein Objekt vorhanden oder nach dem gegenwärtigen Bauleitplan zulässig ist, handelt es sich um eine bestehende Situation. Zweitens legt die Verordnung einen Zielwert für das standortbezogene Risiko weniger gefährdeter Objekte fest. Dieser Begriff bedeutet, zum Beispiel, Arbeitsstätten mit relativ wenigen Angestellten wie Büros für weniger als 50 Mitarbeiter und gewerblich und kommerziell genutzte Räumlichkeiten, von denen keine Gefährdung der Umgebung ausgeht und in denen sich während des größten Teils des Tages im Allgemeinen keine große Zahl von Menschen aufhält. Ein Zielwert ist ein bis zu einem bestimmten Datum so weit wie möglich zu erreichender Standard, der danach so weit wie möglich aufrechtzuerhalten ist. Die Behörden müssen den Zielwert für das standortbezogene Risiko beachten.

Die wichtigste Ermächtigungsgrundlage für die Anlagenumgebungssicherheits-Verordnung ist Kapitel 5 des Umweltmanagementgesetzes. Hierdurch wird der Rat ermächtigt, für bestimmte Umweltbereiche Qualitätsstandards festzulegen. Die Standards können entweder allgemein oder in ausgewiesenen Gebieten Anwendung finden und entweder zwingend erforderliche Grenzwerte oder Zielwerte sein, die so weit wie möglich bis zu einem angegeben Datum zu erreichen sind (weitere Erläuterungen nachstehend). Die Umweltqualitätsstandards richten sich in erster Linie an den öffentlichen Sektor. [Sie sind zu beachten], wenn Umweltgenehmigungen für Betriebe erteilt werden, in denen gefährliche Stoffe gelagert, hergestellt, verarbeitet usw. werden sollen und ebenfalls, wenn Entscheidungen über die Flächennutzung von Gebieten getroffen werden, die an diese Betriebe angrenzen. Zielwerte müssen beachtet und dürfen nur bei Vorliegen zwingender Gründe überschritten werden.

Standortbezogenes Risiko und gemeinschaftsbezogenes Risiko

Die Verordnung umfasst Definitionen des standortbezogenen Risikos und des gemeinschaftsbezogenen Risikos. Das standortbezogene Risiko wird definiert als das für Personen an einem bestimmten Ort bestehende Risiko, ausgedrückt in der jährlichen Wahrscheinlichkeit, in der Nachbarschaft eines Betriebes, in dem gefährliche

Stoffe (legal) gelagert werden, als unmittelbare Folge eines Unfalls vor Ort, an dem diese Stoffe beteiligt sind, zu sterben. Die Verordnung legt Umweltqualitätsstandards in Form von Grenzwerten für das standortbezogene Risiko fest, beispielsweise 10-6 pro Jahr für gefährdete Objekte und für Standorte, die gegenwärtig saniert werden. Beträgt das standortbezogene Risiko für ein bestehendes gefährdetes Objekt zwischen 10-5 und 10-6 pro Jahr, gilt ein Moratorium, d. h. in dem Betrieb dürfen keinerlei Änderungen vorgenommen werden, die das Risiko erhöhen. Diese Grenzwerte sind in erster Linie für die betroffenen Behörden bindend. Neben der Pflicht, Entscheidungen über die Höhe des gemeinschaftsbezogenen Risikos zu erläutern, haben sie auch die Aufgabe, ein Grundniveau des Schutzes gegen Risiken vorzusehen, die mit dem Umgang mit gefährlichen Stoffen in Betrieben einhergehen. Die Pflicht zur Erläuterung von Entscheidungen, die das Niveau des gemeinschaftsbezogenen Risikos in der Nachbarschaft potenziell gefährlicher Betriebe betreffen, bedeutet, dass Behörden den Zugang zu Informationen über ihre in dieser Hinsicht getroffene Wahl ermöglichen. Die Bürger können die Einhaltung der Verordnung gerichtlich erzwingen. Das gemeinschaftsbezogene Risiko berücksichtigt die Bevölkerungsdichte in der nahen Umgebung einer Aktivität mit hohem Risikopotenzial und gibt die Wahrscheinlichkeit einer bestimmten Mindestzahl von Todesfällen aufgrund eines einzigen Unfalls mit gefährlichen Stoffen an. Der Fokus liegt auf der hohen Sterbewahrscheinlichkeit im Falle eines Unfalls. Wenn sie die in der Verordnung beschriebenen Entscheidungen treffen, müssen Provinz- und Kommunalbehörden erklären, wie sie das gemeinschaftsbezogene Risiko im Entscheidungsprozess beachtet haben. Ferner müssen die Gemeinden die Auswirkung der erteilten Umweltgenehmigung oder der entsprechenden Planungsentscheidung auf das gemeinschaftsbezogene Risiko erläutern. Die Erläuterung muss Informationen über Maßnahmen, die die betroffene Bevölkerung zu ihrem Schutz ergreifen kann und über Notfallschutzpläne im Falle eines Unfalls beinhalten. Die Feuerwehr ist befugt, in dieser Hinsicht beratend tätig zu werden.

Bauleitplanung und Sicherheitszone

Da die Verordnung auch für die Verabschiedung von Bauleitplänen und die Erteilung von Ausnahmeregelungen gilt, beruhen einige ihrer Teile auf dem Raumordnungsgesetz. Aufgrund der Verantwortlichkeit der Kommunen für die Bauleitplanung müssen diese auch eine angemessene Sicherheit in der Nähe potenziell gefährlicher Betriebe garantieren. Dies erfolgt größtenteils auch über Bauleitpläne, die festlegen, wie Flächen entwickelt und genutzt werden können. Das Gesetz macht die Kommunen für die Bauleitplanung in ihren Gebieten und somit für die Gewährleistung angemessener Sicherheit in der Nähe potenziell gefährlicher Betriebe verantwortlich. Ein Großteil dieser Arbeiten wird durch Bauleitpläne geleistet, die festlegen, wie die Fläche entwickelt und genutzt werden kann. Eine Kommune kann ein Grundstück für eine Nutzung mit hohem Risikopotenzial in einem in dieser Verordnung erwähnten Betrieb nur ausweisen, wenn das für die Nachbarschaft damit einhergehende Risiko die in der Verordnung festgelegten Grenzwerte nicht überschreitet. Darüber hinaus muss die Kommune darüber befinden, ob diese Risiken für die Gemeinschaft akzeptabel sind.

Die zuständige Behörde kann eine Sicherheitszone um den Standort eines einzelnen Betriebes mit hohem Risikopotenzial oder einer Gruppe solcher Betriebe wie beispielsweise ein Industriegebiet, ausweisen. Die Behörde muss die Ausweisung jeder derartigen Zone auf die derzeit geltende 10^{-6} -Kontur beziehen, die bei der Erteilung von Umweltgenehmigungen für die einzelnen Betriebe berechnet wird. Sie kann jedoch alle relevanten Entwicklungen, mit denen in Zukunft vernünftigerweise zu rechnen ist sowie die mögliche Akkumulation standortbezogener Risiken in Fällen, in denen mehr als ein Betrieb besteht, in Betracht ziehen. Das ermöglicht es der zuständigen Behörde, Gelände für die Erweiterung des Industriegebiets zu reservieren und jede sich daraus ergebende Erhöhung des Risikos zu neutralisieren. Nach Ausweisung der Sicherheitszone im Bauleitplan ist Wohnungsbau beziehungsweise der Bau oder die Errichtung anderer gefährdeter Objekte oder weniger gefährdeter Objekte ohne funktionellen Zusammenhang mit den relevanten Betrieben oder dem Gebiet innerhalb der Zone nicht mehr gestattet. Bei Ausweisung der Sicherheitszone kann die zuständige Behörde zudem das zulässige Risikoniveau der Betriebe absenken und die Umweltgenehmigungen der relevanten Betriebe entsprechend ändern. Nach Ausweisung einer Sicherheitszone gelten in ihr die Grenzwerte für standortbezogene Risiken nicht mehr. Die Voraussetzung dafür ist jedoch, dass mit Ausnahme der Betriebe mit hohem Risikopotenzial, die gegenwärtig in der Zone bestehen oder deren Errichtung genehmigt wurde, bei allen dort zu errichtenden gefährdeten oder weniger gefährdeten Objekten oder Gewerberäumen ein funktioneller Zusammenhang mit einem oder mehreren der Betriebe mit hohem Risikopotenzial in dem relevanten Gebiet oder mit dem von der Sicherheitszone umschlossenen relevanten Gebiet besteht.

3.8 Austria (Vienna)

3.8.1 Question 1: Systematic preliminary question

What is the relationship between binding land use plans and the approval of individual construction projects?

In view of the fact that Vienna is both an Austrian state and a municipal authority, there is no binding land use plan on state level. Only the plans for local land use, the zoning plans (preparatory land use plans) and the actual land use plans issued by decree of the municipal council are legally binding.

In Vienna, the zoning plan alone does not constitute the right to construct buildings. This right is only granted by the binding land use plan (otherwise building is prohibited, with exceptions only granted by means of a special official procedure). However, as a rule in Vienna the zoning plan and land use plan are usually grouped together in one common document that is jointly adopted by the municipal council. Apart from a few small areas, there is one zoning plan and one land use plan to cover the whole of the city area.

3.8.2 Question 2: European Habitat and Protection of Species Law

a) *What is the relationship between national nature conservation law and the building planning law? Are nature conservation law requirements taken into account in development planning or not dealt with until a subsequent approval procedure takes place?*

Laws relevant to nature conservation in Vienna are the Viennese Nature Conservation Law, the National Park Law for the National Park Danube-March-Thaya Meadows, the Viennese Biosphere Park Law for the Viennese part of the Vienna Woods Biosphere Park, the Viennese Tree Protection Law and the Forest Law. Apart from the Forest Law (federal law), these are all state laws.

The Viennese Nature Conservation Law basically regulates both the protection of species and the protection of certain areas or natural objects.

Area and object protection

The nature conservation authority defines

- natural monuments
- ecological development areas

Decrees by the state governments define

- protected parts of the countryside
- landscape protection areas
- nature conservation areas
- European protection areas (Natura-2000 areas)
- the precise demarcation of the Vienna Woods Biosphere Park

The area of the Danube-March-Thaya Meadows was declared a national park under the corresponding state laws.

The Nature Conservation Law (and also the National Park Law and the Biosphere Park Law) defines which measures inside these areas need approval under the nature conservation laws and under which conditions such approval is to be granted. Permits under the Nature Conservation Law must be obtained in addition to those permits required under other legislation (e.g. building laws) (cumulation).

Species protection

The Viennese Nature Conservation Law differentiates between protected and strictly protected species and stipulates the prohibitions resulting from this protection category together with the criteria for exceptions to the prohibitions. The individual species are placed under a protection order by decree of the state government.

Exceptions to the prohibitions stated in the Nature Conservation Law are granted pursuant to a corresponding procedure.

The Viennese planning laws (Viennese Building Code) contain no explicit regulation for taking account of nature conservation issues in the zoning plans and binding land use plans. However, it can be presumed that if these plans could give rise to developments that would have an impact on nature conservation areas or jeopardise the existence of species, then the corresponding impacts have to be presented in the context of a strategic environmental assessment (SEA). Certainly all environmentally relevant departments of Vienna City Council are involved in the consultation process for elaborating the zoning plans and binding land use plans (for partial areas of the city), particularly also the Nature Conservation Authority and the Environmental Advocacy Office to warrant comprehensive consideration of all aspects.

b) Within the scope of the assessment of the implications of a plan or project for a site, how has the exception provided by Article 6 Paragraph 4 of the FFH Directive been transposed into national law in the field of building planning and how is it applied?

No answer

Is there any distinction regarding whether a project is built within or outside an area which is already continuously built up?

Implementation of the SEA Directive in the Viennese Building Code with the corresponding reference to European protection areas also implicitly implements the FFH Directive. However, the Viennese Building Code does not contain any decision-making obligation ensuing from the results of an SEA (regarding the impacts on European protection areas).

In fact, it can be presumed that developments within European protection areas based purely on the regional planning policy of the zoning plans and binding land use plans could only be put into practice if the corresponding permits can be obtained under nature conservation laws. It is therefore improbable that there will be any stipulations made in the zoning plan and binding land use plan whose implementation is clearly ruled out under the Nature Conservation Law.

Developments based on higher legislation (Federal Highways Law, Railways Law) are merely declared as such in the zoning plan and binding use plan and would probably be subject to a project-level environment compatibility review.

The Viennese Building Code makes no explicit distinction as to the position of a project inside or outside a built-up area. But this criterion scarcely seems relevant in the context of European protection areas, as in Vienna these are essentially non-built-up areas.

c) Apart from the conservation area related assessment of implications for the site pursuant to Article 6 Paragraph 3 of the FFH Directive, is an assessment under species protection law also always performed according to Article 12 of the FFH Directive?

The Viennese Nature Conservation Law differentiates between protected and strictly protected species.

Building projects impacting on the habitat of strictly protected species have to undergo a species protection procedure.

c) Is there a need to exempt such construction projects from prohibitions under species protection law which are erected in areas which have already been developed? What role could it play if the protected species had their natural habitat outside the development area and also occurred in greater number there?

The current legislation results (automatically) in areas intended for building development on the basis of planning law provisions or suitable for building development on account of their position in built-up areas usually being "cultivated" in such a way that no fauna and flora relevant to nature conservation will develop there.

From the point of view of Viennese urban planning, it should certainly be asked whether this is a desirable situation, or whether a relaxed approach to species protection in these areas could in the end promote actual biological diversity. However, this would also entail dealing with the Forest Law and the Viennese Tree Protection Law.

In this context, it would be necessary to ascertain exactly what the Directive means with "natural habitat". There are in fact signs that certain species that occurred hitherto in agricultural areas have been greatly decimated here on account of the current agricultural methods used in these areas (in their natural habitats), but that these species frequently move to built-up areas and have now settled there (certain species, such as bats or various bird species, even use actual buildings as their habitat).

d) How is the element of "intention" in Article 12 Paragraph 1 Letter a, b and c of the FFH DIRECTIVE interpreted for construction projects in the light of the legal decisions taken by the European Court of Justice?

The Viennese Natural Conservation Law differentiates between intentional disturbances (e.g. during the reproduction phase) and all other forms of interference (e.g. trapping or killing). Intention (in a disturbance) is deemed to apply when a reproduction base (nest, bats' breeding grounds, etc.) is deliberately destroyed with the intention of getting rid of the protected species. Destruction of a (local) habitat through building development would not constitute intention in this respect. Killing a species through building development would be prohibited in any case.

3.8.3 Question 3: SEA Directive (2001/42/EC)

a) Which levels and instruments of development planning exist and with respect to this, how was the SEA Directive transposed into national law?

In view of the fact that Vienna is both an Austrian state and a municipal authority, there is no binding land use plan on state level but only the plans for local land use (competence area of the municipal authorities). These refer to the zoning plans (preparatory land use plans) and the actual land use plans.

With regard to Viennese land use planning, the SEA Directive has been implemented in the Viennese Building Code ("Viennese Urban Planning and Building Code").

b) How was the SEA Directive transposed into national law?

No answer

c) Are options, especially the option of insignificance within the context of preparing, amending and supplementing the development plans, used to accelerate and simplify procedures for particular cases?

The options in the Directive (slight changes, stipulating the use of small areas on a local level) were not included in the Viennese Building Code (legal advisers found the phrases "slight" and "change" to indefinite). In the case of slight changes, the argument is frequently brought forward that the general framework has already been (basically) staked out so that no (further) SEA is necessary.

Otherwise there is no simplification or acceleration in the legally prescribed period for slight changes (at the most when it comes to ascertaining the basic principles).

d) What is the relationship between the environmental audit and the FFH assessment of implications for the site?

The FFH compatibility assessment is carried out in the framework of the environmental audit and not as a separate review.

3.8.4 Question 4: EU Services Directive (2006/123/EC)

Is it deemed necessary to transpose the EC Services Directive (Directive 2006/123/EC) for the area of national urban planning law and development planning?

No.

3.8.5 Question 5: Renewable Energy Directive (2009/28/EC)

Is it considered necessary to transpose Article 13 of the Renewable Energy Directive (Directive 2009 /28/EC) for the area of national urban planning law and development planning?

No.

3.8.6 Question 6: SEVESO-II-Directive

How has Article 12 Paragraph 1 Subsection 2 Council Directive 96/82/EC of 9 December 1996 on the control of major-accident hazards involving dangerous substances (OJ L 10, 14.1.1997, p. 13; "Seveso-II-Directive") been transposed into national law?

No answer

3.8 Österreich (Wien)

Aufgrund der kompetenzrechtlichen Zuordnung der Gesetzgebung im Bereich des Bau- und Planungsrechts zu den Bundesländern erfolgte für Österreich die Darstellung der Antworten am Beispiel der Stadt Wien.

3.8.1 Frage 1: Systematische Vorfrage

In welchem Verhältnis stehen Bebauungspläne und die Genehmigung einzelner Bauvorhaben?

Da Wien sowohl Bundesland als auch Gemeinde ist, gibt es keine rechtsverbindlichen Raumordnungspläne der Landesebene. Rechtsverbindlich sind nur die per Verordnung des Gemeinderats (im eigenen Wirkungsbereich der Gemeinden) erlassenen Pläne der örtlichen Raumplanung, die Flächenwidmungspläne (Flächennutzungspläne) und die Bebauungspläne.

In Wien begründet der Flächenwidmungsplan alleine nicht das Recht, Bauwerke zu errichten. Dieses Recht entsteht erst durch den Bebauungsplan (ansonsten besteht Bausperre, von der mit einem speziellen Behördenverfahren Ausnahmen gewährt werden können). Allerdings werden in Wien in der Regel Flächenwidmungsplan und Bebauungsplan zu einem gemeinsamen Dokument zusammengefasst und gemeinsam vom Gemeinderat beschlossen. Von wenigen Teilflächen abgesehen besteht für das gesamte Stadtgebiet ein Flächenwidmungsplan und ein Bebauungsplan.

3.8.2 Frage 2: Europäisches Habitat- und Artenschutzrecht

a) In welchem Verhältnis stehen das nationale Naturschutzrecht und das Bauplanungsrecht? Werden die Anforderungen des Naturschutzrechts bereits in der Bauleitplanung oder erst in einem nachfolgenden Genehmigungsverfahren bewältigt?

Für den Wiener Naturschutz relevante Gesetze sind das Wiener Naturschutzgesetz, das Nationalparkgesetz für den Nationalpark Donau-March-Thaya-Auen, das Wiener Biosphärenparkgesetz für den Wiener Teil des Biosphärenparks Wienerwald, das Wiener Baumschutzgesetz sowie das Forstgesetz. Mit Ausnahmen des Forstgesetzes (Bundesrecht) handelt es sich dabei um Landesgesetze.

Das Wiener Naturschutzgesetz regelt grundsätzlich sowohl den Artenschutz als auch den Schutz bestimmter Gebiete bzw. Naturobjekte.

Gebiets- und Objektschutz

Durch Bescheid der Naturschutzbehörde können festgelegt werden:

- Naturdenkmäler

- Ökologische Entwicklungsflächen

Durch Verordnung der Landesregierung können festgelegt werden:

- Geschützte Landschaftsteile
- Landschaftsschutzgebiete
- Naturschutzgebiete
- Europaschutzgebiete (Natura-2000–Gebiete)
- Die genaue Abgrenzung des Biosphärenparks Wienerwald

Der Erklärung von Gebieten der Donau-March-Thaya-Auen zum Nationalpark erfolgte durch eigene jeweilige Landesgesetze.

Im Naturschutzgesetz (bzw. zusätzlich im Nationalparkgesetz und im Biosphärenparkgesetz) ist definiert, welche Maßnahmen innerhalb dieser Gebiete naturschutzrechtlicher Bewilligungen bedürfen und unter welchen Bedingungen diese zu erteilen sind. Bewilligungen nach dem Naturschutzgesetz sind zusätzlich zu den nach anderen Gesetzesmaterien erforderlichen Bewilligungen (z. B. Baurecht) zu erwirken (Kumulation).

Artenschutz

Das Wiener Naturschutzgesetz unterscheidet zwischen geschützten und streng geschützten Arten und regelt die aus dieser Schutzwürdigkeit resultierenden Verbote sowie die Kriterien für Ausnahmen von diesen Verboten. Die Unterschutzstellung der einzelnen Arten erfolgt durch Verordnung der Landesregierung.

Die Bewilligung von Ausnahmen von den im Naturschutzgesetz festgelegten Verboten unterliegt einem entsprechenden Verfahren.

Das Wiener Planungsrecht (Bauordnung für Wien) enthält keine explizite Vorschrift, in welcher Weise naturschutzrechtliche Gegebenheiten in den Flächenwidmungsplänen und Bebauungsplänen zu berücksichtigen sind. Es ist jedoch davon auszugehen, dass dann, wenn durch diese Pläne Entwicklungen ermöglicht werden, die von Einfluss auf Naturschutzgebiete sind oder den Bestand von Arten gefährden, die Auswirkungen im Rahmen einer strategischen Umweltprüfung (SUP) darzustellen sind. Jedenfalls sind bei der Ausarbeitung von Flächenwidmungsplänen und Bebauungsplänen (für Teilgebiete der Stadt) alle umweltrelevanten Dienststellen des Magistrats der Stadt Wien, insb. auch die Naturschutzbehörde und die Umweltanwaltschaft in den Abstimmungsprozess eingebunden, sodass eine umfassende Abwägung aller Aspekte gewährleistet ist.

b) Wie ist im Rahmen der Verträglichkeitsprüfung die Ausnahmeregelung des Artikels 6 Absatz 4 FFH-Richtlinie im Bereich des Bauplanungsrechts in nationales Recht umgesetzt worden und wie wird diese angewandt?

Gibt es eine Differenzierung in Bezug darauf, dass ein Vorhaben innerhalb oder außerhalb eines bereits im Zusammenhang bebauten Gebietes errichtet wird?

Durch die Umsetzung der SUP-Richtlinie in der Wiener Bauordnung und die darin enthaltenen Bezugnahmen auf Europaschutzgebiete ist implizit auch die FFH-Richtlinie umgesetzt. Die Wiener Bauordnung enthält allerdings keine aus dem Ergebnis einer SUP (im Zusammenhang mit Auswirkungen auf Europaschutzgebiete) resultierende Entscheidungspflicht.

Faktisch ist davon auszugehen, dass Entwicklungen innerhalb von Europa-Schutzgebieten, die ihre Grundlage ausschließlich in den raumordnerischen Festlegungen des Flächenwidmungs- und Bebauungsplans hätten, nur dann realisierbar wären, wenn auch entsprechende naturschutzrechtliche Bewilligungen erwirkt werden könnten. Es ist daher unwahrscheinlich, dass es zu Festlegungen des Flächenwidmungs- und Bebauungsplanes kommt, deren Umsetzung aus Naturschutzrechtlicher Sicht eindeutig auszuschließen ist.

Entwicklungen auf der Grundlage übergeordneter Rechtsmaterien (Bundesstraßengesetz, Eisenbahngesetz) werden im Flächenwidmungs- und Bebauungsplan nur deklaratorisch dargestellt und wären wohl in der Regel einer Umweltverträglichkeitsprüfung auf Projektebene zu unterziehen.

Eine ausdrückliche Differenzierung im Hinblick auf die Lage eines Gebiets innerhalb oder außerhalb bebauter Strukturen gibt es in der Wiener Bauordnung nicht. In Bezug auf Europaschutzgebiete erscheint dieses Kriterium aber auch kaum relevant, da es sich bei diesen Gebieten in Wien um weitestgehend bebauungsfreie Flächen handelt.

c) Wird neben der schutzgebietsbezogenen Verträglichkeitsprüfung des Artikels 6 Absatz 3 FFH-Richtlinie bei baulichen Vorhaben auch eine artenschutzrechtliche Prüfung nach Artikel 12 FFH-Richtlinie durchgeführt?

Das Wiener Naturschutzgesetz unterscheidet zwischen geschützten und streng geschützten Arten. Bauvorhaben, die den Lebensraum streng geschützter Arten beeinträchtigen, sind einem Artenschutzverfahren zu unterziehen.

d) Wird ein Bedarf dafür gesehen, solche Bauvorhaben von artenschutzrechtlichen Verboten freizustellen, die in Gebieten errichtet werden, die bereits im Zusammenhang bebaut sind? Welche Rolle könnte es hierbei haben, wenn die geschützte Art ihre natürliche Verbreitung außerhalb des Siedlungsbereichs hat und dort auch in größerer Zahl vorkommt?

Die aktuelle Gesetzeslage führt (zwingt) dazu, dass Flächen, die für eine bauliche Entwicklung aufgrund planungsrechtlicher Bestimmungen vorgesehen sind oder aufgrund ihrer Lage innerhalb baulicher Strukturen infrage kommen, im Regelfall derart „gepflegt“ werden, dass sich keine naturschutzrelevante Fauna und Flora entwickeln kann.

Es wäre aus Sicht der Wiener Stadtplanung durchaus zu hinterfragen, ob dies eine aus dem Gesichtspunkt des Naturschutzes wünschenswerte Situation ist, oder ob nicht ein entspannterer Umgang mit dem Artenschutz in solchen Gebieten letztendlich die tatsächliche biologische Vielfalt fördern könnte. Dies beträfe allerdings auch den Umgang mit dem Forstgesetz und dem Wiener Baumschutzgesetz.

In diesem Zusammenhang wäre zu prüfen, was genau die Richtlinie unter „natürlicher Lebensraum“ versteht. Tatsächlich ist zu beobachten, dass der Bestand von Arten, die bislang in landwirtschaftlich genutzten Bereichen vorkamen, aufgrund der aktuellen Methoden der landwirtschaftlichen Nutzung in diesen Gebieten (ihren natürlichen Lebensräumen) stark dezimiert wurde, diese Arten aber häufig in bebautes Gebiet ausweichen und nunmehr dort heimisch sind (bestimmte Arten, wie beispielsweise Fledermäuse oder verschiedene Vogelarten, nutzen sogar die Gebäude selbst als Lebensraum).

e) Wie wird das Tatbestandsmerkmal der „Absicht“ in Artikel 12 Absatz 1 Buchstabe a, b und c FFH-Richtlinie bei Bauvorhaben im Lichte der Rechtsprechung des Europäischen Gerichtshofs verstanden?

Im Wiener Naturschutzgesetz wird zwischen absichtlichen Störungen (z. B. während der Fortpflanzungszeit) und allen anderen Formen der Beeinträchtigung (z. B. Fangen oder Töten) unterschieden. Absicht liegt (bei einer Störung) dann vor, wenn ein Fortpflanzungsquartier (Nest, Fledermausquartier etc.) bewusst mit dem Zweck, die geschützten Arten damit los zu werden, zerstört wird. Zerstörung eines (lokalen) Lebensraumes durch Bebauung wäre in dieser Hinsicht keine Absicht.

Tötung einer Art durch Bebauung wäre in jedem Fall verboten.

3.8.3 Frage 3: Plan-UP-Richtlinie (2001/42/EG)

a) Welche Ebenen und Instrumente städtebaulicher Planungen bestehen und wie erfolgte diesbezüglich die Umsetzung der Plan-UP-Richtlinie in nationales Recht?

Da Wien sowohl Bundesland als auch Gemeinde ist, bestehen keine rechtsverbindlichen Raumordnungspläne der Landesebene sonder nur die Pläne der örtlichen Raumplanung (eigener Wirkungsbereich der Gemeinden). Dies sind die Flächenwidmungspläne (Flächennutzungspläne) und die Bebauungspläne.

In Bezug auf die Wiener Raumordnung wurde die SUP-Richtlinie in der Bauordnung für Wien („Wiener Stadtplanungs- und Baugesetzbuch“) umgesetzt.

b) Wie wurde die Plan-UP-Richtlinie in das nationale Recht umgesetzt?

Keine Antwort

c) Werden für bestimmte Fälle, insbesondere der Geringfügigkeit im Rahmen der Aufstellung, Änderung und Ergänzung von Bauleitplänen, Möglichkeiten zur Beschleunigung sowie Verfahrenserleichterungen genutzt?

Die Ausnahmetatbestände der Richtlinie (geringfügige Änderungen, Festlegung der Nutzung kleiner Gebiete auf lokaler Ebene) wurden in die Wiener Bauordnung nicht aufgenommen (die Begriffe „geringfügig" und „klein" wurden von Juristen als zu unbestimmt beurteilt). Im Fall der geringfügigen Änderung wird des Öfteren damit argumentiert, dass die Rahmensetzung (im Wesentlichen) bereits erfolgt ist und deswegen keine (neuerliche) SUP erforderlich ist.

Ansonsten gibt es für geringfügige Änderungen keine Erleichterung oder Verkürzung des rechtlich vorgeschriebenen Verfahrens (allenfalls im Bereich der Grundlagenerhebung).

d) In welchem Verhältnis stehen Umweltprüfung und FFH-Verträglichkeitsprüfung zueinander?

Die FFH-Verträglichkeitsprüfung findet in Form der Umweltprüfung statt und stellt keine eigenständige Prüfung dar.

3.8.4 Frage 4: Dienstleistungsrichtlinie (2006/123/EG)

Wird ein Umsetzungsbedarf der Dienstleistungsrichtlinie (Richtlinie 2006/123/EG) für den Bereich des nationalen Städtebaurechts und der Bauleitplanung gesehen?

Nein.

3.8.5 Frage 5: Erneuerbare-Energien-Richtlinie (2009/28/EG)

Wird ein Umsetzungsbedarf hinsichtlich des Artikels 13 Erneuerbare-Energien-Richtlinie (2009/28/EG) für den Bereich des nationalen Städtebaurechts und der Bauleitplanung gesehen?

Nein.

3.8.6 Frage 6: SEVESO-II-Richtlinie

Wie ist Artikel 12 Absatz 1 Unterabsatz 2 der Richtlinie 96/82/EG vom 09.12.1996 zur Beherrschung der Gefahren bei Unfällen mit gefährlichen Stoffen (ABl. L 10 vom 14.01.1997, S. 13; „Seveso-II-Richtlinie") in nationales Recht umgesetzt worden?

Keine Antwort

3.9 Poland

3.9.1 Question 1: Systematic preliminary question

What is the relationship between binding land use plans and the approval of individual construction projects?

According to the Polish Law of 27 March 2003 on spatial planning and development, binding land use plan is the act of local law which sets the purpose and principles of land development. Drawing up a binding land use plan is optional unless it is an obligation arising from any others legal acts. Decision on building permit is granted on the basis of binding land use plan if it exists. In case of the absence of the binding land use plan the planning permission (decision building) or decision on the location of the public investment is granted. Decisions are binding for the issuing authorities.

Furthermore, according to article 35 section 1 point 1 of the act of 7 July 1994 on Building Law, before the decision on building permit is granted, competent authority is obliged to verify the compliance of the construction project with the binding land use plan or with the planning permission in case of the absence of the binding land use plan.

3.9.2 Question 2: European Habitat and Protection of Species Law

a) *What is the relationship between national nature conservation law and the building planning law? Are nature conservation law requirements taken into account in development planning or not dealt with until a subsequent approval procedure takes place?*

Polish Building Law does not contain regulations concerning the procedure of drawing up a binding land use plan. Drawing up and enacting the binding land use plan is based on spatial planning and development law. The obligation to take into consideration the nature conservation issues during the development planning results from:

- 1. Law on spatial planning and development
- 2. Law of 16 April 2004 on nature protection
- 3. Law of 3 October 2008 on the availability of information about the environment and its protection, public participation in environmental protection, and the environmental impact assessment

During the procedure of drawing up and enacting the binding land use plan nature conservation requirements are taken into account. It is a result of:

- Article 1 section 2 point 3 of the Law on spatial planning and development
- Article 3 of the Law on nature protection
- Article 15 section 2 points 3, 7, 9 of the Law on spatial planning and development (in binding land use plan it is mandatory to define: rules on environmental protection, nature and cultural landscape, boundaries and ways of land development or other objects to be protected in the mining areas and exposed to the danger of flooding and risk of subsidence of soil mass, specific conditions of land development and restrictions in land using including non-construction) the processing of the environmental impact assessment, the effects of realization of this plan along with estimate of the environmental impact

Draft of land use plan requires the processing of the environmental impact assessment, the effects of realization of this plan along with estimate of the environmental impact

- Article 46 point 1 of the Law on the availability of information about the environment and its protection, public participation in environmental protection, and the environmental impact assessment
- Article 51 section 1 of the Law on the availability of information about the environment and its protection, public participation in environmental protection, and the environmental impact assessment (competent authority developing draft land use plan prepares an estimate of the environmental impact)
- Article 17 point 4 of the Law on spatial planning and development (draft land use plan is prepared along with an estimate of the environmental impact)

In the absence of a binding land use plan, to obtain a building permit requires the prior determination of the conditions allowing for appropriate land development in the way of a decision on planning permission.

Environmental Impact Assessment should be conducted before the decision on planning permission is granted. In some cases decision on environmental requirements must be obtained.

- According to Article 59 of the Law on the availability of information about the environment and its protection, public participation in environmental protection, and the environmental impact assessment, is required in following cases:

- 1) planned projects likely to have always significant effects on the environment (projects of group I)
- 2) planned projects likely to have potentially significant effects on the environment (projects of group II), if the obligation to assess the impact of environment is based on Article 63 section 1 and is a result of screening.

Implementation of planned projects other than those specified in the above paragraphs 1 and 2 requires an assessment of impacts on the nature of 2000, if

- 1) the project is likely to have significant effects on the Natura 2000, and is not directly related to the protection of this area or does not follow that protection
- 2) obligation to carry out impact assessment on the Natura 2000 has been determined on the basis of Article 63 paragraph 1 of above Act
- Article 72 paragraph 1 point 3 of the Law on the availability of information about the environment and its protection, public participation in environmental protection, and the environmental impact assessment (decision on the environmental conditions follows before the decision on building and land development).

Essentially a procedure for assessing the environmental impact is carried out once. However, in special cases it is possible to re-evaluate the impact of assess the environmental impact if the requirement is at the request of the parties and if it is determined by the competent authority to issue a decision that the dossier does not correspond to the findings the project decision.

- Article 82 paragraph 1 point 4 Law on the availability of information about the environment and its protection, public participation in environmental protection, and the environmental impact assessment (in environmental decision issued after an assessment of environmental impact, competent authority shall submit its position on the need for an assessment of environmental impact and conduct on transboundary environmental impact in the context of the decision on the construction permit, the decision to allow the realization for road investment project or a decision to allow the implementation of the investment in the airport for public use)
- Article 88 section 1 point 1 of the Law on the availability of information about the environment and its protection, public participation in environmental protection, and the environmental impact assessment (assessment of environmental impact in the context of the decision on the construction permit, decision to allow the imple-

mentation of road investment or decisions for investments in the airport for public use, carried out at the request of the entity proposing to undertake the project, filed with the competent authority to issue a decision)

- Article 88 section 1 point 2 of the Law on providing environmental information and its protection, public participation in environmental protecting and the environmental impact assessment (assessment of environmental impact in the context of the issue of the decision on the construction permit, decision to allow the implementation of road investment or decisions for investments in the airport for public use, is also carried out where the competent authority to issue a decision finds that the proposal for a decision which changes were made in relation to the requirements specified in the decision on environmental conditions).

Therefore the implementation of environmental requirements is already setting the stage for development conditions and destination area to be established in the local plan as an administrative decision. Should be noted, however, that there is a chance to re-evaluate the an environmental impact assessment in the context of the decision on the construction permit.

Exception to this rule are made investments in roads under the provisions of the Law on special rules for the preparation and implementation of investment in national roads and airports for public use, which are located under the Law of 12 February 2009 on special principles of preparation and implementation of investment in the airports for public use. Authorization for execution of works is based on the decision to permit the implementation of these investments, which identifies their location and at the same time approve the construction project.

- At the same time it should be noted that in the current state of legal procedures related to the evaluation of their impact on the environment are not associated with the procedures set out in planning law on planning for land development and shall be conducted under a separate law. Rules and procedures in matters of environmental impact assessments are established by law to release information about the environment and its protection, public participation on environmental protection and the environmental impact assessment. The Act provides implementation of community law concerning environmental impact assessment, in particular Council Directive 85/337 of 25 June 1985 on the assessment of the effects of certain public and private projects on the environment and Directive 2003/35 of the European Parliament and the Council of 26 May 2003 providing for public participation in relation to the drawing up of certain plans and programs relating to the environment and amending with regard to public participation and access to justice Council Directives 85/337/EEC and 96/61/WE.

- Please also note that the provisions of the Law on the availability of information about the environment and its protection, public participation in environmental protection, and the environmental impact assessment does not come under the Ministry of Infrastructure.
- Referring to questions b, c, d and e in section 2 enclosed questionnaire, concerning the habitat and species conservation, we inform that the implementation of the Polish law of Council Directive 92/43/EEC of 21 May 1992 on the conservation of natural habitats and of wild fauna and flora, so-called Habitats Directive has been in the provisions of the Law on nature protection.

b) Within the scope of the assessment of the implications of a plan or project for a site, how has the exception provided by Article 6 Paragraph 4 of the FFH Directive been transposed into national law in the field of building planning and how is it applied?

With regard to questions b, c, d and e in paragraph 2 of the enclosed questionnaire on habitat and species protection, please be informed that Council Directive 92/43/EC of 21 May 1992 on the conservation of natural habitats and of wild fauna and flora, the so-called Habitat Directive, was transposed in Polish law in the provisions of the nature conservation law.

Is there any distinction regarding whether a project is built within or outside an area which is already continuously built up?

No Answer

c) Apart from the conservation area related assessment of implications for the site pursuant to Article 6 Paragraph 3 of the FFH Directive, is an assessment under species protection law also performed according to Article 12 of the FFH Directive?

See answer to question 2 paragraph b.

d) Is there a need to exempt such construction projects from prohibitions under species protection law which are erected in areas which have already been developed? What role could it play if the protected species had their natural habitat outside the development area and also occurred in greater number there?

See answer to question 2 paragraph b.

e) How is the element of "intention" in Article 12 Paragraph 1 Letter a, b and c of the FFH Directive interpreted for construction projects in the light of the legal decisions taken by the European Court of Justice?

See answer to question 2 paragraph b.

3.9.3 Question 3: Directive 2001/42/EC of the European Parliament and of the Council on of 27 June 2001 on the assessment of the effects of certain plans and programs on the environment (SEA Directive).

a) Which levels and instruments of development planning exist and with respect to this, how was the SEA Directive transposed into national law?

Polish system is a three-stage planning. The development planning shall be made at national, provincial and local (municipal).

Spatial concept of the country is a document of spatial planning policy makers throughout the country.

Act of the regional government is planning zoning plan population.

Planning document on municipal level are: study of conditions and directions of spatial management of a community and local spatial plan. A study is drawn for the whole area of the municipality and its role is to determine the spatial policy of the municipal authorities and determine the spatial directions of the municipality. The implementation of a spatial policy takes place in a municipal policy on local spatial plan, which sets out the purpose and principles of land management plan. This is the provision of local law which applies to the area covered by the plan and provides the basis for changes in land management plan, including the location of the building. Establishment of a local plan is optional, with some exceptions under the statutory duty.

The Directive 2001/42/EC of the European Parliament and of the Council of 27 June 2001 on the assessment of the effects of certain plans and programs on the environment (SEA Directive) was transposed into force on October 3 of the Law on the availability of information about the environment and its protection, public participation in environmental protection, and the environmental impact assessment

- Article 46 point 1 – carry out strategic environmental impact assessment projects require the concept of spatial development of the country, the study of conditions and directions of spatial of a municipal area, land management plans;
- Article 51 paragraph 1 – body developing a draft concept of national spatial management, the study of conditions and directions of

community planning, zoning plans, draws up an estimate of the environmental impact;

- Article 17 point 4 of the Law on spatial planning and development - the project of the plan is drawn up together with an estimate of the environmental impact;
- Article 41 paragraph 1 point 4 of the Law on spatial planning and development - the draft of region plan is drawn up together with an estimate of the environmental impact.

Furthermore, the draft of the law amending the Law on spatial planning and development and some other laws complements an implementation of EU directives which have been made earlier in the regulations on environmental protection to the provisions of the spatial planning and development. This includes, inter alia, Directive 2001/42/EC.

The proposed changes will allow the addition to the principles of environmental impact assessment and planning acts of investment introduced by the Law on the availability of information about the environment and its protection, public participation in environmental protection, and the environmental impact assessment, by coordinating the various activities required under the environmental and planning procedures.

b) How was the SEA Directive transposed into national law?

Information on this subject can be found in the answer to the question 3 point a).

c) Are options, especially option of insignificance within the context of preparing, amending and supplementing the development plans, used to accelerate and simplify procedures for particular cases?

Law on spatial planning and development does not foresee the possibility of draw up an amended zoning plan under simplified procedures. The mode of preparation of local plans and their changes results from requirements that apply of local law.

- Article 27 of the Law on spatial planning and development - change of study or a local plan shall be made in such a way as they are passed.

d) What is the relationship between the environmental audit and FFH assessment of implications for the site?

According to Article 59 of the Law on the assessment environmental impact, in case of projects that may always be a significant or potentially affect the environment, the impact of projects on the area of the Nature 2000 is considered in the context of the decision on the environmental conditions.

For investments other than those mentioned above, (not covered by the Directive EIA), an assessment of the impact of the project on the area of the Nature 2000 is carried when all the following conditions occur:

- project may have significant effects on the area of the Nature 2000;
- project is not directly related to the protection of the area of the Nature 2000 or not due to its protection;
- competent authority issues a decision required before the project said the potential for significant impact on the area of Nature 2000.

3.9.4 Question 4: EU Services Directive (2006/123/EC)

Is it deemed necessary to transpose the EC Services Directive (Directive 2006/123/EC) for the area of national urban planning law and development planning?

Implementation of the Directive 2006/123/EC does not require changes in the provisions of the Law on spatial planning and development. In Poland there are no limitations, for both EU and third country citizens to perform urban planning services. There are no restrictions for foreigners who would like to establish in Poland a company that will provide these services. However, in Poland there are regulations determining who has the right to pursue the town planner profession. These requirements are established by Law on self-government of professional architects, construction engineers and town planners. According to the Article 6 paragraph 1 of this law, the right to self-design space in the regional and local scale or to lead a team carrying out such a design is entitled only to persons included in the list of members of the professional self-government.

3.9.5 Question 5: Renewable Energy Directive (2009/28/EC)

Is it considered necessary to transpose Article 13 of the Renewable Energy Directive (Directive 2009/28/EC) for the area of national urban planning law and development planning?

It should be noted that the Polish system of spatial planning does not include separate provisions governing the design and cities development. Among the planned changes in legislation there are proposals increasing the importance of energy management in spatial planning and an increase in correlation between the energy sector plans. The proposed amendments to the Law on spatial planning and development provide complementary a catalogue of the establishment of the study of conditions and directions of their use and local spatial plans of local community resources and local plans for the rules of the use of these resources for diversifying supply sources and to create optimal conditions for the development of distributed based on locally available raw materials, including renewable energy.

Furthermore, there has been taken legislative initiative which obliges to include all sectoral development strategies, including the Polish Energy Policy until 2030, when drawing up the study of conditions and directions of spatial development community. Main directions of the Polish Energy Policy until 2030 condition the change in approach to energy policy, inter alia, by changing the structure of exploitation an energy sources. Exploitation of the renewable sources of energy is one of the main directions of the Polish energy Policy. The requirement to consider energy policy objectives in the study of municipal will cause their move from the national to local level.

These proposed regulations are consistent with the provisions of the Directive on the promotion of the use of energy from renewable sources, under the national law of design and urban planning

3.9.6 Question 6: SEVESO-II-Directive

How has Article 12 Paragraph 1 Subsection 2 Council Directive 96/82/EC of 9 December 1996 on the control of major-accident hazards involving dangerous substances (OJ L 10, 14.1.1997, p. 13; "Seveso-II-Directive") been transposed into national law?

No answer

3.9 Polen

3.9.1 Frage 1: Systematische Vorfrage

In welchem Verhältnis stehen Bebauungspläne und die Genehmigung einzelner Bauvorhaben?

Nach dem polnischen Gesetz über Planung und Raumwirtschaft vom 27.03.2003 ist die Erstellung eines bindenden Bebauungsplans ein kommunalrechtlicher Akt, der den Zweck und die Grundsätze der Flächennutzung festlegt. Soweit sie sich nicht verpflichtend aus anderen Rechtsakten ergibt, ist die Erstellung eines verbindlichen Bauleitplans freiwillig. Entscheidungen über Baugenehmigungen werden aufgrund des verbindlichen Bauleitplans getroffen, sofern dieser existiert. Existiert kein verbindlicher Bauleitplan, wird die Planungsgenehmigung (Bauentscheid) oder die Entscheidung über den Standort des öffentlichen Investitionsvorhabens ausgestellt. Die Entscheidungen sind für die erteilenden Behörden bindend.

Nach Artikel 35 Absatz 1 Punkt 1 Baurechtsgesetz vom 7. Juli 1994 ist die zuständige Behörde verpflichtet, vor Ausstellung eines Baugenehmigungsbescheids die Übereinstimmung des Bauvorhabens mit dem verbindlichen Bauleitplan oder, falls ein verbindlicher Bauleitplan nicht existiert, mit der Planungsgenehmigung zu prüfen.

3.9.2 Frage 2: Europäisches Habitat- und Artenschutzrecht

a) *In welchem Verhältnis stehen das nationale Naturschutzrecht und das Bauplanungsrecht? Werden die Anforderungen des Naturschutzrechts bereits in der Bauleitplanung oder erst in einem nachfolgenden Genehmigungsverfahren bewältigt?*

Das polnische Baurecht umfasst keine Regelungen über das bei der Aufstellung eines verbindlichen Bauleitplans zu nutzende Verfahren. Die Aufstellung und Verabschiedung des verbindlichen Bauleitplans beruhen auf dem Gesetz über Planung und Raumwirtschaft. Die Pflichten zur Berücksichtigung von Fragen des Naturschutzes bei der Bauleitplanung ergeben sich aus dem:

- Gesetze über Planung und Raumwirtschaft,
- Gesetz über den Naturschutz vom 16. April 2004
- Gesetz über den Zugang zu Informationen über die Umwelt und ihren Schutz, Öffentlichkeitsbeteilung in Umweltschutzangelegenheiten sowie über Bewertungen der Auswirkungen auf die Umwelt vom 3. Oktober 2008.

Während der Aufstellung und Verabschiedung des verbindlichen Bauleitplans werden Naturschutzerfordernisse berücksichtigt. Dies resultiert aus:

- Artikel 1 Absatz 2 Punkt 3 des Gesetzes über Planung und Raumwirtschaft,
- Artikel 3 des Naturschutzgesetzes,
- Artikel 15 Absatz 2 Punkte 3, 7, 9 Gesetz über Planung und Raumwirtschaft (es ist zwingend erforderlich, in einem verbindlichen Bauleitplan Folgendes festzulegen: Regeln für den Schutz der Umwelt, Natur und Kulturlandschaft; Grenzen und Möglichkeiten der Flächennutzung oder anderer Objekte, die in überflutungsgefährdeten Bergbaugebieten, in denen Gefahr der Bodensenkung besteht, liegen; besondere Voraussetzungen der Flächennutzung und Erschließung des Baugeländes einschließlich Bauverbot), Durchführung der Umweltverträglichkeitsprüfung, die Auswirkungen der Verwirklichung dieses Plans mit der Umweltverträglichkeitsbewertung.

Eine Voraussetzung für die Aufstellung des Bauleitplans ist die Durchführung der Umweltverträglichkeitsprüfung. Sie besteht aus der Ermittlung der Auswirkungen der Verwirklichung dieses Plans und der Umweltverträglichkeitsbewertung.

- Artikel 46 Punkt 1 Gesetz über den Zugang zu Informationen über die Umwelt und ihren Schutz, Öffentlichkeitsbeteiligung in Umweltschutzangelegenheiten sowie die Umweltverträglichkeitsprüfung.
- Artikel 51 Absatz 1 Gesetz über den Zugang zu Informationen über die Umwelt und ihren Schutz, Öffentlichkeitsbeteiligung in Umweltschutzangelegenheiten sowie die Umweltverträglichkeitsprüfung (die für die Aufstellung des Bauleitplanentwurfs zuständige Behörde erarbeitet eine Umweltverträglichkeitsbewertung).
- Artikel 17 Punkt 4 Gesetz über Planung und Raumwirtschaft (der Entwurf des Bauleitplans wird zusammen mit einer Umweltverträglichkeitsbewertung erstellt).

Existiert kein verbindlicher Bauleitplan, ist die vorherige Festlegung der Bedingungen für die angemessene Flächennutzung zwecks Entscheidung über die Baugenehmigung Voraussetzung für die Erteilung der Baugenehmigung.

Die Umweltverträglichkeitsprüfung sollte durchgeführt werden, bevor die Entscheidung über die Baugenehmigung getroffen wird. In einigen Fällen müssen Entscheidungen zu Umweltanforderungen eingeholt werden.

- Nach Artikel 59 Gesetz über den Zugang zu Informationen über die Umwelt und ihren Schutz, Öffentlichkeitsbeteiligung in Umweltschutzangelegenheiten sowie über die Umweltverträglichkeitsprüfung ist Letztere in folgenden Fällen erforderlich:

- 1) Bei Planungsvorhaben, die wahrscheinlich stets erhebliche Wirkungen auf die Umwelt haben werden (Projekte der Gruppe I).
- 2) Planungsvorhaben, die wahrscheinlich potenziell erhebliche Wirkungen auf die Umwelt haben werden (Projekte der Gruppe II), wenn die Pflicht zur Durchführung der Umweltverträglichkeitsprüfung auf Artikel 63 Absatz 1 beruht und Ergebnis des Umweltscreenings ist).

Die Verwirklichung anderer Planvorhaben als den in den vorausgehenden Absätzen 1 und 2 genannten Projekten erfordert die Bewertung der Umweltauswirkungen auf Natura 2000-Gebiete, wenn

- 1) das Vorhaben wahrscheinlich erhebliche Auswirkungen auf ein Natura 2000-Gebiet haben wird und nicht unmittelbar mit dem Schutz dieses Gebietes in Zusammenhang steht oder den Schutz nicht beachtet.
- 2) die Pflicht zur Durchführung der Umweltverträglichkeitsprüfung in dem Natura 2000-Gebiet nach Artikel 63 Absatz 1 des obengenannten Gesetzes festgestellt wurde.
- Artikel 72 Absatz 1 Punkt 3 Gesetz über den Zugang zu Informationen über die Umwelt und ihren Schutz, Öffentlichkeitsbeteiligung in Umweltschutzangelegenheiten und die Umweltverträglichkeitsprüfung (die Entscheidung über die Umweltbedingungen erfolgt vor der Entscheidung über die Bebauung und die Flächennutzung).

Ein Umweltverträglichkeitsprüfungsverfahren wird im Wesentlichen sofort durchgeführt. In Sonderfällen ist es jedoch möglich, eine Neubewertung vorzunehmen, wenn die Parteien dies fordern und wenn die zuständige Behörde entscheidet, dass das Dossier nicht mit den Feststellungen der Projektentscheidung übereinstimmt.

- Artikel 82 Absatz 1 Punkt 4 Gesetz über den über den Zugang zu Informationen über die Umwelt und ihren Schutz, Öffentlichkeitsbeteiligung in Umweltschutzangelegenheiten und die Umweltverträglichkeitsprüfung (in der umweltbezogenen Entscheidung, die nach der Umweltverträglichkeitsprüfung ausgestellt wird, soll die zuständige Behörde ihren Standpunkt zur Notwendigkeit einer Umweltverträglichkeitsprüfung darlegen und im Rahmen der Entscheidung über die Baugenehmigung, die Entscheidung über die Genehmigung der Realisierung von Investitionsvorhaben im Straßenbau oder einer Entscheidung über die Durchführung von Investitionen in einen öffentlich genutzten Flughafen eine grenzüberschreitende Umweltverträglichkeitsprüfung durchführen.)

- Artikel 88 Absatz 1 Punkt 1 Gesetz über den Zugang zu Informationen über die Umwelt und ihren Schutz, Öffentlichkeitsbeteiligung in Umweltschutzangelegenheiten und die Umweltverträglichkeitsprüfung (Bewertung der Umweltauswirkungen im Zusammenhang mit der Entscheidung über die Baugenehmigung, die Entscheidung über die Genehmigung der Realisierung von Investitionsvorhaben im Straßenbau oder einer Entscheidung über die Durchführung von Investitionen in einen öffentlich genutzten Flughafen, die auf Ersuchen des Bauwilligen durchgeführt und bei der für die Entscheidung zuständigen Behörde eingereicht wird.)
- Artikel 88 Absatz 1 Punkt 2 Gesetz über den Zugang zu Informationen über die Umwelt und ihren Schutz, Öffentlichkeitsbeteiligung in Umweltschutzangelegenheiten und die Umweltverträglichkeitsprüfung (die Bewertung der Umweltauswirkungen im Zusammenhang mit der Entscheidung über die Baugenehmigung, die Entscheidung über die Genehmigung der Realisierung von Investitionsvorhaben im Straßenbau oder Entscheidungen über Investitionen in einen öffentlich genutzten Flughafen wird auch durchgeführt, wenn die für die Entscheidung zuständige Behörde feststellt, dass der Vorschlag, über den entschieden werden soll, bezüglich der Anforderungen, die in der Entscheidung über die Umweltbedingungen genannt wurden, geändert wurde).

Daher ist die Durchsetzung der Umweltanforderungen bereits der erste Schritt zur Festlegung der Entwicklungsbedingungen und des als Verwaltungsakt im Kommunalplan festzulegenden Zielgebietes. Es soll jedoch darauf hingewiesen werden, dass die Möglichkeit einer Neubewertung der Umweltverträglichkeitsprüfung im Zusammenhang mit der Entscheidung über die Baugenehmigung besteht.

Ausnahmen von dieser Regelung werden bei Investitionen in den Straßenbau nach den Bestimmungen des Gesetzes über Sonderregelungen für die Vorbereitung und Durchführung von Investitionen in Nationalstraßen und öffentlich genutzte Flughäfen vom 12. Februar 2009 gemacht. Grundlage für die Genehmigung der Ausführung der Arbeiten ist die Entscheidung über die Durchführung dieser Investitionen, in der der Standort festgelegt und das Bauvorhaben gleichzeitig genehmigt wird.

Es sei ebenfalls darauf hingewiesen, dass der gegenwärtige Stand der die Umweltverträglichkeitsbewertung betreffenden rechtlichen Verfahren nicht mit Verfahren in Zusammenhang steht, die im Planungsrecht für die Bauleitplanung beschrieben sind und nach einem separaten Gesetz durchgeführt werden.

Regelungen und Verfahren zu Angelegenheiten der Umweltverträglichkeitsprüfungen sind festgelegt in dem Gesetz über den Zugang zu Informationen über die

Umwelt und ihren Schutz, die Öffentlichkeitsbeteiligung in Umweltangelegenheiten und die Umweltverträglichkeitsprüfung. Das Gesetz sieht vor, das die Umweltverträglichkeitsprüfung betreffendes Gemeinschaftsrecht, insbesondere die Richtlinie 85/337 des Rates vom 25. Juni 1985 über die Umweltverträglichkeitsprüfung bei bestimmten öffentlichen und privaten Projekten und die Richtlinie 2003/35 des Europäischen Parlaments und des Rates vom 26. Mai 2003 über die Beteiligung der Öffentlichkeit bei der Ausarbeitung bestimmter umweltbezogener Pläne und Programme und zur Änderung der Richtlinien 85/337 EWG und 96/61 EG in Bezug auf die Öffentlichkeitsbeteiligung und den Zugang zu Gerichten umzusetzen.

Bitte nehmen Sie auch zur Kenntnis, dass die Bestimmungen des Gesetzes über den Zugang zu Informationen über die Umwelt und ihren Schutz, die Öffentlichkeitsbeteiligung in Umweltschutzangelegenheiten und die Umweltverträglichkeitsprüfung nicht in den Zuständigkeitsbereich des Infrastrukturministeriums fallen.

b) Wie ist im Rahmen der Verträglichkeitsprüfung die Ausnahmeregelung des Artikels 6 Absatz 4 FFH-Richtlinie im Bereich des Bauplanungsrechts in nationales Recht umgesetzt worden und wie wird diese angewandt?

Bezug nehmend auf die Fragen b, c, d und e in Absatz 2 des beigefügten Fragebogens über den Habitat- und Artenschutz teilen wir Ihnen mit, dass die Richtlinie 92/43/EWG des Rates vom 21. Mai 1992 zur Erhaltung der natürlichen Lebensräume und der wildlebenden Tiere und Pflanzen, die sogenannte Habitat-Richtlinie, in den Bestimmungen des Naturschutzgesetzes in polnisches Recht umgesetzt wurde.

Gibt es eine Differenzierung in Bezug darauf, dass ein Vorhaben innerhalb oder außerhalb eines bereits im Zusammenhang bebauten Gebietes errichtet wird?

Keine Antwort

c) Wird neben der schutzgebietsbezogenen Verträglichkeitsprüfung des Artikels 6 Absatz 3 FFH-Richtlinie bei baulichen Vorhaben auch eine artenschutzrechtliche Prüfung nach Artikel 12 FFH-Richtlinie durchgeführt?

Siehe Antwort 2b

d) Wird ein Bedarf dafür gesehen, solche Bauvorhaben von artenschutzrechtlichen Verboten freizustellen, die in Gebieten errichtet werden, die bereits im Zusammenhang bebaut sind? Welche Rolle könnte es hierbei haben, wenn die geschützte Art ihre natürliche Verbreitung außerhalb des Siedlungsbereichs hat und dort auch in größerer Zahl vorkommt?

Siehe Antwort 2b

e) Wie wird das Tatbestandsmerkmal der „Absicht" in Artikel 12 Absatz 1 Buchstabe a, b und c FFH-Richtlinie bei Bauvorhaben im Lichte der Rechtsprechung des Europäischen Gerichtshofs verstanden?

Siehe Antwort 2b

3.9.3 Frage 3: Plan-UP-Richtlinie (2001/42/EG)

a) Welche Ebenen und Instrumente städtebaulicher Planungen bestehen und wie erfolgte diesbezüglich die Umsetzung der Plan-UP-Richtlinie in nationales Recht?

Polen hat ein dreistufiges Planungssystem. Die Raumplanung erfolgt auf nationaler Ebene, auf der Ebene der Woiwodschaften und auf örtlicher (kommunaler) Ebene.

Das nationale Raumordnungskonzept wird von den mit der Raumordnung befassten Politikern ausgearbeitet. Die Erarbeitung des Raumbewirtschaftungsplans ist eine Aufgabe der Regionalregierung.

Zu den kommunalen Planungsdokumenten zählen: Studie der Rahmenbedingungen und der Perspektiven der Raumbewirtschaftung der Gemeinde und der örtliche Raumbewirtschaftungsplan. Die Studie wird für das gesamte Gemeindegebiet durchgeführt; ihre Rolle besteht darin, die Raumordnungspolitik der kommunalen Behörden und die räumliche Ausrichtung der Gemeinde festzulegen. Der örtliche Raumbewirtschaftungsplan konkretisiert die Vorgaben der Studie. Dies ist im Kommunalrecht vorgesehen, das für das von dem Plan erfassten Gebiet gilt. Das Kommunalrecht gilt für das vom Plan erfasste Gebiet und bildet die Grundlage für Änderungen der Studie einschließlich des Gebäudestandortes. Bis auf wenige Ausnahmen, in denen die Pflicht gesetzlich festgeschrieben ist, ist die Aufstellung eines kommunalen Plans freiwillig.

Die Richtlinie 2001/42/EG des Europäischen Parlaments und des Rates vom 27. (UVP-Richtline) wurde durch das Gesetz über den Zugang zu Informationen über die Umwelt und ihren Schutz, die Öffentlichkeitsbeteiligung in Umweltschutzangelegenheiten sowie über die Umweltverträglichkeitsprüfung vom 3. Oktober in Kraft gesetzt.

- (Artikel 46 Punkt 1) Eine strategische Umweltprüfung ist durchzuführen bei der Aufstellung des Raumordnungskonzeptes des Landes, der Studie der Rahmenbedingungen und der Perspektiven der Raumbewirtschaftung der Gemeinde.
- (Artikel 51 Absatz 1) Das Gremium, das den Entwurf des nationalen Raumordnungsplans erarbeitet, die Bedingungen untersucht und die Richtungen der Gemeindeplanung und Zonierungspläne festlegt, erarbeitet die Prüfung der Umweltauswirkungen.

- (Artikel 17 Punkt 4 Gesetz über Planung und Raumwirtschaft) Der Planentwurf wird zusammen mit einer Prüfung der Umweltauswirkungen erstellt.
- (Artikel 41 Absatz 1 Punkt 4 Gesetz über Planung und Raumwirtschaft) Der Entwurf des Regionalplans wird zusammen mit einer Prüfung der Umweltauswirkungen erstellt.

Außerdem ergänzt der Gesetzentwurf zur Änderung des Gesetzes über Planung und Raumwirtschaft und einiger anderer Gesetze die in früheren Umweltschutzregelungen vorgesehene Umsetzung der EU-Richtlinien um die Bestimmungen zur Planung und Raumwirtschaft. Dazu zählt unter anderen die Richtlinie 2001/42/EG.

Die beabsichtigten Änderungen werden es ermöglichen, Investitionsmaßnahmen, die vom Gesetz über den Zugang zu Informationen über die Umwelt und ihren Schutz, die Öffentlichkeitsbeteiligung in Umweltschutzangelegenheiten und die Umweltverträglichkeitsprüfung eingeführt werden, durch Koordinierung der verschiedenen Aktivitäten, die nach den Umweltschutz- und Planungsverfahren vorgesehen sind, in die Grundsätze der Umweltverträglichkeitsprüfung und -planung einzubeziehen.

b) Wie wurde die Plan-UP-Richtlinie in das nationale Recht umgesetzt?

Angaben zu diesem Thema sind in der Antwort auf Frage 3 Punkt a zu finden.

c) Werden für bestimmte Fälle, insbesondere der Geringfügigkeit im Rahmen der Aufstellung, Änderung und Ergänzung von Bauleitplänen, Möglichkeiten zur Beschleunigung sowie Verfahrenserleichterungen genutzt?

Das Gesetz über Planung und Raumwirtschaft sieht keine Möglichkeit vor, einen geänderten Nutzungsplan in einem vereinfachten Verfahren aufzustellen. Die Art der Aufstellung von Kommunalplänen und ihre Änderung ergibt sich aus den Anforderungen des Kommunalgesetzes.

- (Artikel 27 Gesetz über Planung und Raumwirtschaft) Änderungen der Untersuchung oder eines Kommunalplans sind in dem gleichen Verfahren wie bei der Aufstellung vorzunehmen.

d) In welchem Verhältnis stehen Umweltprüfung und FFH-Verträglichkeitsprüfung zueinander?

Nach Artikel 59 Gesetz über die Umweltverträglichkeitsprüfung werden bei Projekten mit erheblichen oder potenziell erheblichen Umweltauswirkungen auf Natura 2000-Gebiete die Auswirkungen im Zusammenhang mit der Entscheidung über die Umweltverträglichkeit betrachtet.

Für andere als die vorstehend genannten Investitionen, (die nicht in der UVP-Richtlinie erfasst sind) wird eine Bewertung der Auswirkung des Projekts auf das Natura 2000-Gebiet bei Vorliegen der folgenden Bedingungen vorgenommen:

- das Projekt kann erhebliche Auswirkungen auf das Natura 2000-Gebiet haben;
- das Projekt steht in keinem direkten Zusammenhang mit dem Schutz des Natura 2000-Gebiets bzw. ist nicht auf seinen Schutz zurückzuführen;

die zuständige Behörde erlässt eine vor dem Projekt erforderliche Entscheidung über die Möglichkeit erheblicher Auswirkungen auf das Natura 2000-Gebiet.

3.9.4 Frage 4: Dienstleistungsrichtlinie (2006/123/EG)

Wird ein Umsetzungsbedarf der Dienstleistungsrichtlinie (Richtlinie 2006/123/EG) für den Bereich des nationalen Städtebaurechts und der Bauleitplanung gesehen?

Die Umsetzung der Richtlinie 2006/123/EC erfordert keine Änderungen der Bestimmungen des Gesetzes über Planung und Raumwirtschaft. In Polen bestehen sowohl für EU-Bürger als auch Bürger von Drittstaaten keine Beschränkungen bezüglich der Erbringungen städtebaulicher Leistungen. Es existieren keine Beschränkungen für Ausländer, die in Polen ein Unternehmen zur Erbringung dieser Leistungen gründen möchten. Es gibt in Polen jedoch Regelungen, die festlegen, wer berechtigt ist, als Stadtplaner tätig zu sein. Diese Anforderungen sind im Gesetz über die Selbstverwaltung von Architekten, Bauingenieuren und Stadtplanern festgelegt. Nach Artikel 6 Punkt 1 dieses Gesetzes ist das Recht, regionale oder örtliche Flächen selbst zu gestalten oder ein Team zu leiten, das diese Planungsaufgaben ausführt, nur für Personen vorgesehen, die in die Liste des beruflichen Selbstverwaltungsgremiums aufgenommen wurden.

3.9.5 Frage 5: Erneuerbare-Energien-Richtlinie (2009/28/EG)

Wird ein Umsetzungsbedarf hinsichtlich des Artikels 13 Erneuerbare-Energien-Richtlinie (2009/28/EG) für den Bereich des nationalen Städtebaurechts und der Bauleitplanung gesehen?

Es wird darauf hingewiesen, dass das polnische Planungssystem keine gesonderten Bestimmungen zur Planung und zum Städtebau umfasst. Zu den geplanten gesetzlichen Änderungen zählen Vorschläge zur Stärkung der Bedeutung des Energiemanagements in der Raumordnung und eine stärkere Abstimmung der Pläne des Energiesektors. Die beabsichtigten Änderungen des Gesetzes über Planung und Raumwirtschaft sehen vor, einen die Untersuchung der Bedingungen und Richtungen der Nutzung kommunaler Bauleitpläne und kommunaler Ressourcen betreffenden ergänzenden Katalog der Vorschriften für die Nutzung dieser

Ressourcen zu erstellen, um eine Diversifizierung der Versorgung sowie optimale Bedingungen für die Erschließung dezentral und örtlich vorhandener Rohstoffe, einschließlich erneuerbarer Energien, zu schaffen.

Außerdem wurde eine Vorlage eingebracht, die vorsieht, alle sektorspezifischen Entwicklungsstrategien, einschließlich der polnischen Energiepolitik bis zum Jahr 2030, bei der Festlegung der Untersuchung der Bedingungen und Richtungen der kommunalen Bauleitplanung zu berücksichtigen. Die Hauptziele der polnischen Energiepolitik bis 2030 sehen eine Änderung des Herangehens an die Energiepolitik vor, was u. a. durch die Änderung der Strukturen zur Nutzung der Energiequellen geschehen soll. Die Nutzung erneuerbarer Energien ist eines der Hauptziele der polnischen Energiepolitik. Da energiepolitische Zielsetzungen in der Untersuchung der Kommunalentwicklung berücksichtigt werden müssen, werden sie künftig nicht mehr auf nationaler Ebene, sondern auf örtlicher Ebene behandelt.

Nach dem nationalen Gesetz über Planung und Städtebau stehen diese beabsichtigten Regelungen im Einklang mit den Bestimmungen der Richtlinie zur Förderung der Nutzung von Energie aus erneuerbaren Quellen.

3.9.6 Frage 6: SEVESO-II-Richtlinie

Wie ist Artikel 12 Absatz 1 Unterabsatz 2 der Richtlinie 96/82/EG vom 09.12.1996 zur Beherrschung der Gefahren bei Unfällen mit gefährlichen Stoffen (ABl. L 10 vom 14.01.1997, S. 13; „Seveso-II-Richtlinie") in nationales Recht umgesetzt worden?

Keine Antwort

3.10 Portugal

3.10.1 Question 1: Systematic preliminary question

What is the relationship between binding land use plans and the approval of individual construction projects?

No answer

3.10.2 Question 2: European Habitat and Protection of Species Law

a) *What is the relationship between national nature conservation law and the building planning law? Are nature conservation law requirements taken into account in development planning or not dealt with until a subsequent approval procedure takes place?*

The domestic legislation which translates the birds and habitats directives rules establish that any building activity inside designated areas under Natura 2000 network is subject to a favourable opinion of nature and biodiversity conservation authority, a decision that can depend on the conclusions of an assessment of environmental incidences, if significant impacts are expected over designated areas. Exception is made to building located in urban perimeters defined in municipalities' master plans. These urban perimeters, legally defined, apart from urbanized areas, contain areas likely to become so and constitute areas of minimal natural value, with a high degree of disturbance.

Yes, the provisions of the concerned domestic legislation are taken into account in urban planning.

b) *Within the scope of the assessment of the implications of a plan or project for a site, how has the exception provided by Article 6 Paragraph 4 of the FFH Directive been transposed into national law in the field of building planning and how is it applied?*

Is there any distinction regarding whether a project is built within or outside an area which is already continuously built up?

The No. 4 of the article 6 of the Habitats Directive was literally translated to the domestic legislation.

There is no distinction between projects located in areas with continuously built up or scattered building. However this feature is evaluated in the assessment process.

c) Apart from the conservation area related assessment of implications for the site pursuant to Article 6 Paragraph 3 of the FFH Directive, is an assessment under species protection law also performed according to Article 12 of the FFH Directive?

The article 12° of the Habitats Directive was totally transposed to internal law. The environmental incidences assessment to which actions, plans or projects are subject, when significant negative impacts are expected over a Natura 2000 network area, take into consideration the legal provisions for species strict protection. When it is expected that a plan, action or project will violate those provisions, the plan, action or project will only be authorised through a license issued by the nature and biodiversity conservation authority and only if all these conditions are satisfied:

a) no satisfactory alternative,

b) the maintenance of populations of the concerned species, in its natural occurrence area, in a favourable conservation state is not affected,

c) and the plan, action or project is oriented to ensure priority public interests, namely of social or economic character.

d) Is there a need to exempt such construction projects from prohibitions under species protection law which are erected in areas which have already been developed? What role could it play if the protected species had their natural habitat outside the development area and also occurred in greater number there?

There are no exemptions. Art 12 of FFH Directive applies in all areas where protected species occurs, including in areas which have already been developed. It must be ensured that no relevant negative changes on conservation status over those species occur due to projects. So art. 12 applies irrespective both of the numbers of the species occurring in a development area or the location of its natural habitat, being dependent on the impact of a project on its conservation status and distribution area.

e) How is the element of "intention" in Article 12 Paragraph 1 Letter a, b and c of the FFH Directive interpreted for construction projects in the light of the legal decisions taken by the European Court of Justice?

The provisions of strict species protection are applied independently of the "intention". However the value of the fine is reduced in a half if we are before a case of negligence.

3.10.3 Question 3: SEA Directive (2001/42/EC)

a) *Which levels and instruments of development planning exist and with respect to this, how was the SEA Directive transposed into national law?*

No answer

b) *How was the SEA Directive transposed into national law?*

No answer

c) *Are options, especially the option of insignificance within the context of preparing, amending and supplementing the development plans, used to accelerate and simplify procedures for particular cases?*

No answer

d) *What is the relationship between the environmental audit and the FFH assessment of implications for the site?*

No answer

3.10.4 Question 4: EU Services Directive (2006/123/EC)

Is it deemed necessary to transpose the EC Services Directive (Directive 2006/123/EC) for the area of national urban planing law and development planning?

No answer

3.10.5 Question 5: Renewable Energy Directive (2009/28/EC)

Is it considered necessary to transpose Article 13 of the Renewable Energy Directive (Directive 2009/28/EC) for the area of national urban planning law and development planning?

Issues relating to Article 13 of the Directive 2009/28/EC will be transposed into national legislation.

However, the question specifically related to the use of renewable energy sources in the planning, design, construction and renovation of urban and industrial areas, has not yet been decided the best way to transpose it.

However, these issues are already expressed in legislation, specifically in the individual approach of construction projects and rehabilitation of housing and services and the management of industrial units through their own regulations (DL 78/2006, 79/2006 e 80/2006 for housing and DL 71/2007 for industries)

3.10.6 Question 6: SEVESO-II-Directive

How has Article 12 Paragraph 1 Subsection 2 Council Directive 96/82/EC of 9 December 1996 on the control of major-accident hazards involving dangerous substances (OJ L 10, 14.1.1997, p. 13; "Seveso-II-Directive") been transposed into national law?

No answer

3.10 Portugal

3.10.1 Frage 1: Systematische Vorfrage

In welchem Verhältnis stehen Bebauungspläne und die Genehmigung einzelner Bauvorhaben?

Keine Antwort

3.10.2 Frage 2: Europäisches Habitat- und Artenschutzrecht

a) In welchem Verhältnis stehen das nationale Naturschutzrecht und das Bauplanungsrecht? Werden die Anforderungen des Naturschutzrechts bereits in der Bauleitplanung oder erst in einem nachfolgenden Genehmigungsverfahren bewältigt?

Die nationale Gesetzgebung, in der die Vogelschutz- und Habitat-Richtlinien umgesetzt werden, legt fest, dass jede Bautätigkeit innerhalb von Gebieten, die nach dem Natura 2000-Netzwerk ausgewiesen sind, einer positiven Stellungnahme der Behörde für Naturschutz und den Schutz der Biodiversität unterliegt. Diese Entscheidung hängt von der Bewertung der Umweltauswirkungen ab, z. B. falls erhebliche Auswirkungen auf die ausgewiesenen Gebiete erwartet werden. Davon ausgenommen ist die Bautätigkeit in den Stadtrandgebieten (urban perimeter), die in den Masterplänen der Kommunen festgelegt sind. Diese gesetzlich festgelegten Stadtrandgebiete umfassen neben den städtebaulichen Flächen auch Gebiete, die in Zukunft wahrscheinlich bebaut werden und wenig wertvolle Naturstandorte mit hohem Störpotenzial aufweisen.

Ja, die Bestimmungen der entsprechenden nationalen Gesetze werden in der Stadtplanung berücksichtigt.

b) Wie ist im Rahmen der Verträglichkeitsprüfung die Ausnahmeregelung des Artikels 6 Absatz 4 FFH-Richtlinie im Bereich des Bauplanungsrechts in nationales Recht umgesetzt worden und wie wird diese angewandt?

Gibt es eine Differenzierung in Bezug darauf, dass ein Vorhaben innerhalb oder außerhalb eines bereits im Zusammenhang bebauten Gebietes errichtet wird?

Gibt es eine Differenzierung in Bezug darauf, dass ein Vorhaben innerhalb oder außerhalb eines bereits im Zusammenhang bebauten Gebietes errichtet wird?

Artikel 6 Nr. 4 Habitat-Richtlinie wurde wörtlich in das inländische Gesetz umgesetzt.

Es gibt keine Differenzierung zwischen Vorhaben in durchgängig bebauten Gebieten oder Streusiedlungsgebieten. Dieses Merkmal wird jedoch im Prüfungsprozess beurteilt.

c) Wird neben der schutzgebietsbezogenen Verträglichkeitsprüfung des Artikels 6 Absatz 3 FFH-Richtlinie bei baulichen Vorhaben auch eine artenschutzrechtliche Prüfung nach Artikel 12 FFH-Richtlinie durchgeführt?

Artikel 12 der Habitat-Richtlinie wurde vollständig in nationales Recht umgesetzt. Die Umweltverträglichkeitsprüfung, der Maßnahmen, Pläne und Projekte unterworfen werden, wenn mit erheblichen negativen Auswirkungen auf ein Gebiet des Natura 2000-Netzwerks gerechnet wird, berücksichtigt die gesetzlichen Bestimmungen für den strengen Artenschutz. Wenn erwartet wird, dass ein Plan, eine Maßnahme oder ein Projekt diese Bestimmungen verletzt, wird der Plan, die Maßnahme oder das Projekt nur durch eine von der Behörde für Naturschutz und Schutz der Biodiversität erteilte Zulassung genehmigt, sofern alle nachstehenden Bedingungen erfüllt sind:

a) es gibt keine befriedigende Alternative,

b) der Fortbestand der Populationen betroffener Arten in ihrem natürlichen Lebensraum und in einem günstigen Erhaltungszustand wird nicht beeinflusst,

c) und der Plan, die Maßnahme oder das Projekt hat die Gewährleistung vorrangiger öffentlicher Interessen sozialer oder wirtschaftlicher Natur zum Ziel.

d) Wird ein Bedarf dafür gesehen, solche Bauvorhaben von artenschutzrechtlichen Verboten freizustellen, die in Gebieten errichtet werden, die bereits im Zusammenhang bebaut sind? Welche Rolle könnte es hierbei haben, wenn die geschützte Art ihre natürliche Verbreitung außerhalb des Siedlungsbereichs hat und dort auch in größerer Zahl vorkommt?

Es sind keine Ausnahmeregelungen vorgesehen. Art. 12 FFH-Richtlinie gilt für alle Gebiete, in denen geschützte Arten vorkommen, einschließlich bereits bebauter Gebiete. Es muss sichergestellt sein, dass es wegen der Projekte nicht zu relevanten nachteiligen Veränderungen des Schutzstatus dieser Arten kommt. Daher gilt Art. 12 ungeachtet der Anzahl der in einem Entwicklungsgebiet vorkommenden Arten und des Standorts ihres natürlichen Lebensraums, die von der Auswirkung eines Projektes auf ihren Schutzstatus und ihr Verbreitungsgebiet abhängig sind.

e) Wie wird das Tatbestandsmerkmal der „Absicht" in Artikel 12 Absatz 1 Buchstabe a, b und c FFH-Richtlinie bei Bauvorhaben im Lichte der Rechtsprechung des Europäischen Gerichtshofs verstanden?

Die den strengen Artenschutz betreffenden Bestimmungen werden unabhängig von der "Absicht" angewandt. In Fällen von Fahrlässigkeit wird die Höhe des Bußgeldes jedoch um die Hälfte gesenkt.

3.10.3 Frage 3: Plan-UP-Richtlinie (2001/42/EG)

a) Welche Ebenen und Instrumente städtebaulicher Planungen bestehen und wie erfolgte diesbezüglich die Umsetzung der Plan-UP-Richtlinie in nationales Recht?

Keine Antwort

b) Wie wurde die Plan-UP-Richtlinie in das nationale Recht umgesetzt?

Keine Antwort

c) Werden für bestimmte Fälle, insbesondere der Geringfügigkeit im Rahmen der Aufstellung, Änderung und Ergänzung von Bauleitplänen, Möglichkeiten zur Beschleunigung sowie Verfahrenserleichterungen genutzt?

Keine Antwort

d) In welchem Verhältnis stehen Umweltprüfung und FFH-Verträglichkeits-prüfung zueinander?

Keine Antwort

3.10.4 Frage 4: Dienstleistungsrichtlinie (2006/123/EG)

Wird ein Umsetzungsbedarf der Dienstleistungsrichtlinie (Richtlinie 2006/123/EG) für den Bereich des nationalen Städtebaurechts und der Bauleitplanung gesehen?

Keine Antwort

3.10.5 Frage 5: Erneuerbare-Energien-Richtlinie (2009/28/EG)

Wird ein Umsetzungsbedarf hinsichtlich des Artikels 13 Erneuerbare-Energien-Richtlinie (2009/28/EG) für den Bereich des nationalen Städtebaurechts und der Bauleitplanung gesehen?

Die Artikel 13 der Richtlinie 2009/28/EG betreffenden Fragen werden in nationales Recht umgesetzt.

Über die beste Art und Weise der Umsetzung der Fragen, die speziell die Nutzung erneuerbarer Energiequellen in der Planung, der Gestaltung, dem Bau und der Sanierung von Stadt- und Industriegebieten betreffen, wurde jedoch noch nicht entschieden.

Diese Fragen wurden jedoch bereits in der Gesetzgebung behandelt, besonders in der individuellen Betrachtungsweise von Bauvorhaben und der Sanierung von Wohngebäuden und der technischen Gebäudeausrüstung sowie der Leitung von Industriebetrieben nach eigenen Regelungen (Richtlinien 78/2006, 79/2006 und 80/2006 für Wohnungsbau und 71/2007 für Industriebauten)

3.10.6 Frage 6: SEVESO-II-Richtlinie

Wie ist Artikel 12 Absatz 1 Unterabsatz 2 der Richtlinie 96/82/EG vom 09.12.1996 zur Beherrschung der Gefahren bei Unfällen mit gefährlichen Stoffen (ABl. L 10 vom 14.01.1997, S. 13; „Seveso-II-Richtlinie") in nationales Recht umgesetzt worden?

Keine Antwort

3.11 Slovakia

3.11.1 Question 1: Systematic preliminary question

What is the relationship between binding land use plans and the approval of individual construction projects?

Individual construction projects must be in accordance with the approved binding land use plans. Buildings can be placed on the basis of the land-use decision. The mere implementation of buildings can be made on the basis of building permission or notice to the building Office.

3.11.2 Question 2: European Habitat and Protection of Species Law

a) *What is the relationship between national nature conservation law and the building planning law? Are nature conservation law requirements taken into account in development planning or not dealt with until a subsequent approval procedure takes place?*

Nature Protection is determinated in Protection of nature and landscape Act No. 543/2002 Coll. Nature conservation authority which is concerned authority in proceedings conducted under the Building Act applies the interests protected by Protection of nature and landscape Act and its interaction is governed by the provisions of § 9 and § 103 clause 7 and 8 of the Protection of nature and landscape Act in relation to the provision of § 126 of Building Act. The interests of nature and landscape conservation are applied by statement to the activity that is permitted under the Building Act.

By land-use proceedings according to § 9 clause 1 letter b) Protection of nature and landscape Act is nature conservation authority the aggrieved authority, which is both legal and required in this case in a form of its statement present to protect the interests of nature and landscape protection.

Building authority define the area for proposed aim in land-use decision and specify the conditions which will ensure the interests protected by specific legislation, that means also the interests of nature and landscape conservation in collaboration with the competent authority for nature conservation, based on its statement.

Land-use decision does not constitute a right to carry out the works. Works can be carry out on the basis of building permission issued at the request of the developer. Building authority in proceeding under the § 62 of Building Act must determine whether the submitted application together with the project documentation of the proposed building meets the conditions specified in the land-use decision, i. e. also the requirements specified in the nature conservation authority statement, which protects the interests of nature and landscape protection.

b) Within the scope of the assessment of the implications of a plan or project for a site, how has the exception provided by Article 6 Paragraph 4 of the FFH Directive been transposed into national law in the field of building planning and how is it applied?

Whole FFH Directive was transposed into law on nature and landscape protection No. 543/2002 Coll. and is applied in a form of statement of nature conservation authority.

Is there any distinction regarding whether a project is built within or outside an area which is already continuously built up?

There is no distinction.

c) Apart from the conservation area related assessment of implications for the site pursuant to Article 6 Paragraph 3 of the FFH Directive, is an assessment under species protection law also performed according to Article 12 of the FFH Directive?

This area falls within the competence of the Ministry environment, Building Act does not deal with it.

d) Is there a need to exempt such construction projects from prohibitions under species protection law which are erected in areas which have already been developed? What role could it play if the protected species had their natural habitat outside the development area and also occurred in greater number there?

Protection of species is considered the same in any area within the Act on the Protection of Nature and Landscape No. 543/2002 Coll.

e) How is the element of "intention" in Article 12 Paragraph 1 Letter a, b and c of the FFH Directive interpreted for construction projects in the light of the legal decisions taken by the European Court of Justice?

The nature conservation authority expresses its statement to individual construction projects.

3.11.3. Question3: SEA Directive (2001/42/EC)

a) Which levels and instruments of development planning exist and with respect to this, how was the SEA Directive transposed into national law?

The basic tools of land-use planning are land-use planning materials, land-use planning documentation and land decision.

b) How was the SEA Directive transposed into national law?

The SEA Directive was transposed into the national law by Act No. 24/2006 Coll. on the assessing of environmental influences and on amending and supplementing certain Acts, as amended by later legislation. In the Annex are specified strategic documents that are subject to assessment. Assessment of the Land-use planning documentation is done by separate annex to this act.

c) Are options, especially the option of insignificance within the context of preparing, amending and supplementing the development plans, used to accelerate and simplify procedures for particular cases?

No, if the document is being reviewed by SEA.

d) What is the relationship between the environmental audit and the FFH assessment of implications for the site?

This area falls within the competence of the Ministry environment, Building Act does not deal with it.

There is no relationship – environmental audit deals with facts in current status, impact assessment predicts status in the future.

3.11.4 Question 4: EU Services Directive (2006/123/EC)

Is it deemed necessary to transpose the EC Services Directive (Directive 2006/123/EC) for the area of national urban planning law and development planning?

EC Services Directive has been transposed in Slovak republic by Act No. 136/2010 Coll on services in the internal market and amending certain acts as amended and Act. No. 138/1992 Coll. of Chartered Architects and Chartered Civil Engineers.

3.11.5 Question 5: Renewable Energy Directive (2009/28/EC)

Is it considered necessary to transpose Article 13 of the Renewable Energy Directive (Directive 2009 /28/EC) for the area of national urban planning law and development planning?

Renewable Energy Directive is now in the process of transposition by Ministry of Economy of the Slovak republic. By our opinion it is necessary to transpose Article 13 for the area of national urban planning law and development planning particularly the letter a):Member States shall, in particular take the appropriate steps to ensure that: a) subject to differences between Member States in their administrative structures and organisation, the respective responsibilities of national, regional and local administrative bodies for authorisation, certification and licensing procedures including spatial planning are clearly coordinated and defined, with transparent timetables for determining planning and building applications.

3.11.6 Question 6: SEVESO-II-Directive

How has Article 12 Paragraph 1 Subsection 2 Council Directive 96/82/EC of 9 December 1996 on the control of major-accident hazards involving dangerous substances (OJ L 10, 14.1.1997, p. 13; "Seveso-II-Directive") been transposed into national law?

This article has been in Slovakia transposed into national law by Act No. 261/2002 Coll. from 20th of March 2002 on the prevention of major industrial accidents and amending certain acts, as amended:

> ***4th Section***
>
> *Land use planning and permitting buildings, facilities and other activities*
>
> ***§ 10Planning and permitting in surroundings of establishments***
>
> *(1) Public authorities that under special regulations 17) procure (supply) or approve development site planning and land-use documentation, or permit buildings, equipment and other activities are required to take into account the need*
>
> *a) meet a sufficient distance between establishments of category A and category B, between establishments of these mentioned categories and other buildings, facilities and activities, mainly from residential settlements and other places regularly populated with more people, territories protected by special regulations, 12), transportation routes, energy and other distribution or*
>
> *b) identify additional technical and other measures so as not to increase the risk of major industrial accidents.*
>
> *(2) Taking into account the requirements set out in clause 1 the competent authorities assumed particular from notice (§ 5), risk assessment (§ 6), and safety reports (§ 9), as well as from statements pursuant to clause 3*
>
> *(3) In proceedings under clause 1 is concerned authority the district nature conservation authority, the interests of preventing major accidents and limiting their consequences are applied in a form of statement.*

3.11 Slowakei

3.11.1 Frage 1: Systematische Vorfrage

In welchem Verhältnis stehen Bebauungspläne und die Genehmigung einzelner Bauvorhaben?

Einzelne Bauvorhaben müssen den genehmigten verbindlichen Bauleitplänen entsprechen. Gebäude können auf der Grundlage der Flächennutzungsentscheidung errichtet werden. Die Realisierung von Gebäuden kann auf der Grundlage der Baugenehmigung oder der Anzeige beim Bauamt erfolgen.

3.11.2 Frage 2: Europäisches Habitat- und Artenschutzrecht

a) *In welchem Verhältnis stehen das nationale Naturschutzrecht und das Bauplanungsrecht? Werden die Anforderungen des Naturschutzrechts bereits in der Bauleitplanung oder erst in einem nachfolgenden Genehmigungsverfahren bewältigt?*

Der Naturschutz ist im Natur- und Landschaftsschutzgesetz Nr. 543/2002 (coll.) geregelt. Die Naturschutzbehörde wird in Verfahren zur Landnutzung beteiligt und vertritt hier die Interessen des Naturschutzes. Die Wechselwirkungen zwischen dem Baurecht wird durch die Bestimmungen des Artikels 9 und des Artikels 103 Absatz 7 und 8 Natur- und Landschaftsschutzgesetz in Bezug auf die Bestimmung des Artikels 126 Baugesetz geregelt. Die Interessen des Natur- und Landschaftsschutzes werden durch eine Stellungnahme der Behörde auf die nach dem Baugesetz gestattete Aktivität angewendet.

Nach dem Bauleitverfahren gemäß Artikel 9 Absatz 1 Buchstabe b) Natur- und Landschaftsschutzgesetz ist die Naturschutzbehörde die geschädigte Behörde, was in diesem Fall zum Schutz der Interessen des Natur- und Landschaftsschutzes sowohl rechtlich als auch tatsächlich die Stellungnahme erforderlich ist.

Die Baubehörde legt in der Flächennutzungsentscheidung Ziele für bestimmte Gebiete fest und beschreibt, wie die Vorgaben des Naturschutzes (neben andere gesetzlichen Vorgaben) sichergestellt werden.

Die Flächennutzungsentscheidung begründet keinen Rechtsanspruch auf Durchführung der Arbeiten. Arbeiten können auf der Grundlage der Baugenehmigung durchgeführt werden, die auf Antrag des Bauträgers ausgestellt wird. In dem Verfahren nach Artikel 62 Baugesetz muss die Baubehörde darüber befinden, ob der gestellte Antrag zusammen mit der Projektdokumentation des geplanten Gebäudes die in der Flächennutzungsentscheidung festgelegten Bedingungen, d. h. auch die in der Erklärung der Naturschutzbehörde, welche die Interessen des Natur- und Landschaftsschutzes wahrnimmt, beschriebenen Anforderungen erfüllt.

b) Wie ist im Rahmen der Verträglichkeitsprüfung die Ausnahmeregelung des Artikels 6 Absatz 4 FFH-Richtlinie im Bereich des Bauplanungsrechts in nationales Recht umgesetzt worden und wie wird diese angewandt?

Die gesamte FFH-Richtlinie wurde im Gesetz Nr. 543/2002 (coll.) über Natur- und Landschaftsschutz umgesetzt und wird in Form einer Erklärung der Naturschutzbehörde angewendet.

Gibt es eine Differenzierung in Bezug darauf, dass ein Vorhaben innerhalb oder außerhalb eines bereits im Zusammenhang bebauten Gebietes errichtet wird?

Nein, es gibt keine Differenzierung.

c) Wird neben der schutzgebietsbezogenen Verträglichkeitsprüfung des Artikels 6 Absatz 3 FFH-Richtlinie bei baulichen Vorhaben auch eine artenschutzrechtliche Prüfung nach Artikel 12 FFH-Richtlinie durchgeführt?

Dieser Bereich fällt in den Zuständigkeitsbereich des Umweltministeriums; das Baugesetz befasst sich nicht mit ihm.

d) Wird ein Bedarf dafür gesehen, solche Bauvorhaben von artenschutzrechtlichen Verboten freizustellen, die in Gebieten errichtet werden, die bereits im Zusammenhang bebaut sind? Welche Rolle könnte es hierbei haben, wenn die geschützte Art ihre natürliche Verbreitung außerhalb des Siedlungsbereichs hat und dort auch in größerer Zahl vorkommt?

Das Gesetz über Natur- und Landschaftsschutz Nr. 543/2002 (coll.) sieht für alle Gebiete denselben Artenschutz vor.

e) Wie wird das Tatbestandsmerkmal der „Absicht" in Artikel 12 Absatz 1 Buchstabe a, b und c FFH-Richtlinie bei Bauvorhaben im Lichte der Rechtsprechung des Europäischen Gerichtshofs verstanden?

Die Naturschutzbehörde gibt ihre Erklärung zu einzelnen Bauvorhaben ab.

3.11.3 Frage 3: Plan-UP-Richtlinie (2001/42/EG)

a) Welche Ebenen und Instrumente städtebaulicher Planungen bestehen und wie erfolgte diesbezüglich die Umsetzung der Plan-UP-Richtlinie in nationales Recht?

Die grundlegenden Instrumente der Bauleitplanung sind die Bauleitplanungsmaterialien, -unterlagen und Flächennutzungsentscheidungen.

b) Wie wurde die Plan-UP-Richtlinie in das nationale Recht umgesetzt?

Die Plan-UP-Richtlinie wurde durch das Gesetz Nr. 24/2006 (coll.) über die Bewertung der Umweltauswirkungen und die Änderung und Ergänzung bestimmter Gesetze in nationales Recht umgesetzt. Im Anhang sind Planunterlagen genannt, die der Prüfung unterliegen. Die Prüfung der Bauleitplanungsdokumente erfolgt nach einem gesonderten Anhang zu diesem Gesetz.

c) Werden für bestimmte Fälle, insbesondere der Geringfügigkeit im Rahmen der Aufstellung, Änderung und Ergänzung von Bauleitplänen, Möglichkeiten zur Beschleunigung sowie Verfahrenserleichterungen genutzt?

Nein, wenn das Dokument einer Plan-UP unterzogen wurde.

d) In welchem Verhältnis stehen Umweltprüfung und FFH-Verträglichkeits-prüfung zueinander?

Dieser Bereich fällt in den Zuständigkeitsbereich des Umweltministeriums; das Baugesetz befasst sich nicht mit ihm.

Es besteht kein Verhältnis – die Umweltprüfung befasst sich mit den gegenwärtigen Fakten, die Verträglichkeitsprüfung sagt den künftigen Zustand vorher.

3.11.4 Frage 4: Dienstleistungsrichtlinie (2006/123/EG)

Wird ein Umsetzungsbedarf der Dienstleistungsrichtlinie (Richtlinie 2006/123/EG) für den Bereich des nationalen Städtebaurechts und der Bauleitplanung gesehen?

Die EG-Dienstleistungsrichtlinie wurde in der Slowakischen Republik durch Gesetz Nr. 136/2010 (coll.) über Dienstleistungen im Binnenmarkt und zur Änderung bestimmter Gesetze in geänderter Fassung und Gesetz Nr. 138/1992 coll. über Architekten und Ingenieure, die dem Berufsverband angehören, umgesetzt.

3.11.5 Frage 5: Erneuerbare-Energien-Richtlinie (2009/28/EG)

Wird ein Umsetzungsbedarf hinsichtlich des Artikels 13 Erneuerbare-Energien-Richtlinie (2009/28/EG) für den Bereich des nationalen Städtebaurechts und der Bauleitplanung gesehen?

Die Erneuerbare-Energien-Richtlinie wird zurzeit vom Wirtschaftsministerium der Slowakischen Republik umgesetzt. Unserer Ansicht nach ist es notwendig, Artikel 13 für den Bereich des nationalen Städtebaurechts und der Bauleitplanung umzusetzen, insbesondere Buchstabe a): Die Mitgliedstaaten ergreifen insbesondere angemessene Maßnahmen, um sicherzustellen, dass: vorbehaltlich der Unterschiede zwischen den Mitgliedstaaten hinsichtlich ihrer Verwaltungsstruktur und -organisation die entsprechenden Zuständigkeiten der nationalen, regionalen und lokalen Verwaltungsstellen für die Genehmigungs-, Zertifizierungs- und Zulassungsverfahren – auch im Hinblick auf die Raumplanung – eindeutig koordiniert und festgelegt sind und transparente Zeitpläne für Entscheidungen über Planungs- und Bauanträge genau bestimmt sind.

3.11.6 Frage 6: SEVESO-II-Richtlinie

Wie ist Artikel 12 Absatz 1 Unterabsatz 2 der Richtlinie 96/82/EG vom 09.12.1996 zur Beherrschung der Gefahren bei Unfällen mit gefährlichen Stoffen (ABl. L 10 vom 14.01.1997, S. 13; „Seveso-II-Richtlinie") in nationales Recht umgesetzt worden?

Dieser Artikel wurde in der Slowakei durch Gesetz Nr. 261/2002 (coll.) vom 20. März 2002 über die Vermeidung schwerer Betriebsunfälle und die Änderung bestimmter Gesetze in geänderter Fassung umgesetzt:

4. Abschnitt

Bauleitplanung und Genehmigung von Gebäuden, Anlagen und anderen Aktivitäten

§ 10 Planung und Genehmigung in der Nachbarschaft von Betrieben

(1) Behörden, die gemäß Sonderregelungen 17) die Planung des Entwicklungsstandorts und die Bauleitdokumentation (liefern) oder genehmigen oder Gebäude, Ausrüstungen und andere Aktivitäten genehmigen, müssen das Erfordernis beachten

a) einen angemessenen Abstand zwischen Betrieben der Kategorie A und der Kategorie B, zwischen Betrieben der genannten Kategorien und anderen Gebäuden, Anlagen und Aktivitäten, vor allem Wohnsiedlungen und andere Orte; durch besondere Regelungen geschützte Gebiete, 12), Verkehrswege, Energie- und andere Versorgungsleitungen einzuhalten oder

b) zusätzliche technische und andere Maßnahmen festzulegen, damit das Risiko schwerer Betriebsunfälle nicht erhöht wird.

(2) Berücksichtigung der in Absatz 1 dargelegten Anforderungen, die die zuständigen Behörden insbesondere dem Bescheid (§ 5), der Risikoabschätzung (§ 6) und den Sicherheitsberichten (§ 9) sowie den Erklärungen gemäß Absatz 3 entnehmen

(3) In Verfahren nach Absatz 1 ist die betreffende Behörde die Kreis-Naturschutzbehörde. Sie wird den Interessen der Vermeidung schwerer Unfälle und der Begrenzung ihrer Auswirkungen in Form einer Erklärung gerecht.

3.12 Spain

3.12.1 Question 1: Systematic preliminary question.

What is the relationship between binding land use plans and the approval of individual construction projects?

Under Spanish Land Act (June 20, 2010), the binding land use plan which defines the use of land plots for building and other usages, and the approval of individual construction projects has different functions. The planning permission is granted if the construction project fulfils the legal requirements. If a binding land use plan exists, the construction project must also be measured against the requirements of the binding land use plan.

3.12.2 Question 2: European Habitat and Protection of Species Law.

a) *What is the relationship between national nature conservation law and the building planning law? Are nature conservation law requirements taken into account in development planning or not dealt with until a subsequent approval procedure takes place?*

Under Spanish nature conservation law, which is primarily defined in the State "Natural Heritage and Biodiversity Act" (December 13, 2007) the environmental protection is prevalent on the land and regional use plans.

The Land Act also says (article 13.4):

> *"… the use of land with environmental, cultural, historical, archaeological, scientific and landscape values which are subject to protection by the applicable legislation, shall remain subject to the preservation of such values at all times, and shall encomapprov solely such acts intended to modify the natural condition of the land expressly authorised by said laws.*
>
> *It shall only be allowed to modify the boundaries of protected natural spaces or spaces included within the Natura 2000 Network, reducing their total surface or excluding pieces of land from such spaces, when the changes occurred on such spaces due to the scientifically proven natural evolution. The alteration must be submitted to public disclosure, which in the case of Natura 2000 Networks shall be made prior to the submission of the delisting proposal to the European Commission and the acceptance by the latter of such delisting.*
>
> *Compliance with the foregoing provisions shall not exempt from complying with the additional protection requirements set forth under the applicable legislation".*

Besides this, the article 15, paragraph 1, demands the "Assessment and monitoring of the sustainability of the urban development" in this terms:

"The spatial and urban planning instruments are submitted to environmental assessment in accordance with the provisions of the laws on the assessment of the effects of plans and programmes on the environment and of this article, without prejudice to the environmental impact assessment of the projects which are required for their execution, as the case may be".

b) Within the scope of the assessment of the implications of a plan or project for a site, how has the exception provided by Article 6 Paragraph 4 of the FFH Directive been transposed into national law in the field of building planning and how is it applied?

The transposition of the Article 6 Paragraph 4 (FFH Directive) into the Spanish law has taken place through the Royal Decree 1421, December 1rst, 2006, about "*Measures to contribute towards ensuring Biodiversity through the Conservation of Natural Habitats and Wild Flora and Fauna*".

This regulation doesn't contain specific rules in the field of land use planning. The general rule is the necessity of adopting countervailing measures to guarantee the global coherence of the *"Natura 2000 network"*. These measures hast to be carried out during the process of environmental assessment of plans and programs, or environmental impact assessments projects, in accordance with the provisions of its applicable law.

In Spain, those measures are specific attributions of the Authonomous Communities.

Is there any distinction regarding whether a project is built within or outside an area which is already continuously built up?

There is no distinction regarding whether a project is built within or outside an area which is already continuously built up in the State legislation. Nevertheless some Authonomous Communities introduce the distinction in this way: the dimensions and limits shall refer to those resulting at the end of the extension, but the environmental agency can consider exceeded those dimensions and limits where, in combination with other activities may affect the same ecological environment. In that case, a new environmental procedure will be necessary.

c) Apart from the conservation area related assessment of implications for the site pursuant to Article 6 Paragraph 3 of the FFH Directive, is an assessment under species protection law also performed according to Article 12 of the FFH Directive?

The species protection is included, in general terms, in the environmental assesment and must be taken into account to obtain a favourable environmental report, after having ascertained that doesn't affect the integrity of the site concerned and if it's necessary, after a public information.

Anyway, the State *"Natural Heritage and Biodiversity Act"* (December 13, 2007 prohibits any land use plan or construction project that can affect a site with natural habitats or protected species, if it's suppose a negative repercussion to listed species endangered.

d) Is there a need to exempt such construction projects from prohibitions under species protection law which are erected in areas which have already been developed? What role could it play if the protected species had their natural habitat outside the development area and also occurred in greater number there?

The "*Natural Heritage and Biodiversity Act*" (December 13, 2007) establish the following rule: When a plan or programme can affect an area containing a natural habitat or a priority specie, or can impact negatively over listed species endangered, only shall be carried out if there isn't another alternative and it can be argued one of the following considerations:

- Human health and public safety.
- Beneficial consequences of primary importance for the environment.
- Other overriding public interests after the necessary consultation with the European Commission.

In any case countervailing measures must be adopted, during the environmental procedure (for plan and programs, but also for projects).

In conclusion, as a general rule, there isn't a need to exempt such construction projects from prohibitions under species protection law which is erected in areas which have already been developed.

e) How is the element of "intention" in Article 12 Paragraph 1 Letter a, b and c of the FFH Directive interpreted for construction projects in the light of the legal decisions taken by the European Court of Justice?

We really don't know how to answer this question.

3.12.3 Question 3: SEA Directive (2001/42/EC).

a) Which levels and instruments of development planning exist?

The Land Acts in Spain (State and regional level) provides local authorities the instrument of two-stage development planning to control urban development. The planning instruments are the Master Plan and the development land use plan. The Master Plan, for the entire municipal territory is the main one, and the framework for the development land use plan in which the future urban development of the entire municipality's territory is shown. The Master Plan is plenty of legal standards.

The development land use plan defines the planning law for certain areas within the local authority in a legally binding way. It's a statute and is approved and issued by the local authority.

b) How was the SEA Directive transposed into national law?

The SEA Directive has been transposed into national law by a SEA State Act (April 9th, 2006), which contains basic provisions for the environmental assessment. Further provisions exist in regional environmental laws, because, in Spain -in environmental matters- the State level is competent to dictate basic protective legislation, but this legislation can be supplemented by the Autonomous Communities.

For building planning has been the Land Act which transposed the Directive, in that an environmental assessment must be performed in all procedures for the preparation of preparatory spatial and urban planning instruments. However, in Spain, the Authonomous Communities has the complete and unlimited attributions in urbanism.

The procedural requirements under EU legislation (such as scoping, environmental report, participation of the public and authorities, consideration in the decision, announcement of the decision and the summary explanation, monitoring) have been integrated in the existing procedural steps for the preparation of a development plan. In addition, the Land Act has included a new requirement:

> *"The documentation of the urban planning instruments of the urbanization actions must include a report on the financial sustainability, where the impact of such developments on the public finances affected by the implementation and maintenance of the necessary infrastructures or the start-up and provision of the resulting facilities, as well as the suitability and the sufficiency of the land allocated for productive purposes shall be specifically considered".*

Further provisions exist in regional Land Acts, because, in Spain the State cannot regulate specific urbanism matters such as those specifically concerning urban planning and urban development projects.

The environmental assessment in the binding land use plan procedure fulfils both all requirements of the SEA Directive and the requirements of an environmental impact assessment (EIA). This applies regardless of whether the binding land use plan is subject to an actual EIA obligation or not.

c) Are options, especially the option of insignificance within the context of preparing, amending and supplementing the development plans, used to accelerate and simplify procedures for particular cases?

It's not provided by the State Land Act, but some Authonomous Communities has included some of this simplify procedures for particular cases (e.g. Madrid). A high percentage of these regional Acts permit to make use of procedural simplifications and accelerations in the preparation of development plans, which only slightly change existing Master plans, in the preparation of binding land use plans which only cover continuously built-up areas and do not significantly change the criterion of permissibility or in the preparation of binding land use plans which define the use of small areas with the aim of making areas reusable, of subsequently increasing density or other inner development measures.

In all cases, it is a prerequisite that the procedure only justifies the permissibility of projects which are not subject to any EIA obligation and which do not put at risk areas of European significance (FFH and bird conservation areas).

d) What is the relationship between the environmental audit and the FFH assessment of implications for the site?

The FFH assessment of implications for the site as defined by Article 6 Paragraph 3 of the FFH Directive, like the environmental assessment provided by the SEA Directive, is integrated as a dependent element in the development planning and approval procedure.

The environmental audit can adopt the function of a supporting procedure, especially for the definition of scope and its evaluation to be undertaken within the context of the FFH assessment of implications. Implementation of the FFH assessment of implications for the site usually is documented as part of the environmental report.

"The objectives of the FFH Directive and the SEA Directive are very different. Anyway, the Spanish Land Act establishes the following system:

> *"The spatial and urban planning instruments are submitted to environmental assessment in accordance with the provisions of the laws on the assessment of the effects of plans and programmes on the environment and of this Article, without prejudice to the environmental impact assessment of the projects which are required for their execution, as the case may be".*

Besides this, *"It shall only be allowed to modify the boundaries of protected natural spaces or spaces included within the Natura 2000 Network, reducing their total surface or excluding pieces of land from such spaces, when the changes occurred on such spaces due to the scientifically proven natural evolution. The alteration must be submitted to public disclosure, which in the case of Natura 2000 Networks shall be made prior to the submission of the delisting proposal to the European Commission and the acceptance by the latter of such delisting".*

Can be said that the assessment of implications for the site as defined by Article 6 Paragraph 3 of the FFH Directive, like the environmental assessment provided by the SEA Directive, is integrated in the development planning and approval procedure".

3.12.4 Question 4: EU Services Directive (2006/123/EC).

Is it deemed necessary to transpose the EC Services Directive (Directive 2006/123/EC) for the area of national urban planing law and development planning?

The Services Directive coordinates national legal and administrative regulations which specifically define the freedom to provide services and the freedom of establishment. This Directive has been implemented in Spanish national law last year (November and December). The Directive says that is not going to be applied in some cases (consideration 9 of the Directive). Among themselves:

> *"...requirements, such as road traffic rules, rules concerning the development or use of land, town and country planning, building standards as well as administrative penalties imposed for non-compliance with such rules which do not specifically regulate or specifically affect the service activity but have to be respected by providers in the course of carrying out their economic activity in the same way as by individuals acting in their private capacity."*

Urban planning provisions concerning the planning law permissibility of construction projects don't define a service activity per se but instead, as abstract general provisions, are aimed at any person if and provided they use a plot of land for building or other purposes. These provisions lack a special service relationship. They are general provisions in the sense of Consideration 9 of the Services Directive. Therefore there is no need for transposition as regards urban planning provisions which concern the planning law permissibility of projects.

3.12.5 Question 5: Renewable Energy Directive (2009/28/EC).

Is it considered necessary to transpose Article 13 of the Renewable Energy Directive (Directive 2009 /28/EC) for the area of national urban planning law and development planning?

With Directive 2009/28/EC of the European Parliament and the Council on the promotion of the use of energy from renewable sources and amending and subsequently repealing Directives 2001/77/EC and 2003/30/EC (RE Directive), the law of renewable energy in Europe is redefined and especially the promotion of all energy sectors is combined in a single legal act. In addition, the Directive contains binding targets for a further development of renewable energy. In this sense, according to Article 3 of the RE Directive, by the target year 2020, energy from renewable sources is to be developed to account for a share of at least 20 % of the Community's gross final consumption of energy. To achieve this target, each member state is obliged to take measures to encourage and promote energy from renewable sources. To this end, each state has been assigned a development target.

The Renewable Energy Directive (2009/28/EC) has not been implemented in Spanish national law, but some Authonomous Communities have approved some

Acts in their respective territories with the same objectives (e. g. Galicia with a Wind Energy Act, Cataluña with a Electric Power Act or Castilla-La Mancha with a Renewable Energy an energy efficiency).

In Spain, the Authonomous Communities have shared powers (with the State) in energy matters, including the authorization of production facilities, the promotion of renewable energies, environmental protection and protected natural areas and, of course, the complete attribution in spatial and urban planning. Some of them are shared also with Local Administration, chiefly in management.

3.12.6 Question 6: SEVESO-II-Directive

How has Article 12 Paragraph 1 Subsection 2 Council Directive 96/82/EC of 9 December 1996 on the control of major-accident hazards involving dangerous substances (OJ L 10, 14.1.1997, p. 13; "Seveso-II-Directive") been transposed into national law?

In Spain Article 12 Paragraph 1 Subsection 2 of the Seveso-II-Directive has been transposed by the following State regulation:

The Royal Decree 1254, July 16th, 1999, about *"Measures of control of major accident hazards involving dangerous substances"*. Article 12 order the Authonomous Communities to take into account the objective of prevention of major-accident hazards involving dangerous substances in their spatial and land use planning, controlling specifically:

- The allocation of new buildings.
- The alterations in some existent buildings.
- The new works, made in territorial scope of influence of the safety study of the building (roads, publicly used areas or residential zones, if the implementation or the works can increase the risk or the consequences of the major-accident).

This regulation also demand to the land-use and/ or other relevant policies and the procedures for implementing those policies take account of the need, in the long term, to maintain appropriate distances between establishments covered by this Directive and residential areas, areas of public use and areas of particular natural sensitivity or interest, and, in the case of existing establishments, of the need for additional technical measures in accordance with Article 5 so as not to increase the risks to people.

The Royal Decree 1196, September 19, 2003, specifically Article 6, about Spatial Plannig, on the same terms. Article 12 permits different methods to evaluate the safety distances that can be chosen by the authority. In general terms, the method established is the one based on the consequences which evaluate those that can derived from a number of foreseeable accidental events, without quantifying its probability. But the competent authority can also decide, after the agreement with the industrial person, to establish an analysis of the consequences of the potential accidents and the estimation of its probability.

Finally, the Land Act (article 12) says the land *"with natural or technological risks, including flooding or serious accidents"* shall be in condition of rural land, in any case, and has to be preserved *"by the spatial and urban planning from being transformed through urbanization"*.

3.12 Spanien

3.12.1 Frage 1: Systematische Vorfrage

In welchem Verhältnis stehen Bebauungspläne und die Genehmigung einzelner Bauvorhaben?

Nach spanischem Bodenrecht (Gesetz vom 20. Juni 2010) haben der verbindliche Flächennutzungsplan, der die Nutzung von Flächen zur Bebauung und für andere Zwecke regelt, und die Genehmigung einzelner Bauvorhaben unterschiedliche Funktionen. Die Planungsgenehmigung wird erteilt, wenn das Bauvorhaben die gesetzlichen Anforderungen erfüllt. Existiert ein verbindlicher Flächennutzungsplan, muss das Bauvorhaben auch gegenüber den Anforderungen dieses Plans geprüft werden.

3.12.2 Frage 2: Europäisches Habitat- und Artenschutzrecht

a) *In welchem Verhältnis stehen das nationale Naturschutzrecht und das Bauplanungsrecht? Werden die Anforderungen des Naturschutzrechts bereits in der Bauleitplanung oder erst in einem nachfolgenden Genehmigungsverfahren bewältigt?*

Nach spanischem Naturschutzrecht, das vorrangig im nationalen „Gesetz zum Naturerbe und zur Artenvielfalt" (13. Dezember 2007) geregelt wird, hat der Umweltschutz Vorrang vor Flächennutzungsplänen und regionalen Nutzungsplänen.

Das Bodenrecht (Artikel 13.4) besagt außerdem:

> *„… die Nutzung von ökologisch, kulturell, historisch, archäologisch, wissenschaftlich und landschaftlich wertvollen Flächen, die durch die geltende Gesetzgebung geschützt sind, unterliegt der Verpflichtung, diesen Wert zu jeder Zeit zu erhalten und darf nur solche Maßnahmen zur Änderung des natürlichen Zustands dieser Flächen umfassen, die in diesen Gesetzen ausdrücklich erlaubt sind.*
>
> *Es ist nur dann zulässig, die Grenzen geschützter natürlicher Räume oder Räume, die Teil von Natura 2000 sind, durch Verkleinerung ihrer Gesamtfläche oder Ausschluss von Teilflächen zu ändern, wenn diese Veränderungen Folge einer wissenschaftlich erwiesenen natürlichen Entwicklung sind. Die Veränderung unterliegt der Offenlegung, die im Falle des Natura 2000-Netzes vor Übermittlung des Vorschlags an die Europäische Kommission, das Gebiet aus der Liste geschützter Gebiete zu streichen, und vor der Genehmigung der Listenstreichung durch letztere erfolgen muss.*
>
> *Die Einhaltung der vorgenannten Bestimmungen befreit nicht von der Einhaltung zusätzlicher Schutzanforderungen, die im geltenden Recht festgelegt werden."*

Darüber hinaus verlangt Artikel 15 Absatz 1 die „*Bewertung und Überwachung der Nachhaltigkeit der Stadtentwicklung*“:

> „*Die Raum- und Stadtplanungsinstrumente unterliegen der Umweltprüfung gemäß den Bestimmungen der Gesetze zu den Umweltauswirkungen von Plänen und Programmen und dieses Artikels, ungeachtet der Umweltverträglichkeitsprüfung der Vorhaben, die je nach Fall für deren Ausführung erforderlich ist.*“

b) Wie ist im Rahmen der Verträglichkeitsprüfung die Ausnahmeregelung des Artikels 6 Absatz 4 FFH-Richtlinie im Bereich des Bauplanungsrechts in nationales Recht umgesetzt worden und wie wird diese angewandt?

Die Umsetzung von Artikel 6 Absatz 4 (FFH-Richtlinie) in spanisches Recht erfolgte über den königlichen Erlass 1421 vom 1. Dezember 2006 zu „Maßnahmen, die zur Gewährleistung der Artenvielfalt durch den Erhalt natürlicher Lebensräume und wild lebender Pflanzen- und Tierarten beitragen“.

Diese Rechtsnorm enthält keine spezifischen Vorschriften zur Flächennutzungsplanung. Die allgemeine Regel ist die Notwendigkeit, Ausgleichsmaßnahmen vorzusehen, um die globale Kohärenz des Natura-2000-Netzes sicherzustellen. Diese Maßnahmen müssen während des Prozesses der Umweltprüfung von Plänen und Programmen oder UVP-Projekten gemäß den Bestimmungen des geltenden Rechts durchgeführt werden.

In Spanien fallen diese Maßnahmen in die besondere Zuständigkeit der autonomen Gemeinschaften (Communidades Autónomas).

Gibt es eine Differenzierung in Bezug darauf, dass ein Vorhaben innerhalb oder außerhalb eines bereits im Zusammenhang bebauten Gebietes errichtet wird?

Das spanische nationale Recht trifft keine Unterscheidung zwischen Bauvorhaben innerhalb oder außerhalb eines bereits zusammenhängend bebauten Gebiets. Dennoch unterscheiden einige autonomen Gemeinschaften (Communidades Autónomas) wie folgt: Maßgeblich sind die Größe und Grenzen, die sich am Ende der Erweiterung ergeben, aber die Umweltbehörde kann die Größe und Grenzen als überschritten betrachten, wenn in Verbindung mit anderen Aktivitäten das gleiche ökologische Umfeld beeinträchtigt wird. In diesem Fall ist ein neues Umweltverfahren erforderlich.

c) Wird neben der schutzgebietsbezogenen Verträglichkeitsprüfung des Artikels 6 Absatz 3 FFH-Richtlinie bei baulichen Vorhaben auch eine artenschutzrechtliche Prüfung nach Artikel 12 FFH-Richtlinie durchgeführt?

Der Artenschutz ist in allgemeiner Form Teil der Umweltprüfung und wird entsprechend im Umweltbericht dargestellt, wenn festgestellt wurde, dass das betreffende Gebiet nicht beeinträchtigt wird. Bei Bedarf muss eine Unterrichtung der Öffentlichkeit stattfinden.

In jedem Fall verbietet das nationale „Gesetz zum Naturerbe und zur Artenvielfalt" (13. Dezember 2007) Flächennutzungspläne oder Bauvorhaben, die sich auf einen Standort mit natürlichen Lebensräumen oder geschützten Arten auswirken könnten, oder wenn negative Auswirkungen für als gefährdet gelistete Tierarten ermöglicht werden.

d) Wird ein Bedarf dafür gesehen, solche Bauvorhaben von artenschutzrechtlichen Verboten freizustellen, die in Gebieten errichtet werden, die bereits im Zusammenhang bebaut sind? Welche Rolle könnte es hierbei haben, wenn die geschützte Art ihre natürliche Verbreitung außerhalb des Siedlungsbereichs hat und dort auch in größerer Zahl vorkommt?

Das "*Gesetz zum Naturerbe und zur Artenvielfalt*" (13. Dezember 2007) legt folgende Regel fest: Wenn ein Plan oder ein Programm ein Gebiet beeinträchtigt, in dem sich ein natürlicher Lebensraum befindet oder eine prioritäre Art lebt, oder sich auf eine als gefährdet gelistete Art auswirken könnte, ist dessen Ausführung nur dann zulässig, wenn es keine Alternative gibt und sie durch eine der folgenden Überlegungen rechtfertigt werden kann:

- Gesundheit des Menschen und öffentliche Sicherheit
- Bedeutende positive Folgen für die Umwelt
- Andere überwiegende öffentliche Interessen nach der erforderlichen Konsultation mit der Europäischen Kommission.

In jedem Fall sind während des Umweltverfahrens Ausgleichsmaßnahmen durchzuführen (für Pläne und Programme, aber auch für Vorhaben).

Als allgemeine Regel kann man schlussfolgernd festhalten, dass nicht die Notwendigkeit besteht, Bauvorhaben in bereits bebauten Gebieten von Verboten nach dem Artenschutzgesetz auszunehmen.

e) Wie wird das Tatbestandsmerkmal der „Absicht" in Artikel 12 Absatz 1 Buchstabe a, b und c FFH-Richtlinie bei Bauvorhaben im Lichte der Rechtsprechung des Europäischen Gerichtshofs verstanden?

Wir wissen nicht, wie diese Frage zu beantworten ist.

3.12.3 Frage 3: Plan-UP-Richtlinie (2001/42/EG)

a) Welche Ebenen und Instrumente städtebaulicher Planungen bestehen und wie erfolgte diesbezüglich die Umsetzung der Plan-UP-Richtlinie in nationales Recht?

Das spanische Bodenrecht (auf nationaler und regionaler Ebene) bietet den lokalen Gebietskörperschaften das Instrument der zweistufigen Bauleitplanung zur Steuerung der Stadtentwicklung. Die Planungsinstrumente sind der Masterplan und der Flächennutzungs- oder Bebauungsplan. Der Masterplan für das gesamte Gebiet der

Kommune ist das Hauptdokument; dieser bildet den Rahmen für den Flächennutzungsplan, in dem die zukünftige Stadtentwicklung des gesamten Gebiets der Kommune aufgezeigt wird.

Der Bebauungsplan legt rechtsverbindlich das Planungsrecht für bestimmte Teilgebiete der lokalen Gebietskörperschaft fest. Er hat den Status einer Satzung und wird von der lokalen Gebietskörperschaft genehmigt und ausgegeben.

b) Wie wurde die Plan-UP-Richtlinie in das nationale Recht umgesetzt?

Die Plan-UP-Richtlinie wurde über ein nationales Plan-UP-Gesetz (9. April 2006) in nationales Recht umgesetzt, das Basisbestimmungen für die Umweltprüfung festlegt. Weitere Bestimmungen sind in regionalen Umweltgesetzen zu finden, da in Spanien die nationale Ebene die Zuständigkeit für die Festlegung der grundlegenden Schutzgesetze hat, diese aber durch die autonomen Gemeinschaften (Communidades Autónomas) ergänzt werden können.

Im Bereich der Bauleitplanung ist dies über das Bodenrecht erfolgt, das die Richtlinie insofern umsetzt, als dass für alle Verfahren zur Aufstellung vorbereitender Raum- und Stadtplanungsinstrumente eine Umweltprüfung vorzunehmen ist. In Spanien haben jedoch die autonomen Gemeinschaften (Communidades Autónomas) im Bereich der Stadtplanung die umfassende und uneingeschränkte Zuständigkeit.

Die verfahrensrechtlichen Anforderungen nach EU-Recht (wie Scoping, Umweltbericht, Beteiligung der Öffentlichkeit und Behörden, Berücksichtigung bei der Entscheidung, Bekanntmachung der Entscheidung, zusammenfassende Erklärung, Monitoring) wurden in die bestehenden Verfahrensschritte zur Aufstellung eines Bebauungsplans integriert. Außerdem beinhaltet das Bodenrecht eine neue Anforderung:

> *„Die Dokumentation der Stadtplanungsinstrumente der stadtplanerischen Maßnahmen muss einen Bericht zur finanziellen Nachhaltigkeit umfassen, in dem die Auswirkungen solcher Vorhaben für die öffentlichen Finanzen eingehend dargelegt werden, die von der Umsetzung und dem Betrieb der notwendigen Infrastruktur oder der Inbetriebnahme und dem Angebot der daraus entstehenden Einrichtungen betroffen sind, sowie über die Eignung und Angemessenheit der für produktive Zwecke ausgewiesenen Flächen."*

Weitere Bestimmungen sind im regionalen Bodenrecht zu finden, da in Spanien der Staat spezifische stadtplanerische Fragen nicht regeln kann, wie zum Beispiel solche, die konkrete Stadtplanungs- und -bebauungsvorhaben betreffen.

Die Umweltprüfung im Flächennutzungsplanverfahren erfüllt sowohl alle Anforderungen der Plan-UP-Richtlinie als auch der Umweltverträglichkeitsprüfung

(UVP). Dies gilt ungeachtet der Frage, ob der Bebauungsplan einer UVP-Pflicht unterliegt oder nicht.

c) Werden für bestimmte Fälle, insbesondere der Geringfügigkeit im Rahmen der Aufstellung, Änderung und Ergänzung von Bauleitplänen, Möglichkeiten zur Beschleunigung sowie Verfahrenserleichterungen genutzt?

Im nationalen Bodenrecht ist dies nicht vorgesehen, aber einige autonome Gemeinschaften (Communidades Autónomas) haben solche vereinfachten Verfahren für besondere Fälle (z. B. Madrid). Ein Großteil dieser regionalen Rechtsnormen lässt die Nutzung erleichterter und beschleunigter Verfahren bei der Aufstellung von Bauleitplänen zu, die bestehende Masterpläne nur geringfügig ändern, bei Aufstellung von Bebauungsplänen, die nur im Zusammenhang bebaute Flächen erfassen und den Zulässigkeitsmaßstab nicht wesentlich ändern, oder bei Aufstellung von Bebauungsplänen, die die Nutzung kleiner Gebiete mit dem Ziel der Wiedernutzbarmachung von Flächen, der Nachverdichtung oder anderer Maßnahmen der Innenentwicklung regeln.

Voraussetzung ist in allen Fällen, dass durch das Verfahren die Zulässigkeit nur solcher Vorhaben begründet wird, die keiner UVP-Pflicht unterliegen und die Gebiete von gemeinschaftlicher Bedeutung (FFH- und Vogelschutzgebiete) nicht beeinträchtigen können.

d) In welchem Verhältnis stehen Umweltprüfung und FFH-Verträglichkeits-prüfung zueinander?

Die FFH-Verträglichkeitsprüfung ist, wie auch die in der Plan-UP-Richtlinie vorgesehene Umweltprüfung als unselbstständiges Element in die Bauleitplanung und das Genehmigungsverfahren integriert.

Die Umweltprüfung kann dabei die Funktion eines Trägerverfahrens übernehmen, insbesondere für die im Rahmen der FFH-Verträglichkeitsprüfung vorzunehmende Grundlagenermittlung und ihre Bewertung. Dokumentiert wird die Durchführung der FFH-Verträglichkeitsprüfung als Teil des Umweltberichts.

Die Zielsetzungen der FFH-Richtlinie und der SUP-Richtlinie sind sehr verschieden. Das spanische Bodenrecht etabliert folgendes System:

Die räumlichen und städtischen Planungsinstrumente werden einer Umweltprüfung in Übereinstimmung mit den Gesetzesvorgaben über die Prüfung der Umweltauswirkungen von Plänen und Programmen unterzogen, ohne Vorwirkung auf die ggf. erforderliche Umweltverträglichkeitsprüfung auf der Realisierungsebene.

Veränderungen im Hinblick auf die Abgrenzung geschützter Naturräume oder von Gebieten innerhalb des Natura 2000 Netzwerks, welche die Gesamtfläche verringern oder Teile hiervon abtrennen sind nur dann zulässig, wenn eine natürliche Entwicklung wissenschaftlich nachgewiesen werden kann. Die Veränderungen müssen öffentlich bekannt gemacht werden. Im Falle des Natura 2000 Netzwerkes muss zunächst der Vorschlag auf Entlassung aus den Gemeinschaftslisten an die Europäische Kommission gemacht und anschließend die Genehmigung auf Entlassung eingeholt werden.

Die Prüfung der Auswirkungen auf Gebiete nach Art. 6 Abs. 3 der FFH-Richtlinie, wie auch die Prüfung nach der Plan-UP-Richtlinie, ist in die Planungs- und Genehmigungsverfahren integriert.

3.12.4 Frage 4: Dienstleistungsrichtlinie (2006/123/EG)

Wird ein Umsetzungsbedarf der Dienstleistungsrichtlinie (Richtlinie 2006/123/EG) für den Bereich des nationalen Städtebaurechts und der Bauleitplanung gesehen?

Durch die Dienstleistungsrichtlinie werden nationale Rechts- und Verwaltungsvorschriften koordiniert, die speziell die Dienstleistungs- und Niederlassungsfreiheit reglementieren. Diese Richtlinie wurde im letzten Jahr (November und Dezember) in spanisches nationales Recht umgesetzt. Die Richtlinie sieht einige Fälle vor, in denen sie keine Anwendung findet (Erwägungsgrund 9 der Richtlinie), darunter:

> *„...Anforderungen wie Straßenverkehrsvorschriften, Vorschriften bezüglich der Stadtentwicklung oder Bodennutzung, der Stadtplanung und der Raumordnung, Baunormen sowie verwaltungsrechtliche Sanktionen, die wegen der Nichteinhaltung solcher Vorschriften verhängt werden, die nicht die Dienstleistungstätigkeit als solche regeln oder betreffen, sondern von Dienstleistungserbringern im Zuge der Ausübung ihrer Wirtschaftstätigkeit genauso beachtet werden müssen wie von Privatpersonen."*

Städtebauliche Regelungen über die planungsrechtliche Zulässigkeit von baulichen Vorhaben regeln nicht eine Dienstleistungstätigkeit als solche, sondern richten sich als abstrakt-generelle Regelungen an jede Person, wenn und soweit sie ein Grundstück in baulicher oder sonstiger Weise nutzt. Diesen Regelungen fehlt ein spezieller Dienstleistungsbezug. Sie sind Jedermann-Regelungen im Sinne des Erwägungsgrunds 9 der Dienstleistungsrichtlinie. Für städtebauliche Regelungen, die die planungsrechtliche Zulässigkeit von Vorhaben betreffen, besteht deshalb kein Umsetzungsbedarf.

3.12.5 Frage 5: Erneuerbare-Energien-Richtlinie (2009/28/EG)

Wird ein Umsetzungsbedarf hinsichtlich des Artikels 13 Erneuerbare-Energien-Richtlinie (2009/28/EG) für den Bereich des nationalen Städtebaurechts und der Bauleitplanung gesehen?

Mit der Richtlinie 2009/28/EG des Europäischen Parlaments und des Rats zur Förderung der Nutzung von Energie aus erneuerbaren Quellen und zur Änderung und anschließenden Aufhebung der Richtlinien 2001/77/EG und 2003/30/EG (EE-Richtlinie) wird das Recht der erneuerbaren Energien in Europa neu geordnet und insbesondere die Förderung aller Energiesektoren, im Rahmen eines einheitlichen Rechtsakts vereint. Außerdem enthält die Richtlinie verbindliche Zielwerte für den weiteren Ausbau der erneuerbaren Energien. In diesem Sinne soll nach Artikel 3 EE-RL bis zum Zieljahr 2020 der Anteil erneuerbarer Energien am gesamteuropäischen Endenergieverbrauch auf mindestens 20 % ausgebaut werden. Zur Erreichung dieses Ziels wird jeder Mitgliedstaat verpflichtet, Maßnahmen zur Förderung von erneuerbaren Energien zu ergreifen. Dazu wird dem jeweiligen Mitgliedstaat ein Ausbauziel zugeordnet.

Die Erneuerbare-Energien-Richtlinie (2009/28/EG) wurde nicht in spanisches nationales Recht umgesetzt, aber einige autonome Gemeinschaften (Communidades Autónomas) haben für ihr jeweiliges Hoheitsgebiet Gesetze mit ähnlichen Zielen verabschiedet (z. B. Galizien mit einem Windenergiegesetz, Katalonien mit einem Stromgesetz oder Castilla-La Mancha mit einem Gesetz über erneuerbare Energien und Energieeffizienz).

In Spanien teilen sich die autonomen Gemeinschaften (Communidades Autónomas) in Energiefragen die Zuständigkeit mit dem Staat, einschließlich Genehmigung von Erzeugungsanlagen, Förderung erneuerbarer Energien, Umweltschutz und geschützte natürliche Räume, haben aber alleinige Zuständigkeit in der Raum- und Stadtplanung. Einige der Zuständigkeiten teilen sie sich auch mit den lokalen Gebietskörperschaften, vor allem im Bereich der Verwaltung.

3.12.6 Frage 6: SEVESO-II-Richtlinie

Wie ist Artikel 12 Absatz 1 Unterabsatz 2 der Richtlinie 96/82/EG vom 09.12.1996 zur Beherrschung der Gefahren bei Unfällen mit gefährlichen Stoffen (ABl. L 10 vom 14.01.1997, S. 13; „Seveso-II-Richtlinie") in nationales Recht umgesetzt worden?

In Spanien wurde Artikel 12 Absatz 1 Unterabsatz 2 der Seveso-II-Richtlinie über folgende nationale Rechtsnorm umgesetzt:

Königlicher Erlass 1254 vom 16. Juli 1999 über *„Kontrollmaßnahmen für schwerwiegende Unfallrisiken mit gefährlichen Substanzen"*. Artikel 12 weist die autonomen Gemeinschaften (Communidades Autónomas) an, das Ziel der Vermeidung schwerer Unfälle mit gefährlichen Substanzen in ihrer Raumordnung und Flächennutzungsplanung zu berücksichtigen, indem sie insbesondere Folgendes kontrollieren:

- Ausweisung neuer Bauflächen
- Änderungen bestehender Bebauung
- Neue Bauvorhaben, die im Einflussgebiet der Sicherheitsstudie für die Bebauung liegen (Straßen, öffentlich genutzte Räume oder Wohngebiete, wenn die Ausführung der Bauarbeiten das Risiko oder die Folgen schwerer Unfälle erhöhen kann).

Diese Rechtsnorm verlangt außerdem, dass die Flächennutzungs- und/ oder andere einschlägige Politiken und die Verfahren zur Umsetzung dieser Politiken die langfristige Notwendigkeit berücksichtigen, einen angemessenen Abstand zwischen von dieser Richtlinie erfassten Einrichtungen und Wohngebieten, öffentlich genutzten Räumen und besonders empfindlichen oder wertvollen Naturräumen einzuhalten und bei bestehenden Anlagen die Notwendigkeit zusätzlicher technischer Maßnahmen nach Artikel 5, um das Risiken für den Menschen nicht zu erhöhen.

Der königliche Erlass 1196 vom 19. September 2003, und vor allem Artikel 6 über die Raumplanung, zur gleichen Frage. Artikel 12 gestattet verschiedene Methoden, um die Sicherheitsabstände zu bewerten, die von der Behörde gewählt werden können. In der Regel basiert die festgelegte Methode auf den Konsequenzen und bewertet die, die sich durch verschiedene vorstellbare Unfallereignisse ergeben können, ohne deren Wahrscheinlichkeit zu beziffern. Die zuständige Behörde kann jedoch nach Einigung mit dem Unternehmer auch beschließen, eine Analyse der Folgen möglicher Unfälle und Abschätzung ihrer Wahrscheinlichkeit vorzunehmen.

Das Bodenrecht (in Artikel 12) besagt schließlich, dass Flächen,

> *„die einem natürlichen oder technologischen Risiko unterliegen, einschließlich der Überflutung oder eines schweren Unfalls“ in ihrem natürlichen Zustand verbleiben sollen und in jedem Fall durch die „Raum- und Stadtplanung vor der Erschließung zu schützen sind“.*

3.13 Hungary

3.13.1 Question 1: Systematic preliminary question

What is the relationship between binding land use plans and the approval of individual construction projects?

This relationship is regulated by a statute in Hungary. The statute is: Act LXXVIII of 1997 on the Formation and Protection of the Built Environment. We enclose the translated statute to this letter.

3.13.2 Question 2: European Habitat and Protection of Species Law

a) What is the relationship between national nature conservation law and the building planning law? Are nature conservation law requirements taken into account in development planning or not dealt with until a subsequent approval procedure takes place?

Architects have to keep the national regulations in connection with environmental protection during planning.

b) Within the scope of the assessment of the implications of a plan or project for a site, how has the exception provided by Article 6 Paragraph 4 of the FFH Directive been transposed into national law in the field of building planning and how is it applied?

This question is regulated by Act LIII of 1996 on the Nature Conservation, a government decree for national professional requirements for town planning, formation and structures, Act CXII of 2000 on the resort of Balaton.

Is there any distinction regarding whether a project is built within or outside an area which is already continuously built up?

There is no distinction regarding whether a project is built within or outside a continuously built up area. The requirements for the plot for buildings are regulated in the binding land use plan.

c) Apart from the conservation area related assessment of implications for the site pursuant to Article 6 Paragraph 3 of the FFH Directive, is an assessment under species protection law also performed according to Article 12 of the FFH Directive?

No answer

d) Is there a need to exempt such construction projects from prohibitions under species protection law which are erected in areas which have already been developed? What role could it play if the protected species had their natural habitat outside the development area and also occurred in greater number there?

Although a development area is already designated, authorities check their requirements during each permission procedure.

The national regulation keeps a balance between built environment and the protection of natural habitat.

e) How is the element of "intention" in Article 12 Paragraph 1 Letter a, b and c of the FFH Directive interpreted for construction projects in the light of the legal decisions taken by the European Court of Justice?

No answer

3.13.3 Question 3: SEA Directive (2001/42/EC)

a) Which levels and instruments of development planning exist and with respect to this, how was the SEA Directive transposed into national law?

This directive is transposed into national law, in a governmental decree.

b) How was the SEA Directive transposed into national law?

This directive is transposed into national law, in a governmental decree.

c) Are options, especially the option of insignificance within the context of preparing, amending and supplementing the development plans, used to accelerate and simplify procedures for particular cases?

In the case of slight changes there is a possibility of procedural simplification with the omission of the environmental assessment. This field is regulated by:

- Act XXI of 1996 on the Regional Development and Regional Planning Act,
- Act LXXVIII of 1997 on the Formation and Protection of the Built Environment,
- Act CXII of 2000 on the Resort of Balaton.

d) What is the relationship between the environmental audit and the FFH assessment of implications for the site?

This lies in the responsibility of the Ministry for the Environment. The building code contains no corresponding regulations. The environment assessment deals with existing circumstances, the implications assessment makes statements on future developments.

3.13.4 Question 4: EU Services Directive (2006/123/EC)

Is it deemed necessary to transpose the EC Services Directive (Directive 2006/123/EC) for the area of national urban planing law and development planning?

We have already transposed the EU Services Directive in connection with land use planning.

3.13.5 Question 5: Renewable Energy Directive (2009/28/EC)

Is it considered necessary to transpose Article 13 of the Renewable Energy Directive (Directive 2009 /28/EC) for the area of national urban planning law and development planning?

We consider necessary to transpose Article 13 of the Renewable Energy Directive, but we have not transposed the Directive yet.

3.13.6 Question 6: SEVESO-II-Directive

How has Article 12 Paragraph 1 Subsection 2 Council Directive 96/82/EC of 9 December 1996 on the control of major-accident hazards involving dangerous substances (OJ L 10, 14.1.1997, p. 13; "Seveso-II-Directive") been transposed into national law?

This directive is transposed into national law, in a governmental decree. (18/2006. (I.26.) governmental decree on the control of major-accident hazards involving dangerous substances)

3.13 Ungarn

3.13.1 Frage 1: Systematische Vorfrage

In welchem Verhältnis stehen Bebauungspläne und die Genehmigung einzelner Bauvorhaben?

Dieses Verhältnis ist in Ungarn gesetzlich geregelt. Bei dem Gesetz handelt es sich um das Gesetz LXXVIII/1997 über die Gestaltung und den Schutz der gebauten Umwelt. Wir fügen die Übersetzung des Gesetzes diesem Schreiben bei.

3.13.2 Frage 2: Europäisches Habitat- und Artenschutzrecht

a) In welchem Verhältnis stehen das nationale Naturschutzrecht und das Bauplanungsrecht? Werden die Anforderungen des Naturschutzrechts bereits in der Bauleitplanung oder erst in einem nachfolgenden Genehmigungsverfahren bewältigt?

Architekten müssen die den Umweltschutz betreffenden nationalen Regelungen bei der Planung einhalten.

b) Wie ist im Rahmen der Verträglichkeitsprüfung die Ausnahmeregelung des Artikels 6 Absatz 4 FFH-Richtlinie im Bereich des Bauplanungsrechts in nationales Recht umgesetzt worden und wie wird diese angewandt?

Diese Frage ist im Gesetz LIII/1996 über den Naturschutz, im Regierungserlass zu nationalen beruflichen Anforderungen für Stadtplanung, Gestaltung und Bauwerke, und dem Gesetz CXII/2000 über das Erholungsgebiet Balaton geregelt.

Gibt es eine Differenzierung in Bezug darauf, dass ein Vorhaben innerhalb oder außerhalb eines bereits im Zusammenhang bebauten Gebietes errichtet wird?

Es gib keine Differenzierung in Bezug darauf, dass ein Vorhaben innerhalb oder außerhalb eines bereits im Zusammenhang bebauten Gebietes errichtet wird. Die Anforderungen an die Bauparzellen sind im verbindlichen Bauleitplan geregelt.

c) Wird neben der schutzgebietsbezogenen Verträglichkeitsprüfung des Artikels 6 Absatz 3 FFH-Richtlinie bei baulichen Vorhaben auch eine artenschutzrechtliche Prüfung nach Artikel 12 FFH-Richtlinie durchgeführt?

Keine Antwort

d) Wird ein Bedarf dafür gesehen, solche Bauvorhaben von artenschutzrechtlichen Verboten freizustellen, die in Gebieten errichtet werden, die bereits im Zusammenhang bebaut sind? Welche Rolle könnte es hierbei haben, wenn die geschützte Art ihre natürliche Verbreitung außerhalb des Siedlungsbereichs hat und dort auch in größerer Zahl vorkommt?

Auch wenn ein Neubaugebiet bereits bestimmt ist, prüfen die Behörden ihre Vorgaben im Zuge jedes Genehmigungsverfahrens.

Die nationale Regelung wahrt das Gleichgewicht zwischen der gebauten Umwelt und dem Schutz der natürlichen Lebensräume.

e) Wie wird das Tatbestandsmerkmal der „Absicht" in Artikel 12 Absatz 1 Buchstabe a, b und c FFH-Richtlinie bei Bauvorhaben im Lichte der Rechtsprechung des Europäischen Gerichtshofs verstanden?

Keine Antwort

3.13.3 Frage 3: Plan-UP-Richtlinie (2001/42/EG)

a) Welche Ebenen und Instrumente städtebaulicher Planungen bestehen und wie erfolgte diesbezüglich die Umsetzung der Plan-UP-Richtlinie in nationales Recht?

Die Hauptinstrumente der Bauleitplanung sind Flächennutzungsplanungsmaterialien und Dokumente und Landentscheidungen.

b) Wie wurde die Plan-UP-Richtlinie in das nationale Recht umgesetzt?

Diese Richtlinie wurde durch einen Regierungserlass in nationales Recht umgesetzt.

c) Werden für bestimmte Fälle, insbesondere der Geringfügigkeit im Rahmen der Aufstellung, Änderung und Ergänzung von Bauleitplänen, Möglichkeiten zur Beschleunigung sowie Verfahrenserleichterungen genutzt?

Im Falle geringfügiger Änderungen besteht die Möglichkeit einer Verfahrenserleichterung durch Verzicht auf die Umweltprüfung. Dieser Bereich wird geregelt durch:

- Gesetz XXI/1996 über das Regionalentwicklungs- und Regionalplanungsgesetz,
- Gesetz LXXVIII/1997 über die Gestaltung und den Schutz der gebauten Umwelt,
- Gesetz CXII/2000 über das Erholungsgebet Balaton.

d) In welchem Verhältnis stehen Umweltprüfung und FFH-Verträglichkeits-prüfung zueinander?

Dies fällt in die Kompetenz des Umweltministeriums. Das Baugesetzbuch trifft hierzu keine Regelung. Die Umweltprüfung behandelt bestehende Sachverhalte, die Verträglichkeitsprüfung trifft Aussagen zur zukünftigen Entwicklung.

3.13.4 Frage 4: Dienstleistungsrichtlinie (2006/123/EG)

Wird ein Umsetzungsbedarf der Dienstleistungsrichtlinie (Richtlinie 2006/123/EG) für den Bereich des nationalen Städtebaurechts und der Bauleitplanung gesehen?

Wir haben die EG-Dienstleistungsrichtlinie bereits im Zusammenhang mit der Bauleitplanung umgesetzt.

3.13.5 Frage 5: Erneuerbare-Energien-Richtlinie (2009/28/EG)

Wird ein Umsetzungsbedarf hinsichtlich des Artikels 13 Erneuerbare-Energien-Richtlinie (2009/28/EG) für den Bereich des nationalen Städtebaurechts und der Bauleitplanung gesehen?

Wir halten die Umsetzung des Artikels 13 Erneuerbare-Energien-Richtlinie für notwendig, haben die Richtlinie jedoch noch nicht umgesetzt.

3.13.6 Frage 6: SEVESO-II-Richtlinie

Wie ist Artikel 12 Absatz 1 Unterabsatz 2 der Richtlinie 96/82/EG vom 09.12.1996 zur Beherrschung der Gefahren bei Unfällen mit gefährlichen Stoffen (ABl. L 10 vom 14.01.1997, S. 13; „Seveso-II-Richtlinie“) in nationales Recht umgesetzt worden?

Diese Richtlinie wurde durch einen Regierungserlass (18/2006) in nationales Recht umgesetzt. (Regierungserlass 18/2006 (I.26.) über die Beherrschung der Gefahren bei schweren Unfällen mit gefährlichen Stoffen).

3.14 Cyprus

3.14.1 Question 1: Systematic preliminary question

What is the relationship between binding land use plans and the approval of individual construction projects?

For any development proposal (i. e. construction project) to be approved a planning permit must be issued by the competent Planning Authority, except for specified, less critical developments, exempted by General and Special Development Orders (e. g. ...). According to the Town and Country Planning legislation, during the examination of a planning permit application, the competent Planning Authority must consider the provisions of any effective (published) statutory (binding) spatial development plan (Local Plan, Policy Statement or Area Scheme, all of which include land use and/ or zoning plans) for the area concerned, as well as any additional "substantial parameters," including environmental considerations. The Planning Authority may **not** issue a permit in departure of the provisions of such plans except through a procedure generally described through a set of relevant Regulations, executed under the authority of the Council of Ministers, which, in such a case, has to consult an independent Council for the Examination of Planning Departures. (DTPH)

3.14.2 Question 2: European Habitat and Protection of Species Law

a) *What is the relationship between national nature conservation law and the building planning law? Are nature conservation law requirements taken into account in development planning or not dealt with until a subsequent approval procedure takes place?*

Concerning nature conservation, two main laws have been enacted at national level, transposing the Birds and Habitats Directives, respectively (for FFH see below for details). The Town and Country Planning Law (on spatial planning and land use) as such had been established prior to the enactment of these two nature conservation laws. However, both the implementation of developments (including construction projects and subdivision of land) and the preparation of land use plans, as prescribed by the spatial planning legislation, **are also governed with explicit references** by the two aforementioned nature conservation laws. Furthermore, in addition to the execution of appropriate environmental impact assessments where required during the development control process (i. e. before the granting of planning permits by the competent Planning Authorities), representatives of the administrative hierarchy of the authorities competent for the Bird and Habitat Directives (i. e. the Ministry of the Interior and the Ministry of Agriculture, Natural Resources and the Environment, respectively) participate in the preparation and amendment of statutory spatial development plans, in their capacity as ex officio members of the **Planning Board**, an independent body with advisory power over large areas of planning policy, which includes the preparation and review of Local Plans and Area Schemes. In this way, the requirements of the two nature conservation laws are incorporated into land use plans during the preparation or amendment stages. (DTPH)

b) Within the scope of the assessment of the implications of a plan or project for a site, how has the exception provided by Article 6 Paragraph 4 of the FFH Directive been transposed into national law in the field of building planning and how is it applied?

The Directive has been transposed into national Law 153(I)/2003. Article 16 of this Law, paragraphs 5, 6, 7 and 8, covers the content of Article 6 paragraph 4 of the Directive and states that if the Environmental Authority rules against a project and in the absence of alternative solutions, if a plan or project must nevertheless be carried out for **imperative reasons of overriding public interest**, including those of a social or economic nature, then the competent authority for approving the project can approve it. However, where the site concerned hosts a priority natural habitat type and/ or a priority species, the only considerations which may be raised are those relating to human health or public safety, to beneficial consequences of primary importance for the environment or, further to an opinion from the Commission, to other imperative reasons of overriding public interest. Moreover, in paragraph 7 of Article 16 it is stated that the Minister of Agriculture, Natural Resources and the Environment, publishes a Decree in which he/ she demands that specific compensatory measures, necessary to ensure the overall coherence of Natura 2000 is protected, be taken. Paragraph 8, states the Commission should be informed of the compensatory measures adopted. (ENV)

Is there any distinction regarding whether a project is built within or outside an area which is already continuously built up?

No, but a project outside the Natura 2000 site may affect the site. Article 16 does not make a distinction and therefore an assessment should be carried out for all projects, within or outside the Natura 2000 site, in order to ensure that no negative impact is caused to the species and habitats found in the site, for which the site has been included in the Natura 2000 network. (ENV)

c) Apart from the conservation area related assessment of implications for the site pursuant to Article 6 Paragraph 3 of the FFH Directive, is an assessment under species protection law also performed according to Article 12 of the FFH Directive?

The appropriate assessment performed according to article 16 of Law 153(I)/2003 includes assessment of how the species will be affected. Also, article 21 of the national law covers the provisions of article 12 of the Directive. (ENV)

d) Is there a need to exempt such construction projects from prohibitions under species protection law which are erected in areas which have already been developed? What role could it play if the protected species had their natural habitat outside the development area and also occurred in greater number there?

If a project is to be constructed in an already developed area it cannot be exempted from the prohibitions if species will be killed or removed. A case by case procedure is followed and a priori exemptions are not supported. If a project will not cause adverse effects then it can be approved. (ENV)

e) How is the element of "intention" in Article 12 Paragraph 1 Letter a, b and c of the FFH Directive interpreted for construction projects in the light of the legal decisions taken by the European Court of Justice?

Article 12 has been transposed in the national law by article 21. Any disturbance by anyone or any type of project is prohibited. (ENV)

3.14.3 Question 3: SEA Directive (2001/42/EC)

a) Which levels and instruments of development planning exist and with respect to this, how was the SEA Directive transposed into national law?

This involves the three-tier hierarchy of statutory spatial development plans prescribed by the Town and Country Planning Law: the Island Plan, which refers to the national territory and the regional distribution of resources and development opportunities but has not yet been prepared; Local Plans, which refer to functional urban areas and urban agglomerations, as well as other areas undergoing intensive development pressures and rapid physical development; and Area Schemes, at the lower end of the hierarchy, which in general refer to areas of a smaller scale and are more detailed and specifically project oriented. For all territory where neither a Local Plan nor an Area Scheme is in force, an additional type of development plan is also considered for SEA, the Policy Statement for the Countryside (PSC), a legally binding document in the form of an adapted regional plan for the promotion and control of development and the protection of the environment in villages and rural areas. Following its transposition into national law (see below for details), the SEA Directive provisions have been integrated into spatial planning procedures, through the definition of their relationship to the various planning stages. Thus, with regard to the development plan review and amendment process (see Addendum 1), the SEA process follows in parallel from the start of stage 3 (elaboration/ discussion of the plan within the Planning Board) to the end of stage 4 (publication of approved plan by Council of Ministers). At the same time, planning process stages have also been incorporated into the SEA scoping and screening procedures and correlated to their various different stages. (DTPH)

b) How was the SEA Directive transposed into national law?

The Directive was fully transposed through Law N. 102(I)/2005, which entered into force on 29.07.2005. Spatial and land use planning is covered by the Law, which requires for a SEA to be carried out, among others, for statutory spatial development plans (Area Schemes, Local Plans, Policy Statements and Island Plan). The provisions of the Law are identical to those of the Directive. (ENV)

c) Are options, especially the option of insignificance within the context of preparing, amending and supplementing the development plans, used to accelerate and simplify procedures for particular cases?

As the first instances of SEA implementation are only now being carried out in Cyprus, within the framework of the current regular review of statutory spatial development plans (2008-2011), there is no adequate relevant experience yet regarding this point. (DTPH)

d) What is the relationship between the environmental audit and the FFH assessment of implications for the site?

An environmental audit should include assessment of the impacts on the habitats and/ or species included in the FFH directive and following the principles of Article 6 of the Directive. (ENV)

3.14.4 Question 4: EU Services Directive (2006/123/EC)

Is it deemed necessary to transpose the EC Services Directive (Directive 2006/123/EC) for the area of national urban planing law and development planning?

It has been deemed that both planning permit and building permit procedures are excluded from the field of application of the Services Directive as they are included among the requirements to which the Directive does not apply (i. e. "*...rules concerning the development or use of land, town and country planning, building standards as well as administrative penalties imposed for non-compliance with such rules...*"), as mentioned under point (9) of the Directive's preamble. Nevertheless, it has been decided that information concerning the planning system and development control procedures, will be provided by the Department of Town Planning and Housing to the national point of single contact. At the same time, within the framework of the current regular review of statutory spatial development plans (2008-2011) as well as with respect to the preparation of any new such plans, relevant provisions concerning the development and use of land are independently under elaboration by the Planning Board in order to be rendered more flexible and facilitate the siting of service-oriented developments. (DTPH)

3.14.5 Question 5: Renewable Energy Directive (2009/28/EC)

Is it considered necessary to transpose Article 13 of the Renewable Energy Directive (Directive 2009/28/EC) for the area of national urban planning law and development planning?

Regarding the transposition of this Directive's provisions into the planning system, this has been done through the amendment of a number of statutory spatial development plans, **with the scope of rendering relevant provisions more flexible**, through the simplification of requirements affecting the siting and construction of renewable energy infrastructure. This process is envisioned to be gradually carried out with all statutory spatial development plans, as they are periodically amended. At the same time, a number of Orders and letters of instructions to Planning Authorities (referred to as *"encyclicals"*) have been issued by the competent Minister of the Interior as follows (some of these refer to Directive 2001/77/EC):

- Order 2/2006 Siting of units for the production of energy from Renewable Energy Sources
- Order 2/2006 Addendum (Statement for the establishment of a wind park)
- Order 2/2006 Addendum (Notification for the establishment of a wind park)
- Instruction -/2008 "Special arrangements for the acceleration of development control procedures regarding planning permit applications for wind parks or individual wind generators"
- Instruction 3/2008 "Photovoltaic system installations for which a planning permit application is not required"

(DTPH)

3.14.6 Question 6: SEVESO-II-Directive

How has Article 12 Paragraph 1 Subsection 2 Council Directive 96/82/EC of 9 December 1996 on the control of major-accident hazards involving dangerous substances (OJ L 10, 14.1.1997, p. 13; "Seveso-II-Directive") been transposed into national law?

The above-mentioned Article has been transposed into spatial planning provisions through relevant Town and Country Planning Regulations [Regulatory Administrative Act 759/2003: The Town and Country Planning (accident hazards involving dangerous substances) Regulations of 2003; and Regulatory Administrative Act 399/2008: The Town and Country Planning (accident hazards involving dangerous substances) (amending) Regulations of 2008], which are taken into consideration by Planning Authorities during relevant planning permit application evaluation processes, where applicable. In these cases, mandatory consultation procedures with the competent authority for the implementation of the Seveso Directive, the

Department of Labour Inspection (DLI), have been established. The Regulations are also taken into consideration during the preparation or review of statutory spatial development plans, with the scope of being gradually incorporated into these plans through forthcoming amendments, in the form of area-specific binding provisions. (DTPH)

Information compiled and questionnaire completed by the Department of Town Planning and Housing, Ministry of the Interior, Republic of Cyprus, after consultation with various competent authorities, 11.3-30.4.2010.

(DTPH) Response provided by the Department of Town Planning and Housing, Ministry of the Interior (Christakis Serghides, Ermis Klokkaris and Phaedon Enotiades, penotiades@tph.moi.gov.cy)

(ENV) Response provided by the Department of the Environment, Ministry of Agriculture, Natural Resources and the Environment (Joanna Constantinidou and Despo Zavrou, dzavrou@environment.moa.gov.cy)

ADDENDUM 1

Statutory spatial development plan review and amendment process – Local Plans and Area Schemes (summary)

> *1. The relevant process is initiated by the publication of a report analysing the territorial state and perspectives of each area under study, within which the vision for the future is incorporated, in consultation with the local authorities involved, as well as a relevant set of strategic objectives. At this stage local authorities convene public meetings to prepare their official comments on the report.*
>
> *2. The Planning Board, an independent body with advisory power over large areas of planning policy, which includes the preparation and amendment of certain types of development plans, is then obliged to consult a Joint Board (a consultative body specifically set up for authorities and local stakeholder groups to express their views during the preparation or amendment of each spatial plan through a report submitted to the Planning Board), as well as the wider public, first through transparent open calls for the submission of opinions and suggestions on each area under study by any interested individual or body, then through structured public hearings, where a representative sample of expressed opinions and suggestions must be heard.*
>
> *3. After acquiring an informed opinion, the Planning Board elaborates and publishes the Local Plan in all its detail, through the Minister of the Interior, and the plan thus published is made available for inspection to all interested parties.*
>
> *4. The public then has the opportunity to file objections on any of the plan's provisions, which must be examined through a prescribed process, before the Planning Board deposits the plan with or without amendments based on the objections examined, through the Minister of the Interior, to the Council of Ministers for approval.*

5. Approved Local Plans will then be monitored and reviewed by the Planning Board for possible amendments at regular intervals through a repetition of the procedures described above.

(DTPH)

3.14. Zypern

3.14.1 Frage 1: Systematische Vorfrage

In welchem Verhältnis stehen Bebauungspläne und die Genehmigung einzelner Bauvorhaben?

Für jeden Erschließungsvorschlag (d. h. Bauvorhaben) ist eine Planungsgenehmigung von der zuständigen Planungsbehörde einzuholen, mit Ausnahme weniger, definierter, kritischer Bauvorhaben, die durch die allgemeinen und besonderen Bauleitverfügungen (General and Special Development Orders) festgelegt werden. Nach Maßgabe des Stadt- und Landplanungsrechts muss die zuständige Planungsbehörde bei Prüfung eines Antrags auf Erteilung einer Planungsgenehmigung die Bestimmungen jedes wirksamen (veröffentlichten) gesetzlich vorgeschriebenen (rechtsverbindlichen) Raumentwicklungsplans (lokaler Plan, Leitbild oder Gebietsprogramm, die alle Flächennutzungspläne und/ oder Bebauungspläne umfassen) für das betreffende Gebiet sowie zusätzliche „wesentliche Parameter", einschließlich von Umweltaspekten, berücksichtigen. Die Planungsbehörde kann **keine** Genehmigungen abweichend von diesen Bestimmungen erteilen, mit Ausnahme von Verfahren, die in der Regel durch die einschlägigen Rechtsnormen festgelegt und unter der Verantwortung des Ministerrats ausgeführt werden. Letzterer muss in diesem Fall einen unabhängigen Rat zur Prüfung von Planungsabweichungen (Council for the Examination of Planning Departures) konsultieren. (DTPH)

3.14.2 Frage 2: Europäisches Habitat- und Artenschutzrecht

a) *In welchem Verhältnis stehen das nationale Naturschutzrecht und das Bauplanungsrecht? Werden die Anforderungen des Naturschutzrechts bereits in der Bauleitplanung oder erst in einem nachfolgenden Genehmigungsverfahren bewältigt?*

Auf nationaler Ebene wurden im Wesentlichen zwei Gesetze zum Naturschutz verabschiedet, die die Vogelschutz- bzw. Habitatrichtlinie umsetzen (zu Einzelheiten zur FFH-Richtlinie siehe unten). Das Stadt- und Landplanungsrecht (zur Raumplanung und Flächennutzung) als solches entstand bereits vor Inkrafttreten dieser beiden Naturschutzgesetze. Die Umsetzung von Erschließungsvorhaben (einschließlich von Bauvorhaben oder der Unterteilung von Land) und die Erstellung von Flächennutzungsplänen, wie sie vom Raumplanungsgesetz vorgeschrieben wird, **unterliegen jedoch ebenfalls ausdrücklich** den beiden vorgenannten Naturschutzgesetzen. Neben der Ausführung angemessener Umweltverträglichkeitsprüfungen während des Erschließungsprüfprozesses, wo diese erforderlich sind, (d. h. vor der Erteilung der Planungsgenehmigungen durch die zuständigen Planungsbehörden) nehmen Vertreter der Verwaltungshierarchie der für die Vogelschutz- und Habitat-Richtlinie zuständigen Behörden (d. h. das Innenministerium bzw. das Ministerium für Landwirtschaft, natürliche Ressourcen und Umwelt) an der Aufstellung und Änderung der gesetzlichen Raumentwicklungspläne teil, da sie kraft ihrer Funktion ex officio Mitglieder des **Planungsgremiums (Planning Board)** sind; dies ist ein unabhängiges Gremium, das Beratungsbefugnisse für große Bereiche der Planungspolitik hat, zu denen die Aufstellung und Überprüfung lokaler Pläne und Gebietsprogramme gehören. Auf diese Weise werden die beiden Naturschutzgesetze während der Aufstellung- oder Änderungsphase in die Flächennutzungspläne integriert. (DTPH)

b) Wie ist im Rahmen der Verträglichkeitsprüfung die Ausnahmeregelung des Artikels 6 Absatz 4 FFH-Richtlinie im Bereich des Bauplanungsrechts in nationales Recht umgesetzt worden und wie wird diese angewandt?

Die Richtlinie wurde in nationales Recht umgesetzt – Gesetz 153(I)/2003. Artikel 16 dieses Gesetzes, Absätze 5, 6, 7 und 8, deckt den Inhalt von Artikel 6 Absatz 4 der Richtlinie ab und besagt, dass im Falle eines negativen Bescheids der Umweltbehörde zu einem Vorhaben und in Ermangelung von Alternativlösungen, wenn der Plan oder das Vorhaben **aus zwingenden Gründen überwiegenden des öffentlichen Interesses** einschließlich solcher sozialer oder wirtschaftlicher Natur dennoch ausgeführt werden muss, die für die Genehmigung des Vorhabens zuständige Behörde das Vorhaben genehmigen kann. Umfasst das betreffende Gebiet jedoch natürliche Lebensräume und/ oder prioritäre Arten, dürfen nur Aspekte im Zusammenhang mit der Volksgesundheit oder öffentlichen Sicherheit, bedeutende positive Folgen für die Umwelt oder (einer Stellungnahme der Kommission folgend) andere zwingende Gründe überwiegenden öffentlichen Interesses angeführt werden. Artikel 16 Absatz 7 besagt weiterhin, dass der/ die Minister(in) für Landwirtschaft, natürliche Ressourcen und Umwelt einen Erlass veröffentlichen wird, der festlegt, dass spezifische Ausgleichsmaßnahmen zur Sicherung der Kohärenz von Natura 2000 durchzuführen sind. Absatz 8 sagt, dass die Kommission über ergriffene Ausgleichsmaßnahmen zu unterrichten ist. (ENV)

Gibt es eine Differenzierung in Bezug darauf, dass ein Vorhaben innerhalb oder außerhalb eines bereits im Zusammenhang bebauten Gebietes errichtet wird?

Nein, aber ein Vorhaben außerhalb eines Natura 2000-Gebiets kann sich auf dasselbige Gebiet auswirken. Artikel 16 trifft keine Unterscheidung und daher sollte für alle Vorhaben innerhalb und außerhalb eines Natura 2000-Gebiets eine Prüfung erfolgen, um sicherzustellen, dass sich keine negativen Folgen für die Arten und Lebensräume ergeben, die in dem Gebiet vorkommen und wegen der es in das Netz „*Natura 2000*" aufgenommen wurde. (ENV)

c) Wird neben der schutzgebietsbezogenen Verträglichkeitsprüfung des Artikels 6 Absatz 3 FFH-Richtlinie bei baulichen Vorhaben auch eine artenschutzrechtliche Prüfung nach Artikel 12 FFH-Richtlinie durchgeführt?

Die zweckdienliche Bewertung nach Artikel 16 von Gesetz 153(I)/2003 umfasst eine Bewertung der Beeinträchtigungen, die sich für die Arten ergeben. Auch Artikel 12 des nationalen Gesetzes deckt die Bestimmungen des Artikels 12 der Richtlinie ab. (ENV)

d) Wird ein Bedarf dafür gesehen, solche Bauvorhaben von artenschutzrechtlichen Verboten freizustellen, die in Gebieten errichtet werden, die bereits im Zusammenhang bebaut sind? Welche Rolle könnte es hierbei haben, wenn die geschützte Art ihre natürliche Verbreitung außerhalb des Siedlungsbereichs hat und dort auch in größerer Zahl vorkommt?

Wird ein Bauvorhaben in einem bereits bebauten Gebiet durchgeführt, kann es nicht von den Verboten ausgenommen werden, wenn Arten getötet oder entfernt werden. Es wird eine Einzelfallprüfung durchgeführt und Ausnahmen a priori werden nicht unterstützt. Hat ein Vorhaben keine negativen Auswirkungen, kann es genehmigt werden. (ENV)

e) Wie wird das Tatbestandsmerkmal der „Absicht" in Artikel 12 Absatz 1 Buchstabe a, b und c FFH-Richtlinie bei Bauvorhaben im Lichte der Rechtsprechung des Europäischen Gerichtshofs verstanden?

Artikel 12 wurde durch Artikel 21 in nationales Recht umgesetzt. Jeglicher Eingriff durch eine Person oder durch jede Art von Vorhaben ist verboten. (ENV)

3.14.3 Frage 3: Plan-UP-Richtlinie (2001/42/EG)

a) Welche Ebenen und Instrumente städtebaulicher Planungen bestehen und wie erfolgte diesbezüglich die Umsetzung der Plan-UP-Richtlinie in nationales Recht?

Dies umfasst die dreistufige Hierarchie der gesetzlich vorgeschriebenen Raumentwicklungspläne, die vom Stadt- und Landplanungsgesetz festgelegt werden: der Inselplan gilt für das nationale Hoheitsgebiet und regelt die regionale Verteilung von Ressourcen und Entwicklungschancen, wurde aber noch nicht aufgestellt; lokale Pläne beziehen sich auf funktionale Stadträume und städtische Verdichtungsräume sowie andere Räume, die unter intensivem Erschließungsdruck stehen und eine schnelle bauliche Entwicklung erleben; die Gebietsprogramme am unteren Ende der Hierarchie beziehen sich in der Regel auf Gebiete kleineren Maßstabs, sie sind detailliert und orientieren sich an Einzelvorhaben. Für Gebiete, für die weder ein lokaler Plan noch ein Gebietsprogramm in Kraft ist, wird eine weitere Art von Bauleitplan auf SUP-Pflicht geprüft: Policy Statement for the Countryside (PSC) – ein Leitbild für die ländlichen Räume, ein rechtsverbindliches Dokument in Form eines adaptierten Regionalplans für die Förderung und Kontrolle der Erschließung und des Umweltschutzes in Dörfern und ländlichen Gebieten. Nach der Umsetzung in nationales Recht (zu Einzelheiten siehe unten) wurden die Bestimmungen der Plan-UP-Richtlinie in die Raumplanungsverfahren integriert – durch die Definition ihrer Beziehung zu den verschiedenen Planungsstufen. Hinsichtlich der Überprüfung und Fortschreibung des Bauleitplans (siehe Anhang 1) verläuft das Plan-UP-Verfahren ab Start von Stufe 3 (Aufstellung/ Erörterung des Plans mit dem Planungsgremium) bis zum Ende von Phase 4 (Veröffentlichung des verabschiedeten Plans durch den Ministerrat) parallel. Gleichzeitig wurden die Stufen des Planungsverfahrens auch in die Scoping- und Screening-Verfahren für die SUP integriert und mit den einzelnen Stufen korreliert. (DTPH)

b) Wie wurde die Plan-UP-Richtlinie in das nationale Recht umgesetzt?

Die Richtlinie wurde durch das Gesetz Nr. 102(I)/2005 vollständig umgesetzt, das am 29.07.2005 in Kraft trat. Der Raum- und Flächennutzungsplan wird durch das Gesetz abgedeckt, das unter anderem verlangt, dass eine SUP für gesetzlich vorgeschriebene Raumentwicklungspläne (Gebietsprogramme, lokale Pläne, Leitbilder und Inselplan) durchgeführt wird. Die Bestimmungen des Gesetzes sind identisch mit der Richtlinie. (ENV)

c) Werden für bestimmte Fälle, insbesondere der Geringfügigkeit im Rahmen der Aufstellung, Änderung und Ergänzung von Bauleitplänen, Möglichkeiten zur Beschleunigung sowie Verfahrenserleichterungen genutzt?

Da erst jetzt die ersten Schritte der SUP-Umsetzung in Zypern im Rahmen der derzeitigen regelmäßigen Überprüfung der gesetzlichen Raumentwicklungspläne erfolgen (2008-2011), gibt es zum gegenwärtigen Zeitpunkt keine einschlägigen Erfahrungen. (DTPH)

d) In welchem Verhältnis stehen Umweltprüfung und FFH-Verträglichkeits-prüfung zueinander?

Die Umweltprüfung sollte eine Bewertung der Folgen für die Lebensräume und/ oder Arten umfassen, die in der FFH-Richtlinie genannt werden, und den Grundsätzen von Artikel 6 der Richtlinie folgen. (ENV)

3.14.4 Frage 4: Dienstleistungsrichtlinie (2006/123/EG)

Wird ein Umsetzungsbedarf der Dienstleistungsrichtlinie (Richtlinie 2006/123/EG) für den Bereich des nationalen Städtebaurechts und der Bauleitplanung gesehen?

Sowohl Planungs- als auch Baugenehmigungen gelten als vom Geltungsbereich der Richtlinie ausgenommen, da sie unter den Anforderungen aufgeführt sind, für die die Richtlinie nicht gilt (d. h. „*Vorschriften bezüglich der Stadtentwicklung oder Bodennutzung, der Stadtplanung und der Raumordnung, Baunormen sowie verwaltungsrechtliche Sanktionen, die für die Nichteinhaltung solcher Vorschriften verhängt werden...*"), wie in Punkt (9) der Präambel zur Richtlinie erwähnt. Nichtsdestotrotz wurde beschlossen, dass die Informationen zum Planungssystem und Bebauungskontrollverfahren von der Abteilung für Stadtplanung und Wohnen an die nationale einheitliche Kontaktstelle geliefert werden. Innerhalb des Rahmens zur regelmäßigen Überprüfung der gesetzlich vorgeschriebenen Raumentwicklungspläne (2008-2011) und hinsichtlich der Aufstellung neuer Pläne dieser Art befinden sich gleichzeitig die relevanten Bestimmungen zur Erschließung und Bodennutzung in der unabhängigen Erarbeitung durch das Planungsgremium, um sie flexibler zu machen und die Standortbestimmung dienstleistungsorientierter Bauvorhaben zu erleichtern. (DTPH)

3.14.5 Frage 5: Erneuerbare-Energien-Richtlinie (2009/28/EG)

Wird ein Umsetzungsbedarf hinsichtlich des Artikels 13 Erneuerbare-Energien-Richtlinie (2009/28/EG) für den Bereich des nationalen Städtebaurechts und der Bauleitplanung gesehen?

Die Umsetzung der Bestimmungen dieser Richtlinie im Planungssystem erfolgte über die Änderung einiger gesetzlicher Raumentwicklungspläne, **um die entsprechenden Bestimmungen flexibler zu gestalten,** und zwar in Form der Vereinfachung der Anforderungen für die Standortbestimmung und den Bau von Infrastruktur für die erneuerbaren Energien. Dieser Prozess soll nach und nach mit allen gesetzlichen Raumentwicklungsplänen durchgeführt werden, die zur regelmäßigen Überprüfung anstehen. Gleichzeitig hat der zuständige Innenminister eine Reihe von Anordnungen und Instruktionen (die *„Enzykliken"* genannt werden) an die Planungsbehörden ausgegeben (einige davon verweisen auf die Richtlinie 2001/77/EG):

- Anordnung 2/2006 Standortbestimmung für Anlagen, die der Energiegewinnung aus erneuerbaren Quellen dienen
- Anordnung 2/2006 Zusatz (Erklärung zum Bau eines Windparks)
- Anordnung 2/2006 Zusatz (Benachrichtigung über den Bau eines Windparks)
- Instruktion -/2008 *„Sondervorschriften für die Beschleunigung von Erschließungskontrollverfahren bei Anträgen für Windparks oder einzelne Windräder"*
- Instruktion 3/2008 *„Fotovoltaikanlagen, für die kein Antrag auf Erteilung einer Planungsgenehmigung erforderlich ist "*

(DTPH)

3.14.6 Frage 6: SEVESO-II-Richtlinie

Wie ist Artikel 12 Absatz 1 Unterabsatz 2 der Richtlinie 96/82/EG vom 09.12.1996 zur Beherrschung der Gefahren bei Unfällen mit gefährlichen Stoffen (ABl. L 10 vom 14.01.1997, S. 13; „Seveso-II-Richtlinie") in nationales Recht umgesetzt worden?

Der oben genannte Artikel wurde durch die einschlägigen Stadt- und Landplanungs-Rechtsnormen in das Raumplanungsrecht überführt [Verwaltungsvorschrift 759/2003: Stadt- und Landplanung (Unfallrisiken mit gefährlichen Substanzen) von 2003; und Verwaltungsvorschrift 399/2008: Stadt- und Landplanungsrecht (Unfallrisiken mit gefährlichen Substanzen) von 2008 (Änderungsvorschrift)]; dieses wird ggf. bei der Prüfung eines Planungsgenehmigungsantrags von den Planungsbehörden berücksichtigt. In diesen Fällen wurden zwingende Konsultationsverfahren mit der für die Umsetzung der Seveso-Richtlinie zuständigen Behörde, der Abteilung für Gewerbeaufsicht (Department of Labour Inspection – DLI), eingeführt. Die Rechtsnormen werden auch während der Aufstellung oder Überprüfung gesetzlicher Raumentwicklungspläne berücksichtigt, mit dem Ziel, diese durch die anstehenden Überprüfungen in Form gebietsspezifischer, verbindlicher Bestimmungen in diese Pläne zu integrieren. (DTPH)

Die Informationen wurden von der Abteilung für Stadtplanung und Wohnen, Innenministerium der Republik Zypern, nach Absprache mit verschiedenen zuständigen Behörden zusammengetragen, 11.3-30.4.2010.

(DTPH) Beantwortung durch die Abteilung für Stadtplanung und Wohnen (Department of Town Planning and Housing), Innenministerium (Christakis Serghides, Ermis Klokkaris und Phaedon Enotiades, penotiades@tph.moi.gov.cy)

(ENV) Beantwortung durch die Umweltabteilung, Ministerium für Landwirtschaft, natürliche Ressourcen und Umwelt (Joanna Constantinidou und Despo Zavrou, dzavrou@environment.moa.gov.cy)

APPENDUM 1

Überprüfungs- und Fortschreibungsverfahren für gesetzlich vorgeschriebene Raumentwicklungspläne – lokale Pläne und Gebietsprogramme (Zusammenfassung)

> *1. Das entsprechende Verfahren wird durch die Veröffentlichung eines Berichts eingeleitet, der den territorialen Zustand und die Perspektiven jedes sich in der Untersuchung befindlichen Gebiets analysiert; in diesem werden auch in Absprache mit den beteiligten Gebietskörperschaften die Vision sowie entsprechende strategische Zielsetzungen dargelegt. In diesem Stadium berufen die lokalen Behörden öffentliche Versammlungen ein, um ihre offiziellen Stellungnahmen zum Bericht vorzubereiten.*
>
> *2. Das Planungsgremium, ein unabhängiger Ausschuss mit beratenden Befugnissen in großen Bereichen der Planungspolitik, zu der auch die Aufstellung und Fortschreibung bestimmter Arten von Bauleitplänen gehört, ist dann verpflichtet, ein ge-*

meinsames Gremium (Joint Board) zu konsultieren (ein Beratungsausschuss, der eigens eingerichtet wird, damit Behörden und lokale Interessenvertretergruppen bei Aufstellung oder Fortschreibung jedes Raumplans über einen Bericht, der dem Planungsgremium unterbreitet wird, ihre Ansichten äußern können) sowie auch die Öffentlichkeit; dies erfolgt zunächst über einen transparenten, offenen Aufruf an alle interessierten Bürger oder Einrichtungen, Meinungen und Vorschläge zu jedem untersuchten Gebiet einzureichen, dann über strukturierte öffentliche Anhörungen, wo ein repräsentativer Querschnitt von Meinungen und Anregungen gehört werden muss.

3. Nach Bildung einer fundierten Meinung erstellt das Planungsgremium den detaillierten lokalen Plan und veröffentlicht ihn über den Innenminister; der so veröffentlichte Plan wird dann für alle interessierten Parteien zur Prüfung offen gelegt.

4. Die Öffentlichkeit hat in diesem Stadium die Möglichkeit, Einsprüche gegen die Bestimmungen des Plans einzulegen, die über das vorgeschriebene Verfahren geprüft werden müssen, bevor das Planungsgremium den Plan mit oder ohne Änderungen aufgrund der geprüften Einsprüche über den Innenminister zur Genehmigung an den Ministerrat überstellt.

5. Verabschiedete lokale Pläne werden dann durch das Planungsgremium überwacht und in regelmäßigen Abständen auf mögliche Änderungen hin überprüft, indem die oben beschriebenen Verfahren erneut durchlaufen werden.

(DTPH)

4. Summary of the written answers

The following overview presents a summary of the answers from the Member States.

4.1 Systematic preliminary question

What is the relationship between construction plans and the approval of individual building projects?

Table 1: Summary question 1

Country	Answer
Estonia	The requirements for the construction of individual projects result from construction and zoning plans.
France	The building permit depends primarily in the local town development plans (PLU) and the municipal maps (CC).
United Kingdom	No statement made
Ireland	The building permit depends primarily on the municipal development plan (DP).
Latvia	Review of compliance with the stipulations in the municipal zoning plan and the building code (part of the zoning plan).
The Netherlands	The construction plan defines the use of land for construction and other uses. The building permit is reviewed for compliance with the construction plan.
Austria **(Vienna)**	The zoning plan and construction plans are the authoritative documents. Building rights only come into being with the construction plan.

Poland	The stipulations of a construction plan (if one exists) are authoritative for the building permit.
Portugal	No statement made
Slovakia	Prime importance goes to how a building project complies with the stipulations in a construction plan.
Spain	The construction plan is authoritative for the building permit (together with other legal requirements).
Hungary	The building permit depends on compliance of a project with the local building codes and zoning maps. If these do not exist, the project has to blend in with its surroundings.
Cyprus	In the framework of planning approval, the stipulations of the regional development plans are reviewed.

Own visualisation

4.2 European habitat and species protection law

Question 2a (European habitat and species protection law)

What is the relationship between national nature conservation law and the building planning law? Are nature conservation law requirements taken into account in development planning or not dealt with until a subsequent approval procedure takes place?

Table 2: Summary question 2a

Country	Answer
Estonia	The requirements of national nature conservation law are considered in the planning procedures.
France	Urban development documents have to take account of the environmental demands in the nature conservation law. Additional environmental demands for the urban development documents result from the "urban solidarity and renewal" law (2000) and the "urbanism and habitat" law (2003). The urban development documents have to undergo a review of the environmental impacts (decree on reviewing the environmental impacts of urban development documents dated 2005). This review also has to include the effects on Natura 2000 areas.
United Kingdom	The requirements of nature conservation law have to be considered in both the planning and the approval of construction projects.
Ireland	All national development plans undergo either a strategic environment assessment or FFH screening. Development plans contain targets for protecting and preserving the environment, also including areas under Community protection. Natural conservation experts influence the draft process by contributing corresponding information which is announced during the procedure for drawing up a plan.

Latvia	Nature conservation law makes protected areas part of a municipal zoning plan. The regulations for preserving and using the areas have to be considered when drawing up land use plans.
The Netherlands	There is no formal link between the building code and the nature conservation law or flora-fauna law. The same can also be said for construction planning. Approval under the flora-fauna law is necessary to obtain a building permit in a habitat. This constitutes an additional assessment for the building permit.
Austria **(Vienna)**	Species protection in Vienna is regulated by Vienna's nature conservation law. This covers both area and building protection. With regard to species protection, a distinction is made between protected and strictly protected species. The impacts on nature conservation areas and the protected species are to be covered by a strategic environment assessment within the planning framework.
Poland	Nature conservation has to be considered when drawing up construction plans. This is based on the law for regional planning and development, the nature conservation law and the law on environment information, public participation in environmental protection and environmental implications assessment. When drawing up a construction plan, the impacts on the environment and also on nature conservation are covered by the environmental implications assessment (EIA). This also applies to the building permit if no construction plan exists.
Portugal	Building projects in areas covered by the Natura 2000 network need approval from a nature conservation or biodiversity authority. The decision is subject to a review of the impacts on the Natura 2000 area. Urban planning has to take account of the national nature conservation law.

Slovakia	Nature conservation is defined by the law for the protection of nature and landscape. In the framework of regional planning, decisions for the use of areas have to comply with the requirements of the corresponding specific legislation. For nature conservation, this consists of an opinion issued by the relevant nature conservation authority. The land use decision does not yet constitute approval to proceed with the project. This is only possible on obtaining the building permit. The building application has to illustrate to what extent nature conservation is taken into account.
Spain	Environmental protection is part of the zoning plan and regional plans. The regional and urban planning instruments undergo an environmental assessment corresponding to the stipulations made in the law for assessing the environmental impact of plans and programs.
Hungary	The national environmental protection requirements have to be heeded by architects in their planning activities.
Cyprus	The nature conservation laws for transposing the FFH and the Bird Protection Directive must be heeded when drawing up zoning plans.

Own visualisation

Question 2b (European habitat and species protection law)

Within the scope of the assessment of the implications of a plan or project for a site, how has the exception provided by Article 6 Paragraph 4 of the FFH Directive been transposed into national law in the field of building planning and how is it applied?

Is there any distinction regarding whether a project is built within or outside an area which is already continuously built up?

Table 3: Summary question 2b

Country	Answer
Estonia	The ruling for exceptions from the FFH implications assessment have been integrated in the law on environmental implications assessment and environ-mental management laws. No distinction is made between built up areas and projects outside such areas.
France	Provisions are made in the environmental law (Article R 414-23) rather than planning laws. Where there are significant harmful implications for conservation of the habitat and species, an evaluation document has to indicate which measures are to be taken to eliminate or diminish the harmful implications. The evaluation document contains a review of alternatives and the justification for proceeding to draw up the planning document. No distinction is made between built up areas and projects outside such areas.
United Kingdom	Transposed by regulation 2010 on the conservation of habitats and species. Accordingly, approval cannot be given for a plan or project that could be detrimental to an area, unless the project developer comes to the conclusion that there is no alternative solution and the plan or project must be carried out for vital reasons in the public interest. In this case, the relevant authority must ensure that all necessary compensation measures

	are taken to protect the overall coherence of the Natura 2000 network. No distinction is made between built up areas and projects outside such areas.
Ireland	Article 6 (4) has been transposed in the habitat regulation. No statement made on transposing the contents of the exception ruling. No distinction is made between built up areas and projects outside such areas.
Latvia	Article 6 (4) FFH Directive has been transposed in the law to conserve specially protected natural areas. This stipulates the need for measures to compensate for intervention; such measures have to be provided by the originator. The law on environment implication assessment stipulates when a EIA has to be carried out in a Natura 2000 area for certain projects (including building projects. No distinction is made between built up areas and projects outside such areas.
The Netherlands	No distinction is made between built up areas and projects outside such areas.
Austria **(Vienna)**	The FFH is transposed by extending the SEA to FFH protection areas. No statement made on transposing the contents of the exception ruling. No distinction is made between built up areas and projects outside such areas.

Poland	The FFH Directive has been transposed in the provisions of the Polish nature conservation law. No statement made on transposing the contents of the exception ruling. No statement made on distinction.
Portugal	Article 6 (4) of the FFH Directive has been adopted word-for-word in the national legislation. No distinction is made between built up areas and projects outside such areas.
Slovakia	The FFH Directive has been transposed fully in the law for the protection of nature and landscape, and is dealt with in the framework of opinions issued by the nature conservation authorities. No distinction is made between built up areas and projects outside such areas.
Spain	Article 6 (4) of the FFH Directive has been transposed in royal decree 1421 "Measures to safeguard biodiversity by preserving natural habitats together with wild flora and fauna". The regulation makes no specific reference to regional planning, but generally stipulates the need for compensatory measures to safeguard coherence of the Natura 2000 network. Corresponding measures have to be taken into account during the planning process. No statement made on distinction.
Hungary	Article 6 (4) of the FFH Directive has been transposed in the 53rd law on nature conservation dated 1996. No statement made on transposing the contents of the exception ruling. No statement made on distinction.

Cyprus	The directive has been transposed in national law. When projects are detrimental to Natura 2000 areas, they can only be implemented if there are no viable alternatives and for vital reasons in the public interest. Further restrictions apply to priority species or habitats. No distinction is made between built up areas and projects outside such areas.

Own visualisation

Question 2c (European habitat and species protection law)

Apart from the conservation area related assessment of implications for the site pursuant to Article 6 Paragraph 3 of the FFH Directive, is an assessment under species protection law also always performed according to Article 12 of the FFH Directive?

Table 4: Summary question 2c

Country	Answer
Estonia	The implications assessment also reviews possible impacts on protected species.
France	No answer
United Kingdom	Where there is a justified probability that protected species may suffer, in the framework of the approval procedure the responsible authority can oblige the project developer to proceed with a corresponding species protection assessment.
Ireland	At the moment there is no mandatory obligation for assessment under the FFH Directive. Destruction or damage to breeding or resting places of species featured in Annex IV of the FFH Directive and harm caused to protected plants constitute a criminal offence. These species are to be identified during the SEA or EIA.

Latvia	Species protection is transposed in the law for the conservation of protected natural habitats. In addition, the demands of species and biotope protection are covered by the law on species and biotope protection.
The Nether-lands	A species protection assessment is necessary in addition to the implications assessment if there is evidence that a building project will harm or damage the habitat of a protected species.
Austria (Vienna)	Distinction between protected and strictly protected species. A species protection assessment has to be carried out if a building project harms the habitat of a strictly protected species.
Poland	No statement made on species protection assessment.
Portugal	Article 12 of the FFH Directive has been transposed fully in national legislation. The FFH implications assessment also covers the strictly protected species.
Slovakia	The building code does not feature any special species protection assessment. No statement made, as this falls under the responsibility of the Ministry for the Environment.
Spain	Species protection is part of the strategic environment assessment and is explained in the environment report. In addition, the law on natural heritage and biodiversity prohibits the use of land if this would have negative implications on natural habitats or protected species.
Hungary	No statement made on species protection assessment.
Cyprus	Law 153(I)/2003 stipulates an appropriate assessment of the implications for species.

Own visualisation

Question 2d (European habitat and species protection law)

Is there a need to exempt such construction projects from prohibitions under species protection law which are erected in areas which have already been developed? What role could it play if the protected species had their natural habitat outside the development area and also occurred in greater number there?

Table 5: Summary question 2d

Country	Answer
Estonia	An exemption can be made if the conservation of a species is safeguarded and the population increases.
France	No statement made
United Kingdom	No need is seen for exemptions as no distinction is made between developments in built up areas and development projects outside such areas.
Ireland	A case of this kind has not occurred hitherto. It would be covered by the exception ruling of Article 6 (4) or Article 16 of the FFH Directive.
Latvia	This depends on each individual case. Both cases have already occurred. No further explanations given.
The Netherlands	A corresponding exemption is seen as helpful for inner-city development projects.
Austria (Vienna)	The problem is acknowledged and dealt with in practical cases.
Poland	No statement made

Portugal	No exemption is provided from species protection bans. Article 12 of the FFH Directive applies to all areas, including those that are already developed.
Slovakia	The areas covered by the law for the protection of nature and landscape also include species protection.
Spain	No need is seen for a ruling on exemptions from the species protection bans.
Hungary	No statement made
Cyprus	No provision is made for exemptions from the species protection bans, even in already developed areas.

Own visualisation

Question 2e (European habitat and species protection law)

How is the element of "intention" in Article 12 Paragraph 1 Letter a, b and c of the FFH Directive interpreted for construction projects in the light of the legal decisions taken by the European Court of Justice?

Table 6: Summary question 2e

Country	Answer
Estonia	No statement made on the element of intention
France	No statement made
United Kingdom	According to the Commission's guidelines for Article 12 of the FFH Directive.
Ireland	According to the requirements for the exemption ruling in compliance with Article 16 of the FFH Directive, any risks regarding the protected species are identified in the SEA or EIA.

Latvia	The law for conserving species and biotopes demands measures to safeguard the requirements corresponding to Article 12 (1) letter a, b and c FFH Directive.
The Netherlands	The element of intention also refers to cases where destruction or killing as a result of an action was not deliberately wanted but known and accepted (reference to the Carette verdict by the European Court of Justice).
Austria (Vienna)	Vienna's nature conservation law distinguishes between intentional destruction and other forms of harm. The destruction of a local habitat through construction does not constitute intention.
Poland	No statement made
Portugal	The stipulations of species protection are implemented regardless of the element of intention.
Slovakia	The responsible nature conservation authority issues an opinion on the building project.
Spain	No statement made
Hungary	No statement made
Cyprus	Article 12 of the FFH Directive has been transposed in Article 21 of the national law which forbids any kind of destruction by a person or project.

Own visualisation

4.3 SEA Directive (2001/42/EC)

Question 3a (SEA Directives (2001/42/EC)

Which levels and instruments of development planning exist and with respect to this, how was the SEA Directive transposed into national law?

Table 7: Summary question 3a

Country	Answer
Estonia	The SEA is to be carried out when drawing up or amending plans on the national, district or local level.
France	Urban development documents are the development plan for conserving territorial coherence (SCOT), the local urban development plans (PLU) and the municipal maps (CC). SEAs are mandatory for the zoning plan for the Ile-de-France region (SDRIF) and particularly the local urban development plans that could cause considerable harm to Natura 2000 areas. Decree 2005-608 dated 25 May 2005 stipulates SEAs for these plans.
United Kingdom	SEAs are required for regional and local plans constituting the framework for projects in urban and land planning.
Ireland	SEAs are required for all urban and county development plans together with development plans for local authorities with more than 10.000 inhabitants. All other statutory building development plans have to undergo preliminary SEAs.
Latvia	Distinction between long/ medium/ short-term development plans (informal plans). Municipal construction plans (mandatory SEA) are drawn up and approved by the local authorities.

The Netherlands	SEAs are required for building development plans that can be drawn up on a local, regional (provinces) or national level.
Austria (Vienna)	On the local level (Vienna), there are zoning plans and construction plans.
Poland	The Polish planning system has three levels, distinguishing between national, regional and local planning. On the local level there are studies for the overall regional development (overall municipality) and local regional development plans and zoning plans. On the local level, SEAs are required for studies on the conditions and guidelines for regional development in the local authorities.
Portugal	No statement made
Slovakia	The instruments for building development planning are building development planning materials, documents and zoning decisions. SEAs are required for the building development planning documents.
Spain	On the local level there is two-stage development planning for controlling urban development. The overall local master plan forms the framework for the land development plan for individual areas. No statement made on mandatory SEAs for the plans.
Hungary	No statement made on mandatory SEAs for the plans.

Cyprus	Three-level structure of the regional planning system with the island plan, local plans (urban and conurbation areas) and regional schemes. These are subject to mandatory SEAs. In addition, SEAs are required for areas that have neither local plans nor regional schemes (policy schemes).

Own visualisation

Question 3b (SEA Directive (2001/42/EC)

How was the SEA Directive transposed into national law?

Table 8: Summary question 3b

Country	Answer
Estonia	The SEA Directive was transposed with the law on the environmental implications assessment and environment management systems (in force since 3.4.2005)
France	The SEA Directive was transposed with ordinance 2004-489 dated 3 June 2004
United Kingdom	The SEA Directive was transposed with the legislation on the environmental assessment of plans and programmes in 2004. This applies to all plans and programmes fulfilling the criteria of the SEA Directive.
Ireland	The SEA Directive was transposed with ordinances (ordinance on the environmental assessment of certain plans and programmes and the ordinance on the strategic environmental assessment).
Latvia	The SEA Directive was transposed in full with the law on the environmental implications assessment. There is a general SEA obligation for plans that could have impacts on Natura 2000 areas.

The Nether-lands	The SEA Directive was transposed with the Dutch environmental management law and the ordinance on the assessment of environmental implications dated 1994.
Austria (Vienna)	The SEA Directive was transposed on the municipal level (regional planning for Vienna) by the Vienna city planning and building code.
Poland	The SEA Directive was transposed by the law on the accessibility of environmental information for environmental protection, public participation in environmental participation and the strategic environmental assessment.
Portugal	No statement made
Slovakia	The SEA Directive was transposed by a law on the assessment of environmental implications. Zoning plan documents also have to undergo the environmental assessment.
Spain	The SEA Directive was transposed with the state law dated 9 April 2006. For building and planning, the SEA Directive was transposed to the extent that all procedures for preparing regional and urban planning instruments have to undergo mandatory SEA.
	The procedural requirements of the SEA were integrated in the procedure for drawing up development plans. The construction plans have to undergo both the SEA and the EIA.
Hungary	The SEA Directive was transposed by means of a decree.
Cyprus	The SEA Directive was transposed with law No. 102(I)2005 dated 2005.

Own visualisation

Question 3c (SEA Directive (2001/42/EC)

Are options, especially the option of insignificance with the context of preparing, amending and supplementing the development plans, used to accelerate and simplify procedures for particular cases?

Table 9: Summary question 3c

Country	Answer
Estonia	No possibilities for accelerating or simplifying procedures.
France	Insofar as local urban development plans are contained in a master plan for territorial coherence, these are not subject to mandatory SEA.
United Kingdom	No statement made
Ireland	When development plans are amended or supplemented, the amendments/ additions are screened to ascertain any possible SEA obligation.
Latvia	Not in the current legal situation. There are plans for simplification when linking planning and project planning procedures and in the advance planning for the designation of planning areas.
The Netherlands	A system of limit values below which assessment is not necessary has been reprimanded by the European Commission and therefore further modified.
Austria (Vienna)	No possibilities for accelerating or simplifying SEA obligations.
Poland	The law on regional planning and development does not provide any possibilities for simplifying the procedures.

Portugal	No statement made
Slovakia	No process simplification on reviewing the document in the framework of an SEA.
Spain	Process simplifications are permitted in certain cases for a few autonomous municipalities insofar as only minor changes are made to a master plan or when drawing up construction plans for the reuse of areas on achieving a higher building density or for inner development.
Hungary	Possibilities for simplifying the procedure exist for minor amendments to plans with omission of the SEA.
Cyprus	Cyprus currently has no experience in this respect.

Own visualisation

Question 3d (SEA Directive (2001/42/EC)

What is the relationship between the environmental assessment and the European species protection in implementation and practical application?

Table 10: Summary question 3d

Country	Answer
Estonia	No statement made
France	No statement made
United Kingdom	No statement made
Ireland	The SEA contains a possibly necessary FFH implycations assessment as distinctive part of the SEA.

Latvia	If an SEA has been carried out, a separate FFH implications assessment is not necessary. The SEA includes covering the implications for Natura 2000 areas.
The Netherlands	The assessment range of the environmental assessment is broader than the FFH implications assessment.
Austria (Vienna)	The FFH implications assessment is carried out in the form of the environmental assessment and does not constitute a separate review.
Poland	If an SEA has been carried out, this also includes assessing the implications for Natura 2000 areas.
Portugal	No statement made
Slovakia	This lies in the responsibility of the Ministry for the Environment. The building code contains no corresponding regulations. The environment assessment deals with existing circumstances, the implications assessment makes statements on future developments.
Spain	FFH implications assessment and SEA are integrated in the planning and approval procedures.
Hungary	No statement made
Cyprus	The environmental report should include the FFH implications assessment.

Own visualisation

4.4 Services Directive(2006/123/EC)

Is it deemed necessary to transpose the Services Directive (Directive 2006/123/EC) for the area of national urban planning law and development planning?

Table 11: Summary question 4

Country	Answer
Estonia	No need for transposition
France	Only insofar as a SCOT or PLU define restrictive criteria for commercial settlement possibilities in stipulations on construction and on the environment.
United Kingdom	No need for transposition in terms of urban planning law.
Ireland	Not applicable to land use plans.
Latvia	Not applicable in terms of planning law.
The Netherlands	No need for transposition with reference to recital 9 of the Directive.
Austria (Vienna)	No need for transposition.
Poland	No need for transposition in planning law.
Portugal	No statement made
Slovakia	No statement made on the need for transposition.
Spain	No need for transposition in terms of urban planning in preparation for the approval of projects under planning law.

Hungary	The Services Directive has already been transposed in connection with development planning.
Cyprus	Planning or building permits are not covered by the Services Directive.
Denmark	No need for transposition (reported by phone)

Own visualisation

4.5 Renewable Energy Directive(2009/28/EC)

Is it considered necessary to transpose Article 13 of the Renewable Energy Directive (2009/28/EC) for the area of national urban planning law and development planning?

Table 12: Summary question 5

Country	Answer
Estonia	The requirements of Article 13 have been adopted in the building laws.
France	No need for transposition
United Kingdom	A national plan of action to transpose the Directive is currently being drawn up.
Ireland	No need for transposition
Latvia	Draft legislation for renewable energies is currently in preparation. There is no need for transposition with regard to planning law.
The Netherlands	No need for transposition, as the Dutch building code already demands adequate energy requirements for the construction, conversion and refurbishment of buildings.
Austria **(Vienna)**	No need for transposition
Poland	No need for transposition. However, in future local planning will also have to consider the aspects of renewable energies, in compliance with the Renewable Energy Directive.

Portugal	Decisions have not yet been taken regarding transposition of the Renewable Energy Directive in the planning, construction and renewal of urban and industrial areas.
Slovakia	The Renewable Energy Directive is currently being transposed in national law. Slovakia sees a need for transposition with regard to planning law.
Spain	The Renewable Energy Directive has not been transposed in national law but has been implemented by various autonomous municipalities.
Hungary	A need for transposition is seen, but the Directive has not yet been transposed.
Cyprus	The requirements of the Renewable Energy Directive have been transposed in the planning system by amending various binding regional development plans. Transposition consisted of simplifying certain requirements regarding the site and construction of plants for renewable energies.

Own visualisation

4.6 SEVESO-II Directive (96/82/EC)

How has Article 12 Paragraph 1 Subparagraph 2 of Directive 96/82/EC of 9 December 1996 the control of major-accident hazards involving dangerous substances (Official Journal of the EC No. L 010 p. 13; "SEVESO-II Directive) been transposed into national law?

Table 13: Summary question 6

Country	Answer
Estonia	In accordance with the requirements of chemical and planning law, local authorities must consider the risks of major accidents in their decisions. This is monitored by so-called rescue centres and the technical monitoring authorities.
France	No statement made
United Kingdom	No statement made
Ireland	No statement made
Latvia	The stipulations of the SEVESO-II Directive are considered in municipal planning.
The Netherlands	Article 12 of the SEVESO-II Directive has been transposed in national law with the ordinance on plant environment safety. This stipulates the environmental requirements to be considered for the surroundings of high-risk plants in the planning and environmental approvals. The ordinance stipulates site-relevant risk limit values for vulnerable objects (residential areas, schools, hospitals, etc.) and distinguishes between existing objects and new planning projects.

	In the framework of development planning, the local authorities are responsible for adequate safety in the surroundings of potentially dangerous facilities in their areas. This refers particularly to the zoning stipulation activities in development planning, with mandatory consideration of the ordinance and its risk limit values. Account can be taken of the risk limit values by designating safety zones around a high-risk facility or group of facilities, such as an industrial estate, for example. Similarly, development planning can reduce the safety risk by placing restrictions on the type of facility.
Austria (Vienna)	No statement made
Poland	No statement made
Spain	Regional planning by the local authorities has to take account of avoiding the risk of major accidents. This refers to sites for new buildings, changes to existing buildings and the development of residential areas and public spaces. Royal decree 1196 dated 19 September 2003 stipulates methods for ascertaining safe distances.
Portugal	No statement made
Slovakia	In the course of regional planning and the approval of buildings, facilities and other activities, the responsible authorities have to consider the requirements for certain distances between facilities and certain uses (particularly residential areas, densely populated areas or areas protected by special regulations). Note: the answer does not indicate whether these requirements result from provisions in the law for transposing the Seveso-II Directive or directly from the planning law or building law which may have been amended in transposition of the Directive.

Hungary	The Directive was transposed by government decree. No concrete statement made on transposition.
Cyprus	The Directive was transposed by amending the planning law, with coordination of the planning process between the planning authorities and the authorities responsible for transposition of the Seveso-II Directive. The provisions have to be considered when drawing up or reviewing binding development plans.

Own visualisation

4. Zusammenfassung der schriftlichen Antworten

In der nachfolgenden Übersicht sind die Antworten aus den Mitgliedstaaten zusammenfassend dargestellt.

4.1 Systematische Vorfrage

In welchem Verhältnis stehen Bebauungspläne und die Genehmigung einzelner Bauvorhaben?

Tabelle 1 Zusammenfassung Frage 1

Land	Antwort
Estland	Die Anforderungen an den Bau von Einzelprojekten ergeben sich aus Bebauungs- und Flächennutzungsplänen.
Frankreich	Maßgeblich für Baugenehmigung sind die Vorgaben der lokalen Stadtentwicklungspläne (PLU) und der kommunalen Karten (CC).
Großbritannien	Keine Aussage
Irland	Maßgeblich für Baugenehmigung ist der kommunale Entwicklungsplan (DP).
Lettland	Prüfung der Übereinstimmung mit den Vorgaben des kommunalen Flächennutzungsplans und der Bauordnung (Teil des Flächennutzungsplans).
Niederlande	Der Bebauungsplan definiert die Nutzung der Grundstücke für Bebauung und sonstige Nutzungen. Für die Baugenehmigung wird die Übereinstimmung mit dem Bebauungsplan geprüft.
Österreich (Wien)	Maßgeblich sind der Flächenwidmungsplan und Bebauungspläne. Baurecht wird erst durch den Bebauungsplan begründet.

Polen	Die Vorgaben eines Bebauungsplans (soweit vorhanden) sind maßgeblich für die Baugenehmigung.
Portugal	Keine Aussage
Slowakei	Maßgeblich ist die Übereinstimmung eines Bauvorhabens mit den Vorgaben eines Bebauungsplans.
Spanien	Der Bebauungsplan ist maßgeblich für Baugenehmigung (neben anderen rechtlichen Anforderungen).
Ungarn	Maßgeblich für die Baugenehmigung ist die Übereinstimmung eines Vorhabens mit den lokalen Bauordnungen (local building codes) und Flächennutzungsplänen (zoning maps). Soweit diese nicht vorhanden sind, muss sich das Vorhaben in die Umgebung einfügen.
Zypern	Im Rahmen der Planungsgenehmigung werden die Vorgaben räumlicher Entwicklungspläne geprüft.

Eigene Darstellung

4.2 Europäisches Habitat- und Artenschutzrecht

Frage 2a (Europäisches Habitat- und Artenschutzrecht)

In welchem Verhältnis stehen das nationale Naturschutzrecht und das Bauplanungsrecht? Werden die Anforderungen des Naturschutzrechts bereits in der Bauleitplanung oder erst in einem nachfolgenden Genehmigungsverfahren bewältigt?

Tabelle 2 Zusammenfassung Frage 2a

Land	Antwort
Estland	Die Anforderungen des nationalen Naturschutzrechts werden in den Planungsverfahren berücksichtigt.
Frankreich	Stadtentwicklungsdokumente müssen den Umweltanforderungen des Naturschutzgesetzes Rechnung tragen. Zusätzliche Umweltanforderungen für die Stadtentwicklungsdokumente ergeben sich aus dem Gesetz „Städtische Solidarität und Erneuerung“ (2000) und „Urbanismus und Habitat“ (2003). Die Stadtentwicklungsdokumente müssen einer Prüfung der Umweltauswirkungen unterzogen werden (Dekret über die Prüfung der Umweltauswirkungen der Stadtentwicklungsdokumente“ aus dem Jahr 2005). Im Rahmen dieser Prüfung müssen auch Auswirkungen auf Natura 2000-Gebiete erfasst werden.
Großbritannien	Die Anforderungen des Naturschutzrechts müssen sowohl im Rahmen der Planung als auch der Genehmigung von Vorhaben berücksichtigt werden.
Irland	Alle Land-Entwicklungspläne müssen entweder einer Strategischen Umweltprüfung oder einem FFH-Screening unterzogen werden. Entwicklungspläne enthalten Ziele zum Schutz und zum Erhalt der Umwelt,

	was auch gemeinschaftsrechtlich geschützte Gebiete umfasst. Naturschutz-Experten beeinflussen den Entwurfsprozess durch entsprechende Informationen, die im Aufstellungsverfahren eines Plans zur Anzeige gebracht werden.
Lettland	Über das Naturschutzrecht werden geschützte Gebiete Bestandteil eines kommunalen Flächennutzungsplans. Die Vorschriften zur Erhaltung und Nutzung der Gebiete müssen bei der Aufstellung von Landnutzungsplänen berücksichtigt werden.
Niederlande	Es gibt keine formale Verknüpfung zwischen dem Baugesetz und dem Naturschutzgesetz oder dem Flora-Fauna-Gesetz. Dies gilt auch für die Bebauungsplanung. Für eine Baugenehmigung in einem Habitat ist eine Genehmigung nach dem Flora-Fauna-Gesetz notwendig. Diese stellt eine zusätzliche Prüfung zur Baugenehmigung dar.
Österreich (Wien)	Das Artenschutzrecht ist für Wien durch das Wiener Naturschutzgesetz geregelt. Dies umfasst sowohl einen Gebiets- als auch einen Objektschutz. In Bezug auf den Artenschutz wird zwischen geschützten und streng geschützten Arten unterschieden. Die Auswirkungen auf Naturschutzgebiete sowie die geschützten Arten sind im Rahmen der Planung durch eine Strategische Umweltprüfung zu erfassen.
Polen	Der Naturschutz muss im Rahmen der Aufstellung von Bebauungsplänen berücksichtigt werden. Grundlage hierfür bildet das Gesetz über räumliche Planung und Entwicklung, das Naturschutzgesetz und das Gesetz über Umweltinformationen, Öffentlichkeitsbeteiligung im Umweltschutz und Umweltverträglichkeitsprüfung.

	Im Aufstellungsverfahren eines Bebauungsplans werden die Auswirkungen auf die Umwelt und damit auch den Naturschutz über die Umweltverträglichkeitsprüfung (EIA) erfasst. Dies gilt auch für die Baugenehmigung, wenn kein Bebauungsplan vorhanden ist.
Portugal	Bei der Zulassung von Bauvorhaben in Gebieten, die zum Netz Natura 2000 gehören, ist die Zustimmung einer Naturschutz oder Biodiversitätsbehörde notwendig. Die Entscheidung ist von einer Prüfung der Auswirkungen auf das Natura 2000 Gebiet abhängig. Das nationale Naturschutzrecht muss bei der Stadtplanung berücksichtigt werden.
Slowakei	Der Naturschutz wird durch das Gesetz zum Schutz von Natur und Landschaft bestimmt. Im Rahmen der räumlichen Planung müssen bei Entscheidungen zur Nutzung von Flächen die Anforderungen der jeweiligen Fachgesetze beachtet werden. Für den Naturschutz erfolgt dies über eine Stellungnahme der zuständigen Naturschutzbehörde. Die Landnutzungsentscheidung ist noch keine Genehmigung für die Ausführung von Vorhaben. Dies wird erst durch die Baugenehmigung möglich. Im Rahmen des Bauantrags muss dargelegt werden, inwiefern der Naturschutz berücksichtigt wird.
Spanien	Der Umweltschutz ist Teil der Flächennutzungspläne und der Regionalpläne. Die räumlichen und städtischen Planungsinstrumente werden einer Umweltprüfung unterzogen, die den Vorgaben des Gesetzes zur Prüfung der Auswirkungen von Plänen und Programmen auf die Umwelt entspricht.
Ungarn	Die nationalen Umweltschutzanforderungen müssen während der Planung von Architekten eingehalten werden.

Zypern	Im Rahmen der Aufstellung von Landnutzungsplänen müssen die Naturschutzgesetze zur Umsetzung der FFH- und der VS-Richtlinie beachtet werden.

Eigene Darstellung

Frage 2b (Europäisches Habitat- und Artenschutzrecht)

Wie ist im Rahmen der Verträglichkeitsprüfung die Ausnahmeregelung des Artikels 6 Absatz 4 FFH-RICHTLINIE im Bereich des Bauplanungsrechts in nationales Recht umgesetzt worden und wie wird diese angewandt?

Gibt es eine Differenzierung in Bezug darauf, dass ein Vorhaben innerhalb oder außerhalb eines bereits im Zusammenhang bebauten Gebietes errichtet wird?

Tabelle 3 Zusammenfassung Frage 2b

Land	Antwort
Estland	Die Ausnahmeregelung von der FFH-Verträglichkeitsprüfung wurde in das Gesetz über Umweltverträglichkeitsprüfung und Umweltmanagementgesetze integriert. Es erfolgt keine Differenzierung zwischen im Zusammenhang bebauten Gebieten und Gebieten außerhalb.
Frankreich	Keine Regelung im Planungsrecht, sondern Regelung im Umweltgesetz (Artikel R 414-23). Soweit signifikante schädliche Auswirkungen auf den Erhaltungszustand des Habitats und der Arten bestehen, muss in einem Evaluierungsdokument dargelegt werden, welche Maßnahmen zur Beseitigung oder Minderung der schädlichen Auswirkungen ergriffen werden. Das Evaluierungsdokument enthält eine Alternativenprüfung sowie eine Rechtfertigung zur Begründung der Aufstellung des Planungsdokuments. Es erfolgt keine Differenzierung zwischen im Zusammenhang bebauten Gebieten und Gebieten außerhalb.

Großbritannien	Die Umsetzung erfolgte über die Rechtsnorm 2010 über den Erhalt von Lebensräumen und Arten. Danach darf ein Plan oder ein Vorhaben nicht genehmigt werden, wenn ein Gebiet beeinträchtigt werden könnte, es sei denn, der Vorhabenträger kommt zu dem Schluss, dass es keine Alternativlösung gibt und der Plan oder das Vorhaben aus zwingenden Gründen des öffentlichen Interesses ausgeführt werden muss. Die zuständige Behörde muss hierbei sicherstellen, dass alle notwendigen Ausgleichsmaßnahmen durchgeführt werden, um die Gesamtkohärenz des Netzes Natura 2000 zu schützen. Es erfolgt keine Differenzierung zwischen im Zusammenhang bebauten Gebieten und Gebieten außerhalb.
Irland	Artikel 6 Abs. 4 wurde in die Habitat-Verordnung umgesetzt. Keine Aussage zur inhaltlichen Umsetzung der Ausnahmeregelung Es erfolgt keine Differenzierung zwischen im Zusammenhang bebauten Gebieten und Gebieten außerhalb.
Lettland	Artikel 6 Abs. 4 FFH-RL wurde in das Gesetz zum Schutz besonders geschützter Naturräume umgesetzt. Hierin wird die Notwendigkeit von Maßnahmen zum Ausgleich von Eingriffen festgelegt, die vom Verursacher zu erbringen sind. Das Gesetz über die Umweltverträglichkeitsprüfung regelt für bestimmte Vorhaben (darunter auch Bauvorhaben) in einem Natura 2000-Gebiet, wann eine UVP durchzuführen ist. Es erfolgt keine Differenzierung zwischen im Zusammenhang bebauten Gebieten und Gebieten außerhalb.
Niederlande	Es erfolgt keine Differenzierung zwischen im Zusammenhang bebauten Gebieten und Gebieten außerhalb.

Österreich (Wien)	Durch den Bezug der SUP auf FFH-Schutzgebiete wird auch die FFH-Richtlinie umgesetzt. Keine Aussage zur Umsetzung der Ausnahmeregelung Es erfolgt keine Differenzierung zwischen im Zusammenhang bebauten Gebieten und Gebieten außerhalb.
Polen	Die FFH-Richtlinie wurde in den Bestimmungen des polnischen Naturschutzgesetzes umgesetzt. Keine Aussage zur Umsetzung der Ausnahmeregelung. Keine Aussage zur Differenzierung
Portugal	Artikel 6 Abs. 4 der FFH-RL wurde wörtlich in die nationale Gesetzgebung übernommen. Es erfolgt keine Differenzierung zwischen im Zusammenhang bebauten Gebieten und Gebieten außerhalb.
Slowakei	Die FFH-Richtlinie wurde vollständig in das Gesetz zum Schutz von Natur und Landschaft umgesetzt und wird im Rahmen der Stellungnahmen der Naturschutzbehörden behandelt. Es erfolgt keine Differenzierung zwischen im Zusammenhang bebauten Gebieten und Gebieten außerhalb.
Spanien	Die Umsetzung des Artikels 6 Abs. 4 der FFH-RL erfolgte über das königliche Dekret 1421 „Maßnahmen zur Sicherstellung der Biodiversität durch Erhaltung natürlicher Lebensräume sowie wilder Flora und Fauna". Die Regelung enthält keinen spezifischen Bezug zur räumlichen Planung, sondern regelt generell die Notwendigkeit ausgleichender Maßnahmen zur Kohärenzsicherung des Netzes Natura 2000. Maßnahmen hierzu müssen während der Planung berücksichtigt werden.

	Keine Aussage zur Differenzierung
Ungarn	Die Umsetzung des Artikels 6 Abs. 4 der FFH-RL erfolgte über das 53. Gesetz über den Naturschutz von 1996. Keine Aussage zur inhaltlichen Umsetzung der Ausnahmeregelung Keine Aussage zur Differenzierung
Zypern	Die Richtlinie wurde in nationales Recht umgesetzt. Bei Beeinträchtigungen von Natura 2000-Gebieten kann das Projekt nur durchgesetzt werden, soweit keine Alternativen vorhanden sind und zwingende Gründe des öffentlichen Interesses bestehen. Weitere Einschränkungen hierfür ergeben sich, wenn prioritäre Arten oder Habitate betroffen sind. Es erfolgt keine Differenzierung zwischen im Zusammenhang bebauten Gebieten und Gebieten außerhalb.

Eigene Darstellung

Frage 2c (Europäisches Habitat- und Artenschutzrecht)

Wird neben der schutzgebietsbezogenen Verträglichkeitsprüfung des Artikels 6 Absatz 3 FFH-Richtlinie bei baulichen Vorhaben auch stets eine artenschutzrechtliche Prüfung nach Artikel 12 FFH-Richtlinie durchgeführt?

Tabelle 4 Zusammenfassung Frage 2c

Land	Antwort
Estland	Im Rahmen der Verträglichkeitsprüfung werden auch die möglichen Auswirkungen auf geschützte Arten geprüft.
Frankreich	Keine Antwort

Großbritannien	Soweit eine begründete Wahrscheinlichkeit besteht, dass geschützte Arten beeinträchtigt werden, kann die zuständige Behörde im Rahmen des Genehmigungsverfahrens den Vorhabenträger dazu verpflichten, eine entsprechende artenschutzrechtliche Prüfung durchzuführen.
Irland	Derzeit besteht keine Verpflichtung zur Durchführung einer Prüfung nach der FFH-Richtlinie. Die Zerstörung oder Beschädigung von Brutstätten oder Ruhestätten von Arten des Anhangs IV der FFH-Richtlinie und die Beeinträchtigung geschützter Pflanzen stellen einen Straftatbestand dar. Die Identifizierung dieser Arten soll im Rahmen der SUP oder der UVP erfolgen.
Lettland	Der Artenschutz wird über das Gesetz zum Schutz besonders geschützter Naturräume umgesetzt. Zusätzlich werden die Belange des Arten- und Biotopschutzes durch das Gesetz über den Arten- und Biotopschutz erfasst.
Niederlande	Eine artenschutzrechtliche Prüfung ist neben der Verträglichkeitsprüfung notwendig, wenn sich Hinweise ergeben, dass durch ein Bauvorhaben der Lebensraum einer geschützten Art beeinträchtigt oder zerstört wird.
Österreich (Wien)	Unterscheidung zwischen geschützten und streng geschützten Arten. Soweit ein Bauvorhaben den Lebensraum einer streng geschützten Art beeinträchtigt, ist ein Artenschutzverfahren durchzuführen.
Polen	Keine Aussage zur artenschutzrechtlichen Prüfung
Portugal	Artikel 12 der FFH-Richtlinie wurde vollständig in die nationalen Gesetze umgesetzt. Die FFH-Verträglichkeitsprüfung erfasst auch die streng geschützten Arten.

Slowakei	Das Baugesetzbuch behandelt keine gesonderte artenschutzrechtliche Prüfung. Keine Aussage, da dies in die Kompetenz des Umweltministeriums fällt.
Spanien	Der Artenschutz ist in der Strategischen Umweltprüfung enthalten und wird im Umweltbericht erläutert. Zusätzlich regelt das Gesetz über das Natürliche Erbe und Biodiversität ein Verbot der Inanspruchnahme von Land, wenn sich negative Auswirkungen auf natürliche Habitate oder geschützte Arten ergeben.
Ungarn	Keine Aussage zur artenschutzrechtlichen Prüfung
Zypern	Das Gesetz 153(I)/2003 regelt eine angemessene Prüfung der Auswirkungen auf Arten.

Eigene Darstellung

Frage 2d (Europäisches Habitat- und Artenschutzrecht)

Wird ein Bedarf dafür gesehen, solche Bauvorhaben von artenschutzrechtlichen Verboten freizustellen, die in Gebieten errichtet werden, die bereits im Zusammenhang bebaut sind? Welche Rolle könnte es hierbei haben, wenn die geschützte Art ihre natürliche Verbreitung außerhalb des Siedlungsbereichs hat und dort auch in größerer Zahl vorkommt?

Tabelle 5 Zusammenfassung Frage 2d

Land	Antwort
Estland	Wenn der Erhalt einer Art gesichert ist und die Population zunimmt, besteht die Möglichkeit einer Ausnahme.
Frankreich	Keine Aussage

Großbritannien	Da keine Differenzierung zwischen einer Entwicklung von bereits im Zusammenhang bebauten Gebieten und der Entwicklung von Gebieten im Außenbereich erfolgt, wird keine Notwendigkeit zur Freistellung gesehen.
Irland	Eine solche Fallgestaltung ist bisher nicht aufgetreten. Diese würden unter die Ausnahmeregelungen des Artikels 6 Abs. 4 oder Artikel 16 der FFH-Richtlinie fallen.
Lettland	Dies ist vom Einzelfall abhängig. Beide Fallgestaltungen sind bereits aufgetreten. Keine weitere Erläuterung hierzu.
Niederlande	Eine entsprechende Ausnahme wird als hilfreich für die Innenentwicklung der Städte angesehen.
Österreich (Wien)	Die Problematik wird erkannt und man setzt sich in der Praxis damit auseinander.
Polen	Keine Aussage
Portugal	Eine Ausnahme von den artenschutzrechtlichen Verboten ist nicht vorgesehen. Artikel 12 der FFH-Richtlinie findet auf alle Gebiete Anwendung – auch solche, die bereits entwickelt sind.
Slowakei	Die nach dem Gesetz zum Schutz von Natur und Landschaft erfassten Gebiete umfassen auch den Artenschutz.
Spanien	Die Notwendigkeit zur Regelung einer Ausnahme von den artenschutzrechtlichen Verboten wird nicht gesehen.
Ungarn	Keine Aussage

Zypern	Eine Ausnahme von den artenschutzrechtlichen Verboten ist auch in bereits entwickelten Gebieten nicht vorgesehen.

Eigene Darstellung

Frage 2e (Europäisches Habitat- und Artenschutzrecht)

Wie wird das Tatbestandsmerkmal der „Absicht" in Artikel 12 Absatz 1 Buchstabe a, b und c FFH-Richtlinie bei Bauvorhaben im Lichte der Rechtsprechung des Europäischen Gerichtshofs verstanden?

Tabelle 6 Zusammenfassung Frage 2e

Land	Antwort
Estland	Keine Aussage zum Begriff der Absicht
Frankreich	Keine Aussage
Großbritannien	Entsprechend dem Leitfaden der Kommission zu Artikel 12 der FFH-Richtlinie.
Irland	Entsprechend den Anforderungen der Ausnahmeregelung in Übereinstimmung mit Artikel 16 der FFH-Richtlinie wird durch die SUP oder die UVP ein Risiko im Hinblick auf die geschützte Art identifiziert.
Lettland	Das Gesetz zur Erhaltung von Arten und Biotopen fordert Maßnahmen zur Sicherstellung der Anforderungen entsprechend Artikel 12 Abs. 1 Buchst. a, b und c FFH-Richtlinie.
Niederlande	Der Begriff der Absicht bezieht sich auch darauf, dass die Zerstörung oder Tötung als Resultat einer Aktion zwar nicht gewollt, aber bekannt ist und hingenommen wird (Verweis auf das Caretta-Urteil des EUGH).

Österreich (Wien)	Im Wiener Naturschutzgesetz erfolgt eine Unterscheidung zwischen absichtlicher Störung und anderen Formen der Beeinträchtigung. Die Zerstörung eines lokalen Lebensraumes durch Bebauung stellt keine Absicht dar.
Polen	Keine Aussage
Portugal	Die Vorgaben des Artenschutzes werden unabhängig vom Begriff der Absicht angewendet.
Slowakei	Die zuständige Naturschutzbehörde gibt eine Stellungnahme zu Bauvorhaben ab.
Spanien	Keine Aussage
Ungarn	Keine Aussage
Zypern	Artikel 12 der FFH-Richtlinie wurde durch Artikel 21 des nationalen Gesetzes umgesetzt. Danach ist jegliche Störung durch eine Person oder ein Projekt verboten.

Eigene Darstellung

4.3 Plan-UP-Richtlinie (2001/42/EG)

Frage 3a (Plan-UP-Richtlinie (2001/42/EG)

Welche Ebenen und Instrumente städtebaulicher Planungen bestehen und wie erfolgte diesbezüglich die Umsetzung der Plan-UP-Richtlinie in nationales Recht?

Tabelle 7 Zusammenfassung Frage 3a

Land	Antwort
Estland	Die SUP ist bei der Aufstellung oder Änderung von Plänen auf Bundes-, Kreis- oder kommunaler Ebene durchzuführen.
Frankreich	Stadtentwicklungsdokumente sind der Leitplan für die Erhaltung der territorialen Kohärenz (SCOT), die lokalen Stadtentwicklungspläne (PLU) und die kommunalen Karten (CC). Umweltprüfungspflichtig sind der Flächennutzungsplan für die Region Ile-de-France (SDRIF) und insbesondere die lokalen Stadtentwicklungspläne, durch die eine erhebliche Beeinträchtigung von Natura 2000-Gebiete ermöglicht wird. Für diese Pläne präzisiert das Dekret 2005-608 vom 27. Mai 2005 die Anwendung der Umweltprüfung.
Großbritannien	Regionale und lokale Pläne, die den Rahmen für die Genehmigung von Vorhaben in der Stadt- und Landplanung vorgeben, erfordern eine SUP.
Irland	Alle Stadt- und Kreis (County) -Entwicklungspläne sowie Entwicklungspläne von Kommunen mit mehr als 10.000 Einwohnern müssen generell einer SUP unterzogen werden. Alle anderen satzungsmäßigen Bauleitpläne müssen einer SUP-Vorprüfung unterzogen werden.

Lettland	Unterscheidung zwischen Lang- und mittel- und kurzfristigen Entwicklungsplänen (informelle Pläne). Kommunale Bebauungspläne (SUP-pflichtig) werden von den Gemeinden erstellt und genehmigt.
Niederlande	Bauleitpläne, die auf örtlicher, regionaler (Provinzen) oder nationaler Ebene aufgestellt werden können, sind einer Umweltprüfung zu unterziehen.
Österreich (Wien)	Auf örtlicher Ebene (Wien) gibt es Flächenwidmungspläne und Bebauungspläne.
Polen	Das polnische Planungssystem besitzt einen dreistufigen Aufbau und unterscheidet nationale, regionale und gemeindliche Planungsebene. Auf gemeindlicher Ebene gibt es Studien für die gesamträumliche Entwicklung (gesamtgemeindlich) und lokale Raumordnungspläne sowie Landnutzungspläne. Der SUP-Pflicht unterliegen (auf lokaler Ebene) die Studien über Bedingungen und Leitlinien der räumlichen Entwicklung der Gemeinden.
Portugal	Keine Aussage
Slowakei	Die Instrumente für die Bauleitplanung sind Bauleitplanungsmaterialien, -unterlagen sowie die Flächennutzungsentscheidung. Der SUP-Pflicht unterliegen die Bauleitplanungsdokumente.
Spanien	Auf lokaler Ebene gibt es eine zweistufige Entwicklungsplanung zur Steuerung der städtischen Entwicklung. Der gesamtgemeindliche Masterplan bildet den Rahmen für den Landentwicklungsplan für einzelne Gebiete. Keine Aussage zur SUP-Pflicht der Pläne

Ungarn	Keine Aussage zur SUP-Pflicht der Pläne
Zypern	Dreistufiger Aufbau des räumlichen Planungssystems mit dem Insel-Plan, lokalen Plänen (städtische Gebiete und Agglomerationen) und Gebietsschemata. Diese unterliegen der SUP-Pflicht. Zusätzlich besteht eine SUP-Pflicht für Gebiete, in denen weder lokale Pläne noch Gebietsschemata vorhanden sind (Policy-Schemes).

Eigene Darstellung

Frage 3b (Plan-UP-Richtlinie (2001/42/EG)

Wie wurde die Plan-UP-Richtlinie in das nationale Recht umgesetzt?

Tabelle 8 Zusammenfassung Frage 3b

Land	Antwort
Estland	Die Umsetzung erfolgte über das Gesetz über die Umweltverträglichkeitsprüfung und Umweltmanagementsysteme (in Kraft seit 3.4.2005)
Frankreich	Die Umsetzung der Plan-UP-Richtlinie erfolgte über die Verordnung 2004-489 vom 3. Juni 2004.
Großbritannien	Die Umsetzung erfolgte mittels der Rechtsvorschrift zur Umweltprüfung von Plänen und Programmen im Jahr 2004. Diese gilt für alle Pläne und Programme, die die Kriterien der Plan-UP-Richtlinie erfüllen.
Irland	Die Umsetzung der Plan-UP-Richtlinie erfolgte über Verordnungen (Verordnung über die Umweltprüfung bestimmter Pläne und Programme sowie die Verordnung über die Strategische Umweltprüfung).

Lettland	Die Plan-UP-Richtlinie ist vollständig über das Gesetz über die Umweltverträglichkeitsprüfung umgesetzt. Eine generelle SUP-Pflicht besteht für Pläne, die Auswirkungen auf Natura 2000-Gebiete haben können.
Niederlande	Die Plan-UP-Richtlinie wurde über das Niederländische Umweltmanagement-Gesetz und die Verordnung über die Prüfung der Umweltauswirkungen von 1994 umgesetzt.
Österreich (Wien)	Die SUP-Richtlinie auf kommunaler Ebene (Wiener Raumordnung) wurde durch das Wiener Stadtplanungs- und Baugesetzbuch umgesetzt.
Polen	Die Umsetzung der Plan-UP-Richtlinie erfolgte über das Gesetz über die Zugänglichkeit zu Umweltinformationen zum Umweltschutz, Öffentlichkeitsbeteiligung im Umweltschutz und die Strategische Umweltprüfung.
Portugal	Keine Aussage
Slowakei	Die Plan-UP-Richtlinie wurde durch ein eigenes Gesetz zur Prüfung der Umweltauswirkungen umgesetzt. Der Umweltprüfung unterliegen auch Flächennutzungsplanungsdokumente.
Spanien	Die Plan-UP-Richtlinie wurde durch Staatsgesetz vom 9. April 2006 umgesetzt. Im Bereich Bauen und Planen wurde die SUP dahin gehend umgesetzt, dass alle Verfahren zur Vorbereitung räumlicher und städtischer Planungsinstrumente einer SUP-Pflicht unterliegen. Die Verfahrensanforderungen der SUP wurden in die Verfahren zur Aufstellung der Entwicklungspläne integriert. Die Bebauungspläne fallen sowohl in den Anwendungsbereich der SUP als auch der UVP.

Ungarn	Die Umsetzung der Plan-UP-Richtlinie erfolgt durch Erlass.
Zypern	Die Umsetzung der Plan-UP-Richtlinie erfolgte durch das Gesetz N. 102(I)2005 aus dem Jahr 2005.

Eigene Darstellung

Frage 3c (Plan-UP-Richtlinie (2001/42/EG)

Werden für bestimmte Fälle, insbesondere der Geringfügigkeit im Rahmen der Aufstellung, Änderung und Ergänzung von Bauleitplänen, Möglichkeiten zur Beschleunigung sowie Verfahrenserleichterungen genutzt?

Tabelle 9 Zusammenfassung Frage 3c

Land	Antwort
Estland	Möglichkeiten zur Beschleunigung oder Verfahrenserleichterungen sind nicht vorgesehen.
Frankreich	Soweit lokale Stadtentwicklungspläne in einem Leitplan zur territorialen Kohärenz enthalten sind, unterliegen diese keiner SUP-Pflicht.
Großbritannien	Keine Aussage
Irland	Bei Änderungen oder Ergänzungen von Entwicklungsplänen erfolgt eine Vorprüfung (Screening) der Änderung/ Ergänzung zur Feststellung der SUP-Pflicht.
Lettland	Nicht in der derzeitigen Gesetzeslage. Geplant ist eine Vereinfachung bei Verknüpfung von Planung und Vorhabenplanung und bei der „Vorplanung“ für Ausweisung neuer Plangebiete.

Niederlande	Ein System aus Grenzwerten, unterhalb deren eine Prüfung nicht notwendig ist, wurde von der Europäischen Kommission gerügt und daher wieder geändert.
Österreich (Wien)	Keine Verfahrenserleichterungen oder Ausnahmen von der SUP-Pflicht.
Polen	Das Gesetz über die räumliche Planung und Entwicklung sieht keine Verfahrenserleichterungen vor.
Portugal	Keine Aussage
Slowakei	Keine Verfahrensvereinfachung, wenn das Dokument im Rahmen einer SUP geprüft wird.
Spanien	Verfahrensvereinfachungen wurden von einigen autonomen Kommunen für bestimmte Fälle vorgesehen, soweit nur geringfügige Änderungen eines Masterplans vorgenommen werden oder bei der Aufstellung von Bebauungsplänen für die Wiedernutzung von Gebieten, wenn zusätzlich eine höhere bauliche Dichte erreicht wird oder für die Innenentwicklung.
Ungarn	Bei geringfügigen Änderungen von Plänen besteht die Möglichkeit von Verfahrenserleichterungen unter Auslassung der SUP.
Zypern	Derzeit bestehen hierzu keine Erfahrungen in Zypern.

Eigene Darstellung

Frage 3d (Plan-UP-Richtlinie (2001/42/EG)

In welchem Verhältnis stehen Umweltprüfung und FFH-Verträglichkeitsprüfung zueinander?

Tabelle 10 Zusammenfassung Frage 3d

Land	Antwort
Estland	Keine Aussage
Frankreich	Keine Aussage
Großbritannien	Keine Aussage
Irland	Die SUP beinhaltet eine ggf. notwendige FFH-Verträglichkeitsprüfung, die einen eigenständig erkennbaren Teil der SUP.
Lettland	Soweit eine SUP durchgeführt wird, ist keine eigene FFH-Verträglichkeitsprüfung mehr notwendig. Die SUP beinhaltet die Erfassung der Auswirkungen auf Natura 2000-Gebiete.
Niederlande	Das Prüfspektrum der Umweltprüfung ist breiter als das der FFH-Verträglichkeitsprüfung.
Österreich (Wien)	Die FFH-Verträglichkeitsprüfung findet in Form der Umweltprüfung statt und stellt keine eigenständige Prüfung dar.
Polen	Soweit eine SUP durchgeführt wird, beinhaltet diese auch die Prüfung der Auswirkungen auf Natura 2000-Gebiete.
Portugal	Keine Aussage

Slowakei	Dies liegt in der Kompetenz des Umweltministeriums. Das Baugesetzbuch trifft hierzu keine Regelungen. Die Umweltprüfung behandelt bestehende Sachverhalte, die Verträglichkeitsprüfung trifft gibt Aussagen zur künftigen Entwicklung.
Spanien	FFH-Verträglichkeitsprüfung und SUP werden in die Planungs- und Genehmigungsverfahren integriert.
Ungarn	Keine Aussage
Zypern	Der Umweltbericht soll die FFH-Verträglichkeitsprüfung enthalten.

Eigene Darstellung

4.4 Dienstleistungsrichtlinie (2006/123/EG)

Wird ein Umsetzungsbedarf der Dienstleistungsrichtlinie (Richtlinie 2006/123/ EG) für den Bereich des nationalen Städtebaurechts und der Bauleitplanung gesehen?

Tabelle 11 Zusammenfassung Frage 4

Land	Antwort
Estland	Kein Umsetzungsbedarf
Frankreich	Nur, soweit in einem SCOT oder PLU einschränkende Kriterien für kommerzielle Niederlassungsmöglichkeiten über Festlegungen zur Bebauung und zur Umwelt bestimmt werden.
Großbritannien	Kein Umsetzungsbedarf im Bereich des Städtebaurechts
Irland	Keine Anwendung bei Landnutzungsplänen
Lettland	Keine Anwendung im Bereich des Planungsrechts
Niederlande	Nein, kein Umsetzungsbedarf unter Verweis auf Erwägungsgrund 9 der Richtlinie.
Österreich (Wien)	Kein Umsetzungsbedarf
Polen	Kein Umsetzungsbedarf im Planungsrecht
Portugal	Keine Aussage
Slowakei	Keine Aussage zum Umsetzungsbedarf
Spanien	Im Hinblick auf städtische Planungen, welche die planungsrechtliche Genehmigung von Projekten vorbereiten, besteht kein Umsetzungsbedarf.

Ungarn	Die Dienstleistungsrichtlinie wurde bereits in Verbindung mit der Bauleitplanung umgesetzt.
Zypern	Keine Erfassung der Planungs- oder Baugenehmigung durch die Dienstleistungsrichtlinie
Dänemark	Kein Umsetzungsbedarf (telefonisch mitgeteilt)

Eigene Darstellung

4.5 Erneuerbare-Energien-Richtlinie (2009/28/EG)

Wird ein Umsetzungsbedarf hinsichtlich des Artikels 13 Erneuerbare-Energien-Richtlinie (2009/28/EG) für den Bereich des nationalen Städtebaurechts und der Bauleitplanung gesehen?

Tabelle 12 Zusammenfassung Frage 5

Land	Antwort
Estland	Die Anforderungen des Artikels 13 werden in das Baurecht übernommen.
Frankreich	Kein Umsetzungsbedarf
Großbritannien	Derzeit wird ein nationaler Aktionsplan zur Umsetzung der Richtlinie aufgestellt.
Irland	Kein Umsetzungsbedarf
Lettland	Der Entwurf eines Gesetzes für Erneuerbare Energien wird derzeit vorbereitet. Ein Umsetzungsbedarf im Bereich des Planungsrechts besteht nicht.
Niederlande	Kein Umsetzungsbedarf, da die niederländische Bauordnung bereits adäquate Energie-Anforderungen für die Errichtung, den Umbau und die Renovierung von Gebäuden verlangt.
Österreich (Wien)	Kein Umsetzungsbedarf
Polen	Kein Umsetzungsbedarf. Zukünftig sollen auch bei Planungen auf örtlicher Ebene die Aspekte der Erneuerbaren Energien berücksichtigen. Dies erfolgt in Übereinstimmung mit der Erneuerbare-Energien-Richtlinie.

Portugal	Die Fragen bezüglich der Umsetzung zur Verwendung Erneuerbarer Energien in der Planung, Konstruktion und Erneuerung von städtischen und industriellen Gebieten sind noch nicht entschieden.
Slowakei	Derzeit erfolgt die Umsetzung der Erneuerbare-Energien-Richtlinie in nationales Recht. Aus Sicht der Slowakei besteht ein Umsetzungsbedarf für den Bereich des Planungsrechts.
Spanien	Die Erneuerbare-Energien-Richtlinie wurde nicht durch nationales Recht umgesetzt, dafür aber durch verschiedene autonome Kommunen.
Ungarn	Ein Umsetzungsbedarf wird gesehen. Die Richtlinie wurde aber noch nicht umgesetzt.
Zypern	Die Anforderungen der Erneuerbare-Energien-Richtlinie wurden in das Planungssystem über die Ergänzung verschiedener verbindlicher Raumentwicklungspläne umgesetzt. Die Umsetzung erfolgte über die Vereinfachung bestimmter Anforderungen betreffend den Standort und die Konstruktion der Anlagen für Erneuerbare Energien.

Eigene Darstellung

4.6 Seveso-II-Richtlinie (96/82/EG)

Wie ist Artikel 12 Absatz 1 Unterabsatz 2 der Richtlinie 96/82/EG vom 09.12.1996 zur Beherrschung der Gefahren bei Unfällen mit gefährlichen Stoffen (ABl. L 10 vom 14.01.1997, S. 13; „Seveso-II-Richtlinie") in nationales Recht umgesetzt worden?

Tabelle 13 Zusammenfassung Frage 6

Land	Antwort
Estland	Entsprechend den Anforderungen des Chemikalien- und Planungsrechts müssen lokale Behörden die Risiken schwerer Unfälle bei ihren Entscheidungen berücksichtigen. Die Überwachung obliegt sog. Rettungscentern und technischen Überwachungsbehörden.
Frankreich	Keine Aussage
Großbritannien	Keine Aussage
Irland	Keine Aussage
Lettland	Die Vorgaben der Seveso-II-Richtlinie werden bei kommunalen Planungen berücksichtigt.
Niederlande	Artikel 12 der Seveso-II-Richtlinie wurde in das nationale Recht über die Verordnung zur Anlagenumgebungssicherheit umgesetzt. Diese bestimmt die Umweltanforderungen, die für die Umgebung von Hoch-Risiko-Anlagen bei Planungen und Umweltgenehmigungen zu berücksichtigen sind. Die Verordnung gibt standortbestimmte Risiko-Grenzwerte für verwundbare Objekte (Wohngebiete, Schulen, Krankenhäuser etc.) vor. Hierbei wird zwischen Bestand und Neuplanung unterschieden.

	Im Rahmen der Bauleitplanung tragen die Gemeinden die Verantwortung, für eine adäquate Sicherheit in der Umgebung potenziell gefährlicher Einrichtungen in ihrem Gebiet Sorge zu tragen. Dies erfolgt insbesondere über die Bauleitplanung bei der Festlegung von Flächennutzungen, wobei die Verordnung mit den darin enthaltenen Risiko-Grenzwerten berücksichtigt werden muss. Den Risiko-Grenzwerten kann durch die Ausweisung von Sicherheitszonen um eine oder eine Gruppe von Einrichtungen mit hohem Sicherheitsrisiko, wie zum Beispiel einem Industriegebiet, Rechnung getragen werden. Ebenso kann im Rahmen der Bauleitplanung das Sicherheitsrisiko über die Einschränkung der Anlagentypen reduziert werden.
Österreich (Wien)	Keine Aussage
Polen	Keine Aussage
Spanien	Die Vermeidung von Gefahren schwerer Unfälle durch gefährliche Substanzen muss bei der räumlichen Planung durch die Kommunen berücksichtigt werden. Dies betrifft den Standort von neuen Gebäuden, die Änderung bestehender Gebäude und die Entwicklung von Wohngebieten und öffentlich genutzten Plätzen. Durch das königliche Dekret 1196 vom 19.9.2003 werden Methoden zur Ermittlung von Sicherheitsabständen vorgegeben.
Portugal	Keine Aussage
Slowakei	Im Rahmen der räumlichen Planung, der Genehmigung von Gebäuden, Einrichtungen und anderen Aktivitäten müssen die zuständigen Behörden die Anforderungen an bestimmte Abstände zwischen Anlagen und bestimmten Nutzungen (insbesondere Wohngebieten, dicht besiedelten Gebieten oder durch besondere Bestimmungen geschützte Gebiete) berücksichtigen.

	Anmerkung: Aus der Antwort kann nicht geschlossen werden, ob sich diese Anforderungen aus Bestimmungen des Fachgesetzes zur Umsetzung der Seveso-II-Richtlinie ergeben, oder direkt aus dem Planungsgesetz bzw. dem Baugesetz, dass ggf. im Rahmen der Umsetzung der Richtlinie geändert wurde.
Ungarn	Die Umsetzung erfolgte durch ein Regierungsdekret. Keine konkreten Aussagen zur Umsetzung
Zypern	Die Umsetzung erfolgt durch Änderungen des Planungsrechts. Hierbei sind Abstimmungen zwischen den Planungsbehörden und den für die Umsetzung der Seveso-II-Richtlinie zuständigen Behörden im Planungsprozess vorgesehen. Die Bestimmungen müssen bei der Aufstellung oder Prüfung verbindlicher Bauleitpläne berücksichtigt werden.

Eigene Darstellung

IV. Conference on Berlin

1. Concept

A conference was held in Berlin from 29 to 30 July 2010 to go into the results in depth. The conference was chaired by Professors Battis, Otto and Mitschang. In addition to the introductory addresses by the chairpersons, representatives from the Member States were asked to give a brief kick-off statement on a certain issue. This opened up the discussions to take a more in-depth look at the various topical areas generated by the written answers, which had been made available to the contacts in the Member States that had answered the survey in the run-up to the conference.

Table 14: Conference programme

1th Conference day 29.7.2010
Part I European Habitat and Species Protection
Habitat and Species Protection in German Planning Law Univ.-Prof. Dr. *Christian W. Otto* Berlin University of Technology Germany
The Exemption clause of the FFH Directive in the member states Mr. *Tom Simpson* Department for Communities and Local Government United Kingdom
Species Protection and Inner City Development Drs. *Hendrik van Sandick*, Ministry of Housing, Spatial Planning and the Environment (VROM) Netherlands

Part II SEA Directive
The Implementation of the SEA Directive in Germany Univ.-Prof. Dr. *Stephan Mitschang* Berlin University of Technology Germany
The Implementation of the SEA Directive in the member states Ms. *Ingūna Urtāne* Ministry of Regional Development and Local Governments Latvia
Simplification proceedings in the SEA / Experiences with the Implementation (Monitoring) Mr. *Tadhg O'Mahony*, Office of Environmental Assessment, Environmental Protection Agency, Ireland
Relation between SEA and Habitat Impact Assessment Ms. *Angela de la Cruz Mera*, Spanish Ministry of Housing Spain
2nd Conference day 30.7.2010
Part III Implementation of further EU Directives
The SEVESO-II-Directive and its Implementation in Germany Univ.-Prof. Dr. Dr. *Ulrich Battis* Humboldt University of Berlin Germany
The Renewable Energy Directive of the European Union Prof. Dr. hab. *Marek Dutkowski* University of Szczecin Poland

The Directive on Services in the Internal Market of the European Union Mr. *Phaedon Enotiades* Ministry of the Interior, Department of Town Planning and Housing Cyprus

Own visualisation

2. Results of the conference

2.1 European habitat and species protection law

2.1.1 Area protection

Area protection as stipulated in the FFH Directive is integrated in the environment and planning laws in all Member States. The area protection requirements are heeded in development planning and planning documents are revised accordingly. Given the clear distinction and thus also identification of the protected areas, there are no problems in applying area protection respectively in performing FFH compatibility assessments in the Member States. The impact of the protection regime stipulated in the FFH Directive with the designation of protected areas and compatibility assessments of regional planning has only started to take effect in the new Member States (Latvia, Poland). In the "older" Member States, corresponding application has meanwhile become extensively established in planning and approval procedures. Conflicts in regional planning can be avoided already in advance by heeding appropriate distances to the protected areas (United Kingdom). As a rule, the Member States integrate the FFH compatibility assessment in the Strategic Environmental Assessment (cf. IV 0.). By taking account of habitat protection in development planning, the subsequent approval procedures are therefore shaped by the previous planning decisions.

2.1.2 Using the exemption of Article 6 Paragraph 4 FFH Directive in habitat protection

The Member States use the exemption of Article 6 Paragraph 4 FFH Directive in the case of major infrastructure projects, such as port extension work (the Netherlands) and coastal defences (United Kingdom), but the exemption has no significance for town planning projects. The conference representatives of the Member States see it as a possibility for enforcing a project under certain circumstances in spite of a negative compatibility assessment. The new Member States (Latvia and Poland) have not yet gained any experience in using the exemption.

2.1.3 Species protection

All Member States have transposed species protection in their corresponding environment and planning law, but this issue is handled differently in terms of application in planning and approval procedures. In some cases it is already given due consideration in the planning procedure (United Kingdom, Spain and Cyprus) so that it is practically rendered insignificant in the approval procedure. But in other cases, species protection only really comes into effect in the approval procedure, playing a minor role in the planning procedure. Insofar as the planning system in the new Member States is not fully developed yet, species protection takes place on every planning and decision-making level, without any clear accentuation or classification being apparent. In Poland in fact it takes place on the project approval level, because the environment requirements resulting among others from species protection mean that compilation of development plans is avoided in planning practice as these are not always necessary for implementing a project.

The "advance impact" of species protection on development planning in terms of enforceability and therefore the need for planning pursuant to Section 1 Paragraph 3 German Building Code, is a specific aspect of application in Germany. The other Member States make no allowance for planning in the "exemption situation". The exemption ruling of Article 16 FFH Directive does not apply to the planning level in the Member States. The reason for this is that in most Member States, the planning process merely includes taking stock of the situation, without any specific view or assessment in terms of species protection.

In contrast to area protection, with clear identification of the protected areas thanks to the designation of the protected area network Natura 2000, the new Members States (Latvia and Poland) in particular see the problem that identifying the impact of projects on the protected species is more difficult than the FFH compatibility assessment. In the Netherlands, a national database is being set up to make it easier to apply species protection law, with the possibility of making specific enquiries about protected species.

2.1.4 Species protection in the inner and outer area

In contrast to Germany, the other Member States see no problem in applying the species protection stipulations of the FFH Directive in the inner and outer areas. There is a general acceptance that projects can be located either inside or also outside already developed areas. However, in contrast to Germany, no distinction is made regarding the legal regulations referring to planning in these areas. Accordingly, application of species protection in inner areas is not perceived to be a specific problem of inner development. However, it is agreed that species protection can in principle hinder inner-city development. Consequently, in practical terms approval procedures for inner areas take scarcely any account of species protection (Latvia, Poland and Spain). In the Netherlands, an integrated approval procedure has been introduced for projects, replacing the building permit by a conservation code of conduct in order to overcome species protection hurdles. Compliance with the stipulations in the code of conduct is monitored by the authorities.

The conference participants from the Member States agree that easement with regard to species protection should be possible for development projects within the cities. They therefore suggest a discussion regarding the value or priority of developing spaces in developed areas compared to the normally still natural spaces in virtually uninhabited areas or on the outskirts of estates. This also has to be seen in the context of introducing the Soil Protection Framework Directive with its intentions for greater re-use of space and for moving away from the designation of new building land. Implementation of this objective could increase the pressure to use free spaces in the city centre areas, which in turn could exacerbate the problem of handling species protection in inner areas.

2.2 SEA Directive (2001/42/EC)

2.2.1 Integration in planning

The SEA Directive of 2004 has been implemented in all Member States. The Strategic Environment Assessment is regularly integrated in development planning. In most member states, the environment assessment takes place parallel to drawing up a plan in the planning procedure. In Latvia, it is only the final plan that undergoes environment assessment. In Poland, as in Germany, the Strategic Environment Assessment replaces environment compatibility assessment on the planning level.

2.2.2 Easement and exemptions

Most Member States have exemptions from the environment assessment (but not Latvia). In this respect, the Spanish system is very similar to the German one. Basically an SEA is carried out on every level of development planning with the possibility of reverting to the results of the preceding assessments (classification). the Netherlands up to now had a system of limit values defining which projects of a certain size are subject to an SEA. However, following a reprimand from the European Commission, this system has to be changed.

2.2.3 Relationship between SEA and FFH compatibility assessment

For most Member States, the stipulation in the SEA Directive that a plan (or programme) potentially affecting a Natura 2000 area has to undergo an SEA constitutes the link to the FFH compatibility assessment.

Although the questions differ, the compatibility assessment can usually be integrated in the environment assessment (United Kingdom, the Netherlands, Poland, Spain, Cyprus). However, the level of integration varies between the Member States. The United Kingdom shows clear similarities to the German system of environment assessment procedures in overall regional planning. Here the environment assessment and FFH compatibility assessment are two assessments featured in separate chapters in the environment report. Application in Latvia forms an exception to the rule, where FFH compatibility assessment is carried out separately from the Strategic Environment Assessment.

In Spain, the species protection assessment is also integrated in the Strategic Environment Assessment in addition to the FHH compatibility assessment. This becomes part of the environment report.

2.2.4 Practical issues (monitoring)

As a rule, the monitoring of unexpected environment impacts in the Member States is carried out by the authority responsible for the plan itself and therefore on the local level. It is only in Poland that the responsible authorities are on the level of the Voivodships.

2.3 Implementation of various directives

2.3.1 SEVESO-II Directive (96/82/EC)

The requirements of the SEVESO-II Directive (96/82/EC) are implemented in the Member States. Discussions with the representatives of the Member States reveal that densely populated countries or regions in particular have conflicts with regard to the allocation of use, both from encroaching development and from the settlement of new uses in the surroundings of existing uses worthy of protection or existing industrial sites. In some cases, declining industrial production or the use of new technologies can also help to defuse conflicts (new Member States).

Application of the national regulations with which the SEVESO-II Directive has been transposed would appear to be extensively unproblematic. As far as regional planning is concerned, the focus is on the allocation of compatible uses respectively the observation of appropriate distances between incompatible uses, with regard to the risk of major accidents. In the Netherlands, the SEVESO-II directive is transposed into both environment and planning law. This system ensures that the requirements of the SEVESO II Directive are met by corresponding stipulations in environment law, if they cannot be fulfilled through regional allocation (planning). In the United Kingdom, compliance with the requirements of the SEVESO-II Directive is not part of planning law. In Latvia, Spain and Poland, conflicts are avoided on the planning level by observing appropriate distances between different uses.

2.3.2 Renewable Energies Directive (2009/28/EC)

Most Member States still have to implement the Renewable Energies Directive (2009/28/EC). However, the representatives of the Member States agree that Article 13 of the Directive does not result in any direct need for implementation affecting the areas of town planning and development planning.

Regional planning can make an indirect contribution to increasing the share of renewable energies and thus to achieving the objective of the Directive, by the way in which sites are designated. In the Polish view, renewable energies do not result in any need for action in terms of regional planning and particularly town planning. In Spain, a national plan of action has set the target of covering more than 20 % of energy consumption from renewable energies by 2020. The development of renewable energies and therefore also the demand for sites stipulated by the planning process depends to a great extent on the profit expectations of the private investors. The different possibilities for generating renewable energies are also reflected in the legislation of the autonomous regions. But this has no direct implications for town planning. In Cyprus, implementation of the Renewable Energies Directive has simplified the procedures involved in obtaining approval for renewable energy plants. Problems arise here particularly with regard to compatibility of tourism use with renewable energy sites, which change or impair the landscape. In the Netherlands they are sceptical about fulfilling the objective of the Directive to increase the share of renewable energy in overall energy consumption to 20 %. The physical impact of renewable energy plants such as wind parks is already evident, both inland and offshore. Latvia is currently reorganising its energy supply policy: the Ministry of Economics is responsible for fossil energy supply and the Ministry of the Environment for renewable energies. Regional planning is supposed to create the prerequisites for setting up plants for generating renewable energies. The United Kingdom is currently setting up a national energy plan. In terms of energy policy, the intention is to boost expertise in nuclear energy, with plans to build 10 new nuclear power stations on existing sites. Achieving the objective of the Directive with regard to the share of renewable energy in overall energy consumption is seen as problematical.

2.3.3 Services Directive (2006/123/EC)

The conference representatives of the Member States unanimously expect that the Services Directive (2006/123/EC) will not apply to the plans and programmes for regional planning and particularly for local development planning.

There may be an indirect influence on approval procedures (the Netherlands), with a possible claim to approval within a period of eight weeks, which makes it necessary to adapt the length of the administrative procedures.

3. Participants

Table 15: List of Participants

Land	Name	Institution
Germany	Prof. Dr. Dr. Ulrich Battis	Humboldt University of Berlin Berlin
Germany	RegDir. Dr. Rolf Blechschmidt	Federal Ministry of Transport Building and Urban Development Berlin
Spain	Ms. Angela de la Cruz Mera	Spanish Ministry of Housing Madrid
Poland	Prof. Dr. habil. Marek Dutkowski	University of Szczecin Szczecin
Cyprus	Mr. Phaedon Enotiades	Ministry of the Interior Department of Town Planning and Housing Nicosia
Germany	Prof. Dr. Stephan Mitschang	Berlin University of Technology Berlin

Germany	Prof. Dr. Christian-W. Otto	Berlin University of Technology Berlin
Germany	MR Dr. Jörg Schliepkorte	Federal Ministry of Transport, Building and Urban Development Berlin
Germany	Tim Schwarz	Berlin University of Technology Berlin
Great Britain	Mr. Tom Simpson	Department for Communities and Local Government Natural Environmental Assessment Planning - Environment Division London
Latvia	Ms. Ingūna Urtāne	Ministry of Regional Development and Local Governments Riga
Netherlands	Drs. Hendrik van Sandick	Ministry of Housing, Spatial Planning and the Environment The Hague

Own Visualisation

The conference was attended on the second day by Ministerialrat Dr. Stefan Lütkes from the Federal Ministry for Environment, Nature Conservation and Nuclear Safety (Department N I 6) as a guest.

IV. Konferenz in Berlin

1. Konzept

Zur Vertiefung der Ergebnisse der schriftlichen Umfrage erfolgte vom 29.-30.7.2010 eine Tagung in Berlin. Die Moderation der Tagung wurde von den Professoren Battis, Otto und Mitschang durchgeführt. Neben den Einführungsreferaten der Moderatoren wurden die Vertreter aus den Mitgliedstaaten gebeten, ein kurzes „Kick-Off-Statement" zu einer bestimmten Fragestellung abzugeben. Dies diente zum Einstieg in die Diskussionen, die zur Vertiefung der verschiedenen Themenkomplexe der schriftlichen Antworten geführt wurden. Im Vorfeld der Tagung wurden den Ansprechpartnern in den Mitgliedstaaten, die den Fragebogen beantwortet haben, die Antworten aus den anderen Mitgliedstaaten zur Verfügung gestellt.

Tabelle 14 Tagungsprogramm

1. Konferenztag 29.7.2010
Teil I Europäisches Habitat- und Artenschutzrecht
Habitatschutz und Artenschutzrecht im deutschen Bauplanungsrecht Univ.-Prof. Dr. *Christian W. Otto* Technische Universität Berlin Deutschland
Die Ausnahmeregelung der FFH-Verträglichkeitsprüfung in den Mitgliedstaaten Mr. *Tom Simpson* Department for Communities and Local Government United Kingdom
Artenschutz und Innenentwicklung Drs. *Hendrik van Sandick*, Ministry of Housing, Spatial Planning and the Environment (VROM) Netherlands

Teil II SUP-Richtlinie
Die Umsetzung der SUP-Richtlinie in Deutschland Univ.-Prof. Dr. *Stephan Mitschang* Technische Universität Berlin Deutschland
Die Umsetzung der SUP-Richtlinie in den Mitgliedstaaten Ms. *Ingūna Urtāne* Ministry of Regional Development and Local Governments Latvia
Erleichterungen bei der SUP/ Erfahrungen bei der Anwendung (Monitoring) Mr. *Tadhg O'Mahony*, Office of Environmental Assessment, Environmental Protection Agency, Ireland
Verhältnis von SUP und FFH-Verträglichkeitsprüfung Ms. *Angela de la Cruz Mera*, Spanish Ministry of Housing Spain
2. Konferenztag 30.7.2010
Teil III Umsetzung verschiedener EU-Richtlinien
Die SEVESO-II-Richtlinie und ihre Umsetzung in Deutschland Univ.-Prof. Dr. Dr. *Ulrich Battis* Humboldt Universität zu Berlin Deutschland
Die Erneuerbare-Energien-Richtlinie der Europäischen Union Prof. Dr. hab. *Marek Dutkowski* University of Szczecin Poland

Die Dienstleistungsrichtlinie der Europäischen Union Mr. *Phaedon Enotiades* Ministry of the Interior, Department of Town Planning and Housing Cyprus

Eigene Darstellung

2. Ergebnisse der Konferenz

2.1 Europäisches Habitat- und Artenschutzrecht

2.1.1 Gebietsschutz

Der Gebietsschutz, der sich aus den Erfordernissen der FFH-Richtlinie ergibt, ist in allen Mitgliedstaaten in die Umwelt- und Planungsgesetze integriert. Die Anforderungen des Gebietsschutzes werden in der Bauleitplanung beachtet und die Planungen entsprechend angepasst. Aufgrund der klaren Abgrenzung und damit auch der Identifizierbarkeit der Schutzgebiete ergeben sich keine Probleme in der Anwendung des Gebietsschutzes bzw. bei der Durchführung der FFH-Verträglichkeitsprüfung in den Mitgliedstaaten. Die Auswirkungen des Schutzregimes der FFH-Richtlinie durch die Ausweisung der Schutzgebiete und der Verträglichkeitsprüfung auf die räumliche Planung kommen in den neuen Mitgliedstaaten (Lettland, Polen) erst in den letzten Jahren voll zum Tragen. In den „älteren" Mitgliedstaaten hat sich die Anwendung in Planungs- und Genehmigungsverfahren weitgehend etabliert. Durch die Einhaltung räumlicher Abstände zu den Schutzgebieten können schon in der räumlichen Planung Konflikte vermieden werden (Großbritannien). Die FFH-Verträglichkeitsprüfung wird in der Regel in den Mitgliedstaaten in die Strategische Umweltprüfung integriert (vgl. IV 0.). Aufgrund der Berücksichtigung des Habitatschutzes in der Bauleitplanung sind die nachfolgenden Genehmigungsverfahren dementsprechend von den vorgelagerten Planungsentscheidungen geprägt.

2.1.2 Anwendung der Ausnahmeregelung des Art. 6 Abs. 4 FFH-Richtlinie beim Habitatschutz

Die Ausnahmeregelung des Art. 6 Abs. 4 FFH-Richtlinie wird in den Mitgliedstaaten bei größeren Infrastrukturprojekten, wie z. B. Hafenerweiterungen (Niederlande) und dem Küstenschutz (Großbritannien) angewendet, besitzt aber keine Bedeutung für städtebauliche Projekte. Sie wird von den teilnehmenden Vertretern der Mitgliedstaaten als Möglichkeit betrachtet, trotz einer negativen Verträglichkeitsprüfung, ein Vorhaben unter bestimmten Voraussetzungen doch realisieren zu können. In den neuen Mitgliedstaaten (Lettland und Polen) liegen noch keine Erfahrungen in der Anwendung der Ausnahmeregelung vor.

2.1.3 Artenschutz

Der Artenschutz ist in allen Mitgliedstaaten in den jeweiligen Umwelt- und Planungsgesetzen umgesetzt, wird aber in Bezug auf die Anwendung in Planungs- und Genehmigungsverfahren unterschiedlich gehandhabt. Teilweise wird dieser bereits im Planungsverfahren berücksichtigt (Großbritannien, Spanien und Zypern), sodass er im Genehmigungsverfahren praktisch keine Bedeutung mehr besitzt (Spanien). Teilweise kommt der Artenschutz aber auch erst im Genehmigungsverfahren voll zum Tragen und spielt dann im Planungsverfahren eine untergeordnete Rolle. Soweit in den neuen Mitgliedstaaten das Planungssystem noch nicht ausgereift ist, findet sich der Artenschutz auf jeder Planungs- und Entscheidungsebene, ohne dass eine klare Akzentuierung oder Abschichtung erkennbar ist. In Polen erfolgt faktisch eine Verlagerung auf die Ebene der Vorhabengenehmigung, da aufgrund der Umweltanforderungen, die sich unter anderem auch aufgrund des Artenschutzes ergeben, in der Planungspraxis die Aufstellung von Bebauungsplänen vermieden wird, da diese auch nicht immer für die Realisierung eines Vorhabens notwendig sind.

Die „Vorwirkung“ des Artenschutzes auf die Bauleitplanung, die sich im Hinblick auf die Vollziehbarkeit und damit die Erforderlichkeit der Planung nach § 1 Abs. 3 BauGB ergibt, stellt sich als Besonderheit der Anwendung in Deutschland dar. Eine Planung in die „Befreiungslage“ ist in den anderen Mitgliedstaaten nicht vorgesehen. Die Ausnahmeregelung des Art. 16 der FFH-Richtlinie findet in den Mitgliedstaaten auf der Ebene der Planung keine Anwendung. Der Grund hierfür ist darin zu sehen, dass im Rahmen der Planung in den meisten Mitgliedstaaten lediglich eine Bestandsaufnahme erfolgt, jedoch keine spezifische artenschutzbezogene Betrachtung oder Prüfung durchgeführt wird.

Im Gegensatz zum Gebietsschutz, der durch die Ausweisung des Schutzgebietsnetzes Natura 2000 eine klare Identifizierung der geschützten Gebiete ermöglicht, wird insbesondere bei den neuen Mitgliedstaaten (Lettland und Polen) die Problematik gesehen, dass sich die Identifizierung sowie der Einfluss von Vorhaben auf die geschützten Arten schwieriger gestaltet als bei der FFH-Verträglichkeitsprüfung. In den Niederlanden soll durch den Aufbau einer nationalen Datenbank die Anwendung des Artenschutzrechts erleichtert werden. Diese soll eine Abfrage von geschützten Arten ermöglichen.

2.1.4 Artenschutz im Innen- und Außenbereich

Das Problem der Anwendung der artenschutzrechtlichen Vorgaben der FFH-Richtlinie im Innen- und Außenbereich stellt sich in den Mitgliedstaaten anders als in Deutschland nicht. So besteht grundsätzlich ein Verständnis dafür, dass Vorhaben im bereits bebauten Bereich oder außerhalb des bebauten Bereichs verortet sein können. Es bestehen jedoch im Gegensatz zu Deutschland keine Unterschiede im Hinblick auf die rechtlichen Regelungen zur Planung in diesen Bereichen. Insofern wird die Anwendung des Artenschutzes im Innenbereich nicht spezifisch als Problem der Innenentwicklung wahrgenommen. Es besteht jedoch Einigkeit darüber, dass der Artenschutz die innerstädtische Entwicklung grundsätzlich hemmen kann. Als Reaktion darauf wird der Artenschutz bei Genehmigungen im Innenbereich faktisch kaum berücksichtigt (Lettland, Polen und Spanien). In den Niederlanden wurde im Rahmen der Einführung einer integrierten Genehmigung für Vorhaben, die Baugenehmigung durch einen naturschutzrechtlichen Verhaltenskodex ergänzt, um dadurch artenschutzrechtliche Hürden zu überwinden. Die Einhaltung der Vorgaben dieses Kodex wird von den Behörden überwacht.

Übereinstimmend sind die Teilnehmer aus den Mitgliedstaaten der Auffassung, dass für die Bebauung innerhalb der Städte Erleichterungen in Bezug auf den Artenschutz möglich sein sollen. Diesbezüglich wird von den Teilnehmern der Konferenz eine Diskussion angeregt zu der Frage der Wertigkeit bzw. Priorität der Entwicklung von Flächen im bebauten Bereich im Verhältnis zu den regelmäßig noch naturhaften Flächen im unbebauten Bereich außerhalb oder am Rande von Siedlungen. Diese muss auch vor dem Hintergrund der Einführung der Bodenschutz-Rahmenrichtlinie gesehen werden, die eine stärkere Wiedernutzung von Flächen und die Abkehr von neuen Baulandausweisungen intendiert. Wird diese Zielsetzung umgesetzt, könnte der Verwertungsdruck auf innerstädtische Freiflächen steigen, was wiederum das Problem des Umgangs mit dem Artenschutz im Innenbereich verschärfen kann.

2.2 Plan-UP-Richtlinie (2001/42/EG)

2.2.1 Integration in die Planung

Die Umsetzung der Plan-UP-Richtlinie aus dem Jahr 2004 ist in allen Mitgliedstaaten erfolgt. Die Strategische Umweltprüfung ist hierbei regelmäßig in die Bauleitplanung integriert. In den meisten Mitgliedstaaten erfolgt parallel zur Aufstellung eines Plans im Planverfahren die Umweltprüfung. Eine Besonderheit in Lettland ist, dass erst der abschließende Plan einer Umweltprüfung unterzogen wird. In Polen ersetzt – ähnlich wie in Deutschland – die Strategische Umweltprüfung die Umweltverträglichkeitsprüfung auf der Planungsebene.

2.2.2. Erleichterungen und Ausnahmen

Die Mitgliedstaaten kennen größtenteils Ausnahmen von der Umweltprüfung (nicht aber Lettland). Das spanische System ähnelt hierbei sehr stark dem Deutschen. Hier erfolgt grundsätzlich auf jeder Ebene der Bauleitplanung eine SUP, wobei auf die Ergebnisse der vorangegangenen Prüfungen zurückgegriffen werden kann (Abschichtung). In den Niederlanden bestand bislang ein System von Grenzwerten, die definieren, welche Projekte einer bestimmten Größe einer SUP zu unterziehen sind. Aufgrund einer Rüge durch die Europäische Kommission muss dieses System jedoch geändert werden.

2.2.3 Verhältnis von SUP und FFH-Verträglichkeitsprüfung

Die Vorgabe der Plan-UP-Richtlinie, dass ein Plan (oder Programm), der Auswirkungen auf ein Natura 2000-Gebiet haben kann, auch einer SUP zu unterziehen ist, stellt in den meisten Mitgliedstaaten das Bindeglied zur FFH-Verträglichkeitsprüfung dar.

Die Verträglichkeitsprüfung kann, trotz unterschiedlicher Fragestellungen, zumeist in die Umweltprüfung integriert werden (Großbritannien, Niederlande, Polen, Spanien, Zypern). Der Grad der Integration ist jedoch in den Mitgliedstaaten unterschiedlich ausgeprägt. Deutliche Ähnlichkeiten zum deutschen System der Umweltprüfverfahren in der räumlichen Gesamtplanung ergeben sich in Großbritannien. Hier bilden Umweltprüfung und FFH-Verträglichkeitsprüfung zwei Prüfungen, die im Umweltbericht in separaten Kapiteln dargestellt werden. Eine Ausnahme bildet die Anwendung in Lettland. Hier erfolgt die FFH-Verträglichkeitsprüfung unabhängig von der Strategischen Umweltprüfung.

In Spanien ist zusätzlich zur FFH-Verträglichkeitsprüfung auch die artenschutzrechtliche Prüfung in die Strategische Umweltprüfung integriert. Diese wird Teil des Umweltberichts.

2.2.4 Praxisfragen (Monitoring)

Das Monitoring zur Überwachung von unvorhergesehen Umweltauswirkungen erfolgt in den Mitgliedstaaten in der Regel durch die planaufstellende Behörde und damit auf kommunaler Ebene. Lediglich in Polen erfolgt dies durch die zuständigen Behörden auf der Ebene der Woiwodschaften.

2.3 Umsetzung verschiedener Richtlinien

2.3.1 SEVESO-II-Richtlinie (96/82/EG)

Die Anforderungen der SEVESO-II-Richtlinie (96/82/EG) sind in den Mitgliedstaaten umgesetzt. In der Diskussion mit den Vertretern der Mitgliedstaaten wird deutlich, dass sich gerade in dicht besiedelten Ländern oder Regionen Konflikte im Hinblick auf die Zuordnung von Nutzungen, sowohl durch heranrückende Bebauung als auch durch die Ansiedlung von neuen Nutzungen in der Umgebung schutzwürdiger Nutzungen oder bestehender Industrieanlagen ergeben. Zum Teil kommt es durch rückläufige Industrieproduktion oder in Folge der Verwendung neuer Technologien aber auch zur Entschärfung von Konflikten (neue Mitgliedstaaten).

Die Anwendung der nationalen Regelungen, mit denen die SEVESO-II-Richtlinie umgesetzt wurden, erscheinen weitgehend unproblematisch. In Bezug auf die räumliche Planung steht hierbei die Zuordnung verträglicher Nutzungen bzw. die Einhaltung von Abständen zwischen unverträglichen Nutzungen im Hinblick auf das Risiko schwerer Unfälle im Vordergrund. In den Niederlanden erfolgt die Umsetzung der SEVESO-II-Richtlinie sowohl im Umwelt- als auch im Planungsrecht. Durch dieses System wird sichergestellt, dass die Anforderungen der SEVESO-II-Richtlinie, wenn sie nicht durch die räumliche Zuordnung (Planung) erfüllt werden können, durch entsprechende Vorgaben aus dem Umweltrecht eingehalten werden. In Großbritannien erfolgt die Einhaltung der Anforderungen der SEVESO-II-Richtlinie nicht über das Planungsrecht. In Lettland, Spanien und Polen werden Konflikte durch die Einhaltung von Abständen zwischen verschiedenen Nutzungen auf der Planungsebene vermieden.

2.3.2 Erneuerbare Energien Richtlinie (2009/28/EG)

Die Umsetzung der Erneuerbare-Energien-Richtlinie (2009/28/EG) steht in den meisten Mitgliedstaaten bislang noch aus. Es besteht jedoch Einvernehmen bei den Vertretern der Mitgliedstaaten, dass sich aus Artikel 13 der Richtlinie kein direkter Umsetzungsbedarf für den Bereich des Städtebaus und der Bauleitplanung ergibt.

Indirekt kann die räumliche Planung über Standortausweisungen einen Beitrag zur Erhöhung des Anteils Erneuerbarer Energien und damit der Zielerreichung der Richtlinie leisten. Aus polnischer Sicht entsteht durch die Erneuerbaren Energien kein Handlungsfeld für die Raumplanung und speziell die städtebauliche Planung. In Spanien sieht ein nationaler Aktionsplan das Ziel vor, bis 2020 über 20 % des Energieverbrauchs aus Erneuerbaren Energien zu decken. Die Entwicklung der Erneuerbaren Energien und damit verbunden die Nachfrage nach Standorten, die durch die Planung festgelegt werden, sind im Wesentlichen abhängig von den Gewinnerwartungen der privaten Investoren. Die unterschiedlichen Möglichkeiten zur

Erzeugung Erneuerbarer Energien schlagen sich auch in der Gesetzgebung der autonomen Regionen nieder. Hieraus ergeben sich jedoch keine unmittelbaren Implikationen für die städtebauliche Planung. In Zypern werden in Folge der Umsetzung der Erneuerbare-Energien-Richtline die Verfahren zur Genehmigung von Anlagen für Erneuerbare Energien vereinfacht. Probleme ergeben sich hier insbesondere im Hinblick auf die Vereinbarkeit touristischer Nutzungen mit den Standorten für Erneuerbare Energien, die zu einer Veränderung bzw. Beeinträchtigung des Landschaftsbildes führen. In den Niederlanden wird die Erreichung des Ziels der Richtlinie, den Anteil Erneuerbarer Energien am Gesamtenergieverbrauch auf 20 % zu steigern, skeptisch gesehen. Die räumlichen Auswirkungen der Anlagen für Erneuerbare Energien, wie zum Beispiel Windparks sind sowohl bei inländischen Standorten als auch bei Offshore-Standorten evident. In Lettland erfolgt derzeit eine Restrukturierung der Energieversorgungspolitik, bei der das Wirtschaftsministerium für die fossile Energieversorgung und das Umweltministerium für die Erneuerbaren Energien zuständig ist. Die räumliche Planung soll die Voraussetzungen für die Ansiedlung von Anlagen für die Erzeugung Erneuerbarer Energien schaffen. In England wird derzeit ein nationaler Energieplan aufgestellt. Energiepolitisch ist eine Steigerung der Kompetenz im Bereich der Kernenergie vorgesehen. Hierfür ist der Bau von 10 neuen Atomkraftwerken an bestehenden Standorten geplant. Die Erreichung des Ziels der Richtlinie im Hinblick auf den Anteil Erneuerbarer Energien am Gesamtenergieverbrauch wird problematisch eingeschätzt.

2.3.3 Dienstleistungsrichtlinie (2006/123/EG)

Die anwesenden Vertreter der Mitgliedstaaten gehen einstimmig davon aus, dass die Dienstleistungsrichtlinie (2006/123/EG) keine Anwendung auf die Pläne und Programme der Raumplanung und speziell der örtlichen Bauleitplanung findet.

Ein indirekter Einfluss wird auf Genehmigungsverfahren gesehen (Niederlande). Gegebenenfalls besteht innerhalb von einer Frist von acht Wochen ein Anspruch auf eine Genehmigung, wodurch eine Anpassung der Dauer von Verwaltungsverfahren notwendig erscheint.

3. Konferenzteilnehmer

Tabelle 15 Teilnehmerliste der Konferenz in Berlin

Land	Name	Institution
Deutschland	Prof. Dr. Dr. Ulrich Battis	Humboldt Universität zu Berlin Berlin
Deutschland	RegDir. Dr. Rolf Blechschmidt	Bundesministerium für Verkehr, Bau und Stadtentwicklung Berlin
Spanien	Ms. Angela de la Cruz Mera	Spanish Ministry of Housing Madrid
Polen	Prof. Dr. habil. Marek Dutkowski	University of Szczecin Szczecin
Zypern	Mr. Phaedon Enotiades	Ministry of the Interior Department of Town Planning and Housing Nicosia
Deutschland	Prof. Dr. Stephan Mitschang	Technische Universität Berlin Berlin
Deutschland	Prof. Dr. Christian-W. Otto	Technische Universität Berlin Berlin
Deutschland	MR Dr. Jörg Schliepkorte	Bundesministerium für Verkehr, Bau und Stadtentwicklung Berlin
Deutschland	Tim Schwarz	Technische Universität Berlin Berlin

Großbritannien	Mr. Tom Simpson	Department for Communities and Local Government Natural Environmental Assessment Planning - Environment Division London
Lettland	Ms. Ingūna Urtāne	Ministry of Regional Development and Local Governments Riga
Niederlande	Drs. Hendrik van Sandick	Ministry of Housing, Spatial Planning and the Environment The Hague

Eigene Darstellung

Als Gast zugegen war am zweiten Tag der Konferenz Herr Ministerialrat Dr. Stefan Lütkes vom Bundesministerium für Umwelt, Naturschutz und Reaktorsicherheit (Referat N I 6).

V. Summary of the results

1. European habitat and species protection law

The role played by species protection in development planning differs in the various Member States. Most of the Member States participating in the survey have to take account of species protection aspects already during the planning stage[21]. Insofar as the planning system is still not yet fully developed in the new Member States, here species protection takes place on every planning and decision-making level, without any clearly obvious accentuation or classification[22]. It is only in the Netherlands that species protection is not considered until the approval procedure, thus playing a minor role in the planning procedure. A separate species protection assessment is carried out in the United Kingdom, Latvia, Austria and the Netherlands. To a certain extent, species protection is also covered by other environment assessment procedures, for example during the Strategic Environment Assessment[23] or FFH compatibility assessment[24].

Area protection is part of local planning in all Member States, where area protection requirements are heeded and planning adapted accordingly. The approval procedure is shaped correspondingly by the preceding planning decisions. If not preceded by planning, the approval procedure takes account of the requirements stipulated in the FFH Directive. All Member States used the exemption ruling of Article 6 Paragraph 4 FFH Directive.

The differentiation between inner area and outer area is less common in the Member States than in Germany. The written answer from the Netherlands and the conference participants also drew attention to the fact that species protection can impede inner-city development. Consequently, in some Member States species protection is scarcely given any consideration in approval procedures for already developed areas[25].

The conference participants in Berlin agreed that easement with regard to species protection should be possible for developments within the cities.

21 Cf. written answers from France, Ireland, Latvia, Austria, Spain, Hungary and Cyprus.
22 Cf. written answers from Estonia, Portugal and Slovakia, as well as Poland (conference).
23 Ireland and Spain.
24 Estonia and Portugal.
25 Cf. Poland and Spain (conference).

2. SEA Directive (2001/42/EC)

The Strategic Environment Assessment has been integrated in development planning in the Member States. Various Member States use exemptions from the SEA to simplify the planning procedures[26]. There is no obvious system of exemptions, even after reviewing the written answers. Some countries operate with exemption catalogues (France), limit values (the Netherlands, where the criteria are under review) or preliminary assessments (Ireland). The Spanish system is very similar to the German one, with the possibility of using the results of previous environment assessments (classification). The New Member States have not developed any exemption rulings yet[27]. The compatibility assessment is mostly integrated in the Strategic Environment Assessment, although the questions differ[28].

3. EC Services Directive (2006/123/EC)

The conference representatives of the Member States unanimously expect that the EC-Services Directive (2006/123/EC) will not apply to the plans and programmes for regional planning and particularly for local development planning, an opinion supported by the written answers given to the survey in the Member States[29].

4. Renewable Energies Directive (2009/28/EC)

Most Member States still have to implement the Renewable Energies Directive (2009/28/EC). In principle, regional planning can help to increase the share of renewable energies and thus to achieve the objective of the Directive, by the way in which sites are designated. However, the representatives of the Member States agree that Article 13 of the Directive does not result in any direct need for implementation affecting the areas of town planning and development planning. This is also confirmed by the written answers from the other Member States[30].

26 Cf. written answers from France, Ireland, the Netherlands, Spain and Hungary.

27 Cf. written answers from Estonia, Latvia (where exemptions are planned), Poland, Slovakia and Cyprus.

28 Cf. written answers from Ireland, Latvia, Austria, Poland and the results of the conference.

29 Cf. written answers from Ireland and Austria (Vienna).

30 Cf. written answers from France, Ireland, Austria/ Vienna, Portugal and Slovakia

5. SEVESO-II Directive (96/82/EC)

The requirements of the SEVESO-II Directive (96/82/EC) are implemented in the Member States. Discussions with the representatives of the Member States reveal that densely populated countries or regions in particular have conflicts with regard to the allocation of use, both from encroaching development and from the settlement of new uses in the surroundings of existing uses worthy of protection or existing industrial sites. However, implementation of the national

Application of the national regulations with which the SEVESO-II Directive has been implemented would appear to be extensively unproblematic.

V. Zusammenfassung der Ergebnisse

1. Europäisches Habitat- und Artenschutzrecht

Die Ausgestaltung des Artenschutzes in der Bauleitplanung der Mitgliedstaaten ist unterschiedlich ausgeprägt. In den meisten der befragten Mitgliedstaaten müssen bereits bei der Planung die Aspekte des Artenschutzes berücksichtigt werden[31]. Soweit in den neueren Mitgliedstaaten das Planungssystem noch nicht ausgereift ist, findet sich der Artenschutz auf jeder Planungs- und Entscheidungsebene, ohne dass eine klare Akzentuierung oder Abschichtung erkennbar ist[32]. Nur in den Niederlanden wird der Artenschutz erst im Genehmigungsverfahren berücksichtigt und spielt infolgedessen im Planungsverfahren eine untergeordnete Rolle. Eine eigenständige artenschutzrechtliche Prüfung erfolgt in Großbritannien, Lettland, Österreich und den Niederlanden. Zum Teil wird der Artenschutz auch im Rahmen anderer Umweltprüfverfahren behandelt, wie zum Beispiel der Umweltprüfung[33] oder der FFH-Verträglichkeitsprüfung[34].

Der Gebietsschutz ist in allen Mitgliedstaaten Bestandteil der örtlichen Planung. Die Anforderungen des Gebietsschutzes werden dort beachtet, die Planung entsprechend angepasst. Das Genehmigungsverfahren ist dementsprechend von den vorgelagerten Planungsentscheidungen geprägt. Sofern eine Planung nicht zuvor erfolgt ist, werden die Anforderungen der FFH-RL im Genehmigungsverfahren beachtet. Die Ausnahmeregelung des Art. 6 Abs. 4 FFH-Richtlinie wurde in allen Mitgliedstaaten umgesetzt.

Die Unterscheidung zwischen Innenbereich und Außenbereich ist in den Mitgliedstaaten nicht so geläufig wie in der Bundesrepublik. In der schriftlichen Antwort aus den Niederlanden sowie durch die Teilnehmer der Konferenz wurde indes darauf hingewiesen, dass der Artenschutz die innerstädtische Entwicklung hemmen kann. Als Reaktion darauf wird der Artenschutz bei Genehmigungen in bereits bebauten Gebieten in einigen Staaten kaum berücksichtigt[35].

Übereinstimmend waren die Teilnehmer der Konferenz in Berlin der Auffassung, dass für die Bebauung innerhalb der Städte Erleichterungen in Bezug auf den Artenschutz möglich sein sollten.

---

31 Vgl. schriftliche Antworten aus Frankreich, Irland, Lettland, Österreich, Spanien, Ungarn und Zypern.

32 Vgl. schriftliche Antworten aus Estland, Portugal und Slowakei sowie Polen (Konferenz).

33 Irland und Spanien.

34 Estland und Portugal.

35 Vgl. Polen und Spanien (Konferenz).

2. Plan-UP-Richtlinie (2001/42/EG)

In den Mitgliedstaaten ist die Strategische Umweltprüfung in die Bauleitplanung integriert worden. Verschiedene Mitgliedstaaten kennen Ausnahmen von der Umweltprüfung zur Vereinfachung der Planverfahren[36]. Ein einheitliches System von Ausnahmen ist auch unter Berücksichtigung der schriftlichen Stellungnahmen nicht erkennbar. Teilweise wird mit Ausnahmekatalogen (Frankreich), Grenzwerten (Niederlande – dort steht die Überarbeitung der Kriterien an) oder mit Vorprüfungen (Irland) operiert. Das spanische System ähnelt sehr stark dem Deutschen, in dem bei Vorliegen bereits erfolgter Umweltprüfungen auf die Ergebnisse dieser Prüfungen zurückgegriffen werden kann (Abschichtung). In den neuen Mitgliedstaaten wurden noch keine Ausnahmeregelungen entwickelt[37]. Die Verträglichkeitsprüfung ist, trotz unterschiedlicher Fragestellungen, zumeist in die (Strategische) Umweltprüfung integriert[38].

3. EG-Dienstleistungsrichtlinie (2006/123/EG)

Die auf der Konferenz anwesenden Vertreter der Mitgliedstaaten gingen einstimmig davon aus, dass die EG-Dienstleistungsrichtlinie (2006/123/EG) keine Anwendung auf die Pläne und Programme der Raumplanung und speziell der örtlichen Bauleitplanung findet. Unterstützt wird diese Auffassung auch durch die schriftlichen Antworten der Befragung in den Mitgliedstaaten[39].

4. Erneuerbare-Energien-Richtlinie (2009/28/EG)

Die Umsetzung der Erneuerbare-Energien-Richtlinie (2009/28/EG) steht in den meisten Mitgliedstaaten bislang noch aus. Grundsätzlich kann die räumliche Planung über Standortausweisungen einen Beitrag zur Erhöhung des Anteils Erneuerbarer Energien und damit der Zielerreichung der Richtlinie leisten. Es bestand jedoch Einvernehmen bei den Vertretern der Mitgliedstaaten, dass sich aus Artikel 13 der Richtlinie kein direkter Umsetzungsbedarf für den Bereich des

---

36 Vgl. schriftliche Antworten aus Frankreich, Irland, Niederlande, Spanien und Ungarn.

37 Vgl. schriftliche Antworten aus Estland, Lettland (Ausnahmeregelung geplant), Polen, Slowakei und Zypern.

38 Vgl. schriftliche Antworten aus Irland, Lettland, Österreich, Polen sowie die Ergebnisse der Konferenz.

39 Vgl. schriftliche Antworten aus Irland und Österreich (Wien).

Städtebaus und der Bauleitplanung ergibt. Auch die schriftlich vorliegenden Antworten der anderen Mitgliedstaaten bestätigen dies[40].

5. SEVESO-II-Richtlinie (96/82/EG)

Die Anforderungen der SEVESO-II-Richtlinie (96/82/EG) sind in den Mitgliedstaaten umgesetzt. Insbesondere in der Diskussion mit den Vertretern der Mitgliedstaaten im Rahmen der Konferenz wurde deutlich, dass sich gerade in dicht besiedelten Ländern oder Regionen Konflikte im Hinblick auf die Zuordnung von Nutzungen sowohl durch heranrückende Bebauung als auch durch die Ansiedlung von neuen Nutzungen in der Umgebung schutzwürdiger Nutzungen oder bestehender Industrieanlagen ergeben. Die Anwendung der nationalen Regelungen, mit denen die SEVESO-II-Richtlinie umgesetzt wurde, erscheint jedoch weitgehend unproblematisch.

40 Vgl. schriftliche Antworten aus Frankreich, Irland, Österreich/ Wien, Portugal und Slowakei.

Berliner Schriften zur Stadt- und Regionalplanung

Herausgegeben von Prof. Dr. Stephan Mitschang

Band 1 Stephan Mitschang (Hrsg.): Umweltprüfverfahren in der Stadt- und Regionalplanung. 2006.

Band 2 Stephan Mitschang (Hrsg.): Stadt- und Regionalplanung vor neuen Herausforderungen. 2007.

Band 3 Stephan Mitschang (Hrsg.): Flächennutzungsplanung – Aufgabenwandel und Perspektiven. 2007.

Band 4 Stephan Mitschang (Hrsg.): BauGB-Novelle 2007. Neue Anforderungen an städtebauliche Planungen und die Zulassung von Vorhaben. 2008.

Band 5 Stephan Mitschang (ed. / Hrsg.): Soil Protection Law in the EU. Bodenschutzrecht in der EU. 2008.

Band 6 Stephan Mitschang (Hrsg.): Innenentwicklung – Fach- und Rechtsfragen. 2008.

Band 7 Stephan Mitschang (Hrsg.): Klimaschutz und Energieeinsparung in der Stadt- und Regionalplanung. 2009.

Band 8 Stephan Mitschang (Hrsg.): Fach- und Rechtsprobleme der Baunutzungsverordnung. 2009.

Band 9 Stephan Mitschang (Hrsg.): Aktuelle Fach- und Rechtsfragen des Lärmschutzes. Bauleitplanung, Fachplanung und Zulassung von Bauvorhaben. 2010.

Band 10 Stephan Mitschang / Gerd Schmidt-Eichstaedt (Hrsg.): Die Umweltprüfung in der Regionalplanung. 2010.

Band 11 Ulrich Battis/Jens Kersten / Stephan Mitschang: Rechtsfragen der ökologischen Stadterneuerung. 2010.

Band 12 Stephan Mitschang (ed. / Hrsg.): Energy Efficiency and Renewable Energies in Town Planning Law. Energieeffizienz und Erneuerbare Energien im Städtebaurecht. 2010.

Band 13 Stephan Mitschang (Hrsg.): Planen und Bauen im Außenbereich. 2010.

Band 14 Stephan Mitschang (Hrsg.): Aktuelle Fragestellungen des Städtebau- und Umweltrechts – Ansatzpunkte für eine BauGB- und BauNVO-Novelle. 2011.

Band 15 Tim Schwarz: Die Umweltprüfung in gestuften Planungsverfahren. Möglichkeiten und Grenzen der Koordination und Abschichtung im Rahmen der Umweltprüfung in der Raumordnung und der Bauleitplanung. 2011.

Band 16 Stephan Mitschang (ed. / Hrsg.): Urban Planning Law under EU-Influence. Städtebaurecht unter EU-Einfluss. 2011.

Venture Capitalist Participation and the Performance of IPO Firms: Empirical Evidence from France, Germany, and the UK

Inauguraldissertation
zur Erlangung des akademischen Grades eines
Doktors der Wirtschaftswissenschaft durch die
Wirtschaftswissenschaftliche Fakultät
der Westfälischen Wilhelms-Universität Münster

vorgelegt von
Georg Rindermann
aus **Frankfurt am Main**

2003

Erster Berichterstatter: Prof. Christian Harm, Ph.D.

Zweiter Berichterstatter: Prof. Dr. Klaus Röder

Dekan: Prof. Dr. Karl-Hans Hartwig

Tag der mündlichen Prüfung: 23. Juli 2003

Venture Capitalist Participation and the Performance of IPO Firms

European University Studies

Europäische Hochschulschriften
Publications Universitaires Européennes

Series V
Economics and Management

Reihe V Série V
Volks- und Betriebswirtschaft
Sciences économiques, gestion d'entreprise

Vol./Bd. 3029

PETER LANG
Frankfurt am Main · Berlin · Bern · Bruxelles · New York · Oxford · Wien

Georg Rindermann

Venture Capitalist Participation and the Performance of IPO Firms

Empirical Evidence from France, Germany, and the UK

PETER LANG
Europäischer Verlag der Wissenschaften

Bibliographic Information published by Die Deutsche Bibliothek
Die Deutsche Bibliothek lists this publication in the Deutsche Nationalbibliografie; detailed bibliographic data is available in the internet at <http://dnb.ddb.de>.

Zugl.: Münster (Westfalen), Univ., Diss., 2003

D 6
ISSN 0531-7339
ISBN 3-631-52076-X
US-ISBN 0-8204-6563-1

www.peterlang.de

Acknowledgements

The completion of this dissertation would not have been possible without the generous assistance of numerous persons. Therefore, I would like to express my gratefulness to everyone who contributed to the successful achievement of this thesis.

First, I would like to thank Professor Christian Harm for accepting me as a doctoral student and supervising my thesis. Likewise, I would like to thank Professor Klaus Röder for his advice and his role as the co-supervisor. I am especially grateful to Professor Douglas Cumming for his hospitality at the School of Business of the University of Alberta. He gave me precious feedback on my research during my stay as a visiting doctoral student in Canada.

Second, I owe a considerable debt to the staff of the chair of International Business at the University of Münster, Germany, for their excellent research assistance. In particular, I thank Christian Bender, Carsten Hahn, Doris Hohmeier, Nicole Neunhöffer, Michael Kruse, and Torsten Engel for their proofreading and valuable comments. I also owe many thanks to Armin Schwienbacher, Abolhassan Jalilvand, Carsten Sørensen, Richard Sweeney, and the participants in the Doctoral Student Seminar of the 2002 FMA European Conference in Copenhagen, Denmark. Likewise, I thank my discussant Allan Zebedee for his comments at the 2003 Annual Meeting of the European Financial Management Association in Helsinki, Finland.

Third, I cannot sufficiently acknowledge Martin Bisicky from Credit Suisse First Boston (CSFB) in London, Anne-Lucie Verdem from the Commission des Opérations de Bourse (COB) in Paris, France, as well as Stefanie Franzke from the Center for Financial Studies and Christa Oberländer from the Deutsche Börse AG, both in Frankfurt, Germany, for providing IPO prospectuses. I am also indebted to Andreas Kümmert from DeMinds, Munich, Germany, for his comments related to the classification methodology of venture capitalists. In addition, I thank Jenny Kunz from Thomson Financial in Frankfurt, Germany, and Jennifer Vandermosten from the European Venture Capital Association (EVCA) in Brussels, Belgium, for providing data on European IPOs and venture capital, respectively.

Finally, I would like to thank my parents, my twin brothers Andreas and Matthias, and my girlfriend Ingrid for their enduring patience and strong encouragement during the completion of my dissertation.

Contents

List of Tables

List of Figures

List of Abbreviations

3i	Investors in Industry
AFIC	Association Française des Investisseurs en Capital (French Venture Capital Association)
AIM	Alternative Investment Market
ANVAR	Agence Nationale pour la Valorisation de la Recherche
BAN	Business Angel Network
BDPME	Banque de Développement pour les PME
BES	Business Expansion Scheme
BJTU	Beteiligungskapital für junge Technologieunternehmen
BNP	Banque Nationale de Paris
BTU	Beteiligungskapital für kleine Technologieunternehmen
BVCA	British Venture Capital Association
BVK	Bundesverband Deutscher Kapitalbeteiligungsgesellschaften (German Venture Capital Association)
CDC	Caisse des Dépôts et Consignations
CEO	Chief Executive Officer
CEPME	Crédit d'Equipement des Petites et Moyennes Entreprises
CEPR	Centre for Economic Policy Research
CFROA	cash flow return on assets
CFROS	cash flow return on sales
COB	Commission des Opérations de Bourse
CSFB	Credit Suisse First Boston
DM	German Mark
DtA	Deutsche Ausgleichsbank
DTI	Department of Trade and Industry
Easdaq	European Association of Securities Dealers Automated Quotation System
ECU	European Currency Unit
ed.	editor
eds.	editors
e.g.	for example, for instance (Latin *exempli gratia*)
EIB	European Investment Bank
EIF	European Investment Fund
EIS	Enterprise Investment Scheme
EMI	Enterprise Management Incentives
Empl	employees
EqR	equity ratio
ERP	European Recovery Program
et al.	and others (Latin *et alii*)
EU	European Union
EUR	Euro
EVCA	European Venture Capital Association
FCPR	Fonds Communs de Placement à Risque
FCPI	Fonds Communs de Placement Innovation
FF	French Franc
FPCR	Fonds Public pour le Capital Risque
GBP	British Pound
GDP	Gross Domestic Product

GM	growth market
HGB	Handelsgesetzbuch (German Accounting Standards)
HMSO	Her Majesty's Stationary Office
IAS	International Accounting Standards
ibid.	ibidem
ICFC	Industrial and Commercial Finance Corporation
IDI	Institut de Développement Industriel
i.e.	that is to say, in other words (Latin *id est*)
IFR	International Financial Review
IMF	International Monetary Fund
incl.	including
int.	international
IPO	Initial Public Offering
IRP	Instituts Régionaux de Participation
IT	Information Technology
Jasdaq	Japanese Association of Securities Dealers Automated Quotation System
JFI	Journal of Financial Intermediation
KBG	Kapitalbeteiligungsgesellschaft
KfW	Kreditanstalt für Wiederaufbau
L	likelihood
LBO	leveraged buy-out
LR	likelihood ratio
LSE	London Stock Exchange
m	million
MBG	Mittelständische Beteiligungsgesellschaft
MBO	management buy-out
MITI	Ministry of International Trade and Industry
mkt. cap.	market capitalization
MoF	Ministry of Finance
n.a.	not available
Nasdaq	National Association of Securities Dealers Automated Quotation System
NL	Netherlands
No.	number
NVC	non venture capital
NVCA	National Venture Capital Association
OECD	Organization for Economic Cooperation and Development
OLS	ordinary least squares
OTC	over the counter
Part	participation ratio
p.	page
pp.	pages
Q	Tobin's Q
R&D	research and development
ROA	return on assets
ROS	return on sales
SBIC	Small Business Investment Company
SCR	Société de Capital Risque
SDR	Société de Développement Régional
SFI	Société Financière d'Innovation
SFLGS	Small Firm Loan Guarantee Scheme

SMART	Small Firms Merit Award for Research and Technology
SME	small and medium sized enterprise
SOFARIS	Société Française pour l'Assurance du Capital Risque
S&P	Standard and Poor's
SPUR	Support for Products under Research
TBG	Technologie-Beteiligungs-Gesellschaft
UBGG	Gesetz für Unternehmensbeteiligungsgesellschaften
UK	United Kingdom
UK GAAP	Generally Accepted Accounting Principles in the United Kingdom
UKLA	UK Listing Authority
URank	underwriter rank
US	United States
USD	US Dollar
US GAAP	Generally Accepted Accounting Principles in the United States
USM	Unlisted Security Market
VC	Venture Capital
VCT	Venture Capital Trust
VIF	variance inflation factor
Vol.	volume
vs.	versus
w.a.	without author
WFG	Deutsche Wagnisfinanzierungsgesellschaft
WR	wealth relative

"Innovations fail to create value when they cannot attract resources required to develop them."[1]

"Indeed the central question running through this whole topic is why does Europe [...] have no Microsoft or Apple? Frequently, the blame for this is laid at the door of financial institutions which view such investments as too risky."[2]

1 Introduction

1.1 Background and Motivation

The contribution of young and fast-growing companies to innovation and economic development has become a topic of interest among academics, practitioners, and politicians. Particularly in the 1990s, the creation of businesses acted as an engine for economic growth in most industrialized countries and therefore garnered considerable attention. Yet, entrepreneurs with innovative ideas of potentially high economic value often find it difficult to finance projects through traditional channels like personal savings and bank loans.[3] This shortcoming is due to several critical factors, such as high uncertainty, asymmetric information, the nature of assets, and the conditions in the relevant financial markets.

In order to address these problems, the institutional form of venture capital[4] has evolved. Venture capitalists are financial intermediaries who specialize in providing capital and management assistance to young and technology-based firms.[5] They are referred to as experts in the field of high-risk company funding and play a critical role as information producers who can assess small business quality.[6] Thus, they are able to cope with the problems related to asymmetric information. Venture capitalists generally aim at spurring the performance of their investments. Therefore, they tend to specialize and concentrate their activities on specific industries and certain stages in the development of firms.[7] The specialization allows venture investors to reduce the risk inherent to portfolios by active control of individual investments instead of mere diversification.[8]

Venture capitalists are also frequent participants in the stock market to exit their investments.[9] They choose the exit channel strategically to build reputation through successful initial public offerings (IPOs).[10] Reputation matters since the future success of venture capitalists

---

1 Gompers and Lerner (2001b), p. 1.

2 Storey and Tether (1996), p. 1.

3 See Moore (1994), Berger and Udell (1998). Binks and Ennew (1996) show that younger and growing firms suffer more from credit constraints than older and non-growing firms.

4 The term "venture capital" refers to a subset of private equity and comprises equity investments made for the seed, start-up, or expansion of a business. Private equity consists of venture capital and buyout investments. See Appendix A 1 for an overview of the different stages of venture capital investments.

5 See Kortum and Lerner (2000), Hellmann and Puri (2002).

6 See Sahlman (1990), Lerner (1995), Fenn *et al.* (1997).

7 See Barry *et al.* (1990).

8 See Norton and Tenenbaum (1993).

9 See Lerner (1994b).

10 See Gompers (1996), Gompers and Lerner (1999c).

crucially depends on their current standing in the market.[11] The role of reputation is twofold as it is important for both raising new funds and drawing firms for future investments. On the one hand, outside investors, such as pension funds, banks, insurance companies, and corporations, usually pay attention to the track record of venture capitalists because the invested capital is typically locked in for several years. On the other hand, investment recipients, such as young technology-based firms, take into account the experience of venture capitalists since they do not want to fail as a result of incompetent management support.

Modern venture capital originated in the United States (US) in the 1950s. It is most developed in Silicon Valley and has been the driving force behind some of the most vibrant sectors of the US economy over the past two decades.[12] Some prominent examples of US venture-backed companies indicate that venture capital has been very successful at spurring the performance of firms with innovative technologies and tremendous growth potential.[13]

Given the economic success of many US venture-backed firms, governments around the globe increasingly consider venture capital as an important source of economic growth. They enviously look at the region of Silicon Valley, trying to imitate its economic structure.[14] Thus, the venture capital approach has been dominant in the political debate, and creating venture capital institutions has become a high priority in many European countries.[15] In the 1990s, various governments pursued the objective to actively promote the development of venture capital.[16] Yet, numerous countries failed to create dynamic venture capital markets to support their high-tech industries, sometimes by lavishly spending their taxpayers' money.[17]

The efforts to duplicate the success of US venture capital industry typically share a common rationale: venture capital has spurred firm performance and innovation in the US and can do so elsewhere.[18] However, even though US venture capital promoted the development of firms and technological innovation it is not evident that the model can be seamlessly transferred abroad.[19] Likewise, it is unclear whether venture capitalists outside the US are able to spur the performance of portfolio firms, as suggested by the US experience. Different finan-

---

11 See Diamond (1989) for an analysis of the role of reputation in financial markets.

12 See Hellmann (2001). Gompers and Lerner (2001b) report that California raised with about 30 percent most venture capital funds generated in any state of the US during the period 1978-1999. Beside California, Massachusetts and New York account for a large fraction of fund-raising.

13 Venture capitalists backed successful companies, such as Apple Computers, Cisco, Compaq, DEC, Federal Express, Intel, Lotus, Microsoft, Sun Microsystems, and Yahoo. These companies developed extremely fast from start-ups to large organizations. Venture-backed companies have historically also created a disproportionate share of new, particularly well-paid and highly skilled jobs. See Gompers and Lerner (2001b).

14 See Saxenian (1994) for an analysis of regional clusters in the US, such as the Silicon Valley and Route 128. Empirical evidence on the importance of regional and industrial clusters in Germany is provided by Bender *et al.* (2002).

15 See European Commission (1995, 1996, 1998).

16 See OECD (1996a).

17 See Schertler and Stolpe (2000). Wang *et al.* (2002) explore the reasons behind the success of venture capital in the US and its failure in other countries. Murray and Marriott (1998) find that European venture capitalists failed to invest in high-tech start-ups because of poor returns in the early stages.

18 See Kortum and Lerner (2000) on venture capital and innovation in the US.

19 See Tykvová (2000) for an examination of the impact of the growing venture capital industry in Germany on the promotion of innovation.

cial systems[20], employment practices, and regulatory policies might hamper the formation of venture capital funds.[21] Also, there is a diversity of national approaches to eliminate the lack of capital and managerial skills for young and fast-growing companies, displaying a stronger allegiance to existing institutional set-ups than to the original model operating in the US.[22] Finally, given the existence of alternative forms of entrepreneurial finance, the demand for and the quality of venture capital financing may vary across countries.[23]

1.2 Research Question

The rising public attention paid to venture capital has triggered a growing number of contributions related to the topic in the economic literature.[24] Several authors have addressed the question whether or not venture capitalists have an impact on the performance of portfolio firms going public.[25] The overall results from US-based research indicate that venture-backed IPO firms outperform non venture-backed ones in terms of operating and long-run market performance.[26] However, the empirical evidence relating to the influence of venture capitalist participation on the performance of European IPO firms is scarce. This shortcoming is in part due to the recent development of the industry. Furthermore, the findings of existing studies are not consistent with the results of prior contributions.[27] The mixed evidence indicates differences across Europe with respect to the quality of venture-backing and emphasizes the necessity for a refined assessment, taking into account the heterogeneity of venture capital in different European countries.

To shed light on this conjecture, the present dissertation investigates the role of European venture capitalists. The main question of the analysis is whether or not venture capitalists in Europe have a positive impact on the operating and long-run market performance of firms they bring public. To examine this issue a hand-collected international data set of venture- and non venture-backed IPOs at the French *Nouveau Marché*, the German *Neuer Markt*, and the British *techMARK* dating from 1996 to 1999 is used.[28]

---

[20] The term "financial system" is typically used to describe the collection of financial intermediaries, such as venture capitalists, investment banks, commercial banks, insurance companies, and financial markets that are collectively responsible for allocating resources and redistributing risks in the economy. See Thakor (1996) for an extensive discussion.

[21] See Black and Gilson (1998), Jeng and Wells (2000). Becker and Hellmann (2003) provide an analysis of the difficulties to create a venture capital market in the German bank-based financial system.

[22] See Kluth and Andersen (1999).

[23] See Schmidt (1988) for a discussion of the supply and the quality of venture capital in Germany in the 1980s.

[24] Much of the literature on venture capital finance has been synthesized in Gompers and Lerner (1999c, 2001a).

[25] See e.g. Barry *et al.* (1990), Megginson and Weiss (1991), Lerner (1994b).

[26] See e.g. Jain and Kini (1995), Brav and Gompers (1997).

[27] See e.g. Bottazzi and Da Rin (2002a, 2002c), Neuhaus and Schremper (2003).

[28] The *Neuer Markt* and the *Nouveau Marché* are growth market segments of the Frankfurt and Paris stock exchanges, respectively. The *techMARK* is a tracking instrument and comprises technology firms being part of other indices of the London Stock Exchange (LSE). While the *Nouveau Marché* and the *techMARK* are still operating, the *Neuer Markt* stopped its activities as separate growth market segment in early 2003.

The considered countries (France, Germany, and the UK) represent the most significant European markets for venture capital.[29] However, they have distinct financial systems and follow different regulatory policies.[30] The business environments are conjectured to have an impact on the development of the domestic venture capital markets, and thus on the skills and experience of venture capitalists. Therefore, the institutional context of the respective countries and their traditional approaches to the financing of innovative fast-growing firms are analyzed with respect to their impact on the development of venture capital.

The empirical section examines the effects of venture-backing on IPO firms by using different accounting profitability and market-based performance measures. Moreover, differences in the issuer and offering characteristics as well as in balance sheet data are explored in more detail. In the refined analysis, the influence of venture capitalists is assessed via a number of variables reflecting the quality of venture-backing, such as pre- and post-issue shareholdings, board membership, age, syndication, organizational form, and the overall participation in IPOs of the sample. Using the international dimension of the investigation, the involvement of venture capitalists in IPOs across countries is considered as an additional proxy for the experience of investors.

Overall, the study provides a contribution to the growing number of international studies in the field of venture capital finance as well as to the literature on the operating and market performance of IPO firms.

1.3 Proceeding and Main Findings

The analysis is structured as follows: *Chapter 1* outlines the motivation of the present work and serves as a general introduction to the research problem.

Chapter 2 provides a comprehensive literature review on venture capital financing. It summarizes prior theoretical and empirical work related to the topic of the dissertation and points out open research questions. The discussion reveals that most existing contributions focus on the role of venture capital in the US. International studies, by contrast, are relatively scarce due to problems of data availability and the later development of venture capital in other countries than the US. Although there are some empirical studies on European venture capital and its impact on the performance of portfolio firms, the findings are partly inconsistent with prior US research. Moreover, earlier studies do neither take into account the effects of venture-backing on the long-run performance of IPOs in Europe nor the heterogeneity of venture capitalists. These shortcomings motivate the focus of the present analysis.

Chapter 3 assesses the institutional environments in France, Germany, and the UK and discusses their impact on the development of the domestic venture capital markets. The analysis shows that the government approaches to promote venture capital in the three concerned countries share several features, such as the provision of grant and subsidized loan schemes to reduce the risks encountered by venture capitalists and entrepreneurs. However, using the characteristics of the different national financial systems, funds are channeled via various

---

[29] The largest venture capital market in Europe is the UK, followed by Germany and France. See EVCA (2000), BVK (2000).

[30] See OECD (1993), Allen and Gale (2000).

ways and institutions to their final recipients. The historical analysis reveals that venture capital developed first in the UK, given the early existence of well developed stock markets providing an exit channel for investments. In Germany, by contrast, the first attempt to establish a venture capital industry failed. It is argued that this initial failure is related, among other things, to the characteristics inherent to the German bank-based financial system. The venture capital industries in France and Germany only developed in the 1990s when stock market segments suitable to young growth firms with technology backgrounds were established, such as the *Nouveau Marché* and the *Neuer Markt*.

Chapter 4 describes the research design of the empirical study. Using the insights of prior contributions on venture capital financing, it provides details on the hypotheses building with respect to the expected performance of IPOs and variations between different types of venture capitalists. Moreover, the characteristics of the employed data set are presented in more detail. Regarding the applied methodology, different performance measures are discussed, such as accounting profitability ratios, Tobin's Q, and wealth relatives of buy and hold returns. Finally, the empirical models to test the derived hypotheses by means of ordinary least squares (OLS) regressions are specified.

The main findings of the empirical study are presented in *Chapter 5*. Consistent with the results of previous contributions, the univariate analysis suggests that both venture- and non venture-backed IPOs in all three considered countries underperform the market in the long run.[31] However, the cross-country comparison shows that the underperformance is more severe for firms going public at the German *Neuer Markt* than for IPOs in the two other countries. Apart from significant variations in the selling behavior of old shareholders in the IPO, the univariate analysis of the offering and firm characteristics does not reveal systematic differences between venture- and non venture-backed issues. Yet, the refined analysis distinguishing between different types of venture capitalists shows that firms backed by international venture capital investors, defined as those backing IPOs in at least two countries of the sample, are on average larger in terms of IPO proceeds and display higher 3-year post-issue wealth relatives than those backed by national venture capital investors. In line with this finding, substantial variations appear in the experience and sophistication of different kinds of venture capitalists, i.e. international venture capitalists are on average older than national venture capitalists, back a larger number of IPOs in the sample, are more often represented on the board, invest with a higher number of syndication partners, and hold larger pre- and post-issue equity positions in portfolio firms. The results of the multivariate analysis suggest that venture-backed IPOs do not generally outperform their non venture-backed counterparts, irrespective of the applied performance measure. Instead, merely international venture capitalists appear to have positive effects on the operating and market performance of portfolio firms going public.

Chapter 6 concludes the main findings of the analysis. Moreover, it provides a summarizing discussion and an outlook with respect to future research. The overall results are interpreted as evidence for the heterogeneity of venture capitalists operating in the young European market. While only a subgroup of international and more sophisticated venture capitalists is able to spur the performance of IPOs, most national venture capital investors appear to lack the necessary skills to assess the quality of young and fast-growing firms. The result that

---

[31] See e.g. Ritter (1991), Loughran and Ritter (1995).

IPOs backed by international venture capitalists are more successful than those backed by national ones is consistent with the theory of the multinational enterprise and suggests that venture capitalists become international if they possess an advantage over nationally operating venture capital firms, such as more investment experience and/or better skills in the assessment of firm profitability, to countervail their restricted knowledge of foreign markets. The variance in the experience of different kinds of venture capitalists explains why earlier studies on European venture capital, not distinguishing between different types of venture investors, have not found significant performance differences between venture- and non venture-backed firms. The results emphasize the lacking degree of maturity and critical size of the European venture capital industry as a whole that is currently undergoing a consolidation process.

In order to identify the context of the research question and to lay the foundation for the overall analysis, the following chapter starts with a thorough literature review on the topic of venture capital financing.

2 Literature Review

2.1 Overview

Many theoretical and empirical studies in finance focus on established large firms. On venture capital, by contrast, there have been relatively few contributions from the viewpoint of financial theory. In the venture capital financing decision, agency costs, unique products, and intangible assets are important considerations. Moreover, venture capital funding is typically related to high uncertainty about future outcomes, considerable asymmetries of information between the contracting parties, as well as outstanding opportunities. Given these specific features, venture capital provides an occasion to study modern financial theory in an extreme setting, and thus to learn more about risk, uncertainty, and contracting technology than under conventional circumstances.

Nevertheless, research on venture capital finance is still at an early stage and the published work in the economic literature has been relatively sparse for many years. Barry (1994) argues that comprehensive and up-to-date data, especially at the micro level, have been lacking for a long time. The lack of publicly available data was due to the very nature of venture capital, such as concerns about confidentiality and privacy, and therefore created a major problem for empirical research. This may explain why venture capital finance has received little attention in empirical literature for an extensive period of time.

Despite these difficulties, over the past two decades, a growing body of literature, both theoretical and empirical, has evolved looking into the details and mechanisms of the venture capital industry.[32] Significant improvements have been made toward understanding the problems of venture capital finance. The progress was possible mainly due to the development of comprehensive databases on venture capital activities in the US, such as *Venture Economics* and *Venture One*.[33] As a result, new theories on venture capital financing, about which very little had been known until the 1980s, have been formulated and empirically tested.

The following sections provide a comprehensive review and synthesis of the main findings in the previous literature on venture capital finance. The contributions are put into perspective in order to derive new directions for future research. In section 2.2, the literature review starts to discuss prior studies on the characteristics of venture capital, such as financial intermediation, management assistance, specialization, monitoring[34], and expertise in dealing with high risk. In section 2.3, the strand of literature focusing on the issue of venture capital contracting is sketched out. Section 2.4 outlines contributions, which assess venture capital in an international perspective and investigate cross-country differences. Subsequently, section 2.5 reviews studies on the role of venture capitalists in the going public process and the perform-

---

32 There are some surveys providing a comprehensive overview of the theoretical and empirical research on venture capital. See e.g. Barry (1994), Gompers and Lerner (1999c, 2001a). The survey of Norton (1993) takes a more managerial orientation.

33 The databases *Venture Economics* and *Venture One* record venture capital investments of professional, institutional venture capital firms. See Fenn and Liang (1998) and Wolken (1998) for an overview of resources for research on small business and venture capital finance.

34 According to Hellwig (1990), the term "monitoring" should be understood in a broad sense as any form of collecting information about a firm, its investment prospects and its behavior.

ance of venture-backed IPO firms. Section 2.6 summarizes the main findings of the literature review and identifies open research questions to motivate the contribution of this dissertation.

2.2 Characteristics of Venture Capital

2.2.1 Financial Intermediation

From a purely financial point of view, venture capitalists are financial intermediaries.[35] They intermediate in the supply of investment funds from other financial intermediaries, such as pension funds, insurance companies, and banks, as well as from foundations, universities, wealthy individuals, corporations, foreign investors, to start-up entrepreneurial firms.[36] The intermediation scheme of the venture capital process is depicted in Figure 1.

Figure 1: An Overview of the Venture Capital Process

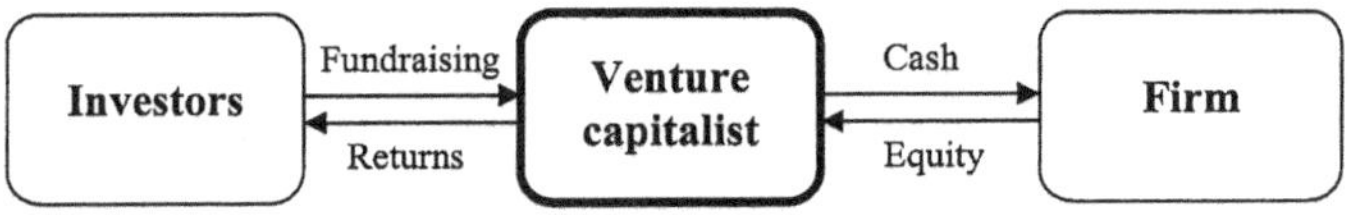

Source: Gompers and Lerner (1999c), p. 9.

Several authors have investigated the positive role of financial intermediaries. The purpose of their research is to provide an explanation for the existence of specialized intermediaries, such as venture capitalists. Since in the world of neoclassic finance theory, there are complete and perfect capital markets, characterized by the absence of information problems, such as moral hazard[37] and adverse selection[38], venture capitalists do not have any special function. In accordance with the central theorems of welfare economics, efficient risk allocations are achieved through a complete *Arrow-Debreu* system of complete contingent-claims markets, in which all agents behave as price takers.[39] In such a system, the allocation of resources is *Pareto*-efficient, and there is no scope for intermediaries to improve welfare, and thus, no need for intermediary institutions, like venture capitalists.[40] Since there is no rationale for financial intermediaries in an ideal world of frictionless and complete financial markets, a

---

35 See e.g. Bhattacharya and Thakor (1993), Greenbaum and Thakor (1995), Allen and Santomero (1996) for comprehensive overviews of the theory of financial intermediation.

36 See Aoki (2000).

37 The term "moral hazard" refers to the form of post-contractual opportunism that arises when actions required or desired under the contract are not freely observable. See Milgrom and Roberts (1992), p. 601.

38 The term "adverse selection" refers to the kind of pre-contractual opportunism that arises when one part to a bargain has private information about something affecting the other's net benefit from the contract and when only those whose private information implies that the contract will be especially disadvantageous for the other party agree to a contract. See Milgrom and Roberts (1992), p. 595.

39 See Arrow and Debreu (1954), Hellwig (1998).

40 The *Arrow-Debreu* model of a complete system of contingent-securities markets is based on the notion that transaction costs are negligible.

consistent theory of financial intermediation justifying the existence of financial intermediaries, such as venture capitalists, must include the assumption of market imperfections.[41]

According to Hellwig (1994), there are two main categories of theoretical approaches on the existence of financial intermediaries. On the one hand, more conventional theories of financial intermediation suggest that the existence of frictions, such as transaction costs, is important in understanding financial intermediation.[42] On the other hand, newer approaches rely on the assumption of asymmetric information. They consider financial intermediaries as information sharing coalitions[43], pools of liquidity[44], and delegated monitors[45]. While the model of Diamond (1984), conceiving financial intermediaries as delegated monitors, will be discussed in the context of contracting issues in section 2.3.2, the theories on financial intermediaries as information sharing coalitions and pools of liquidity are briefly outlined in the following.

Leland and Pyle (1977) consider the benefits obtained by borrowers that form information sharing coalitions in order to justify the existence of financial intermediaries. They assert that financial intermediaries are able to communicate proprietary information about borrowers at a lower cost than borrowers can do individually. In adaptation to the job market theory, as developed by Spence (1973), they assume that firms can signal the quality of their projects by investing more of their own wealth in them. By using the level of retained equity as a signal for investors, firms can partially overcome the problems of adverse selection. However, the signal is costly since firms cannot obtain optimal risk sharing.[46] Provided that financial intermediaries are able to communicate truthfully the quality of projects within the coalition, it is argued that coalitions of borrowers are able to obtain better financing conditions than individuals, since the cost of signaling increases less rapidly than the size of the coalition.

Diamond and Dybvig (1983) consider depositary financial intermediaries, like banks, as pools of liquidity, providing an insurance against idiosyncratic shocks. The authors assume that individuals are initially uncertain as to when they need or want to consume. Therefore, they may not take advantage of the higher rates of return on long-term investments, even if these are expected to be higher than those on short-term investments. Since idiosyncratic shocks may affect the consumption needs of individuals, financial intermediaries can offer them a contingent promise to provide liquidity, when they want to consume. As long as the liquidity shocks affecting individuals are not perfectly correlated, the total cash reserve needed by an intermediary increases less than proportionally with the number of depositors. Thus, by pooling large numbers of depositors, who have uncertain needs for liquidity in the

---

41 Schmidt (1985) claims that an appropriate framework to explain the existence of financial intermediaries is offered by the neo-institutional finance theory, comparing markets with institutions given the presence of market imperfections, such as information asymmetry and transaction costs. See also Hartmann-Wendels (1987).

42 Gurley and Shaw (1955, 1960) assume that financial intermediaries exploit economies of scale in lending and borrowing, and can thus overcome frictions from transaction costs. Benston and Smith (1976) first explicitly develop the argument concerning a transaction cost minimization by financial intermediaries.

43 See Leland and Pyle (1977).

44 See Diamond and Dybvig (1983).

45 See Diamond (1984).

46 This is similar to the solution of the credit rationing problem, in which firms have to provide collateral to obtain credit. See e.g. Stiglitz and Weiss (1981), Bester (1985).

short term, an intermediary can take advantage of the law of large numbers. As the aggregate liquidity needs are more predictable than those of individuals, the intermediary is able to invest in illiquid but more profitable securities while preserving enough liquidity to satisfy the aggregate claims of individual households.

Following Diamond and Dybvig (1983), in providing liquidity financial intermediaries have a comparative advantage over the market. In the model, the assumption of informational asymmetries is crucial for explaining the superiority of banks over financial markets. The market cannot provide insurance against liquidity shocks, because it is not able to distinguish investors with genuine liquidity needs from investors who are trying to make an arbitrage profit.[47] An intermediary, however, can provide liquidity insurance, because it is able to smooth the pattern of return over time. Given that individuals are assumed to be risk-averse, the smoothing of returns provided by financial intermediaries increases their expected utility. All in all, the approach by Diamond and Dybvig (1983) provides a convincing explanation for the existence of depositary financial intermediaries. However, it does not rationalize for the coexistence of depositary and non-depositary financial intermediaries, such as venture capitalists.

Taking up this shortcoming, Chan (1983) investigates theoretically the rationale for the existence of venture capitalists as financial intermediaries in a market with imperfect information. He provides a theory arguing that venture capitalists serve as producers of information. Therefore, he focuses on a venture capital market, where entrepreneurs select the qualities of their projects and their consumption of perks, about which investors are imperfectly informed. Since investors are assumed to have positive search costs, entrepreneurs have an incentive to offer inferior projects. As a result, there occurs a "lemons" problem of adverse selection, as shown by Akerlof (1970), and the capital market may break down. Given this undesirable situation financial venture capitalists evolve acting as informed agents. They can solve the adverse selection problem and allow good projects to come into the market. Thus, by using screening[48] mechanisms, venture capitalist help to mitigate the problems related to asymmetric information and improve the overall efficiency of allocation. However, Chan's (1983) theory does not clarify why venture capitalists typically provide equity rather than debt financing. Also, the characteristics of venture capitalists, producing their *value-added*[49] to portfolio firms, are not modeled.

Considering the problem of adverse selection in venture capital financing, Amit *et al.* (1990) analytically investigate the possible breakdown of the venture capital market. They develop a model where entrepreneurs need capital and want to share risk. Moreover, they assume that entrepreneurs know their own skill levels. The venture capitalists, however, cannot correctly assess the skill level of the entrepreneurs. As a result, in equilibrium, less capable entrepreneurs choose to attract venture capital investment, whereas more capable ones develop their projects without venture capital. As a consequence of this behavior, the market

---

47 See also Allen and Gale (1995), p. 189.

48 The term "screening" refers to activities undertaken by the party without private information in order to separate different types of the informed party along some dimensions. See Milgrom and Roberts (1992), p. 603.

49 See section 2.2.2 on the professionalization and management assistance provided by venture capitalists.

breaks down.[50] However, the model by Amit *et al.* (1990) ignores important aspects of the contract design in venture capital finance, which try to address the problem of adverse selection.

Chan *et al.* (1990) focus on venture capital finance as an alternative financing source beside bank loans and direct financing via capital markets. They suggest a three-tier hierarchy of financing sources, where the most inexperienced borrowers without management skills and reputation choose venture capital financing. By contrast, borrowers with established management skills but without reputation are assumed to prefer bank financing.[51] Further, large firms with both skilled management and reputation for creditworthiness are supposed to choose direct financing via capital markets.[52] In conclusion, the framework of Chan *et al.* (1990) helps to explain why young, risky firms, which often have neither reputation nor experienced managers, tend to rely on venture capital rather than on bank or capital market financing.

In summary, it can be stated that venture capital firms evolve as a specific institutional form of financial intermediaries, which specialize in providing capital to young and growth-oriented companies. Due to the high degree of asymmetric information inherent in the financing of young companies, venture capitalists play a critical role as information producers who can assess small business quality.[53] They are able to overcome the problems related to asymmetric information in financial markets, such as adverse selection and moral hazard, through various activities, such as screening, contracting, and monitoring.[54] Moreover, they provide management assistance and can add value to portfolio firm. In the following sections, these particular features of venture capital are discussed in more detail.

2.2.2 Professionalization and Management Assistance

In the economic literature, it is often emphasized that the relationship between entrepreneurs and venture capitalists entails more than the provision of finance.[55] In addition to the intermediation function, venture capitalists provide consulting expertise that allows entrepreneurs to convert ideas into commercial ventures. For this reason, venture capital is typically based on a close relationship between the entrepreneur and the venture capitalist. A number of contributions have focused on this relationship and examined the activities of venture capitalists in this direction.

Barry (1994) argues that venture capitalists can add value to projects as consultants with a stake in the outcome of the venture. They are often specialists in a particular technology or

---

50 This outcome corresponds to the finding of Akerlof (1970) who provides a theoretical treatment of adverse selection.

51 Diamond (1991) argues that relatively new borrowers without well-established reputation have the most to gain from bank monitoring and hence choose bank loans.

52 Rajan (1992) maintains that banks extract information-related rents for which there is no counterpart in direct borrowing.

53 In the US, venture capital has become a specialized segment of the financial sector geared towards the financing of new high technology firms. See Sahlman (1990), Bygrave and Timmons (1992), Fenn *et al.* (1997) for a description of the industry.

54 See Gompers (1995), Lerner (1995).

55 See e.g. Hellmann (1998a).

stage of development.[56] Hence, in addition to the provision of money, venture capitalists can also offer strategic guidance as a consultant with incentives. The author claims that at least some entrepreneurs with narrow technical specialties may recognize their limitations as managers and therefore welcome an incentivized investor, who can provide assistance and help to guide the business successfully toward a public offering or a trade sale.

Gorman and Sahlman (1989) also document that the role of venture capital is not limited to organizing the financial aspects of portfolio firms. Venture capitalists additionally monitor the activities of the firm, provide consulting services, get actively involved in strategic planning, use their business contacts to introduce the entrepreneur to potential customers and suppliers, and even take part in operational aspects.[57] Thus, venture capitalists become active social constructors and try to shape the future of portfolio companies in ways that improve the outcome of their investments. The degree of investor involvement and the breadth of activities is the most striking characteristic of venture capital.[58] Therefore, venture capital is sometimes referred to as "smart money".

Hellmann and Puri (2002) investigate the impact of venture capital on the development of new firms. They focus on the internal organization and on the professionalization of start-up firms. In particular, they explore how venture capitalists play a role beyond the traditional role of financial intermediaries. The authors provide evidence that obtaining venture capital is related to various organizational milestones for portfolio firms, such as the formulation of human resource policies, the adoption of stock option plans, and the hiring of a vice-president for sales and marketing.[59] Furthermore, they find that firms financed by venture capitalists are more likely and faster to replace the original entrepreneur with an outsider in the position of the Chief Executive Officer (CEO). This effect of venture capital appears to be particular pronounced in the early stages of the development of firms. An important insight of the analysis is that venture capitalists provide more than capital and monitoring and that, from a theoretical point of view, they should be considered as investors gathering information not merely about, but also for firms.[60]

Overall, these studies reveal that venture capital financing is related to the active involvement of investors in their portfolio firms. Hence, venture capitalists provide more than just funding to entrepreneurs. They participate in professionalization issues, offer management assistance, and thus try to affect the success of their ventures. In order to get the capacity of doing so, they are inclined to specialize in certain development stages and industries. This specific feature is discussed in the following section.

---

[56] This particularity will be analyzed in more detail in section 2.2.3.

[57] See MacMillan *et al.* (1988), Sapienza (1992).

[58] Venture capital can be thought as an extreme point on a spectrum from arm's-length to relationship-based finance. The theory of venture capital thus also helps to understand some aspects of relationship-based finance.

[59] A confirmation of these results is provided by Kaplan and Strömberg (2002).

[60] See also Hellmann and Puri (2000).

2.2.3 Specialization by Stage or Industry

Some scholars have documented that venture capitalists tend to specialize by emphasizing a specific industry, geographic area, or a particular stage of development.[61] Lerner (1994a, 1994b, 1995) concentrates on ventures in the domain of biotechnology and provides details of the functioning of venture capital within that specific industry. Moreover, Barry *et al.* (1990) document the industry specialization of venture capital in an analysis of venture-backed IPOs. They suggest that the specialization in a narrow set of industries helps venture capitalists in the monitoring process.

A theoretical explanation for the specialization of equity investors, such as venture capitalists, is offered by Harm (1998). He derives efficient enforcement mechanisms for financial claims by using methodologies applied in institutional economics. It is argued that the choice of debt versus equity depends partially on the personal characteristics of investors, as more educated investors can oversee managerial activity more efficiently by reducing the scope for post-contractual problems. Since debt can achieve greater uniformity and allows for economies of scale due to standardization, it can be considered as an answer to the bounded rationality of the investor.[62] Equity governance, by contrast, typically involves idiosyncrasies and therefore precludes economies of scale from the aggregation. This might explain why depositary financial intermediaries, such as banks, are principally administering the issuance of debt claims. Conversely, non-depositary financial intermediaries administering equity instruments, such as venture capitalists, achieve standardization by specializing in specific industries.[63]

Gupta and Sapienza (1992) test theoretical arguments pertaining to why venture capital firms might systematically prefer varying degrees of industry diversity and geographic scope in their venture investments. They find that venture capitalists specializing in early-stage ventures tend to prefer less industry diversity and a narrower geographic scope relative to those investing in later stage ventures. Venture capital firms, which are subsidiaries of non-financial corporations, also prefer less industry diversity but a broader geographic scope. Larger venture capital firms, by contrast, tend to prefer greater industry and broader geographic scope than smaller ones. The first finding reinforces the notion that venture capitalists are more than just providers of financing. Venture capital investors specializing in early-stage ventures attempt to bring specialized knowledge as well as greater involvement and assistance to portfolio firms.

Norton and Tenenbaum (1993) seek to determine if venture capitalists attempt to control risk through competing portfolio strategies. They argue that venture capitalists can reduce unsystematic risk by investing in a diversified set of firms or industries or, given the risk of liquidation during "cold" IPO markets[64], by containing investments in different stage deals. Alternatively, venture capitalists may seek to specialize, and thus exploit their technical or

---

61 For an overview of different stages of venture capital investments see Appendix A 1.

62 See Simon (1979, 1982, 1992) on bounded rationality.

63 Harm (1998) maintains that the existence of mutual funds is not a counter example to these phenomena. He argues that mutual funds are free-riding on an existing control infrastructure.

64 Ritter (1984) argues that IPO markets follow cycles and distinguishes between as "hot" and "cold" markets. Whereas a "hot" IPO markets are characterized in the literature by an unusually high volume of offerings, severe underpricing, and frequent oversubscription, "cold" IPO markets have lower issuance, less underpricing, and fewer instances of oversubscription.

product expertise in order to control risk. Following the specialization hypothesis, venture capitalists will have less diversified portfolios across companies, industries, and financing stages. Norton and Tenenbaum (1993) report empirical evidence in favor of the specialization hypothesis. Investors in seed deals are less diversified across different industries and firms. Moreover, venture capitalists tend to specialize in certain financing stages rather than stagger in their investments over different financing stages.

All in all, the findings of the present section suggest that venture capitalists control portfolio risk through efforts to specialize. The specialization is vital given the idiosyncrasies of equity governance. Due to the specialization on particular stages and industries, venture capitalists also become specialists in the evaluation of investment opportunities and the monitoring of firms. Therefore, the following section focuses on this particular issue.

2.2.4 Evaluation of Investment Opportunities and Monitoring

A number of authors have investigated the activities of venture capitalists in evaluating new investment opportunities.[65] Tyebjee and Bruno (1984) describe the main contributions of venture capitalists as an orderly process involving five sequential steps. These are:

(1) **deal origination**: the process by which deals enter into consideration as investment prospects;

(2) **deal screening**: a delineation of key policy variables which delimit investment prospects to a manageable few for in-depth evaluation;

(3) **deal evaluation**: the assessment of perceived risk and expected return on the basis of a weighting of several characteristics of the prospective venture and the decision whether or whether not to invest as determined by the relative levels of perceived risk and expected return;

(4) **deal structuring**: the negotiation of the price of the deal, namely the equity relinquished to the investor, and the covenants which limit the risk of the investor;

(5) **post-investment activities**: the assistance to the venture in the area of recruiting key executives, strategic planning, location expansion financing, and orchestrating a merger, acquisition or public offering.

Fried and Hisrich (1994) examine the efforts made by venture capitalists in the evaluation process. They show that venture capitalists spend on average three weeks of full-time effort in evaluating and closing a deal and that this process may take up to almost 100 days. In order to identify the activities that venture capitalists undertake to avoid adverse selection problems, the authors also investigate the investment decision-making process of venture capital organizations. They assert that an intensive screening and evaluation process helps to reduce the problem of adverse selection because it permits to gather substantial amounts of information before investing.

Beyond deal screening and evaluation venture capitalists also perform extensive post-investment activities. Barry *et al.* (1990) analyze the activities of venture capitalists by study-

---

65 See e.g. MacMillan *et al.* (1987), Roberts (1991).

ing IPOs of 433 venture-backed firms from 1978-1987. They find that venture capital investors hold important equity stakes in the firms they finance.[66] Moreover, on average two venture capitalists are represented on the board of directors and 1.8 venture capitalists remain there even one year after the IPO to monitor portfolio firms.

Empirical evidence on the role of venture capitalists as monitors of private firms is presented by Lerner (1995). He investigates the activities of venture capitalists in providing oversight of their portfolio companies and reports that venture capitalists hold more than one-third of the seats on the boards of venture-backed biotechnology firms, and thus more than the number held by management or other outside directors. The author also focuses on the association between involvement of venture capitalists and the need for monitoring. He finds that the representation of venture capitalists on boards of directors is 44 percent higher for firms, in which the CEO has no prior experience in running an entrepreneurial firm, than in firms with experienced CEOs.[67] In addition, he considers the replacements of CEOs as an instance of "organizational crisis" and indication of a need for more intensive monitoring. It is shown that venture capitalists increase the number of board members on average by 1.75 between financing rounds when the CEO is replaced, versus an average increase of 0.24 board members between financing rounds in which the CEO does not change. The results suggest that monitoring activities of venture capitalists seem to intensify in accordance with the dictated needs.

On the whole, the evidence suggests that the deal screening and evaluation process performed by venture capitalists before undertaking an investment is essential to assure the quality of portfolio firms. Moreover, after having invested in a firm, venture capitalists provide extensive oversight by undertaking monitoring efforts. Hence, both pre- and post-investment activities of venture capitalists play an important role with respect to the outcomes of ventures, which typically involve simultaneously high risk and high returns. The relation between high risk and high returns, representing an additional feature of venture capital financing, is scrutinized in the next section.

2.2.5 High Risk and High Return

Venture capitalists provide funding to young, growth-oriented portfolio companies, which are generally considered as risky investments. In order to compensate for the relatively high risk, they are assumed to generate high returns with investments. Huntsman and Hoban (1980) analyze the actual rates of return on investments in new or very young companies made by three venture capital institutions over the period 1960-1975. They demonstrate that the returns depend on outliers. In particular, they find that the annualized rate of return generated by 110 venture investments, treated as a portfolio, over the studied fifteen-year period is almost 20 percent. However, if the top ten percent investments are eliminated, the average return of the portfolio becomes slightly negative. This outcome suggests that the venture capital success is

---

[66] Several contributions in the corporate governance literature have focused on the role of large shareholders in monitoring. Shleifer and Vishny (1986) and Admati *et al.* (1992) argue that block-holders fulfill a beneficial monitoring role. See Harm (2000) for a discussion of empirical findings related to the influence of large shareholders on firms.

[67] See also Baker and Gompers (2003) for an analysis of the determinants of board structures of venture-backed firm at the IPO.

highly dependent on finding some outstanding investments, and that despite the specialization in a narrow set of industries a minimum of diversification is vital.[68]

Martin and Petty (1983) empirically assess the risk-return profiles of publicly traded venture capital companies.[69] They compare the performance of investments in venture capital firms with mutual funds and an index as a benchmark.[70] The findings of the paper indicate that venture capital funds are riskier than either of the comparison standards. However, venture capital funds on average exhibit a higher return than the mutual funds and the compared index. The authors notice that the higher risk does not prevent even risk-averse investors from preferring the return distributions of one or more venture capitalists over some of benchmark investment. In fact, portfolios of venture capital companies combined with mutual funds and individual stocks might offer a source of improved performance.

Gompers (1994) investigates the historical pattern of returns in the US venture capital industry in order to explain why the returns have fallen since the early 1980s. In the first half of the analyzed decade, the annual returns were over 20 percent, which was relatively high for the industry. The author suggests that the early success of venture capital, combined with a "hot" IPO market and financial deregulation led to an increased capital inflow from pension funds into venture capital companies. The large amounts invested in venture capital made new and relatively inexperienced venture capitalists enter the market. Also, the rise in capital under management increased the pressure on venture capital funds to find new deals. Gompers (1994) argues that this pressure gave rise to investments in later-stage investments yielding to earlier results. Yet, since these investments involved less risk, they were also expected to produce lower average returns than early-stage investments.

Overall, the performance analysis shows that venture capital companies take on high risks, which appear to be rewarded on average. Since venture capital firms tend to specialize, as outlined in section 2.2.3, the risk inherent in their portfolios can be characterized as unsystematic in nature. This implies that diversified portfolios, constructed by a variety of venture capital funds, can help investors to reduce the diversifiable risk and to run a much lower risk than with an investment in a single fund.[71]

Summarizing the discussion of the previous sections on the characteristics of venture capital, it can be stated that venture capitalists are specialized non-depositary financial intermediaries in small business financing providing more than merely funding to firms. The main features of venture capital include the issues of professionalization and management assistance, as well as of specialization by stage and industry. Moreover, venture capitalists are experts in evaluating investment opportunities and monitoring. Finally, the investments of venture capitalists typically involve high risk and high returns. Taking into account these findings with respect to the overall research question, IPO firms selected and promoted by venture capital-

---

68 See also Cochrane (2001) on the relation between risk and return in venture capital.

69 Many venture capital firms are not publicly traded. However, the sample of the work by Martin and Petty (1983) is limited to publicly traded firms because private firms do not make results available, nor do they have market prices.

70 The respective benchmarks are a sample of 20 mutual funds with a stated objective of maximum growth potential, and the Standard and Poor's (S&P) 500 stock market index. See Martin and Petty (1983), p. 401.

71 The idea of diversification is based on the assumption that the returns of the individual investments, i.e. venture capital funds, are not perfectly correlated. See Markowitz (1952, 1959), Sharpe (1964, 1970).

ists should be of high quality and are therefore conjectured to outperform non venture-backed firms. An important element that helps venture capitalists to select good firms and to provide incentives to entrepreneurs is the employed financial contracting framework, which is addressed in the following.

2.3 Financial Contracting in Venture Capital

2.3.1 Contracting between Venture Capitalists and Entrepreneurs

Venture capital investing is characterized by a menu of specific contract features. These can be explained as solutions to agency conflicts between entrepreneurs and venture capitalists. Agency problems arise in a venture capital relationship if the entrepreneur expends insufficient effort, exhibits expense preferences behavior, or lacks sufficient information or skill to make optimal production decisions.[72] Numerous scholars have focused on the particularities of contractual arrangements between entrepreneurs and venture capitalists, which are introduced in order to cope with agency problems.

Sahlman (1990) provides a detailed discussion of contracting issues in venture capital. His seminal contribution describes and analyzes the basic structures used in venture capital investments and their economic rationales. He argues that the contracts and operating procedures used in venture capital organizations have evolved in response to the specific agency problems in the venture capital setting being characterized by an environment of substantial uncertainty about payoffs on individual investments and a high degree of information asymmetry between principals and agents. The common procedures and contracts, which have evolved, include investment syndicating, staging the commitment of capital, using convertible securities, basing compensation schemes on value created. Moreover, there are preserving mechanisms and explicit control covenants to force agents to distribute capital and profits. According to Sahlman (1990), these procedures help to sort out skills and intentions of the participants while simultaneously addressing cost and taxation issues.

A comprehensive empirical analysis focusing on the nature of financial contracts in venture capital is provided by Kaplan and Strömberg (2003). They study detailed information on 200 venture capital investments in 118 portfolio companies and compare the characteristics of real world financial contracts to their counterparts in financial contracting theory. According to the main findings, a distinguishing characteristic of venture capital financings is that they allow separately allocating cash flow rights, voting rights, board rights, liquidation rights and other control rights. The contracts observed by Kaplan and Strömberg (2003) are most consistent with control theories, which stress control rights contingent on measures of performance.[73] Also, the observed venture capital contracts are consistent with screening theories.[74] They help venture capitalists to distinguish good entrepreneurs from bad ones.

---

[72] The traditional principal-agent approach, pioneered by Holmström (1979), assumes that the agent's effort is unobservable to the principal.

[73] See Aghion and Bolton (1992), Dewatripont and Tirole (1994).

[74] *Screening* refers to activities undertaken by the party without private information in order to separate different types of the informed party along some dimensions. It may consist in the design of contracts to attract only a desirable portion of applicants, or the design of menus of contracts to sort these parties into distinct groups.

In the following subsections, theoretical and empirical studies on the particular features of the contracting between venture capitalists and entrepreneurs, such as syndication, staging, convertible securities, and specific control rights, are outlined in more detail.

2.3.1.1 Syndication of Investment

Despite intense competition for potential investment opportunities, venture capitalists also cooperate with one another, most often through syndication. The most common reasons for syndication are the size and location of deals. Venture capitalists team up to finance large deals because of restrictions on the percentage a fund can invest in a single deal. A geographic rationale for syndication is related to the value of local monitoring. Moreover, syndication permits the validation of one venture capitalist's judgment by another and is therefore assumed to lead to superior selection of investments.[75] Finally, by allowing other venture capitalists to participate in its deals, a venture capitalist informally obliges others to return the favor in the future, thereby increasing the access to profitable deals.[76]

Admati and Pfleiderer (1994) develop a rationale for syndication in later venture rounds, based on information asymmetries between the initial venture investor and other potential investors. When a venture capitalist is involved in a firm's daily operations he may exploit its informational advantage and overstate the proper price for securities in the next financing round. The authors demonstrate that this kind of opportunistic behavior can be avoided if an inside investor, like a venture capitalist, "holds a fixed fraction contract, where he always receives a fixed fraction of the project's payoff and finances that same fraction of future investment".[77] If the venture capitalist retains a financial stake in the firm, it implies that outside investors are involved in later financing stages. This might explain why later-round financings are syndicated and not fully financed by the lead venture capitalist.

Lerner (1994a) presents evidence that is consistent with Admati and Pfleiderer's (1994) hypothesis. He documents that syndication often insures that the ownership stake of a venture capitalist stays constant in later venture rounds. Beyond, the author finds that syndication is also common in first-round investments. In the first round, experienced venture capitalists primarily offer syndication to other experienced venture capitalists, whereas later rounds also involve less established venture capitalists. The result suggests that multiple evaluations are considered important in the initial decision to finance start-ups. Moreover, venture capitalists seem to be more comfortable with a deal when other venture capitalists of similar experience are willing to invest as well.

Fenn *et al.* (1997) argue that the phenomenon of deal syndication by several established venture capitalists, with one of them acting as a lead manager of the consortium financing, is related to the issue of reputation. The position of lead financier is typically rotated over different projects among experienced and mutually known venture capitalists. Thus, syndication of investment functions as a device to control for possible shirking by the means of reciprocal

---

75 See Sah and Stiglitz (1986).

76 See Fenn *et al.* (1997). The behavior of venture capitalists is similar to syndication among Japanese *keiretsu* banks. See also Prowse (1992), Gilson and Roe (1993), Berglöf and Perotti (1994), Osano (1996).

77 Admati and Pfleiderer (1994), p. 371.

delegation.[78] If a lead venture capitalist shirks due diligence or is incompetent and more than a normal number of financed projects led by that investor fail, his reputation is damaged. Consequently, he loses opportunities for raising additional funds and participating in potentially profitable future projects organized by other venture capitalists.

Generally, the use of syndication among venture capitalists can be considered an approach to deal with the high uncertainty and the problems of asymmetric information which are typically related to venture capital financing. A further distinct characteristic of the financing relationship between venture capitalists and entrepreneurs to handle agency problems is the staging of investments. This subject is discussed in the following subsection.

2.3.1.2 Staging of Investments

A typical feature of venture capital contract design is the use of staged[79] capital commitments.[80] Although the focus of venture capitalists is typically long-term capital gains, they do not provide long-term finance. Instead, they provide capital stage-wised in various financing rounds. Initially, only a fraction of the capital needed for the ultimate development of a project is committed. Subsequent financing is then tied to the successful completion of intermediate objectives. Since staging the capital commitment creates an abandonment option, it increases the value of an investment.[81]

Gompers (1995) further develops the notion of staging and demonstrates how this mechanism addresses agency and monitoring costs in the venture capital industry. He shows that staged capital infusions are the most potent control mechanisms a venture capitalist can employ. Prospects for the firm are periodically reevaluated. The shorter the duration of an individual round of financing, the more frequently the venture capitalist monitors the entrepreneur‘s progress and the greater the need for gathering information. The author argues that the role of staged capital infusion is analogous to that of debt in highly leveraged transactions, such as leveraged buy-outs (LBOs), keeping the owner or manager on a "tight leash" and reducing potential losses from bad decisions.

Overall, by financing projects in a sequence of financing rounds, venture capitalists retain the option to abandon the project at any stage. As a consequence, the entrepreneur may have an incentive to engage in short-termism, that is to bias positively the short-term performance of the project in order to reduce the probability that the project will be liquidated. Hellmann (1993) points out that an unintended consequence of the focus on financing rounds might be myopia. The entrepreneur could lose sight of the big picture by focusing too narrowly on meeting short-term goals. Yet, this problem could be mitigated if short-term performance objectives were essentially designed as intermediate steps towards a longer-term goal. An alter-

---

78 The idea of syndication is analogous to the hostage taking as described by Williamson (1983, 1985).

79 The term "staged" is referred to as the financing in discrete stages over time. See Gompers and Lerner (1999c), p. 130.

80 See Sahlman (1988, 1990), Bergemann and Hege (1998).

81 This idea of value creation via staging is analogous to the concept of real options. See Dixit and Pindyck (1994), Trigeorgis (1996).

native solution is related to the use of convertible securities.[82] This particular contracting feature is addressed in more detail in the next section.

2.3.1.3 Convertible Securities

A mechanism that is typical for venture capital – at least for investments in the US – is the widespread use of convertible securities.[83] Empirical contributions document that, depending on the analyzed data set, venture capitalists use some sort of convertible securities for 20.9 to 94.5 percent of their investments.[84] Motivated by these observations researchers have focused on potential determinants of the financial structure in the venture capital industry and investigated the use of convertibles in venture capital.[85]

Gompers (1999) focuses on the security design and finds that venture capitalists widely use convertible preferred stocks or subordinated debt with conversion privileges. The preferred stock provides venture capitalists with a superior claim to cash flows and distribution in liquidation given that a venture fails. The possibility of conversion in the case of a successful venture provides participation on the upside. Frequently, the conversion price is a function of performance. Thus, if a venture is unsuccessful, there is a better possibility for the venture capitalists to recover the investment. Furthermore, the increased conversion price due to good performance intensifies the incentives of the entrepreneur. The results of Gompers (1999) suggest that convertible securities are important tools to elicit entrepreneurial effort and mitigate the risk-taking of venture capitalists.

Berglöf (1994) provides a theoretical explanation for the predominance of preferred stock and convertible instruments in the case of a firm sale to a private investor. Assuming that an entrepreneur and an external investor want to liquidate their holdings in the future, Berglöf analyzes how both allocate revenues and control among themselves in a venture capital relationship given an incomplete contract framework. He generates contractual arrangements, which closely resemble those observed in venture capital markets. The paper demonstrates how the parties of venture capital arrangements design the capital structure to mitigate distributional conflicts associated with a future firm sale. Berglöf's (1994) findings are related to those of Aghion and Bolton (1992), who demonstrate that when an entrepreneur faces private costs in liquidating his firm, he might lower the cost of outside finance by agreeing to give up control when the firm is doing poorly.

Bascha (2000) also concentrates on a theoretical explanation for the widespread use of convertible securities. He develops a theoretical model that highlights the consequences of an extended moral hazard problem for the optimal contract between an owner-manager (entrepreneur) and an external investor (venture capitalist). Given the possibility of non verifiable project returns in different exit scenarios, he shows that simple contracts, like debt and equity, fail to achieve efficiency. The failure of a first-best solution occurs since pure debt and equity influence the payoff structure of the entrepreneur symmetrically in both good and bad states

---

82 See Cornelli and Yosha (2003).

83 See e.g. Gompers and Lerner (1996), Fenn *et al.* (1997).

84 See Trester (1998), Cumming (2001), Kaplan and Strömberg (2003).

85 See e.g. Gompers (1993), Bergemann and Hege (1998), Marx (1998), Cassamata (1999), Bascha and Walz (2001a), Schmidt (2003).

of nature. However, by writing a financial contract in form of convertible securities, it is possible to provide first-best incentives for the entrepreneur. Hence, in order to assure *Pareto*-efficiency, venture capital contracts should contain combinations of specific equity securities.

Cornelli and Yosha (2003) focus on the role of convertible securities in order to resolve the problem of short-termism. They analyze the function of convertible debt given the problem of an entrepreneur manipulating short-term results for "window-dressing" purposes. The authors show why convertible debt is better than a simple mixture of debt and equity in stage financing situations. Although the "window-dressing"[86] behavior reduces the likelihood of liquidation, it also increases the probability that the venture capitalist converts into equity, reducing the entrepreneur's profit. Cornelli and Yosha (2003) demonstrate that an appropriately designed convertible debt contract succeeds in preventing short-termistic signal manipulation, or "window-dressing", by the entrepreneur.

Despite the theoretical arguments favoring convertible securities in venture capital finance, the existing evidence on the use of these financing instruments is mostly limited to the US. Recent studies by Schwienbacher (2002) and Hege *et al.* (2003) show that convertible securities are less often employed by venture capitalists in Europe. Cumming (2002) provides evidence that financial contracts used by European venture capital firms are quite heterogeneous in terms of both cash flow and control rights. He argues that the different securities are not functional equivalents and interprets the results within the institutional environment, from which the analyzed data were derived. In this line of arguments, Gilson and Schizer (2002) make out a case for taxation issues being partly responsible for the internationally varying use of convertible securities. On the whole, the evidence evokes a need for further theoretical and empirical research to explain the observed differences in the use of convertible securities across countries. Yet, this avenue of research is beyond the scope of the present analysis. In the following section, a final contracting feature of the relationship between venture capitalists and entrepreneurs, i.e. the use of specific control rights, is discussed in more detail.

2.3.1.4 Control Rights

Start-up firms typically represent a unique set of issues for corporate governance[87]. Given the nature of firm assets as well as the high degree of uncertainty and asymmetric information in the financing relationships, venture capitalists differ from other financial intermediaries in a number of ways and usually play a more active governance role. The distinguishing aspects of venture capital are the non-contractible monitoring, support, and control activities by the investor.

Barry *et al.* (1990) point out that venture capitalists often have concentrated stakes in the firms they finance. In some cases the venture capitalist is the sole investor, whereas in others he is the designated lead investor in a small syndicate of investors. The authors report that venture capitalists often hold the largest stake in the project. Moreover, they find that the av-

---

[86] The term "window-dressing" is referred to the manipulation of firm accounting figures in order to generate an optimistic view of future performance.

[87] John and Senbet (1998) define corporate governance deals as the mechanisms by which stakeholders of a corporation exercise control over corporate insiders and management such that their interests are protected. See Shleifer and Vishny (1997) and Harm (2000) for literature surveys on corporate governance.

erage venture-backed IPO has three venture capital investors holding in total 34 percent of the pre-IPO equity in the firm. The overall evidence suggests that venture capitalists play an active role in monitoring, while the syndicating partners tend to remain largely passive.

Sahlman (1990) argues that the incompleteness of contracts[88] might explain why venture capitalists often maintain voting control and board representation in the firms they finance. Venture capitalists typically play a role in the operation of portfolio firms because no contract between an entrepreneur and a venture capitalist can anticipate every possible disagreement. Also, venture capitalists often assume direct control by changing management and sometimes get involved in day-to-day operations.[89] In general, control rights help to protect the interests of venture capitalists and mitigate the problems related to information asymmetries.

Hellmann (1998a) examines the relationship between venture capitalists and entrepreneurs from the control perspective and derives a model looking specifically at the right to appoint the CEO. He asks why entrepreneurs choose contractual arrangements giving venture capitalists strong control rights and afford little protection. It is investigated why, and under which circumstances, control rights are voluntarily relinquished *ex ante* by the entrepreneur, especially if he is liquidity-constrained at the outset. Hellmann (1998a) argues that control rights protect venture capitalists from hold-up by entrepreneurs. Entrepreneurs trade off either giving venture capitalists more favorable financial term or more favorable control terms. If they offer more favorable financial terms they have to give up more equity, increasing the agency costs[90] associated with non-contractible actions. By contrast, if they offer more favorable control terms, there is a higher likelihood of a change of management. Hellmann (1998a) explains the willingness of entrepreneurs to abandon control rights by a trade-off between equity- and debt-induced incentives.

Kirilenko (2001) argues that the disproportionate distribution of control rights in venture capital-financed deals arises in response to adverse selection. He models the source of the problem as a privately observed variable representing the sum of net non-pecuniary benefits that an entrepreneur derives from control, access, and other non-contractible aspects of managing the venture. Since only the entrepreneur knows the value of control, contractual clauses contingent on the value's actual realization cannot be credibly enforced. The author shows that the higher the degree of adverse selection, the more control rights are allocated to venture capitalists. Also, the distribution of control rights affects the investment size, the informativeness of share prices, and the distribution of risk between the parties. As a consequence, a real-

---

88 Economists (and lawyers) use the term "incomplete" to refer to a contract that does not lay out all future contingencies. On incomplete contracts see e.g. Hart and Moore (1988), Aghion and Bolton (1992), or Hart (1995).

89 See also Lerner (1995) who analyzes the determinants of board membership and finds that the presence of venture capitalists on the board increases when CEOs lack entrepreneurial experience. Lerner's (1995) finding evokes a resource-dependent view of the board. The resource-based view of the firm emphasizes the importance of unique and inimitable assets such as tacit knowledge. Prominent contributions to the resource-based view of the firm include Wernerfelt (1984), Barney (1991), and Kogut and Zander (1992), Amit and Schoemaker (1993).

90 According to Jensen and Meckling (1976), agency costs can be defined as the sum of the monitoring expenditures of the principal, the bonding expenditures by the agent, and the residual loss that is induced by the dilution of ownership and the shift out of profits into managerial discretion.

location of control rights can alleviate some commitment and incentive problems, and thus leads to an improved outcome.

Baker and Gompers (1999) analyze the effect of venture capitalists on the CEO equity ownership in young, closely held firms. They consider 1,011 newly public firms and show that equity investments by venture capitalists lower the pre-IPO CEO equity ownership from an average of 35 to 19 percent. Moreover, they explore the active decisions of venture capitalists to increase the post-IPO equity ownership and incentives of the CEO. The evidence reveals that venture capitalists narrow this difference by granting options, reducing secondary sales, and lowering the dilution by primary shares. However, there remains a gap in the post-IPO CEO equity ownership. The findings suggest that venture capitalists, as large outside shareholders, reduce the power of the CEO to control the firm and to extract private benefits. The evidence indicates that venture capitalists try to improve the incentive effects of CEO equity ownership, on the one hand, and to reduce the negative control aspects, on the other.

All in all, the control rights granted to venture capitalists help to protect their interests and to mitigate the problems related to information asymmetries. Summarizing the main insights of the previous sections, the most prominent features concerning the financial relationship between venture capitalists and entrepreneurs comprise syndication, staging, convertible securities, and specific control rights. They can be considered as a general response to the high uncertainty, agency problems, and the incompleteness of contracts. Aside from the studies on financial contracting between venture capitalists and entrepreneurs, there is an area of research concentrating on the contracting between investors and venture capital firms. The main contributions to this strand of literature are presented in the following section.

2.3.2 Contracting between Investors and Venture Capitalists

The research related to the contracting between investors and venture capitalists focuses on the governance structures of venture capital organizations and documents how the compensation and the terms of partnership agreements differ with partnership characteristics, such as age, size, and focus, as well as how they have evolved over time. The research shows how contractual arrangements provide incentives for venture capitalists to select high quality firms. Interestingly, the empirical research on the contracting between investors and venture capitalists pursues a quite different approach to collect data than the research presented in the previous section. Instead of obtaining data from hundreds of partnerships, the researchers contacted a few large investors and gatekeepers who collectively had invested.[91]

Diamond (1984) argues that agency problems in financial contracting may be reduced by delegating the monitoring task to a financial intermediary. If the monitoring activity involves increasing returns to scale, there is no more than one monitor required to undertake this task. However, as a result from the delegation of monitoring to an intermediary, there occur new agency problems, since the information provided by the monitor may not be reliable.[92] Diamond shows that, under certain circumstances, the problems created by the delegation of monitoring are minor due to diversification effects. The agency problems are negligible, if monitoring concerns a large set of diverse borrowers whose aggregate returns are evened out

---

91 See Fenn and Liang (1998).

92 This problem was first discussed by Campbell and Kracaw (1980).

by the law of large numbers. Diamond's (1984) delegated monitoring model helps to explain well-diversified depositary financial intermediaries. Although this prediction seems to be in line with the reality for many intermediaries, its applicability might be limited for venture capital firms because the latter typically provide capital to a relative small number of firms.[93]

Fenn *et al.* (1997) argue that the emergence of the limited partnership as the dominant legal form of venture capital firms in the US is a result of the extreme information asymmetries and potential incentive problems arising in the venture capital market. Partnerships are among the largest and most active shareholders with significant means of both formal and informal control. Thus they are able to direct companies to serve the interests of their shareholders. This is also due to organizational and contractual mechanisms, which align the interests of general and limited partners. Fenn *et al.* (1997) document that about 80 percent of all US venture capital flows through independent limited partnerships, with most of the remaining 20 percent provided by subsidiaries of financial institutions. In partnerships, the general partners usually consist of senior managers of venture capital management firms, and the institutional investors are limited partners.[94] The limited partners typically put up 98 percent or more of the funds and receive 80 percent of the partnership's profits. The general partners receive 20 percent of the partnership's profits plus a fee for managing the fund.[95]

Sahlman (1990) examines the governance structures of venture capital organizations, which aid potential investors to distinguish good venture capitalists from bad ones. He points out that the finite-life nature of venture capital partnership agreements plays a crucial role. It exposes the contract to renewal, and thus helps to address problems of asymmetric information and to align the incentives of the general and limited partners.[96] Moreover, investors often preserve the right to withdraw from funding. Other features include covenants restricting the venture capitalists' management of the fund, mandatory distribution requirements, and other activities, which might be associated with self-dealing. In sum, the contracts and operating procedures that have evolved address three issues simultaneously: sorting good from bad venture capitalists, minimizing the present value of agency costs, and minimizing the present value of operating costs.

Gompers and Lerner (1996) analyze covenants in 140 partnerships establishing venture capital funds. They explore two complementary hypotheses that may help to explain the different use of covenants and restrictions in long-term contracts governing venture funds. Covenant use may be determined by the extent of potential agency problems. Since covenants are costly to negotiate and monitor, they might be employed only when agency problems are severe. Alternatively, the use of covenants may reflect the supply and demand conditions in the venture capital industry. While the price of venture capital services may shift, if the demand for venture funds changes, the supply of fund managers remains fixed in the short run. The evidence of the study suggests that both factors are important determinants of contractual restrictiveness.

---

93 See also section 2.2.3.

94 This is an oversimplification of the precise legal relationship among senior managers, the management firm, and the venture capital funds. See Fenn *et al.* (1997).

95 See Fenn *et al.* (1997) and Gompers and Lerner (1999b) for more details on the specific structures of limited partnerships in the US and associated compensation issues.

96 The typical venture capital fund in the US has a 10-year life span, usually with an option to be extended for two years.

Gompers and Lerner (1999b) investigate the compensation terms specified in 419 venture capital limited partnership agreements and offering memoranda in the US between 1978 and 1992. They produce evidence consistent with the view that reputation is an important motivation for young, unseasoned venture capitalists. Using the age and size of venture capitalists as proxies for reputation, Gompers and Lerner (1999b) find that the compensation of established funds is significantly more sensitive to performance and more variable than that of other funds. Old and large venture capital organizations also have lower base compensations than young ones. The results suggest that venture capital entrants may not have superior information about their investment abilities and be concerned about establishing reputation.

In summary, the financial contracting between venture capitalists and entrepreneurs as well as between venture capitalists and investors is characterized by a number of particularities. On the one side, in the financing relationship between venture capitalists and entrepreneurs, contracting instruments, such as syndication, staging, convertible securities, and control rights, are used to cope with the problems related to high uncertainty and asymmetric information. On the other side, in the relationship between venture capitalists and large investors, performance-sensitive compensation schemes, restrictive covenants, and specialized investment vehicles, such as limited partnerships, are employed to align the incentives of the contracting parties and to handle agency problems.

With respect to the research question, these findings suggest that both venture capitalists and entrepreneurs are highly incentivized to spur the performance of their projects. Accordingly, venture-backed IPO firms can be conjectured to perform better than non venture-backed ones. In order to assess potential differences in the characteristics of venture capital across countries, the subsequent section focuses on studies investigating venture capital in an international context.

2.4 Venture Capital in an International Perspective

The venture capital industry is most developed in the US.[97] However, during the last decade, venture capital has become increasingly important for the funding of entrepreneurial efforts around the world and evolved into an international industry.[98] Kuemmerle (2001) states that "[w]hat is remarkable about the most recent surge of the venture capital industry is its new global reach. Never before has the industry been so active on a global scale, particularly in Western Europe, Japan, Israel, and parts of Asia and South America."[99] Given this trend towards an internationalization of venture capital, it seems interesting to investigate how the venture capital sectors in other countries evolved, how they work, and how they differ from the US.

Yet, relatively little is known about how venture capital markets outside the US are functioning. Studies on venture capital in other countries than the US and international comparative studies are comparatively scarce in the economic literature.[100] The majority of the re-

---

[97] In 1998, US venture capitalists raised and had more capital under management than venture capitalists in any other region of the world. In the same year, professional venture capital firms in the US invested over USD 16 billion. See Venture Economics (1999).

[98] See Stein (2000).

[99] Kuemmerle (2001).

[100] Pfirrmann *et al.* (1996) provide a comparative study on the German and the US venture capital market.

search on venture capital financing concentrates on Anglo-Saxon countries, and in particular on the US, where venture capital has become a well-known form of financing. Due to the history of venture capital in Anglo-Saxon countries, there is a broader set of data available than for continental European countries. For the German or French venture capital industry, for example, only few empirical studies exist so far, which is the consequence of an almost entire lack of appropriate data.[101]

A complication that is related to cross-country comparisons stems from internationally diverging definitions of the notion "venture capital".[102] As a consequence, there often exists confusion in terminology when venture capital is compared across countries. In the US, the term "venture capital" primarily corresponds to early-stage financing.[103] Venture capital is applied more narrowly to equity and equity-like financing of young, growth-oriented companies that is crucial for the development and implementation of their business ideas.[104] Outside the US, however, the term is used in a broader sense and includes also later-stage financing offered to more mature small and medium sized enterprises (SMEs) primarily for capital structure reasons.[105] This more comprehensive definition of the notion "venture capital" is comparable to the American understanding of "private equity" financing[106]. Therefore, it is important to take into consideration the different meanings of venture capital when comparing international figures and interpreting the empirical results of cross-country studies.

Following Kuemmerle (2001), the term "venture capital" is defined as the subset of private equity dealing with the financing of start-up firms, including early-stage and late-stage capital. A consistent and distinctive definition of venture capital is important for the analysis because buy-out and venture capital financing activities require distinctive and separate skills. Moreover, an active venture capital industry is arguably more difficult to create than an active buy-out industry because the former necessitates a fertile technology system and a climate conducive to entrepreneurship aside from a functioning financial system.[107]

Beyond variations in the terminology, there are differences in the institutional, legal, and cultural environment, as well as in the financial and corporate governance systems across countries, which may significantly influence the conduct of business and complicate international comparisons.[108] Despite these difficulties, there is a growing number of international studies on venture capital in the literature. Since the insights of these studies are relevant for the purpose of the present analysis, section 2.4.1 summarizes the main contributions documenting differences in the extent and nature of venture capital across countries. Subsequently, in section 2.4.2, the approaches on the determinants of the different venture capital develop-

---

[101] See Lessat *et al.* (1999), Bascha (2000), and Schefczyk (2000) for empirical investigations of venture capital in Germany based on individual surveys.

[102] See Stedler (1987), Betsch *et al.* (2000), and Balzer (2000) for a discussion in more detail.

[103] Early-stage investments are usually thought of as investments that are made in the initial stages of a company formation. Early-stage capital comprises seed and start-up financing. See also Appendix A 1.

[104] See Hellmann (2000).

[105] Later-stage capital includes expansion, bridge, buy out, and turnaround financing.

[106] Fenn *et al.* (1997) define private equity as professionally managed equity investments in the unregistered securities of private and public companies.

[107] See Kuemmerle (2001).

[108] See Hofstede (1984), Hampden-Turner and Trompenaars (1993) for an analysis of different cultural environments. Schmidt *et al.* (2002) investigate the financial systems of France, Germany, and the UK.

ment patterns are outlined in order to provide an explanation for the observed international variations. Finally, in section 2.4.3, the specific influence of different financial systems and institutional environments, playing a crucial role in the evolution and functioning of venture capital markets, is taken up and discussed.

2.4.1 International Differences in the Extent and Nature of Venture Capital

Several contributions have highlighted the heterogeneity of venture capital across countries. Tyebjee and Vickery (1988) first point out that venture capital in Europe is less developed than in the US. Moreover, they mention that there are persistent differences in the composition of funding sources across countries. Unlike the US, banks are the major source of venture capital funds in Europe. However, venture capital in Europe also shares some characteristics with those observed in the US, such as the investment focus on high-technology and the syndication of funds. The authors also concentrate on differences in the venture capital activity between various European countries. Their findings suggest that European venture capital is most prevalent in the UK, France, and the Netherlands, but less developed in Germany, particularly in relation to the size of its economy. Tyebjee and Vickery (1988) argue that the concept of ownership by venture capital investors is well accepted in the UK and the Netherlands, where the business enterprise has historically been considered as a tradable entity. In France, where public policy has played a strong role in stimulating an interest in stock market investing in general, the government has created attractive fiscal incentives for investors in venture capital funds.[109]

Ooghe *et al.* (1991) review the importance of venture capital in major European countries and look for differences in funding and investment patterns. They document that the most important investors are banks, pension funds, and insurance companies.[110] Moreover, 80 percent of all venture capital funds are raised domestically. Almost half of the European investments are made in the expansion stage, whereas only 14 percent are invested in seed or start-up companies. The findings suggest that the characteristics of more mature venture capital industries are a bigger size, relative to the gross national product of the country, the presence of pension funds and insurance companies as investors in the venture capital industry, the syndication of deals, as well as the absence of the government as investor. Finally, the authors point out that the European venture capital industry cannot be approached as a single, undifferentiated industry. Instead, they emphasize that each country seems to have its own structure, institutions, and policies. Hence, the venture capital industries in different European countries have unique characteristics. Taking into account this important result, the venture capital markets of France, Germany, and the UK, on which the empirical study of this dissertation focuses, are analyzed in chapter 3.

Schertler and Stolpe (2000) compare the venture capital investment patterns in the US, the UK, Germany, and the Netherlands. They document persistent differences not only with respect to size, but also with respect to the concentration across fields of technology and development stages of portfolio firms. Venture capital investments in the US are more heavily concentrated in biotechnology and computer-related industries, as well as in earlier stages of firm development than those in Europe. This finding suggests that US venture capitalists tend to be

---

109 See also section 3.3 on the role of the government in the French venture capital market.

110 The three groups amount for 28, 17, and 12 percent, respectively, in the observed period from 1984 to 1989.

more specialized in particular industries, and might therefore be able to provide more intensive firm oversight and management assistance, than their European counterparts.

Mayer *et al.* (2001) compare the sources of funds and investment strategies of venture capital funds in Germany, Israel, Japan, and the UK. They document significant differences in the sources of funds of venture capital firms across countries, i.e. banks are particularly important in Germany, corporations in Israel, insurance companies in Japan, and pension funds in the UK.[111] Moreover, the investment patterns of venture capitalists differ across countries with respect to the stage and sector of portfolio companies, as well as the geographical focus of investments. While the different sources of finance seem to be related to differences in venture capital activities within countries, they merely account for a small proportion of the differences in venture capital activities across countries. This finding suggests that some of the international differences are due to the demand for funds, which in turn depends on the supply of entrepreneurs, rather than the supply of financial institutions. Likewise, cross-country differences may be due to the availability of alternative sources of entrepreneurial finance.[112]

Jeng and Wells (2000) document that Anglo-Saxon countries, and among them especially the US and the UK, have by far the most developed venture capital markets. Beyond the pure size relative to GDP, venture capital markets in the US and the UK are also better able to finance early-stage and high-tech investments.[113] Table 1 gives an overview of venture capital markets across countries supporting this assessment. The statistics indicate how levels of venture capital and private equity differ across time and countries.[114] The authors also point out that venture capital firms differ internationally with respect to their organizational forms. While firms in the US and the UK are typically organized as limited partnerships, they tend to have different organizational structures in France and Germany with more involvement of banks.[115] Moreover, there are differences in the prevalence of captive funds[116] across countries. Captive funds, which are often subsidiaries of industrial corporations and financial institutions, appear to be very popular in Europe and relatively unimportant in the US.[117] In addition, it is pointed out that there are persistent differences in the management style of venture capitalists across countries. For example, German venture capital firms are not as actively involved in managing investments as their US counterparts.[118]

---

111 Again, these differences might be due to differences in the structure of national financial sectors in the considered countries.

112 Van Osnabrugge (1998) emphasizes the existence of a large business angel market in the UK. The low level of early-stage venture capital financing in the UK may reflect the availability of alternative sources of financing, such as informal venture capital.

113 This finding is consistent with Schertler and Stolpe (2000).

114 To compare the order of the relative importance of venture capital financing across countries, the statistics are normalized by average GDP for each country.

115 See also Lerner (1995).

116 According to EVCA, funds are labeled captive if more than 50 percent of their financing sources derive from one source, i.e. a parent organization.

117 In France, captive funds accounted for 37.1 percent of new funds raised in 1995. See Jeng and Wells (2000), p. 248.

118 Following Jeng and Wells (2000), venture capitalists in Germany have traditionally not held board representation, nor have they been involved in day-to-day management.

Table 1: Country Comparison of Venture Capital Markets

	Venture capital investments as per mil of average GDP			Early-stage investments as per mil of average GDP[1)]			Private equity funds raised as per mil of average GDP[2)]		
Countries	***1986***	***1995***	***Mean[3)]***	***1986***	***1995***	***Mean[3)]***	***1986***	***1995***	***Mean[3)]***
Belgium	1.207	0.628	0.608	0.417	0.038	0.148	2.133	0.946	0.864
Canada	0.206	0.855	0.353	0.082	0.376	0.144	0.483	2.349	0.896
Denmark	0.201	0.132	0.156	0.041	0.031	0.043	0.288	0.253	0.302
Finland	n.a.	0.358	0.202	n.a.	0.096	0.071	n.a.	0.641	0.312
France	0.194	0.336	0.451	0.032	0.027	0.053	0.168	0.799	0.937
Germany	0.031	0.375	0.240	0.007	0.063	0.032	0.036	0.148	0.290
Ireland	0.425	0.557	0.620	0.194	0.026	0.121	0.864	0.360	1.315
Italy	0.021	0.295	0.210	0.011	0.071	0.033	0.204	0.419	0.316
Japan	n.a.	0.216	0.102	n.a.	0.047	0.018	n.a.	0.400	0.240
Netherlands	0.532	1.433	0.743	0.127	0.304	0.119	1.080	1.025	0.717
Portugal	0.004	0.990	0.569	0.004	0.087	0.086	0.036	n.a.	0.924
Spain	0.087	0.425	0.249	0.037	0.056	0.070	0.000	0.450	0.337
Sweden	0.313	0.158	0.202	0.034	0.043	0.027	0.592	3.101	1.056
Switzerland	0.068	0.105	0.157	0.050	0.004	0.034	0.033	0.240	0.263
UK	0.793	1.033	0.969	0.194	0.042	0.136	0.955	2.761	2.581
US	0.556	0.638	0.456	0.058	0.191	0.075	1.539	4.960	2.405

Notes: [1)] Early-stage investments include only seed and start-up investments. [2)] Private equity funds are defined as committed, but not yet paid capital to private equity funds, i.e. including money raised for all stages of investment. [3)] The mean is calculated as the average of all annual figures between 1986 and 1995. Source: Jeng and Wells (2000), pp. 249-251.

Differences in the degree of involvement between venture capital firms in the US and Japan are documented by Hurry *et al.* (1992). They compare US and Japanese venture capital firms and provide evidence that US venture capitalists prefer fewer investments with larger equity stakes than Japanese venture capital firms. They also find that US venture capitalists monitor ventures more frequently than their Japanese counterparts. The findings suggest that the US strategy confers an advantage over the Japanese one since it improves the venture capitalists' ability to focus on managing and understanding portfolio firms. Moreover, there is a greater incentive to monitor for US venture capitalists than for Japanese ones.

Sapienza *et al.* (1996) examine differences in the involvement of venture capitalists between France, the Netherlands, the UK, and the US. They investigate the strategic, interpersonal, and networking roles through which venture capitalists are involved in portfolio companies, and analyze how successful such efforts are. The contribution sheds light on how and when venture capitalists in the four countries expend their effort to provide oversight and value-added assistance to their investment companies. Consistent with prior empirical studies, the results suggest that venture capitalists see strategic involvement as their most important role, i.e. providing financial and business advice, and functioning as a sounding board. Venture capitalists in the US and UK are the most involved in their ventures. In France, by contrast, they are the least involved and add little value. The authors suggest that the different patterns of behavior reflect the manner, in which the different markets operate. Among the European markets, practices in the UK appear to be most like those in the US.

Manigart *et al.* (2000) examine empirically the differences between the information used for investment valuation and the valuation methods used by venture capital investors in five countries (Belgium, France, US, UK, and the Netherlands). They show that, beyond the differences in the relative importance of investment stages and venture capital types, there are important between-country differences with respect to due diligence and valuation methods of venture capital investments. Apparently, similar systems and venture capital markets place varying emphases on different valuation methods. The authors attribute the observed differences in the behaviour of venture capital investors to the dominant corporate governance mechanism and the respective stage of development of the venture capital industry in a country. The findings emphasize the need for venture capital firms entering non-domestic markets to invest considerable effort in understanding the operation of these markets if they intend to fully exploit their perceived competitive advantage.

Manigart *et al.* (2002a) also empirically investigate the determinants of the return required in five countries (Belgium, France, US, UK, and the Netherlands). They focus on the return of different investment stages in order to shed light on the risk-reward trade-off in the venture capital industries across countries. The results indicate that venture capital firms located in the US and the UK have significantly higher required rates of return than those in Belgium, France, and the Netherlands. The authors also find that independent venture capital firms have a higher required rate of return than captive or public funds. Not surprisingly, the required returns for risky early-stage investments are significantly higher than those for other stages.[119]

Bottazzi and Da Rin (2002a) examine the development of European and American venture capital in the 1990s and find a wedge between them to be large and growing. Moreover, they focus on the involvement of venture capital in companies listed on *Euro.nm*.[120] The analysis shows that venture-backed companies raise on average 60 percent more capital than companies without venture-backing. However, it appears that European venture capital has only a limited effect on the ability of the firms to raise equity capital, grow, and create jobs.

Hege *et al.* (2003) compare the conditions in the mature US venture capital market with those in the relatively young European market by focusing on the contractual determinants of success in venture financing. They consider the type of exit as well as the rate of return and report that US venture capital firms perform on average significantly better than their European counterparts. The authors attribute their findings in parts to observed differences in the contractual relationship between venture capitalists and entrepreneurs, and identify three determinants. First, venture capitalists in the US assert more often contingent control rights, i.e. they more frequently seem to use convertible securities and activate contingent control than their European counterparts. Second, they are more effective in using the real options related to stage financing. Third, they seem to be more efficient in the choice between relationship continuity and the advantages of specialization. Overall, the results suggest that venture capitalists in the younger European market are still lagging in their capacity to add value to innovative firms.

---

[119] The finding is consistent with the discussed relationship between risk, return, and stage. See section 2.2.3.

[120] *Euro.nm* was created in 1997 as alliance of the *Nieuwe Markt* (Amsterdam), *Euro.nm* (Brussels), *Neuer Markt* (Frankfurt), and *Nouveau Marché* (Paris). Milan's *Nuovo Mercato* joined the alliance in June 1999.

Finally, Cumming and MacIntosh (2003a) provide an analysis of international differences in venture capital by comparing the exit strategies of venture capitalists in Canada and the US. They find significant differences between the exit behaviors of venture capitalists in the two countries. In particular, they argue that the greater the degree of information asymmetry between the selling venture capitalist and the buyer, the greater the likelihood of a partial exit to signal quality. The results are attributed to a combination of market and regulatory differences between the venture capital industries in Canada and the US. Thus, the findings highlight the impact of legal and institutional factors on exit strategies across countries.

Summarizing the major insights of the present section, the literature contributions discussed above point out significant differences in the extent and nature of venture capital across countries. With respect to the present study, it is important to note that European venture capital is generally classified as less developed than venture capital in the US. Since this delay might affect the ability of European venture capitalists to spur the performance of portfolio firms, the empirical study in the chapters 4 and 5 analyzes the performance-stimulating impact of European venture capitalists on the basis of micro-level data. Given the documented variations between national venture capital industries, the question arises where the observed differences stem from. Therefore, in the subsequent section, the approaches on the determinants of the internationally different venture capital development patterns are discussed in more detail.

2.4.2 Determinants of Different Venture Capital Development Patterns

Several authors have analyzed the driving forces of venture capital markets in order to explain differences across countries. Tyebjee and Vickery (1988) focus on four aspects of country environments to explain the differences between the venture capital sectors in European countries. They analyze the size of the technology sector, the cultural influence on entrepreneurial risk-taking, the government policy to stimulate risk capital and entrepreneurship, and the ability of venture-backed firms to turn to publicly traded markets as a source of future financing. The study concludes that venture capital opportunities depend on the size of the technology sector and the culture of entrepreneurship. If these market forces are not sufficient, the government may instigate venture capital activity by offering tax advantages and loss guarantee programs.[121] A common factor shared by the countries with the highest level of venture capital activities is the presence of a secondary stock market geared to the needs of small, relatively new ventures contemplating an IPO. This finding highlights the necessity of a stock market to function as a viable exit mechanism that allows venture capital investors to liquidate their equity positions.

Taking up the importance of well-developed stock markets and IPOs for venture capital financing, Black and Gilson (1998) consider explanations for the greater vitality of venture capital in stock market- versus bank-centered capital markets. They compare the venture capital industries in the US and Germany, and motivate an informal theory, in which a stock market-centered capital market – present in the US but absent in Germany – is a precondition to a substantial venture capital industry. The authors argue that a well-developed stock market, allowing venture capitalists to exit through an IPO, is critical to the existence of a vibrant ven-

---

[121] The analysis in chapter 3 shows that the governments of France, Germany, and the UK offered tax advantages and designed loss guarantee programs to stimulate the markets for venture capital.

ture capital market. In order to understand the link between the stock market and the venture capital market they stress the role of the contractual arrangements between entrepreneurs and venture capitalists, as discussed in section 2.3.1. It is emphasized that the exit via an IPO allows venture capitalists to enter into implicit contracts with entrepreneurs concerning the future control of start-up firms. Since the stake of venture capitalists typically decreases through the direct sale of shares in an IPO and the sale of new shares of the company to dispersed shareholders, the entrepreneur is able to reassume some of the control formerly ceded to the venture capitalist.[122] In addition, most of the special control rights of venture capitalists typically end at the time of an IPO, and entrepreneurs often finish up with a controlling stock interest.[123] Hence, this implicit contract over future control, which is permitted by the availability of an exit through an IPO and which gives a successful entrepreneur the option to reacquire control, helps to explain the great success of venture capital in stock market-centered systems with liquid IPO markets.

Following Black and Gilson's (1998) theory, the absence of an active IPO market is responsible for the fact that economies other than the US have difficulties imitating the Silicon Valley phenomenon with its well-developed venture capital industry. Although this claim points the way to future research on cross-country differences in the importance of venture capital, Mayer *et al.* (2001) doubt the crucial role of stock markets for venture capital. Comparing venture capital between the UK and Israel, they note that early-stage venture capital investments are common in Israel, even though the IPO market has not been very active.[124] In the UK, by contrast, early-stage investments are relatively uncommon, although the stock market is larger and more liquid than the Tel Aviv stock exchange.

The importance of IPOs in venture capital is also qualified by Aoki (2000). He stresses the advantages of trade sales, depending on the product lines of portfolio firms. The author asserts that successful start-up firms have been increasingly becoming acquisition targets by leading firms operating in the same market rather than going public. Hence, firms which are often themselves grown-up entrepreneurial firms and have taken the leadership in standard setting in their niche markets, aim at acquiring successful firms financed by venture capital to eliminate potential sources of challenges to their set standards. Alternatively, they intend to strengthen their market positions by acquiring and bundling complementary products. This suggests that products might have an influence on venture capitalists in guiding their activities, in particular towards the end of venture capital financing. Therefore, venture-backed firms might prefer an acquisition to an IPO when they have only a single innovative product line.[125]

---

[122] Black and Gilson (1998) do not consider the exit option of a repurchase in their analysis. They argue that a repurchase is generally not feasible for the fast-growing companies that are the typical focus for US venture capitalists.

[123] Gompers (1999) argues that the terms of convertible securities often require conversion into common equity at the time of an IPO. In addition, the covenants granted to venture capital investors typically end with an IPO.

[124] Instead, most high-tech companies opt to go public in the US, on *Nasdaq*. See e.g. Blass and Yafeh (2001). See Rock (2001) for a case study of Israeli high tech firms going public at the *Nasdaq*.

[125] See Hellmann (1998b). Amit *et al.* (1998) argue that if the public is willing to buy the shares of an IPO only at a substantial discount, a trade sale to an informed outside investor, such as an established firm in the industry, may be a more attractive alternative.

Jeng and Wells (2000) empirically investigate the growth determinants in the venture capital industry by using country-level panel data over 10 years for 15 countries. Building on Black and Gilson (1998), they examine the importance of a number of factors for venture capital financing.[126] The authors detect that mainly three institutional factors make a difference in explaining the development of venture capital markets across countries: (1) private pension funds, (2) labor market rigidities, and (3) IPOs. The results indicate that IPOs are the most important determinant of venture capital investing. Another important finding is that different types of venture capital are differently affected by the determinants.[127] The results also suggest that government funded venture capital is not as strongly determined by IPOs as the private one. Yet, it is argued that government funds can act as a catalyst for private sector funding.[128] Although the study provides a first comprehensive empirical cross-country comparison of venture capital markets, it is severely limited by the lack of comparability of internationally aggregated data.

Focusing on the catalyst role of public venture capital, Leleux and Surlemont (2001) analyze the relationship between public and private sources of venture capital in Europe and the industry development in 15 European countries. They find that large public participation is correlated with small venture capital industries. However, the analysis suggests that public venture capitalists are neither acting to seed the industry nor crowding out private funds.[129] Instead, the evidence supports the view that public venture capitalists tend to be latecomers to the industry. Yet, the findings also indicate that public involvement helps greater amounts of money to be invested in the industry as a whole. Accordingly, it is argued that the effects of public intervention, irrespective of their rationale, are beneficial for the venture capital industry as a whole.

Summarizing the main results on the determinants of international venture capital development patterns, the existence of well-developed stock markets turns out to be the most important factor for the maturity of the industry. Yet, the development stage of national stock markets is typically related to the overall financial system of a country. In order to understand whether or not the individual determinants of venture capital development across countries follow systematic patterns, it seems useful to focus on the importance of international differences in the overall financial system and the institutional environment of national innovation schemes. Therefore, this issue is addressed in the following section.

2.4.3 Role of Different Financial Systems and Institutional Environments

Political economists have stressed the historical specificities of national institutional configurations, when analyzing differences between the functioning of financial markets in various

---

126 The respective factors are well-developed stock markets and IPOs, the growth of GDP and market capitalization, labor market rigidities, financial reporting standards, private pension funds, and government programs.

127 For example, labor market rigidities negatively affect early-stage investments, but do not seem to have an impact on later-stage investments. This finding illustrates the necessity of separating venture capital into early- and later-stage investing. Black and Gilson (1998) emphasize that, according to their theory, the success of early-stage venture capital financing is linked to the availability of exit through an IPO.

128 See also Leleux and Surlemont (2001). Empirical evidence against this claim is provided by Cumming and MacIntosh (2002).

129 The findings stand in contradiction to the results of Cumming and MacIntosh (2002) who document the crowding out of private venture capital funds by government-subsidized ones in Canada.

countries.[130] Accordingly, several contributions emphasize that venture capital should be considered as part of a larger financial and innovation system.[131] A systemic perspective helps to determine how venture capital fits into the respective economic system of a country.

Schertler and Stolpe (2000) explain the main differences between national venture capital markets by developing a theory of competition among venture capitalists that emphasizes the role of reputation in the race for capital commitments. The authors claim that reputation helps to overcome information asymmetries between outside and technological investors.[132] The reputation of venture capitalists is typically built through a track record of successful IPOs. It is argued that US venture capitalists have historically enjoyed a more favorable market environment with large stock markets and a fragmented banking system. Therefore, they had more time and opportunity to build reputation than venture capitalists in continental Europe.

Becker and Hellmann (2003) examine the difficulties of creating a venture capital market in a bank-based financial system. They focus on the attempted genesis of venture capital in Germany in the mid-1970s and identify the problem of creating appropriate governance structures to protect investor returns. The analysis exposes the difficulties of established banks and the government to devise venture investment strategies. Moreover, the availability of high quality entrepreneurs willing to apply for venture capital is identified as a critical complement. The results provide a reinterpretation of the hypothesis of Black and Gilson (1998), arguing that the existence of an active stock market is a necessary, but by no means a sufficient condition for the development of venture capital.

Kuemmerle (2001) compares the evolution of the venture capital industries in the US, Germany, and Japan. He claims that "in all three countries the structure of the financial system, and specifically the role of banks and regulators, influenced the path of the venture capital industry."[133] Private initiatives, a competitive science infrastructure, and a favorable legal environment contributed to an early success of the venture capital industry in the US. In particular, some structural elements of the university systems, such as the importance of private research universities as well as some important policy decision like the 1979 change in pension fund regulations, have shaped the venture capital industry in the US. In Germany and Japan, however, early public efforts to create venture capital firms failed.[134] The author notes that both countries started from different positions with respect to their institutional setting. Thus, venture capital firms evolved only when the regulatory environment changed.[135] He

---

130 See Shonfield (1965), Katzenstein (1978), Zysman (1983).

131 See e.g. Nelson (1993) and OECD (1993) for a comparative analysis of national innovation systems.

132 See Harm (1998).

133 Kuemmerle (2001), p. 35.

134 Kuemmerle (2001) refers to the creation of the *Wagnisfinanzierungsgesellschaft* (WFG) in Germany and government efforts, namely by the Ministry of Finance (MoF) and the Ministry of International Trade and Industry (MITI), to create a venture capital industry in Japan in the early 1970s.

135 The venture capital industries in Germany and Japan started to develop actively in the mid-1980s and mid-1990s, respectively. Following Kuemmerle (2001), the reduced regulatory density was a major trigger for the industry development in both countries. The deregulation coincided with the arrival of Anglo-Saxon private equity firms. In Germany, the creation of the law for investment companies (UBGG) in 1987 eliminating certain tax disadvantages represented a first important legal change and public policy measure. In Japan, the reinterpretation of the Anti-Monopoly Law in 1994 allowing private equity firms to hold majority stakes embodied an initial milestone in the development of a national venture capital industry. For more details on legal changes and public policy measures see Kuemmerle (2001), figure 5, p. 49.

points out that the venture capital industry is highly developed in the US and growing rapidly in Germany. Venture capital in Japan, though, has just started to develop.

Kluth and Andersen (1999) examine the making of venture capital markets in France, Germany, and the UK, and focus on the national approaches to the channeling of capital and managerial skills to high-technology entrepreneurs and SMEs. They point out that the approaches to eliminate the lack of capital and management skills for high-technology entrepreneurs and SMEs have been extremely diverse across the three countries and that they display stronger allegiance to existing institutional set-ups than to the model in the US. The authors underline that politics is a vital element of national innovation systems and the dynamics of national institutional development. However, it is claimed that the rigidity of national configurations should not be overestimated.[136]

Chesbrough (1999) compares the national institutional factors, which promote the formation of start-up firms in the US and Japan, by examining the venture capital market, the technical labor market, and the structure of buyer-supplier ties in each country. He points out that the venture capital industry is more developed in the US than in Japan, which he ascribes to, among other things, a long history of regulation and law regarding the access to bank financing, legal constraints placed on bank ownership of firms, taxation of capital gains, pension fund investment regulations, and potential liabilities to which large shareholders are exposed under US law.[137] Chesbrough (1999) emphasizes that complementarities between the individual institutional factors cause them to work as a system. Due to these complementarities the institutional factors in a system must be analyzed together.[138] The overall findings suggest that there may not be a single best way to organize for innovation in different institutional settings. Instead, the efficacy of any organizational structure depends in part upon the environment, in which the organization functions.[139]

Milhaupt (1997) investigates the institutional environment for venture capital in the US and Japan, and focuses on the linkage between corporate governance and innovation. Similar to Black and Gilson (1998), he draws a link between corporate governance systems and the vitality of venture capital markets. In order to explain differences between the US and Japan not only as a function of market liquidity, but also of other differences in key corporate governance traits, Milhaupt (1997) examines data showing that both quantitative and qualitative differences exist between the venture capital markets in the two countries. He suggests that the development of venture capital markets may require not only an active stock market, but the concomitant legal and social practices found in a stock market-oriented corporate culture. He claims that an active venture capital market drives a particular strain of entrepreneurship and innovation that is economically valuable. Thus, certain institutional settings do a better job of facilitating a venture capital market than others. The study concludes that the venture capital market in Japan is constrained by the very institutional framework that has often been the object of envy among American commentators in the 1990s.

---

136 See also Kuemmerle (2001).

137 See Charkham (1994) and Prowse (1996) for discussions.

138 See e.g. Milgrom and Roberts (1990, 1994, 1995), Williamson (1991), Aoki (1994), Hackethal and Schmidt (2000).

139 See Mohnen and Röller (2000) for an analysis of complementarities in innovation policy. For an analysis of the impact of the institutional environment on organizational structures, focusing on changes in the information and communication technology, see Bender (2003).

Crane *et al.* (1995) assert that different financial institutions might fulfill similar functional roles, such as pooling of funds, investing or the execution of payments, across countries. They note, however, that financial institutions generally differ across borders due to differences in size, complexity, and available technology, as well as differences in political, cultural, and historical backgrounds. Even if the names of institutions are the same, the functions they perform might differ substantially. Given that financial functions are more stable than financial institutions, i.e. functions change less over time and vary less across borders than institutions, the authors propose to adopt a functional perspective in order to analyze differences in financial environments across countries.

Allen and Gale (2000) compare the financial systems in five industrialized countries (France, Germany, Japan, the UK, and the US) to shed light on their cross-sectional heterogeneity. They question whether differences in financial systems constitute different ways of doing the same thing. They argue that financial systems have evolved differently in different countries because the timing and nature of economic shocks and financial crises have differed across countries. Also, governments have responded to crises in different ways. Hence, the evolution of financial systems has been shaped by market failures that elicited different types of regulatory responses[140], which then favored either a market or an intermediation orientation.

Summing up, it seems essential to consider venture capital in the overall financial system and institutional context to assess its specific role and development pattern in a country. In order to further elaborate this issue, chapter 3 examines thoroughly the distinct business environments in France, Germany, and the UK with respect to the evolution and functioning of the national venture capital markets. Before doing so, the next section discusses a further particular area of research investigating IPOs, in which venture capitalists have been involved as investors. This strand of literature is most closely related to the empirical study performed in the chapters 4 and 5.

2.5 Venture Capital and Going Public

If projects financed by venture capitalists are successful, the financing often terminates either with an acquisition[141] by another firm or an IPO. Alternative exit vehicles include company buybacks, secondary sales, and write-offs.[142] There is evidence that the prospective appropriateness of different exit choices is considered by venture capitalists to be an important factor in the investment decision.[143] IPOs are considered to be the most attractive possibility of divestment, and therefore play an important role to exit venture capital investments.[144] Though, in specific cases, the purchase of a firm by a strategic buyer may result in the realization of

---

140 The interaction between financial systems and regulatory regime mechanisms is analyzed by Kane (1988).

141 The acquisition exit mode is often referred to as trade sale.

142 See Cumming (2002), Cumming and MacIntosh (2003a, 2003b), and Schwienbacher (2001, 2002) on venture capital exits.

143 See Black and Gilson (1998), Cumming and MacIntosh (2003b).

144 Gompers (1995) reports that US venture capitalists earn an average 60 percent annual return on investments in IPO exits, compared to 15 percent in acquisition exits. See also Sagari and Guidotti (1992), Petty *et al.* (1994), Gompers and Lerner (1997), and Amit *et al.* (1998).

transaction synergies, and thus affect the willingness of the acquirer to pay a higher price.[145] For venture capitalists the successful exit of an investment is critical to ensuring attractive returns for investors and, in turn, to raising additional capital.[146]

Similar to the present analysis, many empirical contributions in venture capital focus on the performance of IPOs, contrasting differences between venture- and non venture-backed firms. This occurrence is primarily due to the fact that public offerings provide a way to find and analyze data on venture capital activities. To identify the main findings of this strand of literature, the following section 2.5.1 discusses relevant contributions on the timing of public offerings by venture capitalists. Then, in section 2.5.2, the effect of reputation and the certification ability of venture capitalists on the underpricing of IPOs are investigated. Finally, in section 2.5.3, a look is taken at studies comparing the operating and market performance of venture-backed IPO firms in relation to non venture-backed offerings.

2.5.1 Timing of Initial Public Offerings

Lerner (1994b) examines the timing of IPOs and private financings by venture capitalists. Using a sample of 350 venture-backed biotechnology firms between 1978 and 1992, he demonstrates that experienced venture capitalists can successfully time the IPO market. In particular, venture capitalists tend to bring portfolio firms public at times when equity valuations are relatively high and employ private financing when market valuations are low. Furthermore, the evidence shows that seasoned venture capitalists are particularly proficient in taking companies public near market peaks. More experienced venture capitalists do not only appear to have a greater ability but also greater incentives to time the market than less experienced ones.[147] The results indicate that venture capitalists provide valuable services in assisting entrepreneurs to time an IPO in the market.

Gompers (1996) investigates potential conflicts of interest between venture capitalists and entrepreneurs relative to the timing of an IPO. Whereas the latter are often fully committed to their venture and tend to overinvest other people's money, the former usually have a portfolio of opportunities and a high opportunity cost of time. Gompers (1996) hypothesizes that, under certain circumstances, venture capitalists may prefer to undertake an IPO earlier than would be optimal for the entrepreneur. He assumes that they might do so to establish reputation and attract capital for new investment funds. Accordingly, less experienced or less reputable venture capitalists without a substantial track record may have incentives to bring a portfolio firm public prematurely to achieve the visibility associated with an IPO. Thereby, they can signal their ability to potential investors. An IPO also permits venture capitalists to mark-to-market an investment previously carried at a lower book value.

Gompers (1996) finds empirical support for his hypotheses. Relatively inexperienced venture capitalists tend to bring portfolio companies public earlier than established venture investors. They also tend to raise new funds sooner after an IPO than more experienced venture

---

145 In some cases, a strategic buyer can be a high valuing purchaser, in particular if he wishes to keep a firm's technology out of the hands of competitors. See Aoki (2000), Cumming and MacIntosh (2003b).

146 Gompers and Lerner (1999c) call this circular relationship the "venture capital cycle".

147 See Gompers (1996).

capitalists. Gompers (1996) calls this phenomenon "grandstanding"[148] in the venture capital industry. Yet, the results are disturbing since they suggest that venture capitalists may have incentives to offer bad advice to their portfolio firms in the matter of timing, i.e. if they prefer to undertake an IPO earlier than it would be optimal for the entrepreneur. Accordingly, the outcome stands to some extend in contradiction to Lerner's (1994b) claim that venture capitalists provide a valuable service in assisting entrepreneurs in timing the market.

Gompers and Lerner (1999c) cite another type of danger with respect to public offerings for entrepreneurs financed by venture capitalists. When entrepreneurs have received financing by a venture capitalist the concerns about exiting can unfavorably affect portfolio firms. The authors mention that less scrupulous investors occasionally encourage portfolio firms to undertake actions boosting the probability of a successful IPO, even if they jeopardize the firm's long-run health.[149] Thus, venture capital investors might exploit their inside knowledge when dissolving their stake in investments. Although this may be in the interest of investors of the fund, it may have harmful effects on the entrepreneurial firm and other shareholders.

Overall, the studies on the role of venture capital in the going public process suggest that skilled venture capitalists can successfully time the IPO market, even though there is evidence that less experienced venture capital investors undertake IPOs earlier than would be optimal for portfolio firms in order to establish reputation. Taking up the issue of reputation, the following section discusses literature contributions which are related to the certification ability of venture capitalists and the underpricing of IPOs.

2.5.2 Reputation, Certification and, Underpricing

As discussed in section 2.2.2, the management support provided by venture capitalists is supposed to create a surplus in venture-backed firms. Several scholars document the important function that venture capitalists play in bringing companies public. Barry *et al.* (1990) develop a comprehensive set of IPOs of venture-backed companies to investigate the influence of venture capitalists in the going public process with respect to underpricing[150]. The evidence indicates that the underpricing is lower when an IPO firm has more experienced venture capitalists as investors. This result indicates that reputation plays a role and that the market recognizes the quality of venture capitalists. It suggests that investors in the capital market require less of a discount to purchase the new shares since the venture capitalist has monitored the quality of the offering.

Megginson and Weiss (1991) focus on the reputation[151] and certification role of venture capitalists in the IPO process. They suggest that venture capitalists can credibly stake their reputation on the quality of issuing firms as they repeatedly bring companies to the public market. Consequently, the presence of venture capitalists in the issuing firms could serve to lower the total costs of an IPO and to maximize the net proceeds to the offering firm. The

---

148 Gompers (1996) refers to Webster's Third New International Dictionary defining the verb "to grandstand" as "to act or conduct oneself with a view to impress onlookers".

149 Gompers and Lerner (1999c) cite the possibility to increase earnings by cutting back research spending.

150 Underpricing, or the initial return, is defined as the spread between the initial offering price and the opening price on the first day of trading. It occurs when outside investors demand a discount for taking the risk of buying new issued stock.

151 Diamond (1989) shows that reputation can be important in accessing debt and equity markets.

authors use a matched sample of IPOs with and without venture-backing to produce evidence that is consistent with the role of venture capitalists as specialized monitors. Similar to other studies pointing out that the quality of the underwriter or auditor can influence an IPO pricing, they find significantly less underpricing in issues with venture capital-backing.[152] Venture-backed firms in the sample are also younger, have greater median book values of assets, and a larger percentage of equity in the capital structure than their non venture-backed counterparts. Finally, the results indicate that venture capitalists retain significant portions of their holdings in the firm after an IPO.

Lin and Smith (1998) focus on insider selling during public issues and examine secondary sales by venture capitalists during IPOs. They assume that venture capitalists have their reputation to protect, since they are likely to return to the IPO market with future offerings. The reputation may influence whether, when, and at what price a company goes public. The authors argue that firms with venture capital are able to come to the market earlier in their development given the backing by venture capitalists. They suppose that the decision to bring a portfolio company public does not only depend on the reputation of venture capitalists but also on the underwriter reputation. The overall findings show that third-party certification by underwriters and venture capitalists reduce the IPO underpricing. The evidence is consistent with Megginson and Weiss (1991) and confirms the importance of reputation as determinant of the organization of the venture capital market.

Francis and Hasan (2001) compare the underpricing of venture- and non venture-backed IPOs in the US during 1990-1993. The objective of the contribution is to determine whether and to which extent the difference in underpricing between the two groups of IPOs is due to deliberate underpricing in the pre-market or to other factors, such as underwriter price support in the aftermarket. The authors find that venture-backed IPOs are managed by more reputable underwriters and are associated with less underwriter compensation. However, the results also indicate that the initial-day returns of venture-backed IPOs are on average higher than in the group without venture-backing. Francis and Hasan (2001) suggest that a significant portion of the initial-day returns is due to deliberate underpricing in the pre-market as a means of compensating investors.

Gompers and Lerner (1999a) focus on venture-backed IPOs to examine the importance of conflicts of interest in the issuance of public securities. Conflicts of interest may arise from the relationship between the underwriter and venture capital investors, i.e. if investment banks simultaneously underwrite an IPO and hold equity in a firm going public through a venture capital subsidiary. Gompers and Lerner investigate how the impact of potential conflicts of interest affects the performance of venture-backed IPOs.[153] The empirical results show that IPOs in which underwriters hold equity investments do not perform worse than offerings in which no underwriter has a venture stake. Though, the findings suggest that outside investors in the market recognize the conflict of interest and require a greater discount to compensate

---

[152] DeAngelo (1981), Beatty and Ritter (1986), Titman and Trueman (1986), Johnson and Miller (1988), and Carter and Manaster (1990) examine how investment bankers and auditors help resolve the asymmetric information inherent in the IPO process. Lin (1996) provides additional evidence that is consistent with the certification role of venture capitalists in IPOs.

[153] The authors scrutinize whether underwriters holding an equity stake in the issuing firms can take advantage of unsuspecting buyers ("naive investor" hypothesis), or whether market participants anticipate the conflict of interest ("rational discounting" hypothesis).

for potential adverse selection. The authors argue that if there is a potential conflict of interest, the reputation of the investment bank seems to reduce some of the required discount.

Ljungqvist (1999) investigates the certification role of venture capitalists and its presupposed relation to lower underpricing amongst venture-backed IPOs. He argues that differences in underpricing are uninformative and possibly misleading when they are not controlled for differences in entrepreneurs' incentives to affect underpricing. Ljungqvist (1999) shows that the entrepreneurs' wealth losses – a measure he supposes to be more suitable than underpricing – are unaffected by the presence of venture-backers. Using data from the 1980s and 1990s, no evidence of venture certification with respect to IPO pricing can be found. The analysis suggests a possible conflict of interest between venture-backers and entrepreneurs.

Hamao *et al.* (2000) investigate venture-backed IPOs in Japan where venture capital funds are often affiliated with major financial institutions. This situation may lead to a potential conflict of interest where the underwriting bank as owner of the issuing company has an interest to set an offer higher than it would be if it were merely acting as financial intermediary.[154] Since investors are assumed to anticipate this conflict of interest, they might demand more underpricing to be compensated. The authors document higher initial returns for IPOs when the lead venture capitalist is also the lead underwriter. Given that affiliations between venture capitalists and underwriting banks exist in other countries as well, it appears important to take this phenomenon into account when comparing venture capital across countries.[155]

Franzke (2001) compares venture-, non venture-backed and bridge-financed IPOs at Germany's *Neuer Markt* between 1997 and 2000. She focuses on potential differences in the issuer and offering characteristics as well as in balance sheet data. In addition, she investigates the role of venture capitalists and underwriters in certificating the quality of a company when going public. The findings reveal that venture-backed firms are less profitable and less strong selling than non venture-backed firms when they go public. However, the analysis suggests that venture-backed IPOs are more underpriced than non venture-backed ones, although the firms backed by prestigious venture capitalists are expected to show less underpricing due to reduced *ex ante* uncertainty.[156] The finding is inconsistent with prior studies and warrants further research on the certification ability of venture capitalists.

Lee and Wahal (2002) provide a possible explanation for the higher underpricing of venture-backed IPOs. They explore a variant of the grandstanding hypothesis as proposed by Gompers (1996). In particular, they investigate the idea that the favorable publicity associated with taking a firm public permits the venture capitalists to raise more money in the future. To test their conjecture cash flow regressions are estimated and indicate that there is a positive relationship between first-day returns of portfolio companies and future inflows to venture capitalists. The findings suggest that the grandstanding offers a compensating benefit to the cost associated with incremental underpricing of venture-backed IPOs.

---

154 See also Gompers and Lerner (1999a).

155 Franzke (2001) considers the affiliation phenomenon in the German venture capital market, but does not find significant differences between affiliated and non-affiliated venture capital-backed IPOs.

156 In similar studies of IPOs at the German *Neuer Markt*, Mayer (2001) and Kraus (2002) document higher initial return for venture-backed than for non venture-backed offerings.

All in all, the evidence of studies investigating the effect of reputation and the certification ability of venture capitalists on the underpricing of IPOs is mixed. While earlier studies from the US document that the market seems to recognize the certified quality of venture capitalists, the findings of more recent non-US studies on the impact of venture-backing on underpricing are not consistent with the results of prior research.[157] The varying evidence emphasizes the necessity for a refined assessment, considering the international heterogeneity of venture capital investors. Therefore, the issue is taken up in the subsequent chapters of this dissertation. Apart from the studies on venture capital and the underpricing of IPOs, a more recent strand of literature has occurred, focusing on the operating and market performance of firms going public. Some prominent contributions related to this area of research are addressed in the following section.

2.5.3 Operating and Market Performance

Jain and Kini (1994) examine changes in the operating performance of IPO firms. They use the operating return on assets as well as the operating cash flows deflated by assets as performance measures and report a significant decline in the operating profitability of firms subsequent to the IPO.[158] The authors find a positive relationship between the post-issue operating performance and the share of equity retained by old shareholders. Finally, they investigate several measures of investor expectations in the market and document post-issue declines in the market-to-book ratio, price-earnings ratio, and earnings per share. Overall, the results suggest that IPO firms are not able to sustain the pre-issue performance levels.

In a similar study, Jain and Kini (1995) focus on the value-added potential of venture capital monitoring in the IPO market. They examine whether the participation of venture capitalists leads to a relative superior post-issue operating and market performance of IPO firms. To test this conjecture, Jain and Kini (1995) compare the post-issue operating performance of venture-backed firms with a matched sample of non venture-backed IPOs. Although both venture- and non venture-backed firms tend to perform worse after the going public when compared to pre-issue levels, the study finds that venture-backed IPO firms exhibit relatively superior post-issue operating performance compared to non venture-backed ones. Furthermore, using market-to-book ratios to measure the market performance of companies, it is shown that in the periods following the IPO the market recognizes the value of venture capital monitoring, as reflected in higher valuations. Finally, the authors demonstrate that different proxies for the quality of venture capitalist monitoring, such as the number of venture investors with equity stakes and the number of board positions held by venture capitalists, are positively correlated to the post-issue operating performance of the venture-backed IPO firms.

Mikkelson *et al.* (1997) investigate the post-issue operating performance of IPO firms over a time period of up to 10 years. They find that the operating performance, measured as the operating income scaled by assets or by sales, exceeds the performance of matched publicly traded firms before the IPO, and after going public declines below the performance level of matched firms. Moreover, it is shown that the size and age of firms are related to operating performance in the first few years after the IPO. Finally, the authors find a relation between

---

157 See e.g. Franzke (2001).

158 The findings are consistent with the results of Degeorge and Zeckhauser (1993) documenting similar declines in the operating profitability for a sample of reverse LBO firms.

operating performance and the sale of shares by incumbent shareholders in the IPO. In particular, the operating performance during the first five years of the post-IPO period decreases more when old shareholders sell shares in the IPO. In line with Jain and Kini (1995), the results suggest that offerings with venture-backing experience a smaller decline in performance between the year prior to and after the IPO than those without.

Aside from contributions focusing on the operating performance of IPO firms, a number of studies have investigated the stock price performance of IPOs in the long-run.[159] In accordance with asset pricing theories, the efficient market hypothesis would suggest that IPOs perform neutrally.[160] Yet, in a seminal study Ritter (1991) examines 1,526 IPOs in the US over the period 1975-1984 and provides evidence that IPO firms after 3 years of trading significantly underperform control firms of similar size and in the same industry. While a control sample of listed firms produces an average buy and hold return of 61.68 percent, the average return of IPOs amounts merely to 34.47 percent over the same holding period. The analysis reveals that the underperformance is more pronounced among small capitalization offers, concentrating among relatively young firms which have gone public in periods with high IPO activity.

In a later study, Loughran and Ritter (1995) investigate the long-run performance of IPO firms between 1970 and 1990. The study shows that IPO stocks underperform relative to non-issuing firms of the same size by 7.4 percent over a 3-year holding period, and by 7 percent over a 5-year holding period. The analysis reveals that the book-to-market ratio accounts only for a small portion of the low returns. The authors find that the underperformance is relatively modest in periods with few IPOs and prominent in periods with many IPOs. Overall, the findings support the notion that firms take advantage of opportunity windows by issuing shares when equity valuations are relatively high.

Replicating Ritter's (1991) study within a venture versus non-venture issuer framework, Brav and Gompers (1997) examine whether venture capitalists affect the long-run performance of newly public firms in a sample of 934 venture-backed IPOs from 1972-1992 and 3,407 non venture-backed IPOs from 1975-1992. They investigate the phenomenon of the long-run underperformance of new issues and find that venture-backed firms outperform non venture-backed ones over a 5-year period. However, the results only hold when returns are weighted equally. Using different benchmarks and the Fama-French (1993) three-factor asset pricing model, it is shown that venture-backed firms do not significantly underperform. Moreover, the findings suggest that the underperformance of firms in the non venture-backed sample is primarily driven by small issuers.[161]

Carter *et al.* (1998) focus on the role of underwriters in the context of underpricing and the long-run performance of IPOs by using different proxies for reputation.[162] They find that each of the reputation proxies is significantly related to the observed initial returns. In particular,

---

159 See e.g. Ritter (1991), Loughran and Ritter (1995), Barber and Lyon (1997a), Kothari and Warner (1997), Fama (1998), Lyon *et al.* (1999). See Jenkinson and Ljungqvist (2001), table 2.2, p. 55, for comparative evidence of IPO long-run performance across countries.

160 See Fama (1970, 1998).

161 Small issuers are defined as firms with a market capitalization of less than USD 50 million.

162 The study employs three different measures of underwriter prestige, developed by Johnson and Miller (1988), Carter and Manaster (1990), and Megginson and Weiss (1991), respectively.

the better the reputation of the underwriter, the less is the underpricing of IPOs. The investigation of market-adjusted long-run returns of IPO firms over a 3-year period reveals that the observed underperformance is less severe for IPOs handled by more prestigious underwriters.[163]

Kutsuna *et al.* (2000) analyze the influence of venture capital involvement on the post-issue market performance of *Jasdaq*[164] companies in Japan. In contrast to prior US studies, they document that the venture capitalist participation in Japanese firms does not favorably affect the post-issue stock price performance. Instead, venture-backed firms going public exhibit lower returns than those without venture capital involvement. Moreover, the study reveals that companies in which venture capital firms sold their equity participations shortly after the IPO underperform relative to companies in which venture capitalists did not invest or maintain their stakes after flotation.[165]

Wang *et al.* (2003) investigate the effects of venture capitalists on the operating and market performance of venture-backed IPO firms in Singapore. They focus on two models concerning the role of venture capitalists with different empirical predictions, i.e. the certification/monitoring model versus the adverse selection/grandstanding model. While the former model predicts that venture-backed IPO firms perform better than non venture-backed ones, the latter forecasts the reverse result. The overall evidence does not support the view that venture-backed firms outperform non venture-backed firms. Wang *et al.* interpret their findings as support for the adverse selection hypothesis. Furthermore, they argue that the evidence supports the grandstanding hypothesis, since firms backed by older venture capitalists perform better than those backed by younger venture investors, the latter achieving the performance levels of non venture-backed firms.

Finally, Neuhaus and Schremper (2003) examine the long-run performance of German IPOs from 1995-2000. They find that IPOs underperform seasoned stocks by more than 50 percent over a 3-year period. The extent of the documented underperformance exceeds the findings of earlier studies on IPOs in the German capital market.[166] The authors also investigate the effects of venture capitalist participation and underwriter reputation on the performance of IPOs. However, while more prestigious underwriters seem to have a favorable impact on the long-run performance, venture capitalists do not reveal any positive influence. Instead, venture-backed firms appear to significantly underperform non venture-backed firms in the long run.[167] It is argued that the venture capital industry in Germany is relatively young and lacks experience in assessing the quality of portfolio firms.

---

[163] See also Michaely and Shaw (1994).

[164] *Jasdaq* stands for the Japanese Association of Securities Dealers Automated Quotation system and is the Japanese equivalent of the *Nasdaq* in the US.

[165] This result is in line with the conjecture that venture capitalists typically have insider information about firms they finance. If venture capitalists sell their equity stakes, the market participants may interpret this action as a signal for the overvaluation of the shares. See also Myers and Majluf (1984).

[166] See e.g. Uhlir (1989), Kaserer and Kempf (1995), Ljungqvist (1997), Stehle and Erhardt (1999), Rehkugler and Schenek (2001).

[167] The significant difference between the performance of venture- and non venture-backed IPOs is observed over a 3-year period after going public but does not apply for the 1- and 2-year post-issue time interval.

Summarizing the main insights of this section, the presence of venture capitalists tends to mitigate the frequently observed post-issue decline in the operating performance of IPO firms. Moreover, at least in the US, there is evidence that venture-backing reduces the underperformance of IPOs relative to secondary stocks in the long-run. Nevertheless, recent non-US contributions do not confirm this pattern and point out potential differences in the quality of venture-backing across countries. To shed more light on this issue, the present dissertation employs an international data set of venture- and non venture-backed IPOs focusing explicitly on the heterogeneity of venture capital in different countries. The subsequent section summarizes the main findings of the literature review and discusses open research questions in order to motivate the contribution of the present analysis.

2.6 Conclusions and Open Research Questions

The literature outlined in the previous sections has covered many questions related to the field of venture capital finance. The cited contributions in section 2.2 have developed theoretical frameworks on the role of venture capitalists as specialized financial intermediaries and producers of information, and given rise to numerous empirical studies focusing on the verification of the models. Given the described characteristics of venture capitalists and their involvement in portfolio firms as well the specific contractual arrangements used in financial relationships, as elaborated in section 2.3, venture-backed IPO firms are generally conjectured to outperform non venture-backed companies in terms of operating and market performance.

Nevertheless, as outlined in section 2.4, there are international differences in the development of venture capital. These are mainly due to the different institutional environments of countries. Given the different maturities of venture capital markets across countries, there might be international differences in the quality of venture-backing. The literature review in section 2.5 pointed out that most of the existing empirical research concentrates on the US and that the findings of some recent non-US contributions are not in line with prior research. Therefore, more empirical evidence is necessary to test the robustness of the theories outside the environment in which they were uncovered. Otherwise, it is hard to ascertain whether the observed regularities also apply for other countries than the US.

There remain several unanswered questions with respect to the impact of venture capitalists outside the US on the performance of IPO firms. Can the findings of previous studies on venture capital be generalized and transferred to countries with a non-US institutional environment? Do venture capitalists outside the US perform a valuable contribution to the firms they finance and spur their operating and market performance? Are there differences in the quality of venture-backing across countries? Do different organizational forms and sources of funds of venture capitalists lead to differences in performance? How severe are the problems of adverse selection related to the provision of venture capital in bank-based financial systems where universal banks often provide services which are close substitutes to those offered by venture capitalists?

A way to take up these questions and to generate new stylized facts on the role of venture capitalists and their impact on firms going public is to employ international data. The data from different countries provide a unique opportunity for the empirical analysis by creating institutional variety. Thus, they offer an independent sample to test the supposed inferences to the extent that other countries are similar to the US. Moreover, the approach relying on international data increases the ability to discriminate among alternative theories to the extent that countries have different institutional structures than the US.[168] However, so far, there are only few international studies on the performance of venture-backed firms.[169] Cross-country studies are scarce due to the limited availability of data and the fact that the rising popularity of venture capital is a recent phenomenon in many countries. As shown in the previous sections, the findings of existing contributions on venture capital in Europe are mixed and point out that additional research is requisite.[170]

Even though there are empirical studies on the impact of venture capitalists on the long-run performance of IPOs, previous research has not yet explored this issue by using international data. Also, the heterogeneity of venture capital across countries, such as the organizational variety of venture capital firms and the importance of captive venture capital funds, which are often subsidiaries of financial institutions, has not been sufficiently considered. In addition, prior studies have not fully analyzed the impact of different sources of capital and management styles of venture capital firms across countries.[171] Therefore, further research should address these issues, controlling for differences in venture capital across countries.

This dissertation investigates whether or not venture capitalists in Europe have a positive impact on the operating and long-run market performance of firms they bring public. Thereby, it provides a contribution to the rising number of international studies on venture capital financing and addresses the shortcomings of prior research. By using a pan-European data set, including IPOs at the French *Nouveau Marché*, the German *Neuer Markt*, and the British *techMARK*, the empirical study assesses the operating and market performance of venture-backed firms in three large industrial countries with the most important European stock markets but distinct financial systems.

The investigation intends to attribute the observed performance figures to different qualities of venture capital involvement, such as the equity stakes and board mandates held by the venture capitalists. Given the recent rise of venture capital in Europe, proxies for the experience and reputation of venture capitalists, such as their age and participation in IPOs of the sample, are examined to elucidate their influence on performance. Furthermore, using the international dimension of the compiled data set, the involvement of venture capitalists in IPOs across countries is considered as an additional proxy for their investment experience. Finally, the empirical analysis explores whether performance differentials can be assigned to variations in the organizational form of venture capital firms relying on different sources of funds.

The analysis based on an international data sample requires analyzing the main differences between the considered countries. Although France, Germany, and the UK are fairly homogenous in their levels of economic development, their institutions, such as the historical role played by banks, financial markets, and the involvement of the government in the economy are different. Likewise, the venture capital markets of the three countries have followed dif-

---

168 See e.g. Rajan and Zingales (1995) for an analogous case in the context of an empirical study on capital structure using an international data set.

169 See e.g. Bottazzi and Da Rin (2002a, 2002c).

170 See e.g. Manigart and Hyfte (1999), Franzke (2001), Bottazzi and Da Rin (2002a, 2002c), Engel (2002).

171 See Mayer *et al.* (2001) for a comparison of the sources of funds and the investment strategies of venture capital funds in Germany, Israel, Japan, and the UK. Wang *et al.* (2002) examine differences in the investment behavior between independent and finance-affiliated venture capital firms in Singapore.

ferent historical paths and vary in their stages of development. To lay the foundation for the empirical analysis in chapter 4 and 5, the following chapter analyzes the institutional environments in France, Germany, and the UK, focusing on the historical developments as well as on the current states of the different national venture capital markets.

3 Venture Capital in Europe

3.1 Overview

Until the 1990s, European venture capital has been relatively underdeveloped in comparison to the US.[172] Given the successful American experience since the 1980s and the important role of venture capital in contributing to the development of technology-oriented growth firms, European policy makers have been looking for adequate policies to promote the venture industry and entrepreneurship.[173] The present chapter briefly portrays the main objectives and milestones of the European Union (EU) policy with respect to venture capital and discusses aggregate statistics of the industry. Afterwards, the developments and particularities of the domestic markets in the UK (section 3.2), France (section 3.3), and Germany (section 3.4) are outlined in more detail to derive implications for the empirical study in chapter 4 and 5.

Traditionally, public decision makers used their ability to influence the supply of investment funds by manipulating legal and fiscal conditions. However, venture capitalists frequently argued that the investment supply is equally linked to the existence of an effective demand side with an adequate quantity of attractive firms.[174] To target at both, supply and demand side issues, the European Commission adopted a number of policy measures, such as the *Action Plan on Innovation* in 1996 and the *Communication on Risk Capital* in 1998.[175] These provided the main policy framework on the EU level, intending to spur entrepreneurial actions and to improve the conditions for venture capital through removing regulatory and fiscal disincentives.[176]

The EU pursued two main strands of support for the development of venture capital and risk capital markets. First, there was a regulatory, legislative strand addressing structural issues related to the functioning of the internal market. The proposals focused on alleviations for pension funds and institutional investors to invest in unquoted companies; also on measures to harmonize and to strengthen accounting and auditing requirements, as well as to facilitate cross-border listings on European stock exchanges; and last but not least on reforming the European patent system.[177] Second, the EU employed its own financial instruments as catalysts to promote private sector commitment to venture funding. A central element of this approach constituted the conversion of the *European Investment Fund* (EIF)[178] into a major investor in venture capital funds. At the end of 1996, the EIF began to provide capital to venture capital firms focusing on new and young companies. Following McGlue (2002), since mid-

---

[172] In 1998, the ratio of the venture capital pool to the larger economy was almost nine times higher in the US than in continental Europe. See Gompers and Lerner (2001b), p. 140.

[173] A main objective of the EU policy consisted in exploiting the ability of innovative companies to create jobs. See Schreyer (2000).

[174] See Locket *et al.* (2002), p. 1013.

[175] See European Commission (1994, 1995, 1996, 1998, 2000a, 2000b, 2000c, 2000d). In addition, there have been several actions undertaken by the national governments, which will be outlined in the sections 3.2 to 3.4.

[176] See Locket *et al.* (2002), p. 1014.

[177] See McGlue (2002), p. 47.

[178] The European Investment Fund was created in 1994 by the European Commission, the European Investment Bank (EIB), and 76 financial institutions as a commercially-oriented financial instrument to facilitate the financing of both large infrastructure projects and SMEs.

2000, the fund has become the EU's specialist provider of funding to this sector with a potential portfolio of more than EUR 2 billion to be invested in venture capital funds.

Figure 2: Funds Raised by Venture Capital Firms in France, Germany, and the UK

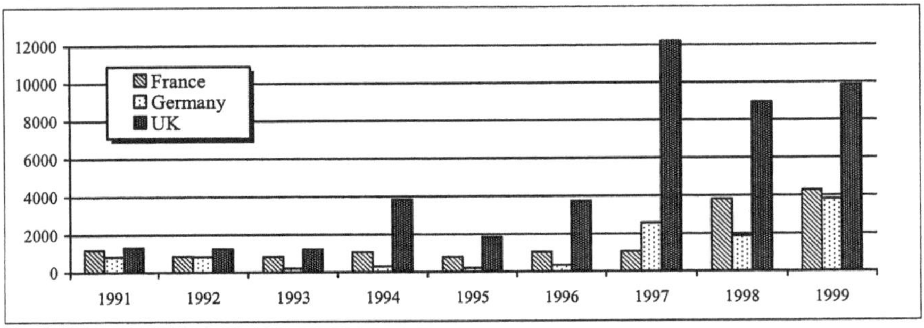

Note: The amounts are expressed in million ECU until 1998 and in million EUR afterwards.
Source: EVCA (2000).

Aside from these measures to spur venture capital and entrepreneurship in the EU, a number of important changes occurred and led to an astonishing rise in the popularity of venture capital at the end of the 1990s, such as the advent of Internet-related technologies and rising stock market valuations. First, there was a general upturn in the European market, and the overall supply of venture capital increased substantially. Between 1995 and 2001, the annual amount invested in young companies in Europe grew from EUR 2.6 to 12.2 billion.[179] In addition, over the same period, the share of venture capital investments of total private equity rose from 47.2 to 50 percent.[180] Figure 2 and Figure 3 illustrate the development of aggregate funds raised and investments made by private equity and venture capital firms in France, Germany, and the UK. The figures indicate that both, funds raised and investments made, increased considerably during the latter part of the 1990s. In addition, the data show that the British market attracted by far the largest amount of capital of the three European countries.

A problem with the aggregated data, as provided by the EVCA (2000), is that they also include funds raised by private equity firms, specializing in management buy-out (MBO) transactions.[181] In order to present the development of venture capital investments in a narrow sense over the last decade, Table 2 excludes replacement capital and buyout transactions. Moreover, the portion of venture capital investments as percentage of private equity investments per country is outlined. Interestingly, the investment figures offer a slightly different picture. Although the investments still increased considerably during the latter half of the

179 See EVCA (2002).

180 Ibid. Gompers and Lerner (2001b) note that historically merely 10 percent of European private equity funds were allocated to early-stage venture capital investments. In the US, by contrast, between 1995 and 2000 the portion of venture capital varied between 86.3 and 97.2 percent of the private equity investments. See also Bottazzi and Da Rin (2002a), p. 241.

181 MBOs are financing operations allowing a management to buy out an existing business from its original owners. MBOs are typically related to established firms in mature industries and therefore quite distinct from venture capital.

1990s, the data reveal that the amount of venture capital invested in the UK was not much higher than in both the other countries. In fact, German venture capital investments in a narrow sense even exceeded the British ones in 1999.

Figure 3: Private Equity Investments in France, Germany, and the UK

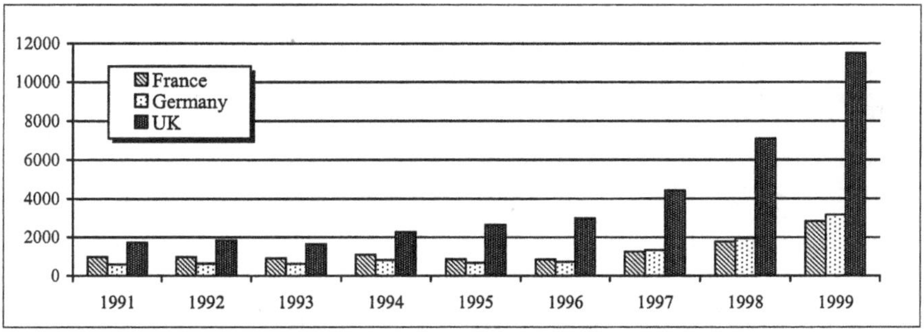

Notes: The amounts are expressed in million ECU until 1998 and in million EUR afterwards. Investments include seed, start-up, expansion, replacement capital, and buyout transactions. Source: EVCA (2000).

Another finding is related to the distribution of investments by stage. The statistics reveal that the British venture capital investments represent merely a small and decreasing fraction of private equity transactions. This outcome reflects the relative importance of later-stage and buyout transactions in the UK, which is discussed in more detail in the sections 3.2.1 and 3.2.3. In France and Germany, by contrast, venture capital investments constituted a bigger share of private equity transactions, amounting up to 58 and 85 percent, respectively.

Table 2: Venture Capital Investments in France, Germany, and the UK

	France		***Germany***		***UK***	
	Amounts[1]	**%[2]**	**Amounts[1]**	**%[2]**	**Amounts[1]**	**%[2]**
1991	575.33	58.36	499.93	85.56	695.20	40.41
1992	438.33	45.29	474.59	75.00	611.01	33.37
1993	422.05	46,76	469.24	74.93	513.93	31.54
1994	528.23	48.55	516.27	64.38	707.42	31.24
1995	333.22	39.13	531.40	79.84	688.33	26.14
1996	461.33	54.36	564.46	78.89	607.54	20.44
1997	544.53	43.65	849.92	64.11	1,174.04	26.52
1998	841.03	47.32	1,311.08	67.29	1,642.77	23.12
1999	1,590.33	56.46	2,583.91	81.70	2,537.24	22.06

Notes: [1)] in million ECU until 1998 and in million EUR afterwards. [2)] Venture capital investments as percentage of all private equity transactions. Investments include seed, start-up, and expansion transactions. Source: EVCA (2000).

The growth of investments in venture capital was complemented by the creation of stock market segments specializing in the launch of technology-oriented growth companies, such as the *Neuer Markt* in Frankfurt, the *Nouveau Marché* in Paris, as well as the *Alternative Investment Market* (AIM) and the *techMARK* in London.[182] The main purpose of these new markets was to help firms with high growth potential to go public, raise equity, and mature.[183] As outlined in Table 3, at the end of 2001, 337 firms were listed at the *Neuer Markt*, 164 at the *Nouveau Marché*, and 246 were included at the British *techMARK*. Table 3 also shows that these three markets represented the most important ones for high-technology firms in Europe in terms of market capitalization in 2001. In 2000, more than EUR 22 billion were raised via IPOs on the "new" stock market segments in Europe.[184] Yet, the major decline in the stock market valuations after March 2000 slowed down the impetus of the IPO markets and reduced the possibility for venture capitalists to exit investments via a going public.[185]

Table 3: European Public Markets for High-Technology Firms

Market	*Year Opened*	*Companies Listed*	*Market Capitalization[1)]*
Nasdaq Europe (formerly Easdaq)	1996	50	8,000
Nouveau Marché (Paris)	1996	164	15,000
Neuer Markt (Frankfurt)	1997	337	50,000
Nieuwe Market (Amsterdam)	1997	7	268
techMARK (London)	1999	246	690,000
Nuovo Mercato (Milan)	1999	45	13,000

Note: [1)] values at the end of 2001, in million EUR. Source: Bottazzi and Da Rin (2002c).

Several authors treat European venture capital as an entity and compare it to the US market.[186] Yet, the institutional environments, such as the legal and market conditions, as well as the sizes of individual domestic markets differ in part significantly among European countries.[187] In addition, the national approaches to eliminate the lack of capital and managerial skills for high-tech entrepreneurs in SMEs have been relatively diverse.[188] Therefore, before investigating the impact of venture capital on a micro level, it seems appropriate to analyze and account for these cross-country differences. The following sections focus on the peculiarities of the national markets. They sketch the main stages of development of the venture capital industries in France, Germany, and the UK, and provide an overview of the varying role of public bodies in the venture capital markets of the three countries.[189]

---

182 See Giudici and Paleari (2001), Giudici and Rosenboom (2002).

183 See Bottazzi and Da Rin (2002b).

184 See McGlue (2002), p. 46.

185 After a number of scandals related to companies listed at *Neuer Markt*, the Deutsche Börse announced on September 29, 2002 to restructure the stock market segments of the exchange in Frankfurt and to close the *Neuer Markt* in 2003.

186 See e.g. Bottazzi and Da Rin (2002a, 2002b, 2002c), Schwienbacher (2002).

187 See e.g. Ooghe *et al.* (1991), Manigart (1994).

188 See Kluth and Andersen (1999).

189 See Appendix A 2 for an overview of the development of the markets for venture capital in France, Germany, and the UK.

3.2 United Kingdom

3.2.1 Pioneering Venture Capital in Europe

The UK played a pioneer role in European venture capital, and there have been early efforts to establish a venture capital market, patterned after the US model.[190] However, several factors impeded the development of an American-style venture capital industry, such as the relatively small size of the British market, the narrowness of the industrial base, and social obstacles. Clark (1987) points out two main types of social barriers in the UK. First, people within commerce and industry were traditionally conservative and little inclined to take risks in new ventures. Second, there was unwillingness in the British society to consider new ideas.

Furthermore, the provision of equity finance for small firms and risky ventures represented a part of the British financial tradition. The first venture capitalists were probably investment trusts investing in the shares of other firms and financing the American railways. In the 1960s and early 1970s, the small business arms of merchant banks, such as Charterhouse, specialized in small company investments and provided equity for new enterprises.[191] But the by far most important source of funds constituted the *Investors in Industry* (*3i*) group, playing a major role in the development of industrial and commercial sectors of the British economy.[192]

Nevertheless, venture capital finance did not generally play an important role in the UK until the end of the 1970s.[193] This situation changed when Margaret Thatcher came into power.[194] The economic policy under her conservative government intended to promote entrepreneurship by means of tax relief and deregulation to fight against unemployment.[195] The overall orientation of the economic policy and the concomitant improvements in the set-up of the corporate environment provided a major contribution to the development of British venture capital. As a result, between 1977 and 1984, the number of venture capital firms in the UK rose from 20 to 110.[196]

In the 1980s, the success of American venture capital attracted widespread attention, and the industry in the UK developed rapidly.[197] Since the beginning of the 1980s, it grew at spectacular rates, expanding from GBP 139.6 million of total funds under management in 1984 to GBP 2.14 billion in 1995.[198] In 1983, the *British Venture Capital Association* (BVCA) was

---

190 See Bishop (1996) for an overview of the British venture capital market.

191 See Clark (1987).

192 A part of *3i* used to be known as *Industrial and Commercial Finance Corporation* (ICFC), founded in 1945 on the initiative of the Bank of England. For more details, see section 3.2.2 on the involvement of the British government in the venture capital market.

193 See Chapman (1986), p. 1.

194 Margaret Thatcher was British prime minister from 1979 to 1990.

195 See Densing (1997), p. 55. For a detailed analysis of the measures undertaken by the government Thatcher see Güllmann (2000).

196 See De and Wimmers (1994), p. 104.

197 See EVCA (1997).

198 See Bank of England (1996), p. 18.

founded, counting 35 full members in its first year.[199] At the end of the 1980s, the number of BVCA full member firms had increased to 124.[200]

While a number of funds was initially investing in early-stage technology firms, many venture capitalists refocused their activities to later stage investments.[201] Locket *et al.* (2002) claims that the few venture capital firms specializing in technology-based investments were generally small, under-capitalized and "regarded with considerable suspicion by institutional investors".[202] Hence, at the start of the 1990s, the UK venture capital industry became increasingly non technology-focused, dominated by funds specializing in MBOs[203] and development finance.[204] Nevertheless, in terms of mere size, the UK turned into the third largest venture capital market in the world, after the US and Japan.[205]

In accordance with Black and Gilson (1998), a factor that favored the pioneering role of the UK in European venture capital was the existence of well-developed stock markets, encouraging IPOs and providing an exit channel for the industry. Prior to the 1980s, the *Official Market* and the unregulated over the counter (OTC) market of the *London Stock Exchange* (LSE) had allowed British companies to raise public equity.[206] In order to enable small firms to raise money, in 1980 the LSE introduced the *Unlisted Securities Market* (USM) with less restrictive requirements than the main market.[207] Moreover, in comparison to most countries in continental Europe, the UK had a relatively well-developed equity culture with sophisticated private and institutional investors.

As a consequence of the favorable institutional environment, British venture capitalists were able to exit portfolio firms via IPOs in the national capital market.[208] Along with the emergence of second tier markets on a pan-European basis, the LSE provided less onerous alternatives than the main market but with greater investor protection than the OTC market.[209] In addition, the LSE launched the *techMARK* as a tracking instrument for innovative technol-

---

199 See Murray (1991), p. 72.

200 Ibid.

201 Murray and Lott (1995) assert that examples of a contrary movement from later-stage to early-stage activity are singularly rare. The Bank of England (1996) states that although the amount invested in early-stage businesses doubled over the same period, investments at this stage merely accounted for 4 percent of the total invested in 1995.

202 Locket *et al.* (2002), p. 1009. See also Murray (1991, 1994).

203 The term MBO is used in a broader sense including all forms of later-stage investment activity, in which buy-out or buy-in investment is made.

204 See Locket *et al.* (2002), Sieber (2002).

205 See Murray (1995).

206 In 1987, the LSE created the Third Market. However, the market was rather informal and companies were typically required to apply for the USM or the *Official Market*. Due to a lack of interest from the investor side the market segment was closed in 1990.

207 In 1986, the USM achieved its greatest popularity with a total of 529 listings of SMEs. See Midland Bank Review (1987) [cited in Goergen (1999)].

208 From 1992-1996, venture-backed IPOs represented almost 40 percent of all going publics in the main market of the LSE. See EVCA (1997).

209 In 1995, the USM was closed and replaced by the *Alternative Investment Market* (AIM), targeting small and young companies. The AIM offered a first step into the public market providing easy listing requirements, such as no minimum trading record, no minimum asset or profit level, no minimum capitalization, and no minimum free float of shares. By the end of 2000, 524 companies with a total market capitalization of GBP 14.5 billion had been admitted on AIM. See Bank of England (2001).

ogy-oriented companies.[210] On the one hand, the *techMARK* aimed to provide firms with a number of benefits, such as a heightened profile and visibility, increased media coverage, investor interest and liquidity, as well as more differentiation within the market as a whole. On the other hand, potential advantages for investors comprised an easier identification of technology stocks, clearer benchmarking of share price performance, as well as higher turnovers and liquidity in technology stocks. Aside from the favorable stock market conditions, the British government took a number of measures to promote venture capital. These are discussed in the following section.

3.2.2 Government Initiatives

The government involvement in the British venture capital market has been insignificant relative to other European countries.[211] Publicly supported investment institutions, such as the *Small Business Investment Companies* (SBICs) in the US, did not exist until the end of the 1990s.[212] However, public initiatives to encourage venture capital date back to the end of World War II. Most noteworthy is the launch of the *Industrial and Commercial Finance Corporation* (ICFC) subsuming later into the *3i* group.[213] The precursor of the *3i* group was created in 1945 on the initiative of the Bank of England in cooperation with major banks, and had the mandate to provide long-term finance to small enterprises in form of equity and loans.[214] Nevertheless, *3i* has never been an instrument of public policy and has not necessarily financed projects favored by the government.[215] Yet, despite being a private sector institution, the investment group differed from independent American-style venture capital firms, as it did not limit investments to firms intending to go public or to offer shares for sale. *3i* dominated the UK market for small firm equity investment until the end of the 1970s, and remained the country's largest investor during the great expansion of the industry in the early 1980s. Today, *3i* is Europe's leading private venture capital firm with operations in more than 12 European countries and a portfolio valuation of more than GBP 5 billion in 2002.[216]

Aside from the funding activities of *3i*, in 1981 the *Small Firm Loan Guarantee Scheme* (SFLGS) was introduced to improve the financial climate for small enterprises. The SFLGS was run jointly by the *Department of Trade and Industry* (DTI) as well as several lending institutions, and differed from similar schemes in other countries, not requiring borrower commitments in form of equity investments or collateral.[217] While this approach intended to overcome the constraints of potential borrowers without assets, it involved the risk of a high

---

[210] 77 companies formerly listed on AIM progressed to the main market of the LSE, whereby over 20 of these firms were included in the *techMARK*. See Bank of England (2001).

[211] See section 3.3 and 3.4 on the government involvement in the French and German venture capital markets, respectively.

[212] See Lerner (1999) for an analysis of the SBICs in the US.

[213] In 1974, the ICFC and the Finance Corporation Industry merged to Finance for Industry. In 1983, Finance for Industry was renamed *3i*. See Bank of England (2001).

[214] Initially, the Bank of England held 15 percent of the ICFC.

[215] See Clark (1987).

[216] See 3i (2002).

[217] The loans guaranteed by the scheme were characterized by a repayment schedule of between 2 and 10 years. The coverage was 75 percent on loans up to GBP 100,000 for firms younger than two years, and 85 percent on loans up to GBP 250,000 for firms older than two years. For loans under GBP 30,000 the banks did not need an approval from the SFLGS authorities. See OECD (1997).

default rate.[218] Nevertheless, in the retrospect the scheme had little effect in stimulating the lending volume to small firms.[219]

In 1999, the government introduced the *Enterprise Fund* to support the provision of early-stage venture capital funding. The scheme was part of several new financing initiatives with the overall objective to build "the knowledge driven economy" and represented the most important public expenditure initiative with respect to start-up finance in the UK.[220] With a three-year budget of about GBP 150 million, the fund absorbed the existing SFLGS, combining public expenditure and private sector funding through two new channels of equity investments. First, a *National Venture Capital Fund*, structured as a fund of funds, invested in early-stage venture capital funds.[221] Second, a network of *Regional Venture Capital Funds* specialized in the provision of small-scale equity to SMEs.[222] Thus, the *Enterprise Fund* offered part-financing to venture capital funds in technological areas and at regional levels.[223]

The government also provided direct financing via grant-based schemes and awards, such as the *Small Firms Merit Award for Research and Technology* (SMART) and the *Support for Products under Research* (SPUR).[224] With grants meeting the costs of expert reviews, the schemes were designed to assist entrepreneurs in the translation of research ideas into commercially-viable products.[225] The receipt of a grant had important implications concerning the attractiveness of a firm in raising additional funds. First, the award was considered as a technology certification. Second, it helped entrepreneurs to advance to a stage where private sector investors could take an interest.[226] However, only relatively few firms receiving grants subsequently obtained venture funding, and there is little evidence that the scheme influenced the attitude taken by venture capitalists to invest in small high-tech firms.[227]

A parallel government initiative, involving a public/private partnership, was the foundation of the *University Challenge Fund* in 1998. The project aimed at encouraging the transfer of science, engineering, and technology from universities into the commercial sector, to demonstrate that early-stage investments can end up being profitable. In order to permit universities to develop their business proposals and start-up firms, the UK government provided funding

---

218 Following the OECD (1997), the default rate varied over the life of the program, reflecting to a large extent the state of the economy. In 1995, the default rate on the guaranteed loans was 22 percent.

219 From 1986 to 1996, only 6 percent of a total of 41,000 loans were granted to technology-oriented firms under the SFLGS. See Bank of England (1996, 2001).

220 See Locket *et al.* (2002), Department of Trade and Industry (1999, 2001).

221 By the end of 2000, the fund had raised GBP 116 million, whereof GBP 20 million stemmed from the public sector. See Bank of England (2001).

222 The Bank of England (2001) defines small-scale equity as sums below GBP 500,000.

223 See McGlue (2002).

224 Interestingly, despite the UK government's strong market ideology, both *SMART* and *SPUR* schemes were allocated by government departments and not commercial investor channels as in similar programs of other countries, such as Germany or the Netherlands. See Murray and Lott (1995).

225 The grant amounts varied from GBP 2,500 to 150,000. See Bank of England (2001).

226 See Moore and Garnsey (1993).

227 See Department of Trade and Industry (1994).

in conjunction with major charities.[228] Thus, 15 consortia of UK universities were granted GBP 45 million to fund the seed venture capital activity in spin-out companies on campus.[229]

The government also supported the industry through legislative measures, without being directly involved in the provision of venture capital funds.[230] There have been several approaches, like the *Business Expansion Scheme* (BES) in 1983, involving the use of tax incentives, such as tax relief on income and capital, to encourage private individuals to invest in unquoted companies.[231] The BES aimed at making individuals invest directly in small unquoted companies by allowing them to claim upfront tax relief at their marginal rate on the amount invested. Yet, despite the success in attracting funds, most of the investments went to property or asset-backed deals, and only relatively small amounts were invested in the type of firms originally targeted.

In the 1990s, the BES was replaced with new complementary schemes: the *Enterprise Initiative Scheme* (EIS) and the *Venture Capital Trusts* (VCTs).[232] The EIS was set up in 1994 providing tax breaks to individuals investing in small unquoted firms.[233] One objective of the scheme was to promote the activity of business angels.[234] Business angels refer to informal private investors who are typically experienced private individuals and prefer to act independently.[235] They often operate close to major population centers and regions with a high density of high-technology firms, as around the research-intensive universities of Cambridge and Oxford, and organize themselves in networks on both national and local/regional scales.[236] Business angel networks (BANs) intend to overcome informational asymmetries between informal investors and entrepreneurs by providing efficient and confidential channels of communication. Several authors stress the crucial role of the informal private investors in the British venture capital market, although the market remains small compared to that in the US.[237] The UK has accumulated great experience with BANs, and there is evidence suggesting that they had a significant impact on informal venture capital activity.[238]

Aside from the EIS, the VCT scheme was established by the Finance Act of 1995.[239] The scheme represented an investment vehicle providing individuals an alternative way to invest

---

228 See Bank of England (2001).

229 See Locket *et al.* (2002).

230 See Hughes (1997) for an overview of recent trends in small business financing in the UK.

231 See Jeng and Wells (2000).

232 In addition, the Capital Gains Tax Re-Investment Relief was introduced.

233 Under EIS, investors were offered an income tax relief equal to 20 percent for investments in newly issued shares, which had to be held for 5 years. In addition, they were exempted from capital tax on disposal of shares held for at least 5 years. Reinvestments in newly issued shares were rewarded with a deferral of tax liability on capital gains. See OECD (1997).

234 See Kluth and Andersen (1999).

235 See Prowse (1998). For a study on informal venture capital in the UK see Harrison and Mason (1996).

236 Reliable information on the activities of business angels is scarce. Yet, recent estimates suggest that there were some 18,000 business angels in the UK in the end of the 1990s whose annual investments amount to about GBP 500 million in some 3,500 businesses. See Harrison and Mason (1999), Bank of England (2001).

237 See e.g. Leopold and Frommann (1998), Kluth and Andersen (1999). Harding (2002) points out that the BAN in the UK is the most developed in Europe. See Sohl (1998) on the informal venture capital market in the US.

238 See OECD (1996b).

239 A VCT is a type of quoted investment, which in turn invests in firms with gross assets of less than GBP 10 million. See OECD (1997).

in small firms and to diversify risk over a number of investments. VCT investors received considerable tax breaks.[240] Yet, VCTs failed to raise as much money as targeted, and merely one in five investments went to emerging high-tech firms.[241] Critics argued that the VCT scheme did not achieve the potential for funding technology-based start-ups because of an overly risk-averse approach.

Finally, a stock option scheme has been established by the legislation to encourage entrepreneurship. Stock options allow the provision of incentives via appropriate remuneration schemes. Thus they help to attract and maintain potential entrepreneurs out of stable employment into young high-tech firms.[242] The 1984 Finance Act reduced the hurdles to a more widespread use of stock options by offering tax incentives.[243] Later in 1999, a new scheme regarding stock options was introduced, known as *Enterprise Management Incentives* (EMI). Under the EMI, firms were able to provide tax-advantaged options to up to 15 key employees.[244] Despite the EMI and the variety of measures taken by the government, British venture capitalists continued to invest in small portions of funds early-stage investments.[245] The following section describes the current state of the British venture capital market and provides a synopsis of its specific features.

3.2.3 Current State and Peculiarities of the British Venture Capital Market

At the end of the 1990s, the venture capital market in the UK was the largest and most developed in Europe.[246] In 1999, British investments accounted for 46 percent of the total invested by the industry in Europe.[247] In addition, as outlined in Table 4, between 1991 and 2000 British venture capitalists represented the largest number of venture capital members of the *European Venture Capital Association* (EVCA). From 1984 to 1999, the member firms of the BVCA invested over GBP 35 billion in close to 19,000 companies.[248] In 1999, a record amount of GBP 7.8 billion was invested in over 1,300 companies.[249]

Although the British venture capital market currently represented the largest and most developed in Europe at the end of the 1990s, only a small percentage of the funds were allocated

---

240 The scheme offered tax relief on dividends and capital gains if investments were held for a minimum of five years. The maximum amount invested by an individual in a VCT was limited to GBP 100,000 per year. See Bank of England (2001).

241 Some VCTs manifested a tendency to prefer investments in property. See Bank of England (2001).

242 See Yermack (1995) on stock options and the principles of agency theory. For a discussion of stock options in the context of corporate governance see Harm (2000), pp. 24-26.

243 The proceeds from stock options were not subject to capital gains tax until exercise date under the prerequisite that they were part of a formal scheme. See Bank of England (2001).

244 According to the Bank of England (2001), shares worth up to GBP 100,000 at the time of the option grant were taxable at more favorable rate than the marginal income tax rate. Following to the 2000 Pre-Budget Report, the British government planned to substitute a total value of shares under EMI options of GBP 2.5 million per company for the limit on the number of employees.

245 See e.g. Murray and Lott (1995), Murray and Mariott (1998).

246 See Güllmann (2000).

247 See Bank of England (2001), p. 9.

248 GBP 29 billion of the amount was invested in the UK. See Bank of England (2001).

249 See Bank of England (2001). GBP 6.1 billion of the amount went to firms in the UK.

to start-up and early-stage investments.[250] Murray and Lott (1995) note that "it is ironic that the UK, as the largest and most experienced venture capital provider in Europe, can in part be characterized by a near universal retreat from investment in new technology-based firms".[251] Between 1989 and 1999, the proportion of start-up and early-stage investments fell from over 15 to about 5 percent.[252] Most of the funds available were invested in the form of expansion capital (20 percent) and MBOs (75 percent).[253] Thus, following the Bank of England (2001), in 1999 merely 260 firms received start-up and early-stage funding of an equivalent amount of GBP 350 million.[254] This trend reflects to a considerable extent the apparent preference of the UK venture capital industry for later-stage deals, requiring larger sums of money than early-stage ones. However, this tendency also suggests that the financing environment for start-ups, where financing difficulties are often particularly acute, has become more difficult.[255]

Table 4: Venture Capital Firms in the EU

Year	*Belgium*	*France*	*Germany*	*Italy*	*NL*	*Spain*	*Sweden*	*UK*
1991	12	30	11	10	21	5	1	42
1992	11	29	15	12	20	6	2	38
1993	12	29	13	11	19	7	3	40
1994	14	27	15	11	17	8	5	42
1995	13	32	18	11	16	8	5	40
1996	13	31	20	13	18	8	4	42
1997	16	32	27	12	19	8	5	52
1998	17	33	36	12	25	10	7	61
1999	23	48	51	16	32	14	11	79
2000	30	59	75	19	33	17	22	90

Note: Venture capital firm members of EVCA. NL - Netherlands. Source: Bottazzi and Da Rin (2002a), p. 244.

Another typical feature of the British venture capital market is the importance of business angels. They help to fill the equity gap in the provision of seed, start-up and early-stage funding for technology-based SMEs.[256] Since they often work in partnership with formal venture capitalists, these providers of informal venture capital play an important role for the industry. According to estimates, the UK had some 18,000 actual and potential business angels invest-

250 Based on a survey of 40 UK venture capitalists Murray and Lott (1995) find that technology projects had to meet more rigorous selection criteria than non-technology projects. For an analysis of the British venture capital market and potential reasons for the industry's preference for later-stage deals see Sieber (2002).

251 Murray and Lott (1995), p. 298.

252 Ibid.

253 According to the Bank of England (2001), 18 percent of the investments of BVCA members in UK companies in 1999 were directed to the high-tech sector. Locket *et al.* (2002) argue that the emergence of technology-focused MBOs, particularly involving entrepreneurial firms in the computing and pharmaceutical sector, has started to confound the simple dichotomy between early-stage and high-tech ventures on the one hand, and later-stage investments on the other.

254 See Bank of England (2001), p. 10.

255 See Garnsey (1995).

256 See Bank of England (2001), p. 3.

ing around GBP 500 million in 1999.[257] Although these figures might appear insignificant when compared to the large business angel market in the US, the business angel network in the UK is by far the most developed in Europe.[258]

Summarizing, there were attempts in the UK to set up a national venture capital system similar to the American model.[259] Although the predecessor of the *3i* group, one of the major players in the contemporary international venture capital, was founded on the initiative of the Bank of England, the British government tended to restrict its role to stimulating networks matching individual investors with entrepreneurs. Since the provision of equity finance for small companies and risky ventures is part of the British financial tradition, and some economic institutions are partly similar to those of the US, such as the well-developed stock markets, venture capital in the UK has grown into the most important and most developed market in Europe.[260] Nevertheless, only a small percentage of the venture capital funds in the UK was allocated to start-up and early-stage investments during the 1990s. Given the industry focus on later-stage investments one might expect venture-backed firms in the British IPO sample to be older and more mature than their counterparts in France or Germany. Moreover, given the tradeoff between industry expertise and diversification, British venture capitalists may on average provide less value-added to firms than their counterparts in the US. To lay the foundations for an international comparison of the British venture capital market with its equivalent in France, the following section reviews the development and the main characteristics of the French venture capital market.

3.3 France

3.3.1 Support for Small Firms in the Tradition of French Industrial Politics

In France, the public decision makers have been concerned to favor the financing of small innovative firms over the last 30 years, irrespective of the political coloring of the administration in power.[261] The approaches followed to some extent the traditional lines of French industrial politics. Thus, in the same way as the science and technology policy was characterized by the dominance of the centralizing state over the past three decades, the market for venture capital developed mainly due to the intervention of public institutions.[262] The government promoted venture capital by providing specifically designed incentive schemes and regulatory frameworks. In addition, the state played an active role using specialized public bodies, such as the financial institution *Caisse des Dépôts et Consignations* (CDC) and the *Agence Nationale pour la Valorisation de la Recherche* (ANVAR), to generate leverage effects and match the needs of small innovative firms.[263]

---

[257] See Bank of England (2001). The BVCA (2000) directory lists some 48 business angel networks in the UK.

[258] According to conservative estimations, the US informal market for venture capital comprises some 250,000 business angels investing around USD 10-20 billion per annum in over 30,000 firms. See Sohl (1998) and Harding (2002).

[259] See Clark (1987), p. 65.

[260] See Güllmann (2000), EVCA (2000).

[261] See Battini (1999).

[262] See Mustar and Larédo (2002) on the innovation and research policy in France and the state.

[263] See Chabbal (1995).

Already in 1955, the government attempted to provide financial support to small but not necessarily innovative firms by creating the *Sociétés de Développement Régional* (SDRs). SDRs received capital from various sources, such as banks, insurance companies, industrial companies as well as regional institutions and individuals, to take interim minority equity participations in small firms.[264] In 1964, the 15 SDRs operating in the French metropolitan area were supplemented by three additional ones in the overseas departments and territories. Over time, the tasks of SDRs were extended, including the possibility to grant subsidized long term loans and to finance real estate investments. At the end of 1993, SDRs administered a portfolio of approximately EUR 580 million[265] and held equity participations in about 3,000 firms.[266] Yet, due to substantial changes in the French financial system during the 1990s, implying the disappearance of subsidized loans and state guarantees on funds hitherto borrowed by the SDRs, the regional development bodies encountered difficulties in their operations.[267] As a consequence of the decreasing profitability most SDRs were either liquidated or acquired by other financial institutions, such as banks and venture capital firms, and do not exist anymore in their original form at the present time.

The point of departure of the French policy focusing on the financing of innovative firms goes back to the end of the 1960s when the state realized the country's international backlog in innovation.[268] In 1970, the government established the semi-public *Institut de Développement Industriel* (IDI) to provide – independently or in cooperation with SDRs – equity financing to SMEs in form of minority and provisional participations.[269] The principal shareholders of the IDI included the state with the major stake of 49.45 percent of the share capital and financial institutions, such as banks and mutual banks.[270] By the end of the 1970s, the government extended the range of activities of the IDI by fostering the collaboration with financial institutions and creating specialized affiliations. Retrospectively the IDI invested more than EUR 0.9 billion[271] and provided funds to approximately 700 firms from its creation until 1998.[272]

Inspired by Anglo-Saxon venture capital firms, in 1972 the public decision makers created the legal form of *Sociétés Financières d'Innovation* (SFIs). The purpose of SFIs was to promote research and the exploitation of inventions by strengthening the equity basis of firms with technological background via minority participations.[273] To derive benefits from fiscal advantages, e.g. the possibility to amortize 50 percent of participations, which had been held for more than three years, and to receive a free of charge risk coverage provided by the state,

---

[264] See Galesne (1999), Stephany (2001).
[265] FF 3.8 billion converted into EUR on the basis of the fixed exchange rate of FF 6.55957 per EUR.
[266] See Galesne (1999).
[267] See Bertero (1994) for an analysis of the changes of the French financial system in the 1990s.
[268] See Stephany (2001), p. 64.
[269] The investments ranged between FF 10 and 15 million per operation and were typically made for an average period of 5 to 7 years. The main targets for the provision of funds were firms with sales between 300 and 600 million FF. However, in particular in the 1980s, the IDI also provided support to bigger firms in financial difficulties.
[270] See Galesne (1999).
[271] FF 6 billion converted with the fixed exchange rate of FF 6.55957 per EUR.
[272] See Galesne (1999).
[273] The statute of SFIs was defined by a specific law (*Loi du 11 juillet 1972*), which was modified in 1985 (*Loi du 11 juillet 1985*).

SFIs had to fulfill a number of constraints.[274] All in all, the establishment of SFIs gave birth to some of the pioneers of French venture capital, such as *Sofinnova* or *Soginnove*, which still play an important role in the contemporary market.[275] Given the fiscal attractiveness, most venture capital firms in France operate currently under the legal statute of SFIs.

Given the concentration of venture capital to the capital region of Paris and the lack of funding accessible to firms operating in the provinces, the government decided to set up the *Instituts Régionaux de Participation* (IRPs) in 1975.[276] Analogous to the IDI, the main purpose of IRPs was to offer equity financing to SMEs through minority and temporary participations.[277] IRPs typically expected to sell participations to the public in later IPOs or back to founders or main shareholders once firms had successfully overcome the critical growth period. Like SFIs, IRPs benefited from a free of charge insurance for invested capital provided by the state. However, after almost 10 years of existence, the legal status of IRPs came to an end with the creation of *Sociétés de Capital Risque* (SCRs) in 1985.[278]

Despite these early initiatives of the state, the funding available to entrepreneurs until the end of the 1970s did not respond adequately to the problems related to the financing of innovation. Stephany (2001) argues that this shortcoming was mainly due to the enduring absence of a financial-industrial and entrepreneurial culture, the lack of mobile capital adapted to the funding of innovation, and the restrictive orientation of the national financial markets towards traditional investments. However, in the 1980s and 1990s, this situation changed with the institutionalization of venture capital, as presented in the subsequent section.

3.3.2 Institutionalization of Venture Capital in the 1980s and 1990s

Some authors refer the main starting point of venture capital in France to the early 1980s, when the state institutionalized venture capital by taking a number of legislative measures.[279] The government initiated a variety of programs channeling capital to small innovative firms at critical moments of their development. A central constituent of these programs was the *Société Française pour l'Assurance du Capital Risque* (SOFARIS), which was created in 1982 to cover financial risks related to granting loans and equity to small firms. SOFARIS was set up with the support of the government, of the financial and banking community, as well as of already existing agencies in order to manage government financed guarantee funds.[280]

---

[274] At least 80 percent of the funds had to be invested in small and innovative firms with sales of less than FF 50 million (or EUR 7.62 million if converted on the basis of the fixed exchange rate of FF 6.55957 per EUR), triennial reinvestments were mandatory, and one third of the invested capital had to be disinvested within a maximum period of six years. For more details see Tabourin (1989).

[275] *Sofinnova* was created by the financial institution *Crédit National* in 1972; *Soginnove* was formed by the *Société Générale* and the *Crédit Hôtelier* in 1973.

[276] See Galesne (1999), Stephany (2001). The creation of these specialized bodies for the funding of SMEs in the French regions accompanied the laws on regionalism that had been passed in 1974.

[277] According to Galesne (1999), the IRPs focused on the provision of funds to SMEs with annual sales between FF 60 and 200 million.

[278] See Dubocage and Rivaud-Danset (2002).

[279] See Stephany (2001), Dubocage and Rivaud-Danset (2002).

[280] SOFARIS was owned 42 percent by the French government. The balance was held by various French financial institutions and insurance companies. See OECD (1997), Kluth and Andersen (1999).

The basic principle of the risk coverage provided by SOFARIS worked in the way that the request for a loan was submitted to the lending bank. Yet, the final decision on granting a loan was subject to the approval of SOFARIS.[281] The venture capital insurance offered by SOFARIS covered up to 50 percent of bank and venture capital funding to SMEs, and up to 70 percent to business start-ups.[282] According to the OECD (1997), in 1995 SOFARIS supported 1,514 SMEs by providing EUR 274 million[283] of guarantees. In the mid-1990s, the activities of SOFARIS were merged with the *Crédit d'Equipement des Petites et Moyennes Entreprises* (CEPME), a state-controlled bank providing funding for SMEs, to the *Banque du Développement*.

In addition to the foundation of SOFARIS, the public bodies set up a legal, fiscal and institutional framework favorable to venture capital investments. By creating the statutes of *Fonds Communs de Placement à Risque* (FCPRs) and SCRs in 1983 and 1985, respectively, the state offered specialized investment instruments adapted to the needs of venture capitalists.[284] Both FCPRs and SCRs were tax exempted if investments represented non-controlling participations and capital gains were reinvested. Taxes had to be paid only once by the beneficiary shareholders at the time when capital gains were distributed.

In 1983, the Paris Bourse launched the *Second Marché* as a specialized capital market segment for SMEs. In order to represent the interests of venture capitalists, the *Association Française des Investisseurs en Capital* (AFIC) was founded in 1984.[285] Yet, despite the numerous measures taken by the public decision makers with respect to the fiscal and institutional framework, venture capital did not prosper until the 1990s. Dubocage and Rivaud-Danset (2002) assert that the industry did not breakthrough until the launch of the *Nouveau Marché* in 1996. By the creation of the stock market segment, aiming to attract the listings of young technology-oriented firms, one major obstacle to the development of venture capital has been removed, i.e. the lack of a liquid stock market for growth companies, allowing venture investors to bring portfolio firms public, and thus to exit investments via IPOs.[286]

More recent measures adopted by the state with respect to the development of venture capital include the creation of *Fonds Communs de Placement Innovation* (FCPI), stock options, and a new law on innovation. FCPIs were initiated in 1997 as specialized mutual funds

---

281 For amounts up to FF 2 million the approval was made by the regional office of SOFARIS. Above this amount, the decision was made by the regional officer of SOFARIS, and in the case of very large proposals by the directorate general. Thus, SOFARIS was a hierarchically ordered organization resembling traditional French state agencies. See OECD (1995), p. 28.

282 The financing covered comprised loans with maturities between 2 and 15 years, leasing agreements, and the funding provided by venture capital companies. The charges for the risk coverage by SOFARIS amounted to 0.6 percent per annum, and in the case of venture capital firms, 0.3 percent and a share of capital gains. The maximum guarantee per company was FF 5 million (or about EUR 0.8 million if converted on the basis of the fixed exchange rate of FF 6.55957 per EUR). See OECD (1997).

283 FF 1.8 billion on the basis of the fixed exchange rate of FF 6.55957 per EUR.

284 See Dubocage and Rivaud-Danset (2002). The organizational type of FCPR is similar to the limited partnerships in the US.

285 See AFIC (1996), Stephany (2001), Dubocage and Rivaud-Danset (2002).

286 See Black and Gilson (1998).

targeting private individuals by offering tax advantages.[287] In 1998, stock options were introduced as a compensation form allowing the founders of start-up firms being unable to offer attractive levels of remuneration to recruit highly competent personnel. Finally, in 1999, the government implemented a law on innovation, aiming at intensifying technology transfers between research institutions and industrial companies. Prior to the adoption of the law, researchers could not found their own firms without resigning from research positions, and thus abandoning both salary and status. Hence, the main idea of the new law was to give more freedom to researchers to start up their own business.[288]

3.3.3 The Role of Public Institutions in the Promotion of Venture Capital

In a synopsis of the development of French venture capital, Dubocage and Rivaud-Danset (2002) stress the essential role played by a small number of public bodies, which were appointed by the state to generate leverage effects and to provide financing as well as assistance to innovative companies. The authors argue that particularly two public institutions played a key role in the development of venture capital in France: (1) ANVAR, and the (2) state-owned CDC. In the following, these institutions are briefly portrayed.

ANVAR was created by the government in 1968 to support the exploitation of the results of public research institutions. Endowed with an annual budget of more than EUR 300 million from the Research and Industry Ministries in 2002, the main purpose of ANVAR still is to generate leverage effects and to provide financing as well as expertise to innovative companies.[289] The public agency provides funding and consultancy services to small firms, laboratories, and individuals involved in research and development activities. It supplies primarily quasi-equity instruments in form of zero-interest loans, which have to be paid back as a fixed percentage of sales revenue if a financed venture turns out to be successful.[290] In addition to the provision of funding and consultancy services, ANVAR often matches firms with venture capitalists. At the end of the 1990s, ANVAR increased its range of services by offering assistance to firms applying for a listing at the *Nouveau Marché*.[291] Looking back, ANVAR allocated approximately EUR 3.35 billion from 1979-1998, most of it for the development of new products, technologies, and services.[292] In 2002, ANVAR held 66 direct participations in young firms.[293]

The second public body that has been actively involved in the promotion of French venture capital is the CDC. The financial institution has been in existence since 1816 and represented a major public investor with approximately EUR 10 billion under management by the end of

---

287 FCPIs offer tax advantages under the condition that at least 60 percent of their capital is invested in innovative French unlisted companies. In 1999, the 14 newly established FCPIs attracted investment capital of approximately EUR 200 million. See Stephany (2001), Dubocage and Rivaud-Danset (2002).

288 See Stephany (2001), Dubocage and Rivaud-Danset (2002).

289 See Ducourtieux (2003).

290 The failure rate in terms of firms being unable to repay has been 10 to 15 percent, whereas successful firms generated a return of six times the amount invested. See Sagari and Guidotti (1992), pp. 30-31.

291 ANVAR guaranteed, for example, a part of the listing expenses. See Bank of England (1996), Dubocage and Rivaud-Danset (2002).

292 See Denis (1999), Ducourtieux (2001).

293 See Ducourtieux (2003).

1998.[294] The CDC conducts activities in both the public interest domain as well as the competitive economy. As outlined in Table 5, the separation between activities in the public interest and the private sector area is reflected in the organizational structure of the venture capital activities of the group. While the affiliate *CDC Participations* operates as a major player in the competitive venture capital market and manages also funds of third parties, *CDC-PME* pursues activities in the public interest domain by offering financing to SMEs in all stages of development. In addition, the CDC promotes investments by managing a *Fonds Public pour le Capital Risque* (FPCR) in cooperation with the *European Investment Bank* (EIB).[295]

Table 5: Venture Capital Compartments of the CDC Group

	CDC Participations	*CDC PME*	*FPCR*
Domain of activity	Competitive market economy	Public interest activity	Public interest activity
Source of funds	80% CDC 20% third parties	100% CDC	66% State 33% EIB
Activity	Management via six investment funds, including *Fonds Invest* and *CDC-Participation* among others	Provision of support at every stage of the venture capital process (seed, expansion)	Funds granted to FPCRs investing in French innovative companies in existence for less than 7 years

Source: Dubocage and Rivaud-Danset (2002), p. 32.

In summary, the most specific feature of French venture capital is the key role of the state, having been the driving force in identifying the problems of capital allocation to small innovative firms and attempting to intervene in the market via specialized public institutions. Stephany (2001) points out that, in contrast to other countries, where venture capital originated from the private sector, the emergence of venture capital in France has to be primarily ascribed to efforts carried out by the government.[296] Some authors interpret the active role of French public agencies in early-stage investments as a substitute for the lack of a market for business angels. Yet, it appears questionable whether government bodies are able to provide the same value added as private business angels.

Despite the favorable legal environment and the awareness of the importance of funding innovation on the part of the government, French venture capital started with difficulties and did not prosper until the latter half of the 1990s when the *Nouveau Marché* was created. Even though the French approach to promote venture capital can be considered as an example of substantial direct governmental participation in the market, some prestigious players in European venture capital, such as *Sofinnova* and *Partech*, originate from France. Nonetheless, given the substantial number of public and semi-public venture investors in the French market, the study controls for performance differences between firms backed by public and private venture capitalists. The following section focuses on the development of venture capital

---

294 See Dubocage and Rivaud-Danset (2002).

295 A first FPCR was formed in 1998, receiving EUR 90 million from the French state, which came partly from the privatization proceeds of *France Télécom* in 1997, and EUR 45 million from the *European Investment Bank* (EIB). In 2000, a second FPCR was set up and received endowments of EUR 150 million. See Dubocage and Rivaud-Danset (2002).

296 See Stephany (2001), p. 64

in Germany, before finally comparing the three European venture capital markets of interest with each other.

3.4 Germany

In Germany, the market for venture capital has been insignificant and underdeveloped for many years.[297] Start-up firms traditionally relied either on internal financing or conventional banks loans to fund the expansion of their businesses. Although German banks generally provided business customers with detailed advice on aspects of running a firm, they were neither specialized in the provision of venture capital nor in the financing of technology-based start-ups. Nevertheless, large German banks heavily influenced the market for SME and technology financing in the past. Since banks feared the competition effects from the operations of venture capitalists on lending activities, they did not promote the venture capital market, and the industry received only little support over many years.[298]

In addition to the lacking support from the supply side, venture capitalists have themselves frequently not been accepted by entrepreneurs who are said to have a strong preference for debt financing to maintain control over firms.[299] Many entrepreneurs belong to SMEs from the *Mittelstand*[300] that played an important role in the reconstruction of Germany during the post-war period. Given the typical industry characteristics of these firms, most entrepreneurs managed to finance growth relying on retained earnings and secured bank loans.[301] Also, entrepreneurs were often unaware of alternative resources or averse to the concept of sharing equity with outside investors. Consequently, venture capitalists encountered difficulties when they intended to work with entrepreneurs.

Despite these problems, there have been substantial efforts by the government to promote venture capital. The following sections first outline briefly the evolution of venture capital in German after World War II and sketch the most important public programs to improve the funding situation for SMEs. Then, the specific problems related to the creation of a venture capital market in a bank-based financial system are discussed on the basis of the unsuccessful effort to start a US-style venture capital industry with the *Deutsche Wagnisfinanzierungsgesellschaft* (WFG) in 1975. Finally, some recent developments, such as the rapid development of the industry during the 1990s and its underlying issues, are addressed.

3.4.1 Government Support Programs and Industry Evolution

Like in other European countries, the German public authorities embarked on numerous measures to promote venture capital and offered a variety of support programs for entrepre-

---

297 A historical overview is provided by Lessat *et al.* (1999).

298 See e.g. Gerke (1996).

299 Bascha and Walz (2001b) underline that German entrepreneurs traditionally favored buy-backs to IPOs and trade sales in order to keep the independence of their firms. Yet, this pattern might have changed after the launch of the *Neuer Markt* in 1997.

300 The term *Mittelstand* is often referred to regional, family-owned businesses, with up to approximately 1,000 employees. See Fiedler and Hellmann (2001).

301 Fiedler and Hellmann (2001) mention that firms from the *Mittelstand* are often specialized in specific industrial niche markets, such as manufacturing machinery and automotive parts. Since these traditional businesses tend to be slow-moving industries, there is no need to grow rapidly.

neurs.[302] One major source of capital for SMEs was the *European Recovery Program* (ERP), also known as the *Marshall Plan*, which was devised for the financing of reconstruction in post-war Europe. The ERP offered investment programs predominantly directed at SMEs, whereby an increasing part of the lending activities was geared toward innovation and venture capital over time.[303] The government mainly employed two public financial institutions to distribute ERP funds: the (1) *Kreditanstalt für Wiederaufbau* (KfW), and the (2) *Deutsche Ausgleichsbank* (DtA).[304] Both institutions granted funds for the refinancing of venture capital firms, offered risk guarantees as well as equity substituting loans to entrepreneurs through banks, and provided equity to companies via venture capitalists.[305]

The KfW is the major German development bank for small firms and was founded in 1948 to administer the financial resources of the ERP.[306] It offers refinancing for venture capital funds at attractive conditions instead of investing directly in portfolio firms.[307] The subsidy of the KfW consists of the difference between the interest rate in the competitive market and the interest that is actually charged.[308] In addition to subsidized interest rates, the KfW covers a part of investment losses if ventures fail.[309] The selective refinancing provided through the KfW is limited to an amount of EUR 1.5 million per firm and offers venture capitalists a way to increase the leverage of investments. An important feature of the KfW operations is the way in which they work within the German banking system. Since entrepreneurs submit the inquiry for KfW funds at banks, the program is handled through the banking system.[310] Moreover, the KfW involvement is limited to co-financings, by which banks make sure that they are not financing the entire amount of investments at stake.[311]

Aside from the KfW, the DtA was involved in the disbursement of ERP funds.[312] The DtA was established in 1950 with the purpose to reintegrate refugees and people from formerly German territories in the East into the economic process.[313] Despite this originally political motivation the role of the bank changed significantly over time. By the end of the 1960s, the focus shifted toward the promotion of small firms, and the DtA became known as the federal government's bank for business start-ups and SMEs.[314] In the following, it turned into the

---

302 See Güllmann (2000), and Schmeisser and Krimphove (2001) for comprehensive overviews of current government schemes.

303 See Kreditanstalt für Wiederaufbau (2002).

304 By the end of 2002, the German government announced its intention to merge both financial institutions. See w.a. (2002).

305 See Bundesministerium für Wirtschaft und Technologie (2000).

306 In 2001, the KfW administered total assets of about EUR 245 billion and had 2,190 employees. See Kreditanstalt für Wiederaufbau (2002) and w.a. (2002).

307 The interest rates of the loans granted by the KfW are set at the lower end of the market rates and fixed over the duration of the loan.

308 The KfW can afford to lend on favorable conditions due to its quasi-government status and a historically low failure rate amongst its customers that is less than one percent. See Bank of England (1996).

309 See Fiedler and Hellmann (2001).

310 See Harm (1992).

311 Ibid.

312 While the mechanism for the distribution of ERP funds via the DtA is similar to that of the KfW, the interest rate charged for the provision of funds by the DtA is slightly lower.

313 The initial name of the bank was *Vertriebenen-Bank* (Bank of Refugees), and changed into *Lastenausgleichsbank* (Bank for the Equalization of War Burdens) in 1954. It was only in 1986, that the bank was renamed to get its current name.

314 See Deutsche Ausgleichsbank (2001).

main provider of official support for small entrepreneurs and contributed, jointly with the KfW, a considerable share to the financing of the market for start-ups.[315] Like the KfW, the main impact of the DtA was an increase in the lending volume to young and small firms. Today, the DtA also offers venture capital financing through specialized subsidiaries, such as the *Technologie-Beteiligungs-Gesellschaft* (TBG), being established in 1989 to promote high-tech start-ups.

The German venture capital industry started to develop in 1965 with the foundation of the first *Kapitalbeteiligungsgesellschaften* (KBGs).[316] KBGs were modeled after the SBICs in the US with the objective to offer equity finance for young firms.[317] Between 1965 and 1972, 33 KBGs were formed, most of them by banks, to foster the growth of SMEs.[318] Kuemmerle (2001) argues that KBGs were profit-oriented and could have evolved into true venture capital firms if they had not been structured as open-ended funds being owned by banks. However, the success of KBGs was relatively modest, and the number of investments in SMEs was quite small. In addition, KBGs mostly invested in established firms and therefore did not qualify as venture capitalists in a narrow sense. Jeng and Wells (2000) note that KBGs provided little management supervision or support although they were often involved with companies up to 8 years. Although 20 of 33 KBGs survived, only 6 of them played a significant role in the venture capital market in the mid-1990s.[319]

Beside KBGs, several *Bundesländer* initiated *Mittelständische Beteiligungsgesellschaften* (MBGs) in the 1960s and 1970s to ensure the provision of financing and management support to small technology-oriented firms.[320] An explanation for the initiative of the federal authorities might be the fact that technology policy represents one of the areas, for which the German federal states are responsible. MBGs were established by regional institutions, such as *Sparkassen* (local savings banks), commercial banks, chambers of commerce, and the Ministry of Economic Affairs of the respective federal state to provide ERP funds in form of silent equity participations to local SMEs.[321] Yet, the MBG investments differed to some extent from the operations undertaken by the private KBGs. First, MBGs pursued rather regional development motives, e.g. the creation of employment rather than pure profit goals. Second, the MBG investment managers faced different incentives, since they were typically bureaucrats with lifetime employment and did not benefit directly from investment gains.[322] Hence, MBGs were not eligible for the role performed by private venture capitalists.

Recent government initiatives comprise the *Beteiligungskapital für junge Technologieunternehmen* (BJTU) program running from 1985-1995. The program pursued a co-

---

315 See Harm (1992).
316 See Fiedler and Hellmann (2001).
317 See Kuemmerle (2001).
318 See Leopold and Frommann (1998).
319 Ibid.
320 See Harrison (1990). For an overview of the technology policy of the German *Länder* see Scherzinger (2002).
321 One important regional venture capital fund operating in Bavaria is the *Bayern Kapital Risikobeteiligungsgesellschaft* that bases on cooperation between the local *Sparkassen*, commercial banks, chambers of commerce, and the Bavarian Ministry of Economic Affairs. See Kluth and Andersen (1999).
322 See Kuemmerle (2001).

investment scheme administered by the DtA subsidiary TBG.[323] The TBG acted as passive investor, matching venture capital investments up to approximately EUR 0.5 million, and gave private investors two options.[324] The investment could either be bought out by private investors at a premium of about 25 percent over the initial investment, or be sold back to the TBG for at least 40 percent of the initial book value, if the venture failed.[325] Since the failure rate of firms financed by the BJTU scheme was relatively low, it was sometimes criticized that the program pursued an overly cautious approach.[326] The BJTU was also subject to debate because of its lengthy and complex application procedure with approvals taking up to one year's time.

In 1995, the BJTU was succeeded by the *Beteiligungskapital für kleine Technologieunternehmen* (BTU) program, representing the main current government scheme for the promotion of venture capital investments.[327] The BTU program offers funds for investments in young high-technology firms[328], containing both a loan and an equity scheme.[329] Under the loan scheme, a publicly supported bank covers partly the losses of venture capitalists by refinancing and providing risk coverage for participations. Under the equity scheme, a publicly supported investor offers co-financing to a firm under the prerequisite that another investor, e.g. a venture capitalist, is involved as lead investor and commits at least the same amount of equity.[330] Similar to the BJTU, the investor can buy out the TBG interest for a 30 percent premium.[331] Moreover, the TBG equity guarantee covers up to 50 percent of the invested capital for the first 5 years of a venture.[332] Since BTU funds are brought in at a level matching private investments, the program generates leverage effects.

3.4.2 The Experience of the Deutsche Wagnisfinanzierungsgesellschaft

Despite the multiple efforts undertaken by the German public authorities, there was no institutional structure available to support risky start-up ventures until the 1970s.[333] In 1975, there was an explicit attempt to create a venture capital industry by the formation of the *Deutsche Wagnisfinanzierungsgesellschaft* (WFG).[334] The WFG was established by 29 German financial institutions with the ambitious goal to fund risky projects in technologically intensive industries.[335] Moreover, the WFG was supported by the German Federal Ministry of Re-

---

323 See Jeng and Wells (2000).
324 See O'Shea (1996).
325 Ibid.
326 According to the OECD (1997), the failure rate of the BJTU scheme was 17 percent in 1995.
327 See Kulicke and Wupperfeld (1996) for a more detailed analysis of the program.
328 The target firms of the BTU program must have a distinctively new product or service under 10 years old, less than 50 employees, and a turnover up to EUR 7 million. See TBG (2001).
329 See Schertler (2002) for an analysis of the BTU program and its predecessors, and a discussion of public loan and equity schemes in venture capital markets.
330 The invested amount offered by the TBG cannot exceed EUR 1.5 billion per company. See TBG (2001).
331 See OECD (1997).
332 Ibid.
333 See Jeng and Wells (2000).
334 See Becker and Hellmann (2003) for a detailed analysis of the WFG.
335 The WFG was initially capitalized with DM 50 million (approximately EUR 25 million) of which 54 percent were contributed by private banks as well as regional public sector banks. See Kuemmerle (2001), Becker and Hellmann (2003).

search. The government agreed to provide risk guarantees covering 75 percent of any capital losses as well as up to 75 percent of annual operating losses.[336]

However, the WFG turned out to be an outright failure. Most investments were unprofitable, and in 1981 the WFG had accumulated losses of roughly EUR 16 million.[337] As a consequence of this financial disaster, in 1984 the WFG was restructured and converted into a private company. After the end of the relationship with the government, five of the largest banks managed the liquidation of existing portfolio and decided to establish a new WFG fund for ensuing investments. Yet, the initial industry focus on high-technology start-ups was abandoned and shifted toward later-stage non-technology investments.[338] After all, in 1988, the main investor Deutsche Bank decided to buy out the remaining shareholders and integrated the WFG as a subsidiary into the group's private equity operations.

Becker and Hellmann (2003) hypothesize that the failure of the WFG was related to the use of inappropriate contract structures for capital disbursement as well as inadequate governance relationships between the venture capitalists and entrepreneurs.[339] Kuemmerle (2001) maintains that the contract structures clearly favored entrepreneurs at the expense of venture capitalists. He points out that the bylaws stated that "the WFG should primarily fund entrepreneurs who could not obtain funding elsewhere".[340] Although the initial intention was to fund risky investments, which had neither been financed by KBGs nor MGBs, this statute evidently created an adverse selection problem.[341]

The case of the WFG revealed the problems inherent to the promotion of a venture capital market in a country with a bank-based financial system. Following Becker and Hellmann (2003), only banks were credible and able to promote venture capital in Germany. Accordingly, the government decided to lean on them to promote venture capital. However, there was a divergence of interests between banks and the government from the very beginning of the project. While the former were primarily concerned about minimizing risks to their reputation, the latter was mostly concerned about the commercialization of new technologies. To prevent a crowding out of the private market, the WFG was set up with the intention to finance only companies which could not obtain financing from other sources. This criterion suited banks fearing competition from the project. Yet, since both the government and banks were more concerned about their interests than the maximization of WFG profits, the project ultimately resulted in inadequate contract structures and poor financial performance.[342]

3.4.3 Recent Developments and the Transformation of the Industry

The WFG collapse caused substantial negative publicity. It was not until 1979 that the first German private venture capital firm was established, adopting the contractual structures and the investment approach of US venture capitalists.[343] In the 1980s, the overall venture capital

---

336 See Albach *et al.* (1986).
337 The losses of the WFG amounted to DM 32 million. See Kuemmerle (2001).
338 See Jeng and Wells (2000).
339 See Becker and Hellmann (2003).
340 Kuemmerle (2001), p. 15.
341 See Mayer and Müller (1991) for a discussion of the selection criteria for portfolio firms of the WFG.
342 See Becker and Hellmann (2003).
343 The first private German venture capital firm was *Hannover Finanz*. See Kuemmerle (2001).

market portfolio grew to roughly EUR 500 million.[344] In 1987, the government established the *Geregelter Markt*, a stock market segment comparable to the OTC market in the US. The market had less stringent listing requirements than the *Amtlicher Handel* (main market) and was geared to appeal small firms. However, instead of attracting the listing of young technology-oriented firms, the segment primarily interested established firms from the *Mittelstand*.[345]

In 1989, the *Bundesverband Deutscher Kapitalbeteiligungsgesellschaften* (BVK) was founded to advocate the interests of venture capitalists.[346] Moreover, in the course of the 1990s, a growing degree of acceptance of venture capital could be observed.[347] Due to several factors, such as the rise in entrepreneurial opportunities and the increasing willingness of individuals to start up their own businesses, which were both related to the advent of the Internet-related technologies, as well as the introduction of the market segment *Neuer Markt* at the Frankfurt stock exchange, German venture capital left its niche existence and witnessed a remarkable rise.[348]

The *Neuer Markt* provided an exit channel and liquidity for venture capital investors. In addition, it aimed from the very beginning at attracting international and institutional investors to a stronger extent than the *Geregelter Markt*.[349] Thus, along with the launch of the market segment for high-technology firms, the German venture capital market did not only attract domestic but also many foreign independent firms, while the former market had primarily been ruled by subsidiaries of established German banks. Between 1990 and 1999, the volume of funds managed by the industry increased from EUR 481 million to 12.81 billion, and at the end of the decade, the annual growth rate reached an all time high with almost 50 percent.[350]

Overall, German venture capital can be characterized by its high level of bank involvement.[351] Since pension funds in Germany were not structured as independent investment entities and played a less important role than in Anglo-Saxon countries, venture capital companies received most of their capital from large banks. Despite the initial reluctance in the promotion of venture capital, depositary financial intermediaries finally played an important role as provider of venture capital funds and continued to be dominant in the market in the 1990s.[352] It seems that banks changed their early attitude and reassessed the financing of small and innovative companies as a field of strategic importance for their core business, being related to the creation of a clientele for future banking services.[353]

The crucial role of banks was recognized and used by the government to channel funds to entrepreneurs. Financial institutions, such as the KfW and DtA, administered publicly subsi-

---

344 See Kuemmerle (2001).

345 See Stedler (1987), Kuemmerle (2001).

346 By 1995, the BVK had already over 100 members.

347 The average annual growth rate of overall funds equaled 22.1 percent in the 1990s. See BVK (1999).

348 See Fiedler and Hellmann (2001) for an analysis of the rapid development of the German venture capital industry in the 1990s.

349 The listing requirements of the *Neuer Markt* stipulate firms to report quarterly earnings in English and to adopt international accounting standards, such as IAS or US GAAP. See Appendix A 3.

350 See BVK (2000).

351 See Kluth and Andersen (1999).

352 According to Kluth and Andersen (1999), banks accounted for 55 percent and insurance companies for 12 percent of Germany's venture capital funds in 1994. See also Table 8 (p. 82) for more recent statistics.

353 See Büschgen (1985), Kluth and Andersen (1999).

dized loans and credit insurance through the banking system, thus succeeding to increase the lending volume to SMEs. Yet, the approaches followed by the KfW and DtA were also subject to debate given their selectivity and the result that many inventions did not make it to the market.[354] Early government efforts to promote venture capital proved to be unsuccessful. The first attempt to establish a venture capital industry by founding the WFG in 1975 was hampered, among other reasons, by the bank-based financial system. The WFG failed because of bad incentive alignments among the involved parties, which were due to conflicts of interest between banks and the government.[355] Thus, the first venture capitalists focused on firms which did not obtain financing from other sources, ending up with adverse selection.[356]

Finally, the considerable transformation and boom of German venture capital during the last decade are noteworthy. Stimulated to a large extent by changes in the regulatory environment and the creation of the *Neuer Markt*, the industry took off in the mid-1990s and turned in a relatively short period into one of the fastest growing and most visible elements of the economy.[357] Therefore, many participants in the market started activities just recently. This trait might be reflected in insufficient skills and experience of many German venture capitalists. The empirical analysis in the following chapters takes up this concern, comparing the performance firms backed by German venture capitalists to those backed by investors of different origins. It remains an open issue whether the strongly grown industry survives its first cyclical crisis and continues to play a key role in the financing of SMEs, or whether the shift in the importance of venture capital during the 1990s represented just a short-lived phenomenon.

3.5 Conclusions

In summary, the analysis of this chapter revealed that the national approaches in France, Germany, and the UK to promote venture capital and to provide funding to innovative SMEs have several features in common. Aside from the creation of specialized venture capital funds on the national or regional levels, public bodies provided grants and subsidized loan guarantee schemes to reduce the risks encountered by entrepreneurs and venture capitalists. In accordance with the findings of the literature review in chapter 2, emphasizing that institutional backgrounds play a crucial role for the development of venture capital, the governments of the considered countries created business environments with adequate legal forms and fiscal incentives for investments of individuals and venture capital firms to foster the development of the industry.

Despite these similarities venture capital developed earlier and more rapidly in the UK than in France and Germany. The pioneering role of the UK in the European venture capital industry was favored by the early existence of liquid IPO markets, tailored to the exit needs of venture capitalists. In France and Germany, by contrast, the expansion of the venture capital

---

354 The Bank of England (1996) notes that only three percent of start-ups in Germany were successful in obtaining long-term finance and that small-ticket equity finance was more difficult to obtain in Germany than in the UK.

355 See Kuemmerle (2001).

356 See Mayer and Müller (1991), Becker and Hellmann (2003). See Schmidt (1988) for a discussion of the quality of venture capital in Germany in the 1980s.

357 See Fiedler and Hellmann (2001), Kuemmerle (2001). Appendix A 2 provides a comparative overview of the evolution of venture capital in France, Germany, and the UK.

industry was not fostered until the mid 1990s by the establishment of stock market segments, suited to the needs of young growth firms with technology backgrounds and providing exit routes for venture capital investments. Hence, the overall findings of the comparative analysis confirm the important function of well-developed stock markets for the development of a venture capital industry, as suggested by the work of Black and Gilson (1998). Nevertheless, it is worth mentioning that only few British venture capital investors specialized in early-stage and technology-oriented activities. This observation underlines that the mere existence of an active IPO market does not represent a sufficient condition for the provision of early-stage venture capital for innovative firms.[358]

As a result of the varying set-ups of the national financial systems in the three countries, different institutions represent different ways of transferring funds to recipient firms. In Germany, with its bank-centered financial system, the government relied for the most part on the involvement of private banks to promote venture capital and support entrepreneurs. In France, where publicly owned banks played a dominant role in the financial sector, public or semi-public financial institutions were mainly involved in the distribution of funds to venture capitalists and entrepreneurs. Finally, in the UK, with its well developed financial markets, the government limited its function largely to taxation measures influencing the supply of private venture capital as well as the creation of networks, such as BANs, matching individual investors with entrepreneurs.

Given the different financial systems and paths of development of the venture capital markets in France, Germany, and the UK, the skills and business experience of the venture capitalists operating in the respective national markets may vary significantly. In particular, taking into account the earlier development of venture capital in the UK, British venture capitalists are supposed to be on average more experienced than their counterparts in France and Germany. Moreover, in the German market, with the traditionally strong influence of banks, venture capitalists might still encounter severe adverse selection problems, which were also related to the first failure to establish a venture capital industry in Germany, as argued by Becker and Hellmann (2003).

Overall, it remains an empirical question whether or not the differences in the business environments and the conjectured cross-country variations in the expertise of venture capitalists affect their ability to spur firm performance, as suggested by prior theoretical contributions and empirical evidence from the US. Therefore, the following chapters investigate the involvement of venture capitalists in firms going public in France, Germany, and the UK on the micro level. The empirical analysis explores cross-country variations in the impact of European venture capitalists on the performance of portfolio firms. Moreover, to account for the different participation of banks and public institutions in the considered venture capital markets, as discussed in the previous sections, potential variations in the venture-backing quality among various types of venture capital investors are scrutinized.

Before presenting the main findings of the empirical analysis in chapter 5, the subsequent chapter outlines the research design of the international study on venture- and non venture-backed IPOs. It includes details on the hypotheses building as well as on the characteristics of the employed data set and the methodology.

---

[358] See e.g. Mayer *et al.* (2001).

4 Design of Analysis

4.1 Overview

In this chapter, the research design of the empirical analysis is outlined in more detail. The study considers the impact of venture capitalists on the operating and long-run market performance of firms going public in France, Germany, and the UK, by using a hand-collected data set of venture- and non venture-backed IPOs at the French *Nouveau Marché*, German *Neuer Markt*, and British *techMARK* dating from 1996 to 1999. In section 4.2, the theoretical and empirical insights of earlier studies on the operating and market performance of venture-backed firms are discussed and used to formulate empirically testable hypotheses.[359] Section 4.3 then describes the international data set of IPOs, which is used to test the derived hypotheses. It provides details on the collection criteria as well as the data sources. Moreover, the sample composition and selection issues are discussed. Finally, section 4.4 describes the methodology and the different variables employed throughout the analysis.

4.2 Hypotheses

4.2.1 Long-Run Performance of IPOs

Various empirical contributions have focused on the performance of IPOs.[360] As pointed out in section 2.5.3, an anomaly that has generated much research in the IPO literature is the finding that the shares of firms going public appear to suffer long-run underperformance relative to seasoned companies in the aftermarket. Theoretical explanations for this observation can be divided into two main groups: (1) approaches related to the underpricing literature, such as signaling approaches, theories of book-building[361], legal liability[362], price support[363], and agency costs[364]; (2) approaches based on a behavioral finance background, such as heterogeneous expectations[365], fads and timing[366], and window-dressing.[367] Alternatively, measurement problems are discussed in the IPO literature. They are indicated as a more trivial cause for the observed performance figures.[368]

Jenkinson and Ljungqvist (2001) discuss the consistency of empirical studies with theoretical explanations. They maintain that the price support argument provides a promising avenue for future research related to the documented underperformance of IPOs in the long run.

---

359 The specific performance measures employed in the analysis are described in more detail in section 4.4 as well as in Appendix A 6.

360 Ritter (1991) and Loughran and Ritter (1995) investigate the long-run performance of IPOs. Brav and Gompers (1997) focus on the long-term underperformance of venture-backed firms in comparison to non venture-backed firms. Jain and Kini (1994) and Mikkelson *et al.* (1997) explore the operating performance of IPOs.

361 See Hanley (1993).

362 See Hughes and Thakor (1992).

363 See Ljungqvist (1995).

364 See Jensen and Meckling (1976).

365 See Miller (1977).

366 See Ritter (1991), Loughran *et al.* (1994), Rajan and Servaes (1994), Loughran and Ritter (1995).

367 See Jenkinson and Ljungqvist (2001), pp. 139-161, for an extensive overview of possible explanations.

368 See Fama (1998), Barber and Lyon (1997a), Kothari and Warner (1997), Brav (2000), Mitchell and Stafford (2000).

The widespread practice by underwriters to support the initial trading price of IPOs in the aftermarket may spuriously lead to the observation of abnormally low returns. Jenkinson and Ljungqvist (2001) also emphasize that it is difficult to control accurately for risk over long time horizons. Thus, no adequate estimates of systematic risk are available.

Since the present study deals with the performance of IPOs, the analysis first focuses on the issue of long-run performance, before investigating the particular impact of venture capitalists on firms going public. Based on the theoretical arguments as stated above and in previous empirical findings, the following performance-related hypothesis is derived:

H1: The stocks of IPO firms underperform relative to seasoned stocks in the long-run.

To test this conjecture, different measurement techniques for the long-run performance of IPOs, such as buy and hold returns and wealth relatives are investigated over a 3-year post-issue interval in section 5.1.2.2. A detailed description of the variables, which are used to assess the long-run stock price performance in the aftermarket, is outlined in section 4.4.2.2.

4.2.2 Screening, Monitoring, and Certification of Venture Capitalists

The literature review in chapter 2 reveals that venture capitalists can be viewed as specialized financial intermediaries who provide much more than capital to young technology-based firms.[369] Venture capitalists also provide intensive oversight and often share some of the private information about the prospects of portfolio firms.[370] Besides high-risk funding, they can add value to firms through pre-investment screening, monitoring, and management support.[371] Moreover, the reputation and certification role of venture capitalists play an important role in the context of bringing firms to the capital markets. In the following, the theoretical arguments related to the characteristics of venture capital are employed to derive different hypotheses with respect to the performance of venture-backed IPO firms.

First, venture capitalists typically screen a relatively large number of investment opportunities and seek to concentrate on a manageable number of ventures.[372] Since venture capital firms typically have small staff, they merely invest in a fraction of deals, which come to their attention.[373] Also, the investment decision is often contingent on the identification of a syndication partner who agrees about the attractiveness of the investment.[374] Depending on the investment stage focus, the screening criteria reflect a tendency to limit investments to areas, with which venture capitalists are familiar, specifically in terms of technology, product, and market scope.[375] Although there is evidence that venture capitalists in different countries tend

---

369 See e.g. Kortum and Lerner (2000).

370 See Lerner (1995).

371 See Berger and Udell (1998), Gompers and Lerner (1999c, 2001b), Hellmann and Puri (2002), Kaplan and Strömberg (2003).

372 Gompers and Lerner (2001b) mention that venture capitalists employ various deal identification strategies and techniques. See Gompers and Lerner (2001b), pp. 46-47.

373 Lerner (2002) states that historically only one percent of firms submitting business plans to venture organizations have been funded. Gebhard and Schmidt (2001) report that in Germany on average 5 percent of all offered projects are financed with venture capital.

374 See e.g. Sah and Stiglitz (1986), Lerner (1994a).

375 See Tyebjee and Bruno (1984), Hellmann and Puri (2000). See also section 2.2.3.

to use different contracts,[376] the stringent contractual arrangements between entrepreneurs and venture investors typically help to resolve sorting problems.[377] Given these characteristics venture capitalists should be able to select high quality firms. Accordingly, their involvement in IPOs is conjectured to have a positive influence on the post-issue operating and market performance of portfolio firms, as stated by the following hypotheses:

H2a: Venture-backed firms outperform non venture-backed firms in terms of post-issue operating performance as a consequence of the selection ability of venture capitalists.

H2b: Venture-backed firms outperform non venture-backed firms in terms of market performance as a consequence of the selection ability of venture capitalists.

To investigate the soundness of *H2a* and *H2b*, the post-issue operating and market performance of IPOs is empirically analyzed in section 5.2.1, using a dummy variable for the involvement of venture capitalists (*VC*) that equals one if an IPO is venture-backed. To account for slope effects of venture-backing, the individual explanatory variables, which are outlined in section 4.4, are multiplied by the venture capital dummy variable.[378]

Second, venture capitalists provide valuable services in form of third-party monitoring. As discussed in section 2.2.3, they tend to concentrate on particular industries and development stages, where their value-adding potential is the greatest due to their expertise and monitoring ability.[379] Given the active involvement in portfolio firms, venture capitalists often have profound knowledge about a company's history as well as insider information about its management quality and financial situation. Also, the reputation at stake represents an important contributory factor for continued monitoring. Lerner (1994a) mentions that venture capitalists frequently form syndicates and represent partners on the board.[380] As the venture capital industry is a small community, individual performance is closely monitored.[381]

Third, venture capitalists involved in IPOs can push the post-issue firm performance by building relationships with top-tier financial institutions which permit to overcome informational asymmetries. Therefore, the investment behavior of venture-backed firms is expected to be less dependent upon internally generated cash flows than that of non venture-backed issues.[382] Since venture capitalists tend to hold significant ownership as well as board positions, and continue to be involved in firms after a going public, they can provide access to capital even in the post-issue period.[383] Venture capitalists also tend to put effective management structures in place, which help firms to perform better in the long run.[384] Taking into account

---

376 See e.g. Cumming (2001), Schwienbacher (2002). Gilson and Schizer (2002) stress the link between taxation and the varying use of convertible securities across countries.

377 Sahlman (1990) emphasizes that contractual provisions employed by venture capitalists help to select the best entrepreneurial firms. See also Gompers (1999) and Kaplan and Strömberg (2002, 2003) on venture capital contracts.

378 See section 4.4.5 for more details on the applied testing methodology. On the use of interaction variables see Jobson (1991), pp. 317-321.

379 See e.g. Norton and Tenenbaum (1993), Jain and Kini (1995).

380 See also Manigart *et al.* (2002b).

381 See e.g. Sahlman (1990).

382 See Brav and Gompers (1997).

383 See Barry *et al.* (1990), Megginson and Weiss (1991).

384 See Brav and Gompers (1997).

these arguments, the monitoring provided by venture capitalists as well as their involvement in portfolio firms are expected to yield a superior performance of venture-backed firms relative to non venture-backed ones. In view of that, the following hypotheses are stated:

H3a: Venture-backed firms outperform non venture-backed firms in terms of post-issue operating performance as a consequence of the monitoring and value-adding of venture capitalists.

H3b: Venture-backed firms outperform non venture-backed firms in terms of market performance as a consequence of the monitoring and value-adding of venture capitalists.

To assess the impact of venture capital monitoring and value-adding on performance differences between the two firm groups, the influence of the post-issue equity stakes held by venture capitalists on the operating and market performance is investigated in section 5.2.2.1.[385] Gompers and Lerner (1998) argue that the shareholdings are related to the incentive to monitor and to the quality of information about a firm. Moreover, venture capitalists are supposed to hold large post-offering stakes to signal the assurance of continued monitoring.[386] Also, the board representation of venture capitalists is considered to test the relationship between monitoring and IPO performance. Board membership is typically associated with great access to inside information, which is crucial for monitoring.[387]

Lastly, the performance of venture-backed firms might be related to the certification role of venture capitalists. Venture capital investors intervene in the capital markets repeatedly to exit investments. Given the importance of reputation, venture capitalists have an incentive to only bring firms of high quality to the market in order to signal their skills and generate future deal flows. In line with this argument, one might suppose that venture-backed firms going public are higher valued at the time of the IPO than non venture-backed ones:

H4a: Venture-backed firms undergo higher market valuations at the time of the IPO than non venture-backed firms as a consequence of the certification role of venture capitalists.

To assess the impact of venture capital certification on valuation, the approximated value of Tobin's *Q* at the time of the IPO is compared between venture- and non venture-backed firms in section 5.2.1.2. In the refined analysis, the age of lead venture capitalists (*VC_Age*), defined as the investor with the largest equity stake in the firm at the IPO, is taken into account as a proxy for reputation.[388] In order to detect whether or not the certification of venture capitalists is related to the monitoring and value-adding hypothesis (*H3b*), in section 5.2.2.1 the analysis also controls for the effect of the pre-issue equity held by venture capitalists on Tobin's *Q*.

---

385 The equity stakes have been used in previous research as a proxy for monitoring. See e.g. Barry *et al.* (1990).

386 See Megginson and Weiss (1991).

387 See Gompers and Lerner (1998).

388 Previous studies, such as Barry *et al.* (1990) and Gompers (1996), also use the age of venture capitalists as a proxy for reputation. They conjecture that older and more experienced venture capitalists can better certify firm quality at an IPO.

Jain and Kini (1995) and Wang *et al.* (2003) argue that the certification role of venture capitalists reduces the need of IPOs for excellent operating performance to impress investors in the capital markets. Accordingly, venture capitalists are assumed to bring firms public, which are less profitable than firms without venture-backing. To test this additional certification-related conjecture the following hypothesis is stated:

H4b: Venture-backed firms are less profitable than non venture-backed firms at the time of the IPO as a consequence of the certification role provided by venture capitalists.

To check *H4b*, in section 5.2.1.1 different measures for the operating profitability of venture-backed firms at the time of the IPO are considered and tested with respect to significant differences to firms without venture-backing. In the refined analysis, the age lead venture capitalist is examined regarding potential effects on the operating performance at the time of the going public. To determine whether or not the monitoring and value-adding ability of venture capitalists (*H3a*) has an impact on the certification at the time of IPO, in section 5.2.2.1 the effect of the pre-issue equity-holding of venture capitalists on the operating performance is considered.

4.2.3 Financial Systems and Adverse Selection

The set of the analyzed countries is of particular interest because it includes three major European economies with different financial systems.[389] In the economic literature, Germany is characterized as a bank-based system, whereas the UK is referred to as a market-based system.[390] The case of France is more subtle. At the beginning of the 1980s, the French financial system was strongly dominated by state influence, and publicly owned banks dominated the financial sector.[391] However, although the country continues to have a strong banking sector, there is evidence that the system has undergone some fundamental changes and has adopted increasingly several characteristics of a market-based financial system.[392] Therefore, following Schmidt *et al.* (2002), since the 1990s the French financial system has been difficult to classify.

First, as discussed in the previous chapter, the different financial systems have influenced the development of the national venture capital markets and their participants in the three considered countries.[393] The different historical paths of development are conjectured to have an impact on the expertise and reputation of venture capitalists operating in the contemporary markets. Moreover, as many German banks have a long history of relationship financing and focus on the provision of informed capital to the corporate sector, the services of German

---

389 See appendices A 2 and A 4.

390 See e.g. Allen and Gale (2000). The practice of distinguishing between bank- and capital market-dominated financial systems was introduced by Rybczynski (1984).

391 See e.g. Gueslain (1992), Swary and Topf (1992), de Boissieu (1993). See also Rindermann (1996).

392 Based on the assessment of quantitative measures, such as bank-related intermediation ratios and financing patterns, Schmidt (2001) claims that the French financial system changed from being "more German than the Germans" – or having a strongly bank-dominated system – to being "more British than the British". See Schmidt (2001), p. 6. See also Schmidt *et al.* (1999, 2002). For an analysis of the transformation of the French financial system, see Zerah (1993), Bertero (1994), and Faugère and Voisin (1994).

393 See Black and Gilson (1998). Hyytinen and Pajarinen (2002) compare the financial systems and venture capital in Scandinavian countries.

banks are often close substitutes to those offered by venture capitalists.[394] Therefore, the economic benefits of the venture capital industry in the German bank-based system might be smaller than those of their counterparts in a market-based financial system, where the services provided by venture capital are unique.[395]

Second, the differences in financial systems and established ways of financing SMEs in the considered countries may have an influence on the type of firms looking for venture capital. As discussed in section 3.4, in Germany with its bank-based financial system, many entrepreneurs do not request venture capital, because they traditionally rely on bank financing to keep outside shareholders out of their firm.[396] A study by the *Deutsche Bundesbank* and the *Banque de France* based on comprehensive compiled balance sheet data documents that German SMEs rely more heavily on financing via bank loans than French ones.[397] Following Friderichs *et al.* (1999), the reliance of small German firms on this source of financing even increased in the first half of the 1990s. Given the historical predominance of bank financing for small firms, venture capitalists in Germany are conjectured to encounter relatively severe adverse selection problems, i.e. less profitable firms decide to involve venture capitalists, whereas more profitable firms tend to develop without applying for venture capital.[398]

Against this line of reasoning one might put forward that the types of financing provided by banks and venture capitalists are not perfect substitutes.[399] Ueda (2000) makes a case that venture capitalists have typically more technological expertise than banks which allows them to assess the profitability of firms more accurately. Also, evidence from the US suggests that banks are incapable of adequately financing innovative firms and that venture capital proved to be the superior form of finance in innovative industries.[400] However, some authors argue that the uniqueness of the financial services provided by for venture capitalists may not generally apply in the context of the German banking system.[401] Given some specific aspects of the German banking industry, such as the importance of relationship financing and the universal banking system allowing banks to hold equity, the financial services offered by banks and venture capitalists share some similarities.[402] As a consequence, in the context of the German financial system the adverse selection argument, as developed before, might apply for the type of firms being scrutinized in this study.

---

394 See e.g. Gerschenkron (1962) for a historical analysis of the role of banks in Germany.

395 See Schäfer (2002).

396 See also Leopold and Frommann (1998).

397 See Friderichs *et al.* (1999).

398 See Amit *et al.* (1990) for a general theoretical discussion of adverse selection problems in venture capital. Empirical evidence supporting the adverse selection hypothesis in the case of venture capital in Singapore is provided by Wang *et al.* (2003). Hyytinen and Pajarinen (2002) argue that the credit boom in the 1980s in Nordic countries may have worsened the adverse selection encountered by the infant venture capital industry because only very bad projects did not receive financing from banks. See also section 3.4.2 for a discussion of the adverse selection problems related to the creation of the WFG in Germany.

399 See e.g. Audretsch and Lehmann (2003).

400 See Gompers and Lerner (2001a).

401 See e.g. Schäfer (2002).

402 See Hellmann (1993). He cautiously notes that there still are some important differences between venture capital and banks, and that the German *Hausbank* system has undergone considerable changes. Harm (1992) convincingly demystifies some of the features of the German *Hausbank* system.

Several authors counter that the contracting framework employed by venture capitalists enables them to cope with potential problems of adverse selection, as discussed in section 2.3.[403] Yet, as the European markets for venture capital are relatively young in comparison to the US, many players of the recently developed industry are inexperienced. There is evidence that contract structures employed by European venture capitalists differ from those of US investors.[404] Although these differences might stem from different tax legislations, as advocated by Gilson and Schizer (2002), there is also evidence supporting the view that the use of different contracts across countries is related to the performance of venture-backed firms.[405] These findings do not exclude the possibility that European venture capitalists are less experienced in handling adverse selection problems via contracting than US venture capitalists.

Table 6: Comparison of the Capital Markets in France, Germany, and the UK

	France			*Germany*			*UK*		
Year	Listed firms[1]	Mkt. Cap.[1,2]	Mkt. Cap. to GDP[3]	Listed firms[1]	Mkt. Cap.[1,2]	Mkt. Cap. to GDP[3]	Listed firms[1]	Mkt. Cap.[1,2]	Mkt. Cap. to GDP[3]
1990	873	311.7	26.2	649	355.3	23.1	n.a.	850.0	81.3
1991	839	373.4	28.9	665	392.5	20.3	n.a.	986.1	92.0
1992	786	349.6	27.1	665	346.9	17.8	n.a.	928.4	102.9
1993	726	455.5	37.2	664	460.8	24.7	n.a.	1,150.6	127.0
1994	724	452.1	32.2	666	499.3	22.8	n.a.	1,145.3	114.6
1995	710	500.0	31.5	678	577.4	23.5	1,971	1,346.6	121.7
1996	702	587.0	38.7	681	664.9	28.8	2,091	1,642.6	128.7
1997	740	676.3	49.6	700	825.2	40.5	2,064	1,996.2	150.9
1998	914	985.0	64.9	741	1,086.8	48.2	1,957	2,372.7	168.4
1999	968	1,496.9	111.1	933	1,432.2	72.1	1,826	2,855.4	199.1
2000	1,021	1,446.6	109.6	1,065	1,270.2	66.6	1,929	2,612.2	187.0

Notes: [1)] Domestic firms; [2)] market capitalization in billion USD (converted at the average exchange rate of the respective year); [3)] in percent. Sources: Deutsche Börse (2001), Euronext (2001), IMF (2001), World Federation of Exchanges (2001).

Third, there might be differences in the quality and type of firms considering an IPO, given the different sizes of domestic stock markets and their varying importance for the funding of SMEs across countries. Table 6 exposes that the stock market capitalizations of domestic firms, both in absolute and relative terms, differed substantially between Germany and France on the one side, and the UK on the other, between 1990 and 2000.[406] The statistics also reveal that less domestic firms are listed in France and Germany than in the UK. Aside from these differences, several studies point out variations among the fraction of listed firms and their

---

403 See e.g. Sahlman (1990), Gompers (1995).

404 See Cumming (2002), Kaplan *et al.* (2002), Schwienbacher (2002).

405 See Hege *et al.* (2003).

406 A common way to measure the relative importance of a country's stock markets consists in dividing the market capitalization by the GDP.

average sizes across countries.[407] Thus, in the early 1990s, firms listed in France and Germany tended to be larger in terms of total assets than firms in the UK.[408]

The differences between the domestic stock markets are partly reflected in their IPO activities. Table 7 shows that the number of IPOs in the UK exceeded those of the French and German capital markets in every year between 1995 and 2000. However, after the opening of the *Neuer Markt* and the sharp rise in the number of IPOs in 1999, the amounts of capital raised in Germany surpassed those in the UK, both in absolute and relative terms. In France, where the *Nouveau Marché* was launched in 1996, there was a substantial increase in the amounts of capital raised between 1996 and 1998. In 1998, the proceeds even exceeded the British and German figures, irrespective of employing absolute or relative measures.

Table 7: IPO Markets in France, Germany, and the UK

	France			*Germany*			*UK*		
Year	No. of IPOs[1]	Capital raised via IPOs[1,2]	Capital raised as % of GDP	No. of IPOs[1]	Capital raised via IPOs[1,2]	Capital raised as % of GDP	No. of IPOs[1]	Capital raised via IPOs[1,2]	Capital raised as % of GDP
1995	22	0.2	0.0	20	4.9	0.2	285	4.7	0.4
1996	59	1.2	0.1	20	13.5	0.6	347	16.7	1.4
1997	82	8.2	0.6	35	2.9	0.1	217	11.6	0.9
1998	152	13.7	0.9	67	3.7	0.2	169	6.6	0.5
1999	68	7.3	0.5	134	13.7	0.7	166	7.4	0.5
2000	76	10.7	0.8	135	23.3	1.3	366	14.5	1.0

Notes: [1]) Domestic firms; [2]) in billion USD (converted at the average exchange rate of the respective year).
Sources: Deutsche Börse (2001), Euronext (2001), IMF (2001), World Federation of Exchanges (2001).

Given the low numbers of IPOs as well as the prevalence of large and established firms listed in France and Germany until the opening of new market segments for small and growth-oriented companies, SMEs undertaking an IPO in these countries might have been subject to a more pronounced screening by the market participants than those entering the British capital markets. Hence, as illustrated in Figure 4 by means of an assumed density function of firms according to their profitability, the threshold (S_2) for a public listing in terms of firm profitability is conjectured to be higher in France and Germany than in the UK (S_1), where IPOs, including those of small firms, are more common.

However, as a result of the flourishing IPO markets at the end of the 1990s, allowing particularly the German and French markets to make up ground vis-à-vis the British ones, the distance between the thresholds for a listing (*d*) might have decreased over time and systematic differences in the screening quality as well as their impact on the type of firm considering an IPO might have disappeared. The analysis of IPOs by yearly cohorts in the growth market segments of France, Germany, and the UK indicates that firms going public at the *Neuer Markt* became smaller both in terms of sales and assets from 1997-1999, whereas no particu-

---

407 See e.g. Rajan and Zingales (1995). Edwards and Fischer (1994) estimate that listed firms accounted for 30.5 percent of total sales in the UK, while their listed counterparts in Germany accounted for just 10.6 percent.

408 See Rajan and Zingales (1995), table 1, p. 1425.

lar changes were observed in France and the UK.[409] Moreover, the operating profitability of the average IPO at the *Neuer Markt*, measured by the operating cash flow return on assets at the time of the going public, fell from 22 percent in 1997 to almost 0 percent in 1999. Taking into account these observations, the regression analysis controls for differences in the firm performance between the considered domestic markets as well as for variations over time, by employing country and calendar year dummy variables.

Figure 4: Stylized Profitability Threshold for IPOs in France, Germany, and the UK

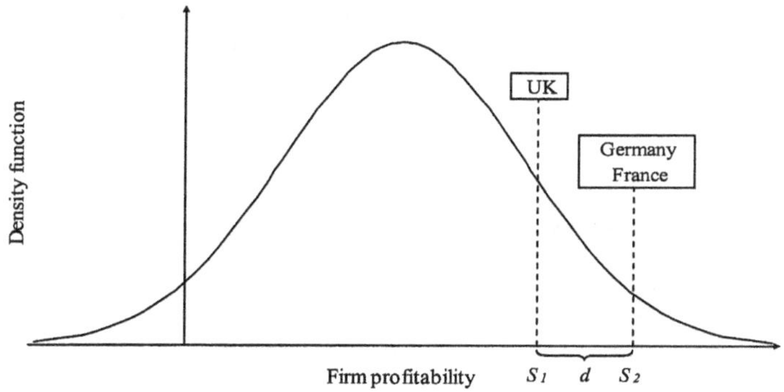

Source: own figure.

Summarizing the arguments related to the influence of different financial systems on the type of firms looking for venture capital, it is conjectured that entrepreneurs in Germany reveal a stronger tendency to rely on bank financing than their equivalents in France or the UK. German firms may less frequently consider venture capital as an option for funding than firms in the two other countries. As a result, there might be severe adverse selection problems for venture capital financing to the extent that venture-backed IPOs in Germany perform worse than IPOs without venture-backing.

This adverse selection effect may be fortified by the listing threshold argument, as illustrated in Figure 4, suggesting that relatively few firms went public at the less developed German stock markets, at least in the early years of the examined period. Hence, firms making it to the public markets without venture capital were typically the most profitable and promising ones. Based on this line of reasoning, there is an important implication for the empirical analysis, as considered by the following hypothesis:

H5: *In Germany, venture-backed firms underperform non venture-backed firms in the aftermarket due to adverse selection caused by the bank-based financial system.*

---

[409] In the firm sample, which is employed in the empirical analysis, the median sales at the time of the IPO decreased from EUR 26.3 to 12.6 million between 1997 and 1999. Likewise, over the same period, the median amount of assets diminished from EUR 20.1 to 9.6 million.

To investigate *H5*, the analysis is performed country-wise. Moreover, in order to detect significant cross-country differences, a dummy variable for venture-backed firms going public in Germany is employed in the full sample analysis in section 5.2.1.3. To make sure that the results are not driven by the hype during the period that several authors refer to as the "Internet bubble", the analysis examines firm performance of the complete and a reduced sample, the latter omitting the IPOs of 1999.[410]

4.2.4 Organizational Forms and Sources of Venture Capital Funds

As discussed in section 2.4.1 and in parts of chapter 3, the differences between the financial systems of France, Germany, and the UK are also reflected in the organizational variety and different sources of venture capital funds across countries. Following the aggregated statistics of the EVCA (2000), as illustrated in Table 8, banks were the main providers of funds in France and Germany in 1999, with 25 and roughly 40 percent, respectively.

Table 8: Raised Funds by Type of Investor

	France		*Germany*		*UK*	
	m EUR	**%**	**m EUR**	**%**	**m EUR**	**%**
Corporate investors	196	4.6	407	10.6	916	9.3
Private individuals	192	4.5	439	11.5	430	4.3
Government agencies	118	2.8	549	14.4	179	1.8
Banks	1,072	25.0	1,520	39.9	2,564	26.0
Pension funds	395	9.2	331	8.7	3,105	31.5
Insurance companies	632	14.8	434	11.4	1,395	14.1
Fund of funds	193	4.5	121	3.2	137	1.4
Academic institutions	9	0.2	0	0.0	50	0.5
Realized capital gains	1,018	23.8	11	0.3	0	0.0
Other	454	10.6	0	0.0	1,094	11.1
Total	4,279	100.00	3,812	100.00	9,870	100.00

Note: figures for 1999. Source: EVCA (2000).

In the UK, private pension funds were the most important investors, contributing nearly one third of the raised funds.[411] Nonetheless, with a share of 26 percent banks also played a considerable role in the provision of financing sources in the UK. In Germany, government agencies represented the second largest source of capital with 14.4 percent. However, as the aggregated statistics do not distinguish between funds raised by private and public financial institutions, the mere focus on the share of government agencies might be misleading to assess the overall contribution of public funds.

---

410 See e.g. Ljungqvist and Wilhelm (2003). More details on the employed sample are provided in section 4.3.

411 Mayer *et al.* (2001) document that the sources of finance are related to differences in venture capital activities within countries. An interesting extension of the analysis by Mayer *et al.* (2001) might be to include details on covenants and their effects on the activities financed by venture capitalists.

Beyond differences in the sources of funds, there are different organizational forms of venture capital across countries. While venture capital firms in the UK are frequently independent and organized as limited partnerships, there are many captive and semi-captive venture capital funds in Germany and France, respectively (see Table 9).[412] Previous research underlines that captive funds, such as corporate venture capitalists, are more likely to provide product market support to complementary companies, thereby increasing the entrepreneurial firms' likelihood of a successful exit.[413] Yet, in France and Germany most captive venture capital firms are subsidiaries of financial institutions, such as major universal banks.

Table 9: Distribution of Venture Capital Investments by Investor Type

	France		*Germany*		*UK*	
	m EUR	%	m EUR	%	m EUR	%
Independent	1,709	60.7	1,330	42.1	9,023	78.4
Captive	227	8.0	1,204	38.1	595	5.2
Semi-captive	865	30.7	126	4.0	1,864	16.2
Public	16	0.6	499	15.8	19	0.2
Total	2,817	100.00	3,159	100.00	11,501	100.00

Note: figures for 1999. Source: EVCA (2000).

Hamao *et al.* (2000) argue that venture capitalists, which are subsidiaries of financial institutions being involved in the underwriting process, can present a conflict of interest with respect to the pricing of an offering.[414] However, as far as the long-term performance of IPOs is concerned, the presence of venture capitalists affiliated with financial institutions might be of advantage for venture-backed firms, given the proximity to financing sources.[415] Thus, venture capitalists associated with financial institutions, such as major universal banks in France and Germany, might benefit the post-issue performance of firms because they can help to overcome asymmetric information and alleviate the accessibility of external financing.[416] Based on this line of reasoning, one might conclude that IPOs backed by bank- or finance-affiliated venture capital firms outperform issues backed by non finance-affiliated venture capitalists.[417] Accordingly, the following hypothesis is stated regarding the impact of the organizational form of venture capital firms on the performance of IPOs:

H6a: Firms backed by bank-affiliated venture capitalists in France and Germany outperform firms backed by non bank-affiliated venture capitalists in the aftermarket.

---

412 The EVCA (2000) defines independent, semi-captive, and captive venture capitalists as firms which are owned less than 20%, between 20% and 50%, and more than 50% by a parent organization, respectively.

413 See e.g. Santhanakrishnan (2002).

414 See also Gompers and Lerner (1999a).

415 Petersen and Rajan (1994), and Berger and Udell (1998) provide evidence that the quality of bank-firm lending relationships is an important determinant for financing conditions of SMEs in the US.

416 Harhoff and Körting (1998) provide evidence for lending relationships in Germany and find that lending in the SME segment is typically concentrated on one or two financing institutions. Elsas and Krahnen (1998) examine the role of the house-bank status in the provision of loans to German SMEs and show that house-banks provide more finance and obtain more collateral.

417 In the following, the terms "bank-affiliated" and "finance-affiliated" are employed interchangeably.

To test *H6a*, dummy variables for firms backed by bank-affiliated venture capitalists in France (*Bank*France*) and Germany (*Bank*Germany*) are employed in the multivariate analysis based on the venture-backed IPO sample in section 5.2.2.2.

Hellmann *et al.* (1999), however, point out that bank-affiliated venture capitalists are not the most aggressive ones and tend to behave more conservatively than independent venture investors. Bank-related venture capital firms typically have a wide range of industries, in which they are willing to invest. Moreover, their activities are less geographically concentrated than those of independent venture capitalists.[418] These findings suggest that bank-affiliated venture capitalists are less specialized and dedicate less time to individual portfolio firms. Accordingly, it is conjectured that bank-affiliated venture capitalists bring less risky firms to the capital markets than independent venture capitalists. This entails firms supported by bank-affiliated venture capitalists to be less sensitive to overall market movements, i.e. they outperform firms financed by independent venture capitalists when markets go down, and vice versa.

Wang *et al.* (2002) investigate the nature of different forms of venture capitalists and find that independent venture investors add more value to their portfolios than their finance-affiliated equivalents. The authors relate the results to differences in four main characteristics of independent and finance-affiliated venture capital funds, as outlined in Table 10. First, Wang *et al.* (2002) focus on differences related to the source of funding. While finance-affiliated venture capitalists typically rely on internal funding and invest the capital of a single dominant investor[419], independent venture capital firms tend to receive capital from various institutional investors, such as pension funds and insurance companies, and therefore have to compete for the provision of funds in the market.[420] In order to signal their ability to potential investors and attract future investments, independent venture capitalists attach more importance to building up reputation and bringing only the most promising firms public.[421]

Second, different investment objectives might have an impact on the performance of venture capital funds. Independent venture capitalists usually wish to maximize the returns on the invested capital and do not pursue strategic investment motives. Finance-affiliated venture capitalists, however, may also follow strategic objectives on behalf of their parent institution.[422] As a result, the mere financial performance of the backed company might be of secondary importance for finance-affiliated venture capital firms.[423]

---

418 See Hellmann (1997), Hellmann *et al.* (1999).

419 The main investor is frequently the parent company of a venture capital firm.

420 See e.g. Norton (1995), Zemke (1995), van Osnabrugge and Robinson (2001).

421 See Wright and Robbie (1996). See Gompers (1996) on IPOs and reputation building. In the literature on corporate governance it is argued that large block shareholders - present in finance-affiliated and absent in independent venture capital firms - have a positive effect on firm performance. However, since affiliated venture capitalists are less dependent on funding from the market, they might have fewer incentives to build up reputation via successful IPOs. See Lin (1996) for an analysis of venture capitalists as large block shareholders.

422 For example, bank-affiliated venture capitalists may intend to benefit from "cross-selling" opportunities, i.e. tying backed companies to the bank for follow-on business relations. See e.g. Büschgen (1985).

423 This is analogous to the behavior of corporate venture capital firms which habitually pursue strategic objectives. See e.g. Hardymon *et al.* (1983), Siegel *et al.* (1988), Schween (1996), Brody and Ehrlich (1998), and Hellmann (1998c) for contributions on corporate venture capital.

Table 10: Comparison of Independent and Finance-Affiliated Venture Capital Funds

	Independent VCs	*Finance-Affiliated VCs*
Source of funding	Mainly public institutional investors	Parent companies and associates
Investment objective	Investment return	Not purely for return, must also bring in business for parental company or have other strategic purposes
Organizational structure	Limited partnerships	Corporate
Staff background	Experienced professionals, entrepreneurs, or industry managers	Staff from parents without technical background

Note: VCs – venture capitalists. Source: Wang et al. (2002), p. 65.

Third, there are differences in the organizational structures between independent and finance-affiliated funds. Independent funds are traditionally organized as limited partnerships and stimulate fund managers by means of incentive structures to work hard and take risks.[424] Finance-affiliated funds, by contrast, are often managed as normal investment institutions and provide fewer incentives to take risks. Given the high debt ratios employed by financial institutions and the reduced risk-taking capacity, finance-affiliated funds tend to be more conservative and unwilling to invest in early-stage high-tech companies.[425]

Finally, different staff backgrounds can influence the behavior of venture capitalists. Independent venture capital firms often employ personnel with technical and industrial backgrounds. Thus, they are better skilled to assess the feasibility of start-up business plans and to manage early-stage investments.[426] Finance-affiliated venture capital funds, by contrast, tend to rely on employees from parent companies, who typically do not possess technical expertise. Accordingly, evaluations are often more conservative and similar to credit analyses in banks, focusing on visible profits and tangible assets.

Based on these arguments, it is assumed that independent venture capitalists are more likely to invest in early stages and high-tech industries than finance-affiliated venture capital firms. Accordingly, independent venture capitalists are more probable to syndicate investments with other venture capital firms in order to share the high risks.[427] Since employees of independent venture capitalists are typically more skilled and technically experienced, it is also more probable that they are involved in the management of portfolio firms. Accordingly, independent venture capital firms are conjectured to have a stronger influence on the performance of the firms they bring public than finance-affiliated investors:

H6b: Firms backed by independent venture capitalists outperform firms backed by captive venture capitalists in the aftermarket.

To shed light on the validity of *H6b*, a dummy variable for independent venture capitalists (*Indep*) is used in the multivariate analysis of the venture-backed sample in section 5.2.2.2.

---

424 See also section 2.3.2.
425 See Wang *et al.* (2002).
426 Wang *et al.* (2002) argue that staff with technical knowledge can better assess the value of intangible assets, such as patents and software, than staff without this type of knowledge.
427 See Lerner (1994a).

4.2.5 International Venture Capitalists and the Role of Government Agencies

Early theories of the multinational enterprise claim that foreign firms must possess a countervailing advantage over local firms to make international activities viable and that the market for the sale of this advantage must be imperfect.[428] More recent approaches on the existence of multinational enterprises rely on the transaction cost theory of internalization and draw from the work of Buckley and Casson (1991).[429] They posit that multinational enterprises grow by replacing imperfect, or non-existent, markets by internal ones. Thus, they suppose that running a wholly-owned subsidiary involves internalizing a market activity within the firm and creates the multinational firm.[430]

Transferring the theoretical arguments concerning the multinational enterprise to the venture capital business, it is conjectured that there exist imperfections in the venture capital market and that internationally operating venture capitalists possess a particular advantage over purely national venture capitalists, such as more investment experience and better skills to assess the profitability of innovative firms. According to the theory, these advantages are necessary to balance the restricted knowledge of international investors about foreign markets. If venture capitalists operating in more than one country have more business experience than national venture investors, it can be conjectured that they are on average older and more sophisticated than less experienced national venture capitalists. Accordingly, international venture capitalists are expected to converge in their management styles toward the genuine US model, i.e. they hold larger equity stakes, syndicate investments with more syndication partners, are represented on the board of portfolio firms more often, and back more IPOs in the sample than national venture investors.[431] The following hypothesis accounts for the conjecture with respect to differences in experience and sophistication between nationally and internationally operating venture capitalists:

H7: International venture capitalists, defined as those backing IPOs in at least two of the considered countries, are more experienced and therefore more involved in portfolio firms, providing more management support, than national venture capitalists.

To test *H7*, in section 5.1.3 the age, the number of syndication partners, the board representation, the backing frequency in sample, as well as pre- and post-issue shareholdings of venture capitalists are analyzed with respect to systematic differences between internationally and nationally operating investors.

Finally, the role of government agencies in energizing venture capital markets differs between countries.[432] Public organizations are often directly or indirectly involved in the fund-

---

[428] See e.g. Hymer (1970, 1976), Kindleberger (1969).

[429] See also Rugman (1980), Teece (1986), Casson (1992). Seminal literature contributions on the transaction cost approach include Coase (1937) and Williamson (1975, 1985).

[430] A discussion of theories of the multinational firm is provided by Calvet (1981) and Dunning (1993).

[431] Several authors emphasize the heterogeneity in management styles and degrees of strategic involvement of venture capital investors across countries. See e.g. Sapienza *et al.* (1996), Jeng and Wells (2000). Bascha and Walz (2001b) report that German venture capitalists often take participations in form of silent partnerships, which offer limited control and monitoring rights.

[432] See Leleux and Surlemont (2001) for a pan-European analysis of the interrelation between public and private venture capital.

ing of young companies.[433] However, publicly sponsored or managed venture capital funds tend to follow different objectives and to behave in different manners, when compared to private investors.[434] In particular, they are often less concerned to make profits than to strengthen the entrepreneurial sector.[435] Additionally, public fund managers are often civil servants or government employees, and might therefore lack the necessary skills to evaluate entrepreneurial firms.

Wang *et al.* (2002) argue that government-funded venture capital firms often syndicate with independent venture capitalists due to the lacking expertise in selecting and monitoring investment firms. Thus, public venture capitalists, such as the German *Technologie-Beteiligungs-Gesellschaft* (TBG) in Germany, require an outside lead investor in all promoted investments, and typically co-invest as silent partners in portfolio firms, matching the funds of other investors.[436] However, if government funds act as the sole or lead venture capital investor in firms going public, one might expect them to have a less favorable influence on performance than private venture capitalists given the missing incentives and investment skills. This conjecture is considered by the following hypothesis:

H8: *Firms backed by private lead venture capitalists outperform firms backed by public lead venture capitalists in the aftermarket.*

To test *H8*, in section 5.2.2.2 a dummy variable (*Public*) is employed that equals one, if a lead venture capitalist is directly or indirectly owned by a public shareholder. To get an overview of all testable conjectures, Table 11 summarizes the derived hypotheses. Details on the data set used in the empirical analysis are described in the following section.

---

433 See e.g. Lerner (1999).

434 See e.g. Cumming and MacIntosh (2002).

435 For example, the investments of government-funded venture capital conventionally aim at building up local high-tech industries or accelerating the growth of local enterprises rather than acquiring high financial returns. Bascha and Walz (2001b) report that public venture capitalists tend to have lower return claims than private ones.

436 See Fiedler and Hellmann (2001) and section 3.4.1.

Table 11: Summary of Hypotheses

H1	Long-run IPO performance	The stocks of IPO firms underperform relative to seasoned stocks in the long-run.
H2a	VC selection and operating performance	Venture-backed firms outperform non venture-backed firms in terms of post-issue operating performance as a consequence of the selection ability of venture capitalists.
H2b	VC selection and market performance	Venture-backed firms outperform non venture-backed firms in terms of market performance as a consequence of the selection ability of venture capitalists.
H3a	VC monitoring and operating performance	Venture-backed firms outperform non venture-backed firms in terms of post-issue operating performance as a consequence of the monitoring and value-adding of venture capitalists.
H3b	VC monitoring and market performance	Venture-backed firms outperform non venture-backed firms in terms of market performance as a consequence of the monitoring and value-adding of venture capitalists.
H4a	VC certification and IPO valuation	Venture-backed firms undergo higher market valuations at the time of the IPO than non venture-backed firms as a consequence of the certification role of venture capitalists.
H4b	VC certification and firm profitability	Venture-backed firms are less profitable than non venture-backed firms at the time of the IPO as a consequence of the certification role provided by venture capitalists.
H5	VC and adverse selection	In Germany, venture-backed firms underperform non venture-backed firms in the aftermarket due to adverse selection caused by the bank-based financial system.
H6a	Bank-affiliated VCs in France and Germany	Firms backed by bank-affiliated venture capitalists in France and Germany outperform firms backed by non bank-affiliated venture capitalists in the aftermarket.
H6b	Independent vs. captive VCs	Firms backed by independent venture capitalists outperform firms backed by captive venture capitalists in the aftermarket.
H7	Experience and involvement of international vs. national VCs	International venture capitalists, defined as those backing IPOs in at least two of the considered countries, are more experienced and therefore more involved in portfolio firms, providing more management support, than national venture capitalists.
H8	Public vs. private VCs	Firms backed by private lead venture capitalists outperform firms backed by public lead venture capitalists in the aftermarket.

4.3 Data Sample

4.3.1 Data Collection and Sources

In the empirical analysis, the data from a sample of IPOs in France, Germany, and the UK are employed. The attention is limited to the stock markets of the three largest European economies with a sufficient number of IPOs to make comparisons meaningful.[437] The analysis uses data from non-financial corporations going public in the respective countries between 1996 and 1999. Like in previous studies, financial institutions are excluded from the sample due to their different balance sheet structures.[438]

The data include IPOs at the *Nouveau Marché* in Paris, the *Neuer Markt* in Frankfurt, as well as the AIM and the Official Listing of the LSE. The growth segments of the Paris and Frankfurt stock exchanges started their activities in 1996 and 1997, respectively. They were created with the common objective to attract listings of innovative companies in high-growth industries. The LSE, however, does not have a separate growth segment.[439] In order to obtain a fairly homogenous sample of "venturable" companies with a high-tech industry background, the IPOs of firms included in the *techMARK* sector[440] are integrated in the sample. Details of the selection procedure and the final sample composition are provided in Appendix A 5.

The final sample consists of 303 IPOs in France, Germany, and the UK. It includes only IPOs, and does not contain relistings, transfers from other stock markets or market tiers, as well as mergers or demergers of previously listed firms. Similarly, listings not accompanied by an equity offer remain unconsidered. IPOs which were delisted or changed to another market segment in the observed post-IPO period are included until the delisting date to prevent an inadvertent survivorship bias. All firms without consistently reported data for the periods of interest around the IPO, i.e. one fiscal year before and two fiscal years after the going public, are omitted. 21 firms are eliminated from the sample, which changed the level of consolidation during the observation period, or presented financial statements on periods with varying time-spans, thus limiting the comparability of successive accounting measures. Likewise, the data of 19 companies remain unconsidered because of liquidations, mergers, and acquisitions by other companies, as well as missing data.[441] The final sample used in the regressions includes only firms with available data for at least two consecutive financial years around the IPO, i.e. one fiscal year before and after the going public.

The data of the sample were collected from various sources. Information about the offering terms and characteristics of each IPO, such as the size and per-share price of the offering as

---

437 Therefore, the analysis does not consider IPOs at the growth market segments of smaller stock exchanges, such as the *Nieuwe Markt* (Amsterdam), *Euro.nm Brussels* (Brussels), or *Nouvo Mercato* (Milan). Bottazzi and Da Rin (2002a) report that no company listed at these segments was backed by venture capitalists. See Bottazzi and Da Rin (2002a), p. 248.

438 Previous studies investigating companies listed at Germany's Neuer Markt excluded firms from the financial sector for the same reason. See e.g. Fischer (2000), Franzke (2001).

439 Although the AIM was established in order to allow small companies to have their shares listed before acquiring a full listing, it cannot be compared to other growth markets in Europe, being a mixed segment that also contains pubs and football clubs besides technology shares.

440 The *techMARK* is a tracking instrument and comprises technology firms being part of other indices of the LSE.

441 See Appendix A 5 for details on the firms excluded from the sample.

well as the name of the underwriters, were gathered from Deutsche Börse, Euronext/Bourse de Paris[442], and the LSE. The IPO data were checked and, if necessary, completed with information from the offering prospectuses and the *IFR Platinum New Issues* database. The aftermarket stock prices of the firms going public, such as the first-day closing price and the monthly closing prices over different time intervals after the IPO, as well as the values of different benchmark indices, which are used for market adjustments of the stock price performance of individual firms, were obtained from *Thomson Financial Datastream*. To assess the reputation of underwriters, annual statistics on the IPO numbers and proceeds of investment banks in France, Germany, and the UK over the sample period were taken from the annual league tables of the *IFR Thomson Financial* database.

Firm specific information, such as the date of the firm creation and incorporation, pre- and post-IPO accounting data, ownership structure, and board membership were obtained from the offering prospectuses, annual reports, and Internet websites of the firms going public. For companies listed at the *Neuer Markt* most of the IPO prospectuses were available as downloads on the websites of the stock exchange and the individual firms. Prospectuses of firms listed at the *Nouveau Marché* were obtained from corporate websites or photocopied at the *Commission des Opérations de Bourse* (COB) in Paris, the French stock exchange regulator. Most of the prospectuses of firms in the UK sample, which were not available online, were sent directly by the companies or obtained via investment banks involved in the public offering.[443]

The IPO sample is divided into venture- and non venture-backed issues. The Panels A to C in Appendix A 8 list the names and origins of the venture capital firms, which are the most frequently involved in IPOs in each country sample. To characterize the quality of venture capital organizations, their age at the time of the IPO as well as their international involvement in firms outside their country of origin was determined. Moreover, the shares of equity held by venture capitalists before and after the IPO were collected to explore whether or not the quality of monitoring and certification is related to how much of the company is owned by venture investors. In addition, in order to compute the selling intensity, the percentage of equity sold by venture capital investors in the IPO was assessed. The detailed description of the most important variables used throughout the empirical analysis is reported in Appendix A 7.

4.3.2 Sample Characteristics

Table 12 provides an overview of the distribution of the sample by country and year, both in terms of the number of IPOs and gross proceeds. It shows that the number and value of IPOs are not evenly distributed over the sample period. Instead, the time series figures illustrate that the annual number of going publics increased continuously over time and achieved a peak with a total of 152 IPOs in 1999.[444] Thus, 50.2 percent of all IPOs, and 62.1 percent of the proceeds, were carried out during the last year of the covered time interval. The statistics also reveal that more than half of the IPOs, both in terms of the number of issues and the gross

---

[442] *Euronext* was created in September 2000 by a merger of the exchanges in Paris, Amsterdam, and Brussels.

[443] I am grateful to Martin Bisicky from CSFB for the provision of firm prospectuses, which were neither available from the firms going public nor from the investment banks involved in the IPOs.

[444] The *Neuer Markt* in Germany started its activity only in March 1997, which explains the missing item in 1996 and the low number of IPOs in 1997.

proceeds, took place at the German *Neuer Markt*. Yet, as discussed in section 4.2.3., the fact that the market segment of the Frankfurt stock exchange attracted more than 76 percent of its proceeds in 1999 alone might also be interpreted as an indicator of bad screening and overall hype in the market. To elucidate this conjecture, the empirical analysis explores changes in the screening quality of the considered IPO markets over time, by including calendar year dummies.

Table 12: Distribution of IPOs by Year

	France		*Germany*		*UK*		*All IPOs*	
	No. IPOs	Proceeds	No. IPOs	Proceeds	No. IPOs	Proceeds	No. IPOs	Proceeds
1996	13 (14.4)	197.7 (15.4)			18 (29.0)	530.5 (12.3)	31 (10.2)	728.2 (5.6)
1997	13 (14.4)	109.1 (8.5)	8 (5.3)	373.7 (5.1)	18 (29.0)	811.3 (18.8)	39 (12.9)	1,294.2 (10.0)
1998	37 (41.1)	427.5 (33.3)	32 (21.2)	1,361.3 (18.4)	12 (19.4)	1,118.6 (25.9)	81 (26.7)	2,907.3 (22.4)
1999	27 (30.0)	549.5 (42.8)	111 (73.5)	5,658.9 (76.5)	14 (22.6)	1,866.2 (43.1)	152 (50.2)	8,074.5 (62.1)
All	**90** **(100.0)**	**1,283.8** **(100.0)**	**151** **(100.0)**	**7,393.9** **(100.0)**	**62** **(100.0)**	**4,326.6** **(100.0)**	**303** **(100.0)**	**13,004.2** **(100.0)**

Note: proceeds in million EUR; percentage of IPOs in parentheses. Source: own calculations.

Table 13 shows that the overall composition of the sample, in regard to venture- and non venture-backed IPOs, is relatively equalized. In total, 154 firms, or 50.8 percent of the IPO sample, are venture-backed. Similarly, the patterns in the different subsamples by country appear fairly balanced. The average portion of venture-backed IPOs varies between 43 percent in Germany and 60 percent in France. The figures by year (Panel A) reveal that about 75 percent of the IPOs taking place at the British *techMARK* in 1998 were venture-supported. Likewise, a relatively high portion of IPOs in France in 1999 was venture-backed (74.1 percent). This finding is consistent with the argument of Lerner (1994b) that venture capitalists are able to time IPOs and tend to bring firms public, when stock market valuations are relatively high.

The industry distribution figures (Panel B) expose that the majority of both venture- and non venture-backed IPOs operate in information technology (IT), software, and Internet-related businesses. Firms from the technology sector form the second largest industry in both firm groups, representing about one fifth of each subsample. The comparison between the venture- and non venture-backed IPOs reveals that in all three countries the majority of firms from the biomedical sector is venture-backed.[445] Interestingly, a relatively high portion (15.6 percent) of firms going public at the *Nouveau Marché* belongs to traditional industries.[446] At the *Neuer Markt*, only 8 firms representing 5.3 percent of the going publics in the German growth market segment are related to traditional businesses. The British IPO sample does not

[445] See e.g. Lerner (1994a, 1994b, 1995) for analyses of the functioning of venture capital in the domain of biotechnology.

[446] For example, some of the French IPO firms operate in the beverage, grocery or textile industries.

contain any firm with traditional activity background. This is due to the focus of the *techMARK* that favors firms from non-traditional industries.[447]

Table 13: Venture- and Non Venture-Backed IPOs

Panel A: Time Series Distribution

	France		*Germany*		*UK*		*All IPOs*	
	VC	NVC	VC	NVC	VC	NVC	VC	NVC
1996	8	5			9	9	17	14
	(61.5)	(38.5)			(50.0)	(50.0)	(54.8)	(45.2)
1997	6	7	5	3	10	8	21	18
	(46.2)	(53.8)	(62.5)	(37.5)	(55.6)	(44.4)	(53.8)	(46.2)
1998	20	17	16	16	9	3	45	36
	(54.1)	(45.9)	(50.0)	(50.0)	(75.0)	(25.0)	(55.6)	(44.4)
1999	20	7	44	67	7	7	71	81
	(74.1)	(25.9)	(39.6)	(60.4)	(50.0)	(50.0)	(46.7)	(53.3)
All	**54**	**36**	**65**	**86**	**35**	**27**	**154**	**149**
	(60.0)	**(40.0)**	**(43.0)**	**(57.0)**	**(56.5)**	**(43.5)**	**(50.8)**	**(49.2)**

Panel B: Industry Distribution

	France		*Germany*		*UK*		*All IPOs*	
	VC	NVC	VC	NVC	VC	NVC	VC	NVC
Biomed	7	5	7	3	8	4	22	12
	(13.0)	(13.9)	(10.8)	(3.5)	(22.9)	(14.8)	(14.3)	(8.1)
ITSINT	18	13	31	41	22	13	71	67
	(33.3)	(36.1)	(47.7)	(47.7)	(62.9)	(48.2)	(46.1)	(45.0)
Media	4	6	6	13	1	0	11	19
	(7.4)	(16.7)	(9.2)	(15.1)	(2.9)	(0.0)	(7.1)	(12.8)
Techno	11	5	17	19	3	3	31	27
	(20.4)	(13.9)	(26.2)	(22.1)	(8.6)	(11.1)	(20.1)	(18.1)
Telecom	6	1	2	4	1	7	9	12
	(11.1)	(2.78)	(3.1)	(4.7)	(2.9)	(26.9)	(5.8)	(8.1)
Traditional	8	6	2	6	0	0	10	12
	(14.8)	(16.7)	(3.1)	(7.0)	(0.0)	(0.0)	(6.5)	(8.1)
All	**54**	**36**	**65**	**86**	**35**	**27**	**154**	**149**
	(100.0)	**(100.0)**	**(100.0)**	**(100.0)**	**(100.0)**	**(100.0)**	**(100.0)**	**(100.0)**

Note: VC – venture-backed IPOs; NVC – non venture-backed IPOs; ITSINT stands for firms belonging to the IT, software or internet sector; percentages of IPOs are in parentheses. Source: own calculations.

Table 14 displays the frequency distribution of the venture-backed IPOs by type of the lead venture capitalists. The statistics show that 59.7 percent of the lead venture capitalists backing

[447] Although *techMARK* is theoretically open to all companies, it focuses on innovative technology companies. Therefore, according to the statement of the *techMARK* brochure, companies from the following industry sectors, immediately identifiable as areas of technological innovation, benefit from a streamlined application process: computer hardware, semiconductors, telecommunication equipment, computer services, Internet, and software. See London Stock Exchange (2001).

IPOs are independent, whereas 26 percent of them are captive organizations. Only 14.3 percent of the venture-supported firms received funding from a public lead venture capitalist. A look at the figures by country reveals that the frequency distribution by venture capital type varies considerably across the national subsamples. In the UK, almost all venture capitalists intervening in the IPO market to exit their investments are independent. By contrast, in Germany independent venture capitalists backed 61.5 percent and in France only one third of the venture-funded IPOs. Instead, captive venture capitalists play a more important role in both countries. This outcome is in line with the aggregated figures provided by the EVCA (2000). Captive venture funds are involved in 42.5 and 24.6 percent of the venture-backed transactions at the *Nouveau Marché* and *Neuer Markt*, respectively.

Table 14: Venture Capital Type Distribution of Venture-Backed IPOs

	France		*Germany*		*UK*		*All IPOs*	
	No.	%	No.	%	No.	%	No.	%
Independent	18	33.3	40	61.5	34	97.1	92	59.7
Captive	23	42.6	16	24.6	1	2.9	40	26.0
Public	13	24.1	9	13.9	0	0.0	22	14.3
All	**54**	**100.0**	**65**	**100.0**	**35**	**100.0**	**154**	**100.0**

Source: own calculations.

Finally, Table 14 also exhibits that the portion of public venture capital firms involved in IPOs varies significantly across countries. In France, their contribution is the largest (24.1 percent), whereas in Germany, public venture capitalists are involved in 13.9 percent of all venture-backed IPOs. In the UK, however, public venture capitalists do not participate in any IPO. This finding is consistent with the analysis of chapter 3, stating that the British state plays a less important role as provider of venture funds than the government authorities in France and Germany.

4.3.3 Selection Issues

There are at least two potential sources of bias in the collected data to worry about. First, looking at venture-backed IPOs has the obvious limitation of ignoring those firms remaining private, being sold in a trade sale, or failing. The aggregated EVCA (2000) statistics in Table 15 shows that in France, Germany, and the UK, as well as in the European market as a whole, divestments via IPOs represent a smaller fraction of the exits chosen by venture capitalists than trade sales. Considering the percentage of divested amounts and numbers of investments in 1998 and 1999, most venture capitalists in Europe prefer trade sales to IPOs as exit vehicle.[448]

Venture-backed firms going public are typically among the most successful ones since the rewards for venture capitalists are the highest both in reputation and pecuniary terms. Previous studies underline that IPOs are considered as the most attractive possibility of divestment

---

[448] Only in the UK, the portion of exited amounts via IPO exceeded with 26 percent the share of trade sales (22.1 percent) in 1999, which might be related to that year's generally high stock market valuations.

by venture capitalists.[449] As a consequence, the impact of venture-backing on performance might be overestimated by looking only at firms, which were exited by a going public. However, given that firms undertaking an IPO are in general the most promising and profitable ones, the potential bias towards more successful firms may also apply to the control sample of non venture-backed issue. Taking into account the listing threshold argument, as developed in section 4.2.3, the bias of non venture-supported firms may be even stronger in countries with relatively underdeveloped IPO markets, such as France and Germany during the earlier sample period, where less profitable firms found it difficult to go public without the certification provided by venture capitalists.

Table 15: Comparison of Exit Modes of Venture Capitalists in Europe

The table shows the percentages of the amounts divested via IPO and trade sale relative to all divestments per year. The figures in parentheses represent the percentages of the numbers of firms divested via IPO and trade sale relative to all divestments per year. Aside from trade sales and IPOs, the unreported divestment modes recorded by the aggregated EVCA (2000) statistics include write-offs, repayments of preference shares/loans, sales to other venture capitalists and financial institutions, and divestments by other means.

	France		*Germany*		*UK*		*Europe*	
Year	Trade Sale	IPO	Trade Sale	IPO	Trade Sale	IPO	Trade Sale	IPO
1998	72.0 (18.9)	16.8 (25.3)	20.4 (12.6)	12.5 (5.9)	51.4 (28.8)	18.9 (25.8)	53.6 (21.5)	18.8 (18.2)
1999	58.5 (28.6)	16.4 (23.0)	23.6 (12.7)	17.8 (10.9)	22.1 (27.5)	26.0 (25.0)	36.6 (26.4)	20.7 (17.9)

Source: EVCA (2000).

Second, venture-backing is not randomly distributed but represents typically an endogenous choice by entrepreneurs and venture capitalists. Venture capital financing is not desired by all entrepreneurs, and not all of them receive it. Hence, there might be a selectivity bias in the receipt of venture capital funding.[450] The endogenous choice in providing financing is also reflected in the eventual exit of a venture, and might lead to a non-random distribution of venture-backed IPOs. In particular, the preference of venture capitalists to finance specific types of firms from a particular range of industries might be reflected in the IPO characteristics, such as the size and age of firms going public.[451] However, as outlined in Table 13 of section 4.3.2, the sample composition indicates that the selected stock market segments and IPOs provide appropriate control samples for the purpose of a comparison between venture- and non venture-backed firms. The statistics show that the overall frequency distribution of venture- and non venture-backed IPOs appears relatively well balanced regarding the represented industries. Nevertheless, in order to capture effects, which might be due to the preferences of venture capitalist for particular firm types and industries, the multivariate analysis in section 5.2 employs control variables for age, size, and industries of the IPOs.[452]

---

449 See e.g. Gompers (1995), Gompers and Lerner (1997).

450 See e.g. Lee and Wahal (2002).

451 Lee and Wahal (2002) find that venture-backed IPOs in the US tend to concentrate on software firms and firms conducting commercial biological research. They also document that venture-backed IPOs take place in younger, smaller firms with lower book value of equity.

452 Details on the individual control variables are provided in section 4.4.4.

4.4 Methodology

4.4.1 Accounting Performance Measures

In the empirical analysis, different performance measures are used as dependent variables.[453] First, differences in the accounting performance between venture- and non venture-backed IPOs are explored. To assess the operating performance of firms, two different kinds of profitability ratios are employed, which are widely used as accounting performance measures: (1) operating return on assets, and (2) operating cash flow return on assets.[454] Both ratios are efficiency measures on how a firm is being run and provide information on how much returns are generated by each unit of assets. In addition, both variables measure flows on a pre-tax and pre-interest basis, and avoid the mechanical effect of leverage on the results. Thus, the effects due to differences in the capital gain taxes across the considered countries are controlled for.

The operating return on assets is defined as the operating income before interest, taxes, and extraordinary items divided by total assets. Accordingly, the operating cash flow return on assets is defined as operating cash flow before interest, taxes, and extraordinary items normalized by total assets, whereby the operating cash flow is computed as the operating income plus depreciation, amortization, and provisions. The possible advantage of cash flow-related performance ratios is the elimination of several accruals. Therefore, the operating cash flow return should be less sensitive to manipulation by managers and exhibit more variability than the first measure. However, operating cash flows can be computed in many different ways. Thus, for robustness purposes, different cash flow definitions are employed to calculate the operating cash flow return on assets.[455]

Figure 5: Time Intervals and Reference Points of Accounting Performance Ratios

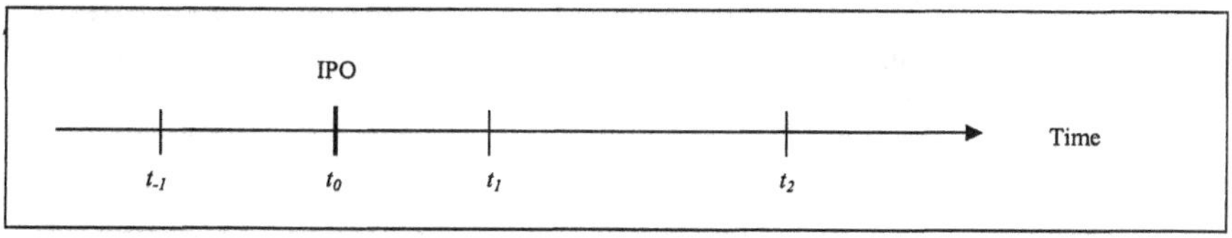

Source: own figure.

Since an IPO is typically related to a substantial increase of total assets, the accounting profitability measures scaled by assets might impart a downward bias after a going public.[456] Therefore, the operating income and operating cash flow scaled by sales are examined as alternative accounting profitability ratios. Analogous to the previously defined performance ratios, the return on sales equals the operating income before interest, taxes, and extraordinary items divided by total sales. Likewise, the operating cash flow to sales ratio is computed as the operating cash flow before interest, taxes, and extraordinary items divided by total sales.

---

[453] The individual performance measures are recapitulated in Appendix A 6.

[454] These two measures of operating performance are employed in previous studies. See e.g. Holthausen and Larcker (1996). See also in Appendix A 7.

[455] See Appendix A 6 for more details on the definition of the different cash flows.

[456] See e.g. Mikkelson *et al.* (1997), Bottazzi and Da Rin (2002a, 2002b).

As depicted in Figure 5, all accounting performance ratios are computed annually at three different reference points in time, i.e. at the end of the fiscal year prior to the IPO (t_{-1}) and two fiscal years subsequent to the IPO (t_1, t_2) in order to assess the pre- and post-issue accounting performance, respectively. Further details on the individual accounting performance measures are recapitulated in Panel A of Appendix A 7.

In the empirical analysis, the relationship between operating performance and the involvement of venture capitalists in IPOs is tested, by estimating multivariate ordinary least squares (OLS) regressions. Based on the specification of Mikkelson *et al.* (1997), the following model is used for the estimation:

$$(1) \qquad CFROA_i = \beta_0 + \beta_1 VC_i + \beta_2 \ln Size_i + \beta_3 \ln Age_i + \beta_4 Part_i + \beta_5 URank_i + \varepsilon_i ,$$

where the dependent variable *CFROA* denotes the operating cash flow return on assets. The explanatory variables include a dummy that equals one if an IPO is venture-backed (*VC*), the natural logarithm of assets (ln*Size*), the natural logarithm of firm age (ln*Age*), calculated in years at the time of the IPO, the participation ratio of old shareholders (*Part*), defined as number of secondary shares sold in the IPO divided by the number of shares outstanding before flotation, and the rank of the underwriter reputation (*URank*).[457]

In line with Mikkelson *et al.* (1997), size and age are expected to have a positive influence on operating performance. Small and young firms usually display low accounting performance due to small scales of operations, low volumes of sales, one-time start-up costs and high initial operating costs, as well as high production and selling costs because of inexperience. Moreover, they often price products at smaller margins over costs than large established firms in order to attract new clients. The portion of secondary sales in the IPO is measured by the participation ratio, representing a frequently used proxy in the IPO literature.[458] Mikkelson *et al.* (1997) argue that firms undertaking an IPO including secondary sales must have favorable prospects. Accordingly, the participation ratio is assumed to be positively correlated with the post-offering operating performance.

4.4.2 Market Performance Measures

The operating performance of venture-backed firms tends to be low or negative in the first years of the corporate history. They often have negative cash flows at the time of the IPO, and the exit from a venture is typically the primary way for venture capitalists to realize a positive return. Therefore, the mere focus on accounting profitability ratios might be misleading in assessing the performance of venture-supported IPOs. Since the value of high-tech firms is related to growth opportunities and expectations of future profitability, it seems appropriate to examine if the market appraises the value-added potential of venture capital participation in firms going public. Therefore, it is analyzed whether the market has higher expectations of future earnings performance from venture-backed than from non venture-backed issues.

---

[457] Details on the rank assignment for the reputation of underwriters are outlined in section 4.4.4 as well as in the Appendices A 6 and A 7.

[458] See e.g. Barry (1989), Habib and Ljungqvist (2001).

4.4.2.1 Tobins's Q

To assess the expectations of investors in the stock market with respect to the certification role of venture capitalists in IPOs, an approximation of Tobin's Q for both, venture- and non venture-backed issues, at the end of the first trading day is explored as a valuation proxy. Numerous empirical studies in the financial economics literature apply Tobin's Q to categorize firms according to their relative performance.[459] The measure is used as a proxy for companies' future investment opportunities and as an indicator of intangible value. Accordingly, a high value of Tobin's Q indicates that investors place a high valuation on the future growth opportunities of the company.

According to Tobin and Brainard (1968), and Tobin (1969), Tobin's Q is defined as the ratio of the market value of outstanding financial claims to the current replacement costs of assets.[460] The definition assumes that the replacement costs are a logical measure for the value of the alternative use of assets. Since Tobin's Q represents the present value of future cash flows divided by the replacement costs of intangible assets, no risk adjustment is necessary to compare the ratio across firms.[461] Moreover, as the numerator of Tobin's Q contains market values, it is less sensitive to discretion of managers and superior to pure accounting-based measures.[462] In addition to accounting-related information, long-term improvements are taken into account, such as growth opportunities and events, which do not immediately affect the accounting statements or cash flows of a firm.

Nevertheless, on the basis of the collected data, the precise Tobin's Q measure of equity at replacement costs is not available. Therefore, an approximation of the ratio is computed, defined as the market value of equity plus the book value of debt divided by the book value of total assets.[463] This way to calculate Tobin's Q, which is sometimes referred to as "simple Q", has been used in previous studies.[464] Furthermore, there is evidence that the approximation tends to yield similar values for Tobin's Q.[465] In the present analysis, an approximation of the "simple Q" ratio is computed at the first trading day for each IPO. Based on the model specifications employed in previous studies on Tobin's Q, such as Mørck *et al.* (1990) and Cho (1998), the following regression is estimated:

$$(2) \qquad Q_i = \beta_0 + \beta_1 VC_i + \beta_2 \ln Size_i + \beta_3 EqR_i + \beta_4 S\_A_i + \beta_5 Alpha_i + \varepsilon_i\,,$$

where the dependent variable is the approximated value of Tobin's Q at the first trading day, as defined above. On the right-hand side of the equation, a dummy variable (*VC*) for the involvement of venture capitalists in the IPO is used. The firm size (ln*Size*), calculated as the natural logarithm of the market capitalization at the end of the first trading day, and the equity

---

459 See e.g. Lindenberg and Ross (1981), Mørck *et al.* (1988), Lang *et al.* (1991), Lang and Stulz (1994), Himmelberg *et al.* (1999).

460 See Lewellen and Badrinath (1997) for a discussion on the measurement of Tobin's Q.

461 See Lang and Stulz (1994), p. 1249.

462 It is often claimed that managers attempt to "window-dress" accounting figures and carry out "creative accounting" practices.

463 The calculation method is explained in more detail in Appendix A 6.

464 See e.g. Loderer and Martin (1997).

465 Chung and Pruitt (1994) show that the correlation between the "simple Q" ratio and a measure of Tobin's Q employing market values is close to 0.97.

ratio (*EqR*), defined as the book value of equity divided by the book value of total assets at the fiscal year prior to the IPO, are included as explanatory variables. The equity ratio represents a widely used proxy for financial risk.

Additionally, a variable controlling for the importance of the firms' soft capital, quantified by the ratio of sales over assets (*S_A*), is included in the regression setup. Himmelberg *et al.* (1999) argue that firms with higher sales to capital ratios are more difficult to monitor. To account for ownership issues, the fraction of the firms' equity retained by the old shareholders immediately after the going public (*Alpha*) is employed as an explanatory variable. Ideally, the regression should also control for the firms' R&D and advertising expenditures, but unfortunately, these figures are not accessible for the majority of firms in the sample.[466]

4.4.2.2 Buy and Hold Returns and Wealth Relatives

In addition to Tobin's *Q*, the buy and hold returns and wealth relatives are calculated for each IPO over different time intervals in the aftermarket. These measures have been employed in many studies on the long-term performance of IPOs.[467] The aftermarket performance covers the period from the first day of trading onwards, and the buy and hold return (*BHR*) for each firm is calculated by compounding monthly returns up to 36 months after the IPO date:[468]

$$(3) \qquad BHR_{i,T} = \left[\prod_{t=1}^{T}(1+R_{i,t})\right]-1,$$

where $R_{i,t}$ denotes the monthly return of firm *i* in month *t* over the time interval *T*. If a firm going public was delisted before the end of the considered post-issue period, the return is compounded until the delisting date. The monthly closing prices, which are used to calculate the returns, are adjusted for dividends, stock splits, and new share emissions.[469]

Given the movements of the general market over time, the raw buy and hold returns of IPOs have to be adjusted to reflect the excess returns over the market. To measure the relative market performance of the IPOs, buy and hold excess returns can be computed relative to an appropriate benchmark, such as a market index. The buy and hold excess return (*BHER*) of a single firm is defined as the difference between the buy and hold return of the IPO and the buy and hold return of an adequate benchmark portfolio over the same time interval:[470]

$$(4) \qquad BHER_{i,T} = \prod_{t=1}^{T}(1+R_{i,t}) - \prod_{t=1}^{T}(1+R_{m,t}),$$

---

466 Bottazzi and Da Rin (2002c) point out the lack of comparable data on research and development intensity for firms listed at Euro.nm.

467 See e.g. Ritter (1991), Loughran and Ritter (1995), Brav and Gompers (1997).

468 Some empirical studies on the long-run stock price performance of IPOs, such as Loughran and Ritter (1995) and Brav and Gompers (1997), compute 5-year buy and hold returns. However, since many firms of the employed sample went public very recently, buy and hold returns can only be calculated for a maximum post-IPO period of three years. 3-year buy and hold returns are also calculated by Ritter (1991).

469 Details on the calculation of monthly returns are provided in Appendix A 6.

470 See Ritter (1991), Loughran and Ritter (1995). See also Röder (1999), p. 44.

where $R_{m,t}$ denotes the monthly return of the market, or benchmark portfolio, in month t. Accordingly, the buy and hold excess return of a portfolio of N firms can be calculated as follows:

$$(5) \qquad BHER_T = \frac{1}{N} \cdot \sum_{i=1}^{N} BHER_{i,T} = \frac{1}{N} \cdot \sum_{i=1}^{N} \left[\prod_{t=1}^{T} (1 + R_{i,t}) - \prod_{t=1}^{T} (1 + R_{m,t}) \right],$$

The measurement of the relative market performance against an adequate benchmark, such as a market index of the corresponding growth market segment, may eliminate some of the distorting effects of the high stock valuations of European companies listed in the growth market segments at the end of the 1990s. Yet, many authors point out that long-run returns display right-skewed distributions and may therefore lead to inference problems of long-term security performance and test statistics.[471]

An alternative way to compare the buy and hold returns of IPOs with different benchmarks, e.g. the market index of the corresponding growth market segment, represents the calculation of wealth relatives. Based on Ritter (1991), Loughran and Ritter (1995), and Brav and Gompers (1999), wealth relatives (*WR*) are computed as follows:

$$(6) \qquad WR = \frac{(1 + BHR_{i,T})}{(1 + BHR_{m,T})} = \frac{\prod_{t=1}^{T} (1 + R_{i,t})}{\prod_{t=1}^{T} (1 + R_{m,t})}.$$

Wealth relatives greater than one can be interpreted as IPOs outperforming the overall market, whereas wealth relatives smaller than one indicate IPO underperformance to the employed benchmark. However, using the returns of market indices as benchmark implies the assumption that all IPOs have an average systematic risk, or that their betas are equal to one. If betas are not equal to one, this measurement technique could seriously bias the long-term performance estimates.[472] Moreover, using index-adjusted returns implicitly bases on the assumption that systematic risk does not alter over time.[473] Yet, as most of the studies trying to account for time-varying beta risk still find that IPOs underperform in the long run, this issue is not further addressed in the present analysis.[474]

While monthly returns tend to be normally distributed, the wealth relative method typically leads to log-normally distributed and right-skewed long-run return figures.[475] To account for this particularity, Jakobsen and Sørensen (2001) propose a methodology which considers the distribution properties of long-run returns. Using wealth relatives as a data transformation leading to log-normally distributed returns, they decompose the expected buy and hold returns of wealth relatives into a mean and a volatility component:

---

471 See e.g. Barber and Lyon (1997a), Kothari and Warner (1997), Fama (1998), Lyon *et al.* (1999).

472 See Jenkinson and Ljungqvist (2001), pp. 161-162.

473 Clarkson and Thompson (1990) document non-stationary betas for IPO firms in the US which decline after high initial levels.

474 See Jenkinson and Ljungqvist (2001), table 6.3, p. 163, for an overview of studies which try to address the issue of time-varying betas.

475 See e.g. Jakobsen and Sørensen (2001), Jakobsen and Voetmann (2003).

$$(7) \qquad E(WR) = \underbrace{\exp(\mu_T \cdot T)}_{mean} \cdot \underbrace{\exp(0.5 \cdot \sigma_T^2 \cdot T)}_{volatility},$$

where μ_T describes the drift parameter and σ_T^2 the volatility parameter of a geometric Brownian motion. Both components can be tested for significance against alternatives. Moreover, the geometric Brownian motion model represents the marginal dynamics of the cross-sectional long-run returns. Equation (7) shows that the expected value of wealth relatives depends on both the mean and the volatility parameter. If there is no noise and the volatility parameter equals zero, the wealth relative is purely described by the drift parameter in the exponential function. Yet, given the presence of noise, the volatility parameter contributes positively to the average value of wealth relatives, depending on the length of the time horizon (T).[476]

In the empirical analysis, returns are calculated in several ways, as described above. Yet, given the typical log-normality as well as the decomposition possibility related to the wealth relative method, the reported tests in chapter 5 are based on the use of wealth relatives.[477]

Several contributions, such as Barber and Lyon (1997a, 1997b), Gompers and Lerner (1999a), and Lyon *et al.* (1999), advocate the use of portfolios of the same firm size and book-to-market ratios to compute wealth relatives. Yet, given the relatively small number of IPOs per country, the construction of such portfolios is beyond the scope of the present study.[478] Instead, the analysis employs the stock market indices of the different domestic growth markets as benchmark portfolios. The distinct benchmarks used for the calculation of wealth relatives include the (1) *Nouveau Marché* index, (2) *Nemax All Share* index, and (3) *techMARK All Share* index, for IPOs at the *Nouveau Marché*, *Neuer Markt*, and *techMARK*, respectively. There are at least two concerns related to the consistency and adequacy of the employed market indices, which are discussed in the following.

First, the constitution of the *techMARK All Share* index includes also firms going public prior to the covered period, and therefore differs from the composition of the *Nemax All Share* and *Nouveau Marché* indices. To examine the effects of the different index constitution, in the Panels A to C of Appendix A 10 equally weighted market indices are calculated for each country on the basis of the firms in the sample and compared to the official growth market indices. In line with expectations, the synthetic indices of the French (Panel A) and the German sample (Panel B) closely follow the official market indices of the respective growth markets, albeit with slightly higher volatilities. The observed difference might be due to the relatively lower number of firms in the sample as well as to different weights used for the index calculation.[479] As shown in Panel C, the synthetic market index in the UK clearly outperforms the *techMARK All Share* index between August 1999 and March 2001. Again, this observation might be related to the relatively lower number of firms in the sample than in the stock market index as well as the use of different weights.

---

476 See Jakobsen and Sørensen (2001), pp. 404-407.

477 A similar procedure is followed by Brav and Gompers (1997).

478 See also Burghof and Kraus (2003), p. 24.

479 While the synthetic sample-based index is equally weighted, official market indices, such as the *techMARK All Share*, are typically weighted with the market capitalizations of the constituent firms.

Second, in the French and German sample, there are concerns for a simultaneity bias related to the use of the official growth market indices as benchmark. Since the calculation of the *Nemax All Share* and *Nouveau Marché* indices started simultaneously with the launch of the growth market segments, they include almost all IPOs of the respective country sample. To make sure that the calculated wealth relatives are not due to a simultaneity bias or systematic differences in the composition of the market indices, the *Nasdaq Composite* index is employed as an alternative benchmark to adjust for the overall market development.[480] Although the *Nasdaq Composite* index does not include any of the firms going public at the regarded European stock markets, Figure 6 shows that it is highly correlated with the indices of the analyzed growth markets and therefore represents an adequate proxy for the overall valuation levels of technology stocks over the sample period.

Figure 6 outlines the evolution of the different stock market indices, based on monthly observations between 1996 and 2002. In order to allow a comparative assessment, the indices are standardized at the value of 100 at March 15, 1997, i.e. the time when the *Neuer Markt* was launched. The figure shows that the *techMARK All Share* and *Nasdaq Composite* indices follow widely similar paths over time. Moreover, they were less volatile than the *Nemax All Share* index that clearly outperformed the other benchmark indices between June 1997 and September 2001. The *Nouveau Marché* index, by contrast, underperformed the other indices between May 1996 and May 2000 as well as after June 2001, although it exhibits shortly a superior performance to the *techMARK All Share* and the *Nasdaq Composite* indices in early 2000.

Figure 6: Evolution of Different Benchmark Indices over Time

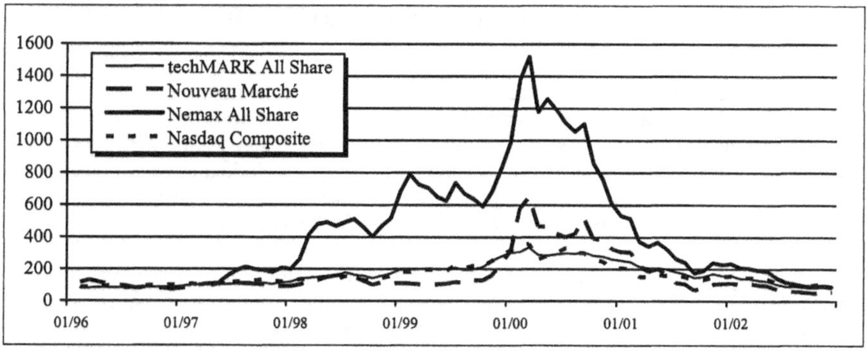

Note: monthly observations (100 = March 15, 1997). Source: Thomson Financial Datastream.

To assess the market performance of the IPOs in the cross-section, based on Brav and Gompers (1997) and Carter *et al.* (1998), the following model specification is employed in the empirical analysis:

---

[480] Fama (1998) points out that benchmark issues might be related to the observed performance anomalies of IPOs in the aftermarket.

(8) $$WR_i = \beta_0 + \beta_1 VC_i + \beta_2 Q_i + \beta_3 \ln Size_i + \beta_4 URank_i + \varepsilon_i,$$

where the dependent variable *WR* denotes the natural logarithm of the 3-year wealth relative, using the respective country's growth market index or the *Nasdaq Composite* index as benchmark.[481] The explanatory variables include a dummy variable (*VC*), taking on the value of one if a firm is venture-backed, the approximated value of Tobin's *Q* (*Q*), and the natural logarithm of the market value of equity (ln*Size*), both measured at the end of the first trading day, as well as the reputation rank of the underwriter (*URank*).

By controlling for firm size, the analysis follows previous studies, such as Fama and French (1992), who document that the measure has an influence on the cross-sectional variation of stock returns.[482] Size is assumed to have a negative effect on stock returns. The rationale behind this expected relationship is that investors require a discount, or higher stock returns, for small firms to compensate for the high portfolio risks related to liquidity, information access, and other factors. Brav and Gompers (1997) argue that asymmetric information is more likely to be a problem for small firms, as it might not pay for sophisticated investors to do research on these firms.

The approximated value of Tobin's *Q* at the end of the first trading day controls for the relation between firm valuation at the time of the IPO and the long-term performance in the stock market. Previous studies, such as Fama and French (1992) or Brav and Gompers (1997), use the book-to-market ratio after the IPO as explanatory variable in the analysis of stock returns. They argue that the observed explanatory power of book-to-market for stock returns is related to the risk of financial distress.[483] Accordingly, firms with higher book-to-market ratios tend to have higher returns because their risk of financial distress is higher. However, as the inverse of the book-to-market ratio is per definition highly correlated with Tobin's *Q*, only the latter is considered in the analysis.

Due to the certification role of venture capitalists, it is assumed that investors value venture-backed issues higher than non venture-backed firms in terms of Tobin's *Q* at the time of the IPO, as graphically shown by the difference ΔQ between Q_1 and Q_2 in Figure 7. As a consequence of the relatively lower initial valuation of non venture-backed issues and the declining information asymmetry regarding the firm quality after flotation, the performance angle of non venture-backed firms (α_2) is expected to be steeper in the aftermarket than the one of venture-backed issues (α_1). This relationship implies that the mere stock price performance of venture-backed firms in the aftermarket (P_1) rests below the performance of non venture-backed firms (P_2), and that it is necessary to consider Tobin's *Q* in the overall performance analysis. To control for the conjectured relationship between Tobin's *Q* and the stock price performance in the aftermarket, as well as for significant differences between venture- and non venture-backed issues, the intercept variable *Q* and an interaction term *VC***Q* is included in the estimated equations in section 5.2.1.3.

---

481 The natural logarithm is employed to account for the distribution properties of wealth relatives, which tend to be right-skewed and log-normally distributed, as pointed out by Jakobsen and Sørensen (2001).

482 See also Fama and French (1993, 1995). Davis (1994), Chan *et al.* (1995), and Barber and Lyon (1997b) show that stocks of small firms with high book-to-market ratios earn high rates of return.

483 See also DeBondt and Thaler (1985).

Figure 7: Stylized Relationship between Tobin's *Q* and Market Performance of IPOs

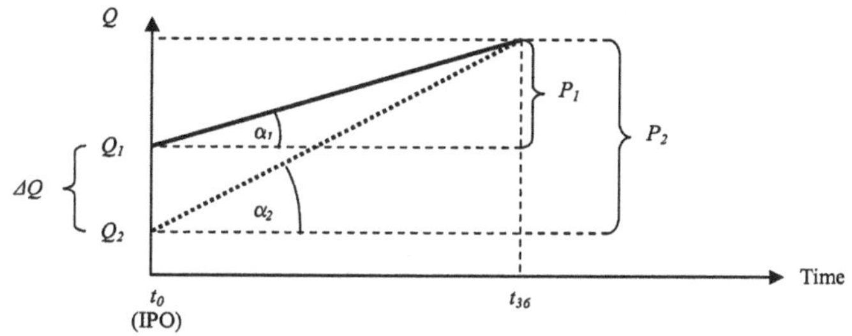

Note: Q_1 (P_1) and Q_2 (P_2) denote the Tobin's Q (market performance) of venture- and non venture-backed IPOs, respectively. Source: own figure.

Finally, the quality of the underwriting investment bank is considered in the regressions. Carter *et al.* (1998) show that the long-term underperformance of IPOs handled by more prestigious underwriters is less severe. Accordingly, the prestige of the underwriting body, measured by the underwriter rank, is conjectured to have a positive effect on the performance of IPOs in the long run. More details on the methodology used to assign ranks for the reputation of underwriters are outlined in section 4.4.4. The following section describes the variables for the quality of venture-backing employed throughout the empirical analysis.

4.4.3 Quality of Venture-Backing

There might be differences in the effectiveness of venture-backing. Therefore, several proxies for the quality of venture capital monitoring are used as controls. These include the representation of venture capitalists on the board of directors (*Board*) as well as the pre- and post-issue equity shares held by all venture investors (*PrEq, PosEq*).[484] Since the use of syndication typically involves complementary skills of additional venture capitalists, providing a second opinion in the process of project selection, the number of venture capitalists having equity positions in the IPO (*NVC*) may also have an impact on the quality of venture-backing.[485] If these monitoring proxies are related to the quality of oversight, a positive sign of the coefficients in the regressions on the operating and market performance of IPOs is expected.

Similar to the study of Gompers (1996), the age of the lead venture capitalist at the time of the IPO (*VC_Age*) is taken into account as a proxy for reputation and certification.[486] If established venture capitalists are able to provide stronger certification than younger ones, the variable is expected to be positively related to the IPO valuation in terms of Tobin's *Q*. Moreover,

[484] Barry *et al.* (1990) investigate the relation between various proxies for monitoring quality and the underpricing of venture-backed IPOs.
[485] The same proxy is also employed by Jain and Kini (1995). See Sah and Stiglitz (1986) on syndication.
[486] In line with Ljungqvist (1999), the lead venture capitalist is defined as the venture capitalist with the biggest stake. Following Franzke (2001), the biggest stake typically corresponds with the longest investment horizon within the portfolio company. See also Barry *et al.* (1990), Lin and Smith (1998).

the number of IPOs in the sample (*No_IPOs*), in which a venture capitalist is involved as financier, is employed as a proxy for experience and reputation. To control for differences in the experience between internationally and nationally operating venture capital investors, a dummy variable (*Int_VC*) is codified for venture capitalists being involved in IPOs in at least two countries of the sample. Finally, the percentage of venture shareholdings sold in the IPO (*Sale*) is considered to find out whether or not venture capitalists try to take advantage of investors by "cashing out".[487] The higher the fraction of equity sold at the time of the IPO, the poorer should be the post-issue performance of the firms.

As discussed in section 4.2.4, there might be differences between the venture-backing of different organizational forms of venture capitalists. To control for systematic variations between the quality of venture-backing provided by independent and captive venture capitalists, a dummy variable for firms backed by independent venture capitalists (*Indep*) is included in the analysis. Based on the distinction made by the EVCA (2000), venture capitalists are defined as independent if less than 20 percent of their equity is held by a single shareholder.

Given the conjecture that public venture capitalists tend to employ staff with little investment skills and to have few incentives to be profitable, as argued in section 4.2.5, an additional dummy variable (*Public*) is employed for firms backed by public venture capital organizations. A venture capitalist is characterized as being public, if the ownership structure indicates that public bodies have a direct or indirect influence on the investments of the firm.[488] Finally, dummies for the national origins of venture capital firms are used to control for the effects of different management styles across countries on the performance of venture-backed IPOs, as pointed out by prior studies.[489]

4.4.4 Control Variables

To avoid specification errors due to omitted variables, the empirical analysis includes a number of control variables.[490] In analogy to previous studies on IPO performance, the underwriter reputation of investment banks is controlled for.[491] Carter and Manaster (1990), Carter *et al.* (1998), and Loughran and Ritter (2003) account for the quality of underwriters by using a classification methodology based on their relative positions in the tombstone of an offering.[492] Yet, the assigned ranks following this methodology, as listed by Loughran and Ritter (2003), merely consider IPOs in the US, and therefore do not adequately reflect the reputation of prestigious investment banks operating primarily in the European markets.

Taking this shortcoming into account, a distinct ranking is employed. It regards the lead management market share of investment banks in the domestic capital markets of France,

---

487 See Gompers and Lerner (1999a).

488 As pointed out by Bascha and Walz (2001b), banks and financial institutions, which are controlled by public authorities, tend to be interested in the promotion of regional business structures and employment.

489 See e.g. Sapienza *et al.* (1996).

490 See Gujarati (1988), pp. 401-404.

491 See Carter and Manaster (1990), Megginson and Weiss (1991), Carter *et al.* (1998), Loughran and Ritter (2003).

492 The resulting rankings range on a scale from 0 to 9, whereby a higher number denotes a higher ranking. Generally, a ranking of 8 and above is considered to be a high rank for prestigious national underwriters.

Germany, and the UK from 1995-1999, based on the underwritten IPO proceeds.[493] On a scale of 1 to 3, the top rating is 1, and the lowest rating equals the value of 3. Investment banks with a market share of equal to or higher than 5 percent are assigned the best value of 1, those with a market share between 1 and 5 percent the value of 2, and those with a market share of lower than 1 percent are assigned the value of 3. To account for the reputation of internationally operating high quality underwriters, which are more active in the US than in the European IPO markets, investment banks classified as prestigious underwriters by Loughran and Ritter (2003) are assigned the value of 1.

For an overview, the ranks of the top investment banks involved in IPOs of each country sample are outlined in the Panels A to C of Appendix A 9. The panels also compare the ranking scores of the method applied in the present analysis to the rankings calculated by Carter *et al.* (1998) and Loughran and Ritter (2003).

To account for industry-related effects, industry dummy variables are employed for six different aggregated industry sectors, based on the classification of the *Financial Times*.[494] In addition, calendar year dummies are included in the analysis, to check whether IPO markets, and in particular the German *Neuer Markt* with its considerable increase in the number of IPOs, attracted firms of low quality and provided ineffective screening in the late years of the sample period.

In order to control for effects related to cross-country differences, country dummy variables are included in the analysis. Dummy variables are also codified for different accounting standards used by the firms going public, such as International Accounting Standards (IAS), French, UK or US GAAP. As required by the listing rules, IPOs at the German *Neuer Markt* report consolidated financial statements either under IAS or US GAAP.[495] At the *Nouveau Marché*, only two firms report under US GAAP while the majority reports under French accounting principles. Information on the applied accounting standards is typically provided in the offerings prospectuses and annual reports of the firms going public.

Finally, an additional dummy variable is employed for firms in the British sample, which originally went public at the *Alternative Investment Market* (AIM) before changing to the *Official Market* (*techMARK*). An overview of all control variables used in the empirical analysis is provided in Appendix A 7. The subsequent section focuses on the applied testing methodology to detect significant differences between venture- and non venture-backed firms.

4.4.5 Testing for Differences between Venture- and Non Venture-Backed Firms

In the univariate analysis, the means and medians of the two firm groups are compared with regard to potential differences. Since the means are sensitive to extreme values, the reported medians may help control for this shortcoming. To identify statistically significant differences

---

493 Methodologies that assess the quality of investment banks on the basis of their market shares as lead underwriters in the IPO market are also used by Megginson and Weiss (1991) and Franzke (2001). See Carter *et al.* (1998) for a comparative analysis of different proxies for underwriter reputation.

494 A similar classification is employed by Bottazzi and Da Rin (2002b, 2002c). More details on the industry classifications are outlined in Appendix A 6.

495 Some of the firms going public at the *Neuer Markt* provided their first post-issue financial statements exceptionally under German HGB accounting standards.

between the means and medians of the two firm groups, Student's t-tests and non-parametric Mann-Whitney-tests are conducted, respectively.

In the multivariate analysis, a qualitative explanatory variable for the involvement of venture capitalists in IPOs, i.e. a venture capital dummy (*VC*), is employed. First, to detect potential effects of venture-backing, a simple intercept dummy variable for the participation of venture capitalists is included in the estimated regressions. Second, in order to reveal slope effects of venture-backing, different interaction dummy variables are used.[496] In the present analysis, interaction dummy variables are the product of the original regressor of each explanatory variable and the binary venture capital dummy variable. To avoid a bias in the estimates of interaction terms, it is vital to use a venture capital intercept dummy in the regressions as well.[497]

Alternatively, the *Chow*-test is applied to identify whether the regressions of venture- and non venture-backed firms are different concerning their structures.[498] The *Chow*-test is a frequently used method of testing for differences between two or more regressions, representing a special application of the restricted least squares method.[499] It compares the results of a pooled regression over all observations with those of separate regressions for different subgroups.[500] The significance of the *Chow*-test is assessed by comparing the impact of the pooled regression with individual estimations for each subgroup by an F-test statistic:

$$(9) \qquad F = \frac{(RSS_{pooled} - RSS_{individual})/k}{RSS_{individual}/(n_1 + n_2 - 2k)},$$

where k is the number of the estimated parameters, n_1 and n_2 are the numbers of observations of each subgroup, and RSS_{pooled} and $RSS_{individual}$ are the residual sums of squares of the pooled and individual regressions, respectively.

Based on the overall research design, as described in the previous sections, the following chapter presents the main findings of the empirical study and provides interpretations with respect to the stated hypotheses as well as the overall research question.

---

496 For the use of intercept and interaction dummy variables see Jobson (1991), pp. 317-321, and Studenmund (1997), pp. 232-241.

497 See Studenmund (1997), p. 236.

498 See Chow (1960), pp. 594-595.

499 See Gujarati (1988), pp. 443-446.

500 See Jobson (1991), p. 307.

5 Empirical Evidence

This chapter presents the main results of the empirical study regarding the impact of venture-backing on IPOs. First, in section 5.1, the findings of the univariate analysis are presented. It focuses on the offering and firm characteristics at the time of the going public (section 5.1.1), the post-issue operating and market performance (section 5.1.2), the experience of venture capital investors and their involvement in firms (section 5.1.3), as well as on the quality of the lead underwriters (section 5.1.4). Section 5.2 then investigates the determinants of the firms' different operating and market performance measures in the cross-section, and in particular the influence of venture-backing, by using a multivariate analysis. Finally, section 5.3 discusses the main findings and addresses the limitations of the study as well as alternative explanations for the obtained results.

5.1 Univariate Analysis

5.1.1 Firm and Offering Characteristics at the Time of the IPO

Before testing the individual hypotheses, the firm and offering characteristics at the time of the going public are investigated. Table 16 provides the descriptive statistics for the IPOs in the different country samples, comparing the means and medians of the characteristics of venture- and non venture-backed firms. To account for the sensitivity of means regarding extreme values, the medians are reported in parentheses. The offering characteristics include the market capitalization at the end of the first trading day (*Mkt_Cap*), the gross IPO proceeds (*IPO_Size*), the participation of old shareholders in the offering (*Part*), measured by the percentage of secondary shares relative to the number of shares outstanding before flotation, the portion of equity held by old shareholders immediately after the IPO (*Alpha*), and the approximated Tobin's *Q* at the first day of trading (*Q*). In addition, several firm characteristics are investigated, such as the total sales (*Sales*), assets (*Assets*), net income (*Income*), equity ratio (*EqR*), the number of employees (*Empl*) at the end of the fiscal year prior to the going public, and the firm age (*Age*) at the IPO.

Comparing the participation ratios between the two firm groups indicates that the old shareholders of venture-backed firms in all three countries sold significantly higher portions of shares in the IPO than their non venture-backed counterparts. Moreover, the differences in the *Alpha* ratios show that old shareholders of venture-backed firms in France and Germany held significantly less equity after the IPO than their equivalents in firms without venture-backing. The remaining firm and offering characteristics reported in Table 16 do not indicate any further significant differences between venture- and non venture-backed issues, which apply for all three considered countries. Although venture-backed firms in all three countries tend to be lower leveraged than non venture-backed ones, these differences remain statistically insignificant.[501] However, it is surprising how old venture-backed firms generally are, relative to the results of previous US studies. Following Gompers (1996), US venture-backed firms are on average between 3 and 4 years old when they go public. Yet, in the present analysis, the median age ranges from 7 years in the French sample to 10 years in the German sample.

---

[501] See Audretsch and Lehmann (2003) for an analysis of the leverage of venture- and non venture-backed IPOs at the German Neuer Markt.

Table 16: Firm and Offering Characteristics at the IPO

The values for venture- and non venture-backed firms are expressed as means (medians in parentheses). To detect significant differences between the means [medians] of the two firm groups, Student's *t*-tests [non-parametric Mann-Whitney tests] were conducted. Bold values indicate statistically significant differences at the 1% or 5% confidence levels. The market capitalization of equity (*Mkt_Cap*) is expressed in million EUR and related to the first day of trading. *IPO_Size* represents the gross proceeds of the IPO incl. exercised overallotments. *Part* is the participation ratio (in percent), defined as number of old shares sold divided by the number of shares outstanding before flotation. *Alpha* is the percentage of the firm's equity retained by old shareholders after the IPO. Tobin's *Q* is the approximated value at the end of the first trading day, defined as the sum of the market value of equity and book value of debt, divided by the book value of total assets. Total sales (*Sales*), assets (*Assets*), and net income *(Income)* are expressed in million EUR and related to the end of the fiscal year prior to the IPO. Likewise, the number of employees *(Empl)* is related to the end of the fiscal year prior to the IPO. The equity ratio (*EqR*) is expressed in percent, defined as the book value of equity divided by the book value of total assets at the end of the fiscal year prior to the IPO. The firm age (*Age*) is expressed in years at the IPO.

	France			***Germany***			***UK***		
Variable	VC-backed	non VC-backed	*t*-test [*z*-score]	VC-backed	non VC-backed	*t*-test [*z*-score]	VC-backed	non VC-backed	*t*-test [*z*-score]
Mkt_Cap	53.3	46.9	0.5	223.2	251.1	-0.8	231.1	166.6	0.9
	(37.2)	(26.0)	[-0.7]	(164.3)	(184.9)	[-0.9]	(132.0)	(58.6)	[-1.7[c]]
IPO_Size	16.6	10.7	1.8[c]	46.7	50.7	-0.6	75.4	62.5	0.4
	(10.7)	(7.4)	[-1.9[c]]	(36.3)	(35.2)	[-0.3]	**(35.1)**	**(13.6)**	**[-2.3[b]]**
Part	**10.0**	**5.8**	**2.2[b]**	**15.4**	**9.3**	**3.2[a]**	**20.6**	**10.7**	**2.4[b]**
	(9.6)	(6.3)	[-1.9[c]]	**(11.7)**	**(6.7)**	**[-2.4[b]]**	**(18.5)**	**(9.0)**	**[-2.1[b]]**
Alpha	**72.9**	**79.0**	**-3.4[a]**	**62.2**	**68.6**	**-3.2[a]**	62.0	68.4	-1.9[c]
	(72.3)	**(79.4)**	**[-3.2[a]]**	**(66.1)**	**(70.7)**	**[-2.3[b]]**	(65.1)	(70.0)	[-1.4]
Q	2.3	2.6	-1.4	4.7	4.9	-0.5	3.8	4.2	-0.9
	(2.3)	(2.3)	[-1.1]	(3.9)	(4.0)	[-0.3]	(3.4)	(4.1)	[-1.0]
Sales	18.9	18.4	0.1	25.0	42.3	-1.2	87.4	33.5	0.9
	(10.4)	(9.8)	[-0.5]	(9.2)	(16.1)	[-1.7[c]]	(15.7)	(7.3)	[-1.1]
Assets	16.4	11.8	1.2	29.0	26.5	0.2	39.6	43.2	-0.1
	(10.7)	(8.4)	[-1.6[c]]	(11.4)	(12.4)	[-1.5]	**(13.7)**	**(5.1)**	**[-2.4[b]]**
Income	-0.1	0.4	-0.9	-0.3	0.2	-0.3	1.6	-1.2	0.8
	(0.3)	(0.4)	[-1.3]	**(0.0)**	**(0.5)**	**[-3.1[a]]**	(1.1)	(0.4)	[-0.2]
EqR	30.3	26.6	0.9	30.2	-10.6	1.7[c]	18.5	-32.3	1.5
	(26.7)	(23.3)	[-1.5]	(32.2)	(21.4)	[-1.0]	(33.6)	(20.7)	[-1.3]
Empl	148	121	0.7	162	218	-1.2	349	162	1.4
	(85)	(94)	[-0.1]	(82)	(214)	[-1.0]	**(113)**	**(53)**	**[-2.1[b]]**
Age	9.4	9.0	0.3	13.0	14.5	-0.7	9.8	11.6	-0.5
	(7.0)	(7.5)	[-0.4]	(10.0)	(10.0)	[-0.9]	(8.0)	(5.0)	[-0.8]

Note: significance levels are denoted by [a)] for 1%, [b)] for 5%, and [c)] for 10%. Source: own calculations.

The refined univariate analysis, distinguishing between IPOs backed by national and international venture capitalists, whereby the latter are defined as those backing firms in at least two of the three considered countries, shows a slightly different pattern. The comparison, which is for brevity outlined in Appendix A 11, indicates that the median firm backed by international venture capitalists has significantly larger IPO proceeds than its equivalent backed by national venture investors, irrespective of the considered country. This finding suggests that venture capitalists in the considered countries are quite heterogeneous, and when classified as international, are able to raise higher amounts of equity in the capital markets than national venture investors.

In order to detect significant differences across countries, Table 17 outlines the Student's *t*-test and non-parametric Mann-Whitney *z*-scores for international differences in the means and medians of the offering and firm characteristics. The comparison shows that venture- as well as non venture-backed firms in the French sample are significantly smaller than those in Germany and the UK, both in terms of market capitalization and issue size. In addition, the test-scores of the participation and alpha ratios indicate that the old shareholders of venture-backed firms in France sold significantly lower portions of shares in IPOs, and held more equity shares after flotation, than their counterparts in Germany and the UK.

Table 17: International Comparison of the Firm and Offering Characteristics at the IPO

Values for venture- and non venture-backed firms are Student *t*-tests statistics (non-parametric Mann-Whitney *z*-scores in parentheses) to detect significant differences between the means (medians) of firm groups in different countries. Bold values indicate statistically significant differences at the 1% or 5% confidence levels. The market capitalization of equity (*Mkt_Cap*) is expressed in million EUR and related to the first day of trading. *IPO_Size* represents the gross proceeds of the IPO incl. exercised overallotments. *Part* is the participation ratio (in percent), defined as number of old shares sold divided by the number of shares outstanding before flotation. *Alpha* is the percentage of the firm's equity retained by old shareholders after the IPO. Tobin's *Q* is the approximated value at the end of the first trading day, defined as the sum of the market value of equity and book value of debt, divided by the book value of total assets. Total sales (*Sales*), assets (*Assets*), and net income *(Income)* are expressed in million EUR and related to the end of the fiscal year prior to the IPO. Likewise, the number of employees *(Empl)* is related to the end of the fiscal year prior to the IPO. The equity ratio (*EqR*) is expressed in percent, defined as the book value of equity divided by the book value of total assets at the end of the fiscal year prior to the IPO. The firm age (*Age*) is expressed in years at the IPO.

Comparison	***France vs. UK***		***Germany vs. UK***		***Germany vs. France***	
Variable	VC-backed	Non VC-backed	VC-backed	Non VC-backed	VC-backed	Non VC-backed
Mkt_Cap	**-3.9**^{a}	**-3.1**^{a}	-0.2	1.6	**6.3**^{a}	**4.8**^{a}
	(-5.3^{a})	**(-2.6^{a})**	(-1.0)	**(-3.1^{a})**	**(-7.4^{a})**	**(-7.4^{a})**
IPO_Size	**-4.1**^{a}	**-2.3**^{b}	-1.9^{c}	-0.7	**4.8**^{a}	**5.2**^{a}
	(-5.0^{a})	**(-2.5^{b})**	(-0.3)	**(-3.0^{a})**	**(-6.7^{a})**	**(-7.6^{a})**
Part ratio	**-3.6**^{a}	-1.9^{c}	-1.5	-0.7	**2.3**^{b}	**2.2**^{b}
	(-2.5^{b})	(-1.0)	(-1.2)	(-0.3)	(-1.9^{c})	(-1.7^{c})
Alpha	**4.4**^{a}	**4.6**^{a}	0.1	0.1	**-4.5**^{a}	**-6.0**^{a}
	(-3.5^{a})	**(-4.0^{a})**	(-0.4)	(-0.2)	**(-3.9^{a})**	**(-5.8^{a})**
Q	**-5.2**^{a}	**-4.4**^{a}	1.9^{c}	1.1	**7.1**^{a}	**4.7**^{a}
	(-4.8^{a})	**(-3.5^{a})**	(-1.8^{c})	(-0.7)	**(-6.7^{a})**	**(-5.8^{a})**
Sales	-1.7^{c}	-1.2	-1.7^{c}	0.4	1.1	1.3
	(-0.7)	(-0.7)	(-0.6)	(-1.8^{c})	(-0.1)	**(-2.7^{a})**
Assets	-1.9^{c}	-1.5	-0.6^{c}	-1.1	1.2	**2.1**^{b}
	(-1.5)	(-1.3)	(-1.2)	**(-2.3^{b})**	(0.0)	**(-2.0^{b})**
Income	-1.3	0.5	-1.6	0.5	-0.5	-0.1
	(-1.5)	(-0.1)	(-1.5)	(-0.5)	(-0.9)	(-0.9)
EqR	1.2	-1.9^{c}	1.7	0.5	0.0	-1.1
	(-0.5)	(-0.4)	(-0.1)	(-0.1)	(-0.6)	(-0.7)
Empl	**-2.1**^{b}	-0.8	**-2.2**^{b}	0.8	0.4	1.7^{c}
	(-1.1)	(-1.3)	(-1.2)	(-1.9^{c})	(-0.2)	(-0.7)
Age	-0.2	-0.8	-1.4	0.9	1.7^{c}	**2.3**^{b}
	(-0.5)	(-0.8)	(-1.2)	**(-2.4^{b})**	(-1.8^{c})	**(-2.3^{b})**

Note: Significance levels are denoted by $^{a)}$ for 1%, $^{b)}$ for 5%, and $^{c)}$ for 10%. Source: own calculations.

Finally, the significantly lower market valuations of firms at the *Nouveau Marché* compared to the two other country sample in terms of Tobin's Q are noteworthy. Since French accounting principles are rather conservative and do not lead to higher book equity valuations than IAS, UK or US GAAP, the finding suggests that IPOs in the French sample underwent systematically lower market valuations than firms in the two other countries.[502] The refined analysis by yearly cohorts between the firm groups and across countries, as shown in the Appendices A 12 and A 13, reveals that the observed differences in Tobin's Q between France and Germany stem from the significantly higher valuations of firms at the *Neuer Markt* in 1998 and the generally low valuations of venture-backed issues at the *Nouveau Marché* in 1999.[503] Yet, the significant differences in valuation between IPOs in France and the UK last over the period 1997 to 1999. A possible explanation for this finding is the higher liquidity available in the UK with its larger capital markets compared to France.

Table 18: Accounting Profitability at the IPO

Values for venture- and non venture-backed firms are expressed as means (medians in parentheses). To detect significant differences between the means [medians] of the two firm groups, Student's t-tests [non-parametric Mann-Whitney tests] were conducted. Bold values indicate statistically significant differences at the 1% or 5% confidence levels. All accounting profitability ratios are expressed in percent and based on accounting figures at the end of the fiscal year prior to the IPO. The operating return (cash flow return) on assets is defined as the operating income (operating cash flow) divided by total assets. The operating return (cash flow return) on sales is defined as the operating income (operating cash flow) divided by sales.

	France			*Germany*			*UK*		
Variable	VC-backed	non VC-backed	*t*-test [*z*-score]	VC-backed	non VC-backed	*t*-test [*z*-score]	VC-backed	non VC-backed	*t*-test [*z*-score]
ROA	**4.3**	**14.0**	**-2.6[a]**	-4.3	-2.0	-0.1	-15.8	-34.7	0.6
	(7.4)	**(9.7)**	**[-2.0[b]]**	**(1.3)**	**(11.6)**	**[-3.8[a]]**	(8.4)	(16.8)	[-1.4]
CFROA	**10.4**	**21.7**	**-3.1[a]**	5.3	9.1	-0.3	-9.7	-29.9	0.7
	(14.7)	**(17.6)**	**[-2.3[b]]**	**(7.3)**	**(20.7)**	**[-3.9[a]]**	(13.7)	(20.6)	[-1.2]
ROS	-121.7	35.0	-1.1	-22.3	-53.3	0.5	-2,874.4	-394.8	-0.8
	(6.5)	(8.0)	[-1.1]	**(1.4)**	**(7.5)**	**[-3.6[a]]**	(4.8)	(9.2)	[-1.0]
CFROS	-98.6	44.8	-1.1	-8.8	-25.9	0.4	-2,763.6	-386.6	-0.8
	(11.9)	(13.3)	[-1.7[c]]	**(6.3)**	**(15.3)**	**[-3.3[a]]**	(8.2)	(10.4)	[-0.9]

Note: Significance levels are denoted by [a)] for 1%, [b)] for 5%, and [c)] for 10%. Source: own calculations.

In the following, the descriptive statistics of different accounting profitability ratios of venture- and non venture-backed firms at the time of the IPO are examined. Table 18 compares the operating performance of the two firm groups, measured by the operating return on assets (*ROA*) and the operating cash flow return on assets (*CFROA*), as defined in section 4.4.1. Moreover, the profitability ratios normalized by total sales (*ROS*, *CFROS*) are investigated. Yet, merely in Germany systematic differences appear between the two firm groups which apply for all defined operating profitability ratios, i.e. the median venture-backed IPO is less profitable than its non venture-backed counterpart. In the French sample, venture-backed firms are significantly less profitable in terms of *ROA* and *CFROA* than non venture-backed

[502] See Ordelheide and KPMG (1995) for an analysis of different accounting practices in countries of the EU.

[503] The statistics in Appendix A 12 indicate that venture-backed IPOs at the *Nouveau Marché* in 1999 underwent significantly lower valuations in terms of Tobin's Q than non venture-backed issues.

issues, whereas in the UK, no systematic variations between the accounting profitability of the firm groups are detected.

The differences between the means and medians of the operating performance measures, which are the most striking for the ratios normalized by sales of British IPOs, indicate that the sample contains a number of extreme values. In fact, a closer look at the data set shows that some firms exhibit particularly large return on sales ratios, caused by very low sales, which are not unusual for young firms during the first years of operations. Given the skewness of the profitability ratios scaled by total sales, the operating performance measures normalized by assets are employed in the multivariate analysis. The international assessment in Table 19 does not identify systematic differences across countries. Although the comparison of the mean returns on assets indicates that French venture-backed issues are more profitable than British IPOs, this finding is not confirmed by the Mann Whitney z-scores for differences in medians. Likewise, the refined comparison between the accounting profitability of firms backed by international and national venture capitalists, as stated in Appendix A 11, does not provide any additional insights regarding systematic differences between the two groups.

Table 19: International Comparison of Accounting Profitability at the IPO

Values for venture- and non venture-backed firms are Student t-tests statistics (non-parametric Mann-Whitney z-scores in parentheses) to detect significant differences between the means (medians) of firm groups in different countries. Bold values indicate statistically significant differences at the 1% or 5% confidence levels. All accounting profitability ratios are expressed in percent and based on accounting figures at the end of the fiscal year prior to the IPO. The operating return (cash flow return) on assets is defined as the operating income (operating cash flow) divided by total assets. The operating return (cash flow return) on sales is defined as the operating income (operating cash flow) divided by sales.

Comparison	***France vs. UK***		***Germany vs. UK***		***Germany vs. France***	
Variable	VC-backed	Non VC-backed	VC-backed	Non VC-backed	VC-backed	Non VC-backed
ROA	**2.3[b]**	1.8[c]	1.3	1.1	-1.8[c]	-0.8
	(-0.4)	(-0.2)	(-0.5)	(-0.4)	(-1.9[c])	(-0.2)
CFROA	**2.3[b]**	2.0[c]	1.6	1.4	-1.0	-0.6
	(-0.5)	(-0.3)	(-0.2)	(-0.6)	(-1.4)	(-0.2)
ROS	1.2	1.5	1.4	1.7[c]	0.9	-1.1
	(-1.3)	(-0.5)	(-0.1)	(-0.4)	**(-2.1[b])**	(-0.1)
CFROS	1.2	1.5	1.4	1.9[c]	1.0	-1.2
	(-1.8[c])	(-1.6)	(-0.6)	(-1.8[c])	(-1.3)	(0.0)

Note: Significance levels are denoted by [a)] for 1%, [b)] for 5%, and [c)] for 10%. Source: own calculations.

In sum, the univariate analysis of the IPO and firm characteristics does not reveal many systematic differences between the two firm groups. First, it was shown that venture-backed IPOs in all three considered countries are characterized by higher secondary sales than non venture-backed issues. This finding can be explained with the need of venture capitalists to return capital to investors and to raise new funds, even though prestigious venture capitalists do typically not sell many shares in IPOs to avoid the impression of "cashing out".[504] Second, the refined analysis, focusing on firms backed by different kinds of venture capitalists, indi-

[504] See e.g. Gompers and Lerner (1999a), p. 18.

cates significant variations with respect to the issue size. The findings support the idea that international venture capitalists can bring bigger issues to the markets than national ones. This result points out the necessity to account for the heterogeneity of venture capitalists, when testing the hypotheses in the multivariate analysis. In the following, the descriptive statistics of the post-issue operating and market performance figures are analyzed.

5.1.2 Post-Issue Performance

5.1.2.1 Accounting Performance

To compare the pre- and post-issue operating performance figures of the sample firms, the average accounting profitability ratios of the two fiscal years subsequent to the IPO are presented in Table 20. As in the fiscal year prior to the going public, venture-backed firms in the German sample continue to display lower accounting performance figures than non venture-backed firms after flotation. Yet, the significance of the cash flow-related performance ratios is only weakly significant. In the two other countries, the overall results do not indicate any systematic differences between the means and medians of the two firm groups. In analogy to the previous analysis, the descriptive statistics in Appendix A 14 compare the post-issue operating performance between firms backed by international and national venture capitalists. However, this refinement of the univariate analysis does not provide any additional insights with regard to significant differences in the post-issue performance.

Previous contributions have shown that firms typically undertake an IPO after a period of apparently strong performance, but then fail to sustain the growth rates.[505] Although it is possible that old shareholders float firms after periods of genuinely strong performance, which they – unlike investors – know cannot be sustained in the future, it is also probable that pre-issue performance is deliberately overstated.[506] Comparing the medians of different profitability ratios of the sample firms over time, it appears that both firm groups display lower post- than pre-IPO performance figures. This pattern is consistent with the idea that managers attempt to "window-dress" their accounting numbers prior to going public, i.e. actively encourage optimistic views of future performance. Yet, the post-issue accounting profitability might be biased by IPO costs, which are frequently treated as an expense in the fiscal year following the issue.[507] The treatment of these costs as expenses creates a considerable bias in the post-issue accounting profitability, and thus limits the comparability of the performance figures over time.[508] Since the nature of the employed data set does not allow controlling for the precise influence of IPO costs on the post-issue accounting performance, the "window-dressing" hypothesis is not further explored.

---

505 DeGeorge and Zeckhauser (1993) emphasize that managers have extraordinary incentives to make their firms "shine" before flotation. For empirical evidence on the post-issue operating performance of IPO firms see Jain and Kini (1994).

506 See Jenkinson and Ljungqvist (2001), pp. 157-160.

507 An empirical study of IPOs at Germany's *Neuer Markt* between 1997 and 1999 by Küting (2000) shows that most firms reporting under IAS treated IPO costs as expenses. See Küting (2000), pp. 69-86.

508 The IPO costs typically vary between 3 and 10 percent of the proceeds, depending on the size of an issue. See e.g. Fessler *et al.* (2000), Kiwitz and Melzer (2001).

Table 20: Post-IPO Accounting Profitability Ratios (2-Year Averages)

Values for venture- and non venture-backed firms are expressed as means (medians in parentheses). To detect significant differences between the means [medians] of the two firm groups, Student's *t*-tests [non-parametric Mann-Whitney tests] were conducted. Bold values indicate statistically significant differences at the 1% or 5% confidence levels. All accounting profitability ratios are averages of two fiscal years following the IPO and expressed in percent. The operating (cash flow) return on assets is defined as the operating income (cash flow) divided by total assets. The operating (cash flow) return on sales is defined as the operating income (cash flow) divided by sales.

	France			*Germany*			*UK*		
Variable	VC-backed	non VC-backed	*t*-test [*z*-score]	VC-backed	non VC-backed	*t*-test [*z*-score]	VC-backed	non VC-backed	*t*-test [*z*-score]
ROA	-5.6	3.0	-1.9[c]	**-15.7**	**-3.3**	**-2.2[b]**	-7.0	-18.8	1.0
	(1.6)	(3.2)	[-0.8]	**(-0.5)**	**(3.5)**	**[-1.8[b]]**	(5.9)	(-1.1)	[-0.2]
CFROA	**-0.3**	**8.3**	**-2.0[b]**	-1.5	4.6	-1.8[c]	1.2	-13.1	1.4
	(7.1)	(6.3)	[-0.7]	(5.5)	(8.3)	[-1.8[c]]	(10.0)	(4.9)	[-0.6]
ROS	-50.5	-60.9	0.17	**-43.9**	**-15.0**	**-2.3[b]**	-1,237.5	-193.4	-0.8
	(0.8)	(3.8)	[-0.9]	**(-2.1)**	**(3.8)**	**[-2.0[b]]**	(5.2)	(-0.9)	[-0.6]
CFROS	-38.8	-17.1	-0.6	-14.5	0.2	-1.6	-1,148.1	-179.0	-0.8
	(8.3)	(8.3)	[-0.5]	(5.7)	(8.8)	[-1.8[c]]	(7.7)	(5.9)	[-0.6]

Note: Significance levels are denoted by [a)] for 1%, [b)] for 5%, and [c)] for 10%. Source: own calculations.

Overall, the univariate analysis of the post-issue operating performance does not reveal any systematic differences between venture- and non venture-backed issues in the considered countries. Likewise, significant differences do not appear between firms backed by different types of venture capitalists. Although the comparison of the pre- and post-issue profitability ratios indicates a potential decline of the operating performance after the IPO, the sort of the employed data does not allow a validation of this finding. Given the limited insights of the accounting figures with respect to differences between venture- and non venture-backed issues, the following section focuses on the market performance of the IPOs.

5.1.2.2 Market Performance

In order to test the hypothesis that IPOs underperform seasoned stocks (*H1*), buy and hold returns as well as wealth relatives are calculated by compounding monthly returns for the 3-year post-issue period.[509] Wealth relatives in the aftermarket are computed by taking the ratio of one plus the IPO portfolio return over one plus the return on the chosen benchmark. Figure 8 plots the mean cumulative wealth relatives of all IPOs by country to the respective growth market index (figure on the left-hand side) and the *Nasdaq Composite* index (figure on the right-hand side) for a period of 36 months subsequent to the going public.

The wealth relatives on the left-hand side of Figure 8 do not suggest that the IPOs systematically underperform the market in the long run. Yet, the comparison of the figures on the left- and right-hand side reveals that the results are sensitive to the employed benchmark. Using the *Nasdaq Composite* index as an alternative benchmark leads to generally lower wealth relatives three years after the IPO than the respective growth market indices, even though the ranking of the portfolios by country does not change. Between 10 and 17 months after going

509 See e.g. Ritter (1991), Loughran and Ritter (1995), Brav and Gompers (1997).

public, French and German firms attain relatively high wealth relatives of about 2.3 and 1.5, respectively. However, towards the end of the observed period the German IPOs appear to underperform the benchmark, ending up with an average wealth relative of merely 0.75.

Figure 8: Mean Wealth Relatives of IPOs by Country

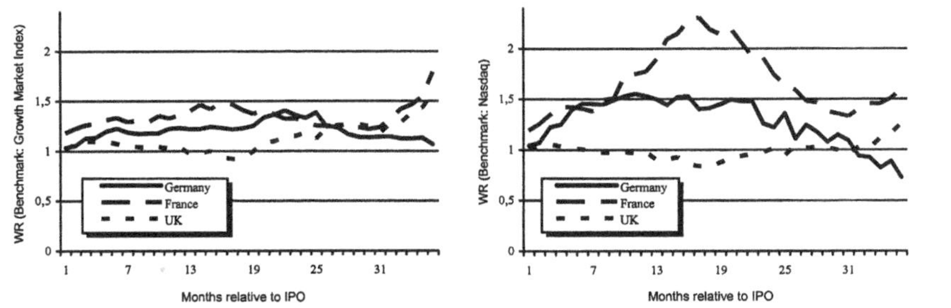

Note: WR – wealth relative. Source: own calculations.

A closer look at the distribution of the wealth relatives reveals that they are highly skewed and that the means of the country portfolios are driven by outliers with relatively high wealth relatives.[510] Therefore, the medians which are less sensitive to outliers are plotted in Figure 9. It illustrates that the median IPO in all three countries tends to underperform seasoned stocks, independent of the applied benchmark, with wealth relatives ranging from 0.2 to 0.8. If the respective country's growth market index is applied as benchmark (figure on the left-hand side), the French IPOs display a better market-adjusted performance in terms of wealth relatives (0.85) than firms in the UK (0.66) and Germany (0.44). Yet, if the *Nasdaq Composite* index is used to adjust for the return of the market (figure on the right hand side), the IPOs in France end up with a considerably lower wealth relative of 0.69, which is only slightly above the value of the IPOs in the UK (0.65). Finally, with very low wealth relatives of only 0.2, the German IPOs perform clearly worse than the market as well as the issues in the two other countries.

Considering the distribution properties of long-run returns expressed in terms of wealth relatives, the application of the method proposed by Jakobsen and Sørensen (2001) widely confirms the findings based on the medians.[511] As documented in the Panels A, D, and G of Appendix A 15, the decomposition of the wealth relatives into mean and volatility components shows that the mean components of the IPOs in all three countries underperform the overall market in the three years following the going public, irrespective of the employed benchmark.

---

[510] Three years after the IPO, the German firm EM.TV ends up with a wealth relative of 19.5 (46.0) against the *Nemax All Share* (*Nasdaq Composite*) index. In the UK, Colt Telecom represents an outlier with a wealth relative of 12.8 (10.2).

[511] The decomposition method used by Jakobsen and Sørensen (2001) necessitates that wealth relatives are log-normally distributed. The examination of the logarithm of wealth relatives by means of Kolmogorov-Smirnov-tests indicates that they are log-normally distributed.

Figure 9: Median Wealth Relatives of IPOs by Country

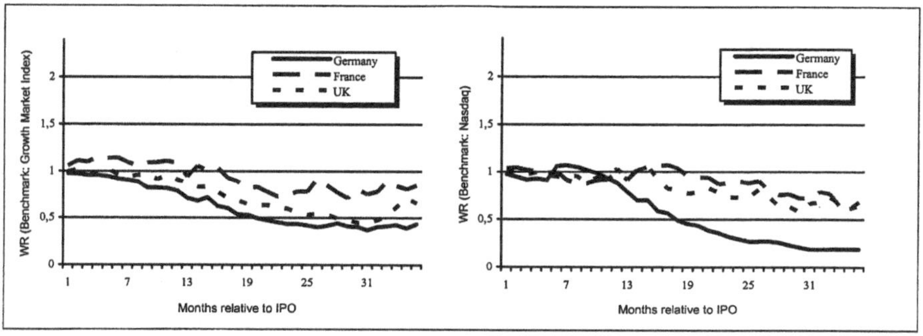

Note: WR – wealth relative. Source: own calculations.

Ordinary *t*-test statistics of the mean component of the wealth relatives, as shown in Table 21, indicate that the null hypothesis of no underperformance to the market over the 3-year post-IPO period can be rejected at confidence levels between 1 and 5 percent, depending on the country and the employed benchmark. Hence, consistent with *H1*, in all three countries, the average IPO of the sample underperforms seasoned stocks in the long run, irrespective of the employed benchmark. The international comparison confirms the order of the previous analysis: 36 months after the going public, IPOs in Germany display the worst and IPOs in France the best market performance.

Table 21: Mean Component of Cumulative Wealth Relatives of Different Time Intervals

Based on the method employed by Jakobsen and Sørensen (2001), the table shows the mean component of the average cross-sectional market performance of the IPOs in levels of wealth relatives (*WR*) over different time intervals using different benchmarks. The values of Student *t*-tests statistics to detect significant differences from one are reported in parentheses. Bold values indicate statistically significant differences at the 1% or 5% confidence levels.

	France		***Germany***		***UK***	
Post-IPO period	WR to Growth Market	WR to Nasdaq	WR to Growth Market	WR to Nasdaq	WR to Growth Market	WR to Nasdaq
12 months	1.07 (1.0)	1.10 (1.0)	**0.83** **(-2.9[a])**	0.95 (-0.7)	0.87 (-1.5)	**0.81** **(-2.2[b])**
24 months	**0.77** **(-2.4[b])**	0.90 (-0.8)	**0.55** **(-6.0[a])**	**0.40** **(-8.9[a])**	**0.61** **(-3.1[a])**	**0.56** **(-3.8[a])**
36 months	**0.74** **(-2.2[b])**	**0.68** **(-2.9[a])**	**0.42** **(-7.5[a])**	**0.18** **(-13.9[a])**	**0.64** **(-2.3[b])**	**0.53** **(-3.4[a])**

Note: Significance levels are denoted by [a)] for 1%, [b)] for 5%, and [c)] for 10%. Source: own calculations.

In the following, the returns and wealth relatives of venture- and non venture-backed IPOs are inspected with regard to systematic differences over the 3-year post-issue period. Similar to Brav and Gompers (1997), Table 22 reports the buy and hold returns as well as wealth relatives of both firm groups against different benchmarks. To account for the influence of outliers, both means and medians are documented. The results show that the venture-backed

IPOs in Germany underperform their non venture-backed equivalents in terms of mean and median IPO returns and wealth relatives, independent of the employed benchmark. However, the observed differences are not significant at conventional statistical levels.[512] Most striking is the fact that both firm groups report highly negative median returns and relatively low median wealth relatives, even though the bad performance of the median firm going public in Germany was already indicated by the time-series figures.

Table 22: 3-Year Post-IPO Returns and Wealth Relatives versus Different Benchmarks

3-year equal weighted returns on IPOs are compared with alternative benchmarks (medians in parentheses). The buy and hold returns (*BHR*) and benchmark returns are calculated by compounding monthly returns for 36 months subsequent to the IPO date. Wealth relatives (*WR*) are computed by taking the ratio of 1 plus the IPO return over 1 plus the return on the chosen benchmark over the same period.

	Venture-Backed IPOs			***Non Venture-Backed IPOs***		
Benchmarks	BHR (%)	Benchmark Return (%)	WR	BHR (%)	Benchmark Return (%)	WR
			France (Nouveau Marché)			
Nouveau Marché	80.5 (4.0)	27.9 (-12.6)	2.06 (1.02)	76.8 (-36.1)	55.3 (-4.2)	1.39 (0.64)
Nasdaq Composite	80.5 (4.0)	28.8 (4.9)	1.79 (0.73)	76.8 (-36.1)	48.4 (11.7)	1.32 (0.55)
			Germany (Neuer Markt)			
Nemax All Share	-61.2 (-88.3)	-28.7 (-72.9)	0.78 (0.35)	47.1 (-86.3)	-54.3 (-73.9)	1.28 (0.51)
Nasdaq Composite	-61.2 (-88.3)	-17.0 (-32.1)	0.39 (0.15)	47.1 (-86.3)	-27.5 (-38.0)	0.98 (0.20)
			UK (techMARK)			
techMARK All Share	194.3 (7.3)	94.6 (134.0)	1.18 (0.66)	447.8 (36.2)	100.9 (144.9)	2.12 (0.65)
Nasdaq Composite	194.3 (7.3)	151.8 (195.1)	0.94 (0.53)	447.8 (36.2)	162.6 (231.1)	1.71 (0.52)

Source: own calculations.

In France, venture-backed firms tend to outperform non venture-backed issues regarding returns and wealth relatives. Contrary to the German sample, both mean and median returns of venture-backed firms are positive. Moreover, venture-supported firms outperform the overall market in terms of wealth relatives, when the *Nouveau Marché* index serves as benchmark. Finally, in the British sample no clear pattern with respect to systematic performance differences can be detected. Although both mean and median returns of non venture-backed issues are higher than those of venture-backed firms, the wealth relatives of the two firm groups are nearly identical. Noteworthy are the relatively high mean returns for both venture- and non venture-backed IPOs, indicating the existence of extremely high return figures for some individual firms.

---

512 For brevity, the test statistics for comparisons in means and medians are not reported.

The decomposition of wealth relatives into mean and volatility components, as advocated by Jakobsen and Sørensen (2001), sheds additional light on the performance figures. The Panels B and C in Appendix A 15 show that the venture-backed IPOs in France end up with a substantially higher mean component of wealth relatives when compared to their non venture-backed counterparts, independent of the applied benchmark.[513] By contrast, the opposite relationship prevails in Germany. The Panels E and F in Appendix A 15 demonstrate that venture-backed IPOs finish up with a mean component of wealth relatives of merely 0.32 (0.15), whereas non venture-backed issues realize a level of 0.51 (0.21) 3 years after the IPO, when the *Nemax All Share* (*Nasdaq Composite*) index is employed as benchmark. In the UK, as illustrated in the Panels H and I, the mean performance figures of venture-backed IPO wealth relatives are slightly worse than those of non venture-backed issues but the differences remain statistically insignificant.[514]

Table 23: International Comparison of the Post-IPO Returns and Wealth Relatives

Values for venture- and non venture-backed firms are non-parametric Mann-Whitney *z*-scores to detect significant differences between the medians of the firm groups in different countries. Bold values indicate statistically significant differences at the 1% or 5% confidence levels. 3-year equally weighted buy and hold returns on IPOs are compared with alternative benchmarks. The IPO returns are calculated by compounding monthly returns for 36 months after the IPO date. Wealth relatives (*WR*) are computed by taking the ratio of one plus the IPO portfolio return over one plus the return on the chosen benchmark over the same period. *GM* denotes the index of the country's growth market, i.e. Nouveau Marché, *Nemax All* Share, and *techMARK All Share* index for *Nouveau Marché*, *Neuer Markt*, and *techMARK*, respectively. *Nasdaq* denotes the *Nasdaq Composite* index.

Comparison	***France vs. UK***		***Germany vs. UK***		***Germany vs. France***	
Variable	VC-backed	Non VC-backed	VC-backed	Non VC-backed	VC backed	Non VC-backed
IPO return	-0.6	-1.4	**-5.4**[a]	**-4.5**[a]	**-6.2**[a]	**-5.3**[a]
WR to GM	-1.1	-0.4	**-2.2**[b]	-1.0	**-3.6**[a]	-0.6
WR to Nasdaq	-1.4	-0.1	**-3.9**[a]	**-2.9**[a]	**-5.5**[a]	**-3.5**[a]

Note: Significance levels are denoted by [a)] for 1%, [b)] for 5%, and [c)] for 10%. Source: own calculations.

The cross-country comparison of IPO returns and wealth relatives underlines the relatively bad performance of the German venture-backed IPOs. Given the skewness of the returns and wealth relatives, Table 23 reports the Mann-Whitney *z*-scores to discover significant cross-country differences between the medians of the firm groups. It shows that German venture-backed IPOs performed significantly worse than venture-backed issues in France and the UK, irrespective of the employed benchmark. But also non venture-backed firms in Germany display worse performance figures than their equivalents in the two other countries, in terms of buy and hold returns as well as wealth relatives against the *Nasdaq Composite* index.

---

[513] Using the *Nouveau Marché* (*Nasdaq Composite*) index as a benchmark, the mean component of the wealth relatives of the venture-backed IPOs equals 0.83 (0.75), whereas non venture-backed IPOs end up with a mean component of 0.62 (0.58).

[514] For brevity, the correspondent *t*-test statistics are not reported.

Previous studies report that the underperformance of IPOs is most severe in years with the highest IPO activity.[515] Therefore, in the refined analysis of Appendix A 16, the returns and wealth relatives of venture- and non venture-backed issues are investigated by yearly cohorts. Although both means and medians are presented, an emphasis is put on the latter, given that the mean returns are highly sensitive to outliers. Yet, the overall results do neither support the notion that the underperformance is particularly severe in years with high IPO activity, nor do they provide any additional insights regarding systematic differences between venture- and non venture-backed issues.

Finally, taking up the conjecture that there are variations in the quality and experience of different types of venture capitalists, Table 24 contrasts the average wealth relatives of firms backed by international and national venture capitalists. It shows that firms backed by international venture capitalists display significantly higher 3-year wealth relatives than issues backed by national venture investors, irrespective of the employed benchmark. Moreover, the median firm backed by international venture investors does not underperform seasoned stocks, if wealth relatives are calculated against the index of the respective growth market.[516] Overall, these findings support the notion that international venture capitalists select better firms and provide venture-backing of higher quality than national venture investors.

Table 24: Wealth Relatives of International and National Venture Capitalists

The table compares the average 3-year wealth relatives of sample firms backed by international and national venture capitalists. Wealth relatives (*WR*) are computed by taking the ratio of one plus the IPO portfolio return over one plus the return on the chosen benchmark over the same period. *GM* denotes the index of the country's growth market. The values are expressed as means (medians in parentheses). To detect significant differences between the means [medians] of the two firm groups, Student's *t*-tests [non-parametric Mann-Whitney tests] were conducted. Bold values indicate statistically significant differences at the 1% or 5% confidence levels.

Variable	*IPOs backed by international venture capitalists*	*IPOs backed by national venture capitalists*	*t-test [z-score]*
WR to GM	**2.10 (1.03)**	**0.70 (0.39)**	**2.7^{a} [-3.8^{a}]**
WR to Nasdaq	**1.56 (0.58)**	**0.57 (0.22)**	**2.7^{a} [-3.7^{a}]**

Note: Significance levels are denoted by $^{a)}$ for 1%, $^{b)}$ for 5%, and $^{c)}$ for 10%. Source: own calculations.

Summarizing the main findings, there is evidence supporting the view that the average IPO in the three considered countries underperforms seasoned stocks over the 3-year post-issue period. This finding is consistent with *H1* and the results of prior studies documenting the long-run underperformance of IPOs. However, as the present study employs only stock market indices for the calculation of market-adjusted returns, it cannot be ascertained whether the underperformance persists when portfolios of the same firm size and book-to-market ratios are used, as advocated by a number of authors.[517] Second, the market performance figures indicate that German IPOs performed particularly bad vis-à-vis issues in the two other countries. Yet, as the differences between venture- and non venture-backed firms are statistically insignificant, further examinations in the multivariate analysis are warranted. Finally, there is

[515] See e.g. Ritter (1991), Loughran and Ritter (1995).

[516] Given the skewness of wealth relatives, the analysis puts an emphasis on the comparison of medians.

[517] See e.g. Barber and Lyon (1997a, 1997b), Lyon *et al.* (1999).

evidence pointing out considerable variations in the experience of different types of venture capitalists and the quality of their venture-backing. To elucidate the characteristics of different kinds of venture capitalists, the following section examines a number of proxies for their experience and involvement in IPO firms.

5.1.3 Experience and Involvement of Different Types of Venture Capitalists

In order to highlight potential heterogeneities in the quality of venture-backing, as hypothesized in section 4.2.5 (*H7*), the present section explores different types of venture capitalists and their involvement in portfolio firms. As shown earlier, a promising method to capture differences in the experience of venture capitalist consists in distinguishing between nationally and internationally operating investors.

Figure 10: Participation of Different Types of Venture Capitalists in the Sample

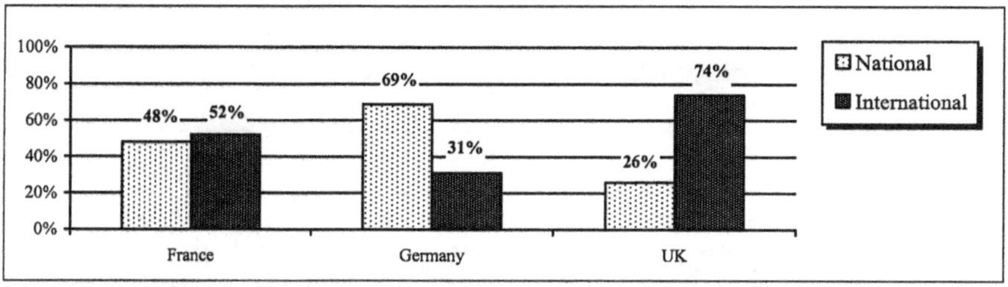

Source: own calculations.

To examine the participation of different types of venture capitalists in firms of the sample, Figure 10 illustrates the composition of IPOs in those backed by national and international venture capital investors for each country. It shows that international venture capitalists were the most active in the British IPO sample and the least involved in the German one. In Figure 11, the composition of different lead venture capitalists involved in IPOs of the sample is outlined by nationality. It shows that most national venture capitalists are German (50 percent) and French (33 percent) investors. By contrast, the majority of the international venture capitalists is British (65 percent). The finding that the more experienced international venture capitalists are mostly British and more frequently involved in IPOs in the UK than in the two other countries is in line with the overall results of chapter 3, pointing out that the British market for venture capital represents the most sophisticated in Europe. In Germany and France, though, where the markets for venture capital developed later than in the UK, less experienced national venture capitalists play a more important role.

To produce more stylized facts regarding variations in the experience of different types of venture capitalists, Figure 12 outlines the distribution of national and international venture capitalists by their founding dates.[518] It shows that most internationally classified venture capitalists (70 percent) were founded before 1980. Moreover, none of the international inves-

---

[518] A number of contributions on venture capital financing employs the age as a proxy for experience and reputation of venture capitalist investors. See e.g. Barry *et al.* (1990), Gompers (1996).

tors started its activities after 1995. This observation suggests that international venture capitalists have extensive business experience in backing portfolio firms. National venture capitalists, by contrast, reveal a higher variance concerning the number of years in business. Only 15 percent of them were established before 1980, whereas 66 percent began operating between 1980 and 1985. Finally, about one fifth of the national venture capitalists has only little business experience and has started investing after 1995. This result indicates that considerable variations exist in the backing experience among different types of venture capitalists.

Figure 11: National Origin of Different Types of Venture Capitalists

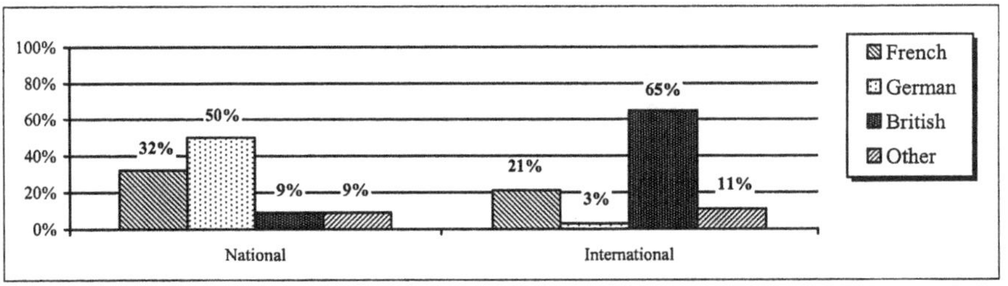

Source: own calculations.

Table 25 compares a number of alternative measures for the experience and sophistication of venture capitalists. These measures include the (1) age of the lead venture investor, (2) board representation at portfolio firms, (3) number of syndication partners, and (4) number of backed IPOs in the sample.

Figure 12: Founding Dates of National and International Venture Capitalists

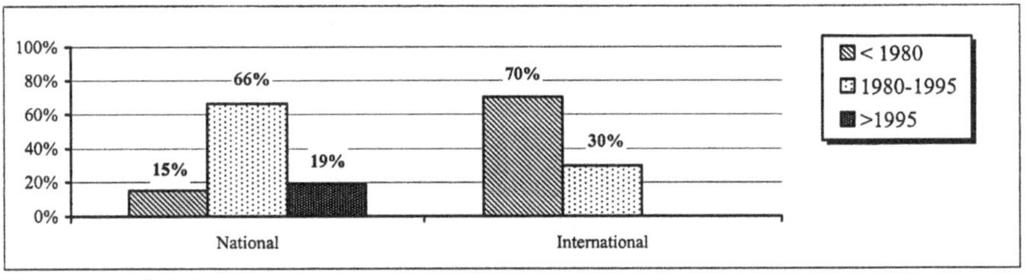

Source: own calculations.

First, the results show that the internationally operating lead venture capitalists are significantly older than their equivalents of nationally venture-backed firms at the time of the IPO. Second, both types of venture investors are represented on the boards of portfolio firms, even though European venture capitalists are often said to employ a hands-off management policy.[519] However, board representation is used significantly more often by international venture capitalists than by national ones. Third, although syndication is a common feature applied by

[519] See Gompers and Lerner (2001b), p. 197.

both types of investors, international venture capitalists tend to invest with a larger number of partners than their national counterparts. Finally, international venture capitalists back a significantly higher number of IPOs in the sample than national ones, suggesting that the former are more experienced in bringing firms to the capital markets. Overall, the findings highlight the existence of considerable differences in the experience and sophistication between international and national venture capitalists, which might influence the quality of venture-backing and thus the performance of portfolio firms.

Table 25: Experience and Involvement of Venture Capitalists in IPOs

The table compares international and national venture capitalists with respect to the age of the lead venture capitalist at the time of the IPO (*VC_Age*) and the percentage of sample firms, in which they were represented on the board (*Board*). Moreover, the total number of venture capitalists having invested in the same portfolio firm (*NVC*) and the number of backed IPOs in the sample (*No_IPOs*) are compared between the two types of venture capitalists. The values are expressed as means (medians in parentheses). To detect significant differences between the means [medians] of the two firm groups, Student's *t*-tests [non-parametric Mann-Whitney tests] were conducted. Bold values indicate statistically significant differences at the 1% or 5% confidence levels.

Variable	*International VCs*	*National VCs*	*t-test [z-score]*
VC_Age	**31.6 (29.0)**	**12.2 (12.0)**	**10.0ª [-7.7ª]**
Board	**76%**	**60%**	**2.1ᵇ**
NVC	**3.3 (3.0)**	**2.1 (2.0)**	**4.6ª [-4.1ª]**
No_IPOs	**11.4 (7.0)**	**3.1 (2.0)**	**8.4ª [-6.9ª]**

Note: significance levels are denoted by a) for 1%, b) for 5%, and c) for 10%. Source: own calculations.

In the following, the venture shareholdings in portfolio firms are investigated to detect statistically significant differences in the involvement between different kinds of venture capitalists.[520] Table 26 outlines the pre- and post-issue equity stakes held by international and national investors at the time of the going public. The ownership figures show that both types of venture capitalists held substantial pre- and post-issue shares of equity. On average, all venture capitalists together owned about 45 percent of the equity before flotation if an IPO was backed by international investors, and roughly 27 percent if it was backed by national ones. In line with US studies, these findings underline that both groups of venture capitalists have significant incentives to participate in the corporate governance of portfolio firms.[521] However, the comparison of different venture capitalists reveals statistically significant differences in their involvement in IPOs, i.e. international venture investors systematically hold more equity in firms they bring public than national ones.

Like all venture capitalists as a group, the international (national) lead manager, defined as the one with the largest equity position in the firm, represents a significant stockholder, owning on average 24.9 (18.5) percent before and 14.4 (10.0) percent after flotation. Yet again, there appear considerable variations in the involvement of lead investors between different types of venture capitalists. In line with the earlier findings for venture capitalists as a group,

---

520 Gompers and Lerner (1998) emphasize that the size of equity stakes held by venture capitalists are related to the incentive to monitor and the quality of information about a firm.

521 See e.g. Barry *et al.* (1990). Shleifer and Vishny (1986) point out that large-block shareholders, who are willing to incur the costs of monitoring management, can play an important role in enhancing firm value.

the lead financier of internationally venture-backed IPOs holds a larger equity stake than its matching part in nationally venture-backed issues.

Table 26: Shareholding of Venture Capitalists in IPOs

The table compares national and international venture capitalists with respect to their involvement in IPO firms. *AVC (LVC) PrEq* and *PosEq* denote the percentage of equity owned by all venture capitalists (the lead venture capitalist) before and immediately after the IPO, respectively. *AVC (LVC) Sale* denotes the percentage of shares sold by all venture capitalists (the lead venture capitalist) at the IPO. The values are expressed as means (medians in parentheses). To detect significant differences between the means [medians] of the two firm groups, Student's *t*-tests [non-parametric Mann-Whitney tests] were conducted. Bold values indicate statistically significant differences at the 1% or 5% confidence levels.

Variable	*International VCs*	*National VCs*	*t-test [z-score]*
AVC PrEq	**44.8 (40.7)**	**27.3 (22.0)**	**5.0^a [-4.9^a]**
AVC PosEq	**26.7 (24.8)**	**15.3 (11.8)**	**5.0^a [-4.4^a]**
AVC Sale	40.0 (35.2)	40.1 (33.3)	0.0 [-0.2]
LVC PrEq	**24.9 (22.3)**	**18.5 (14.4)**	**2.4^b [-3.7^a]**
LVC PosEq	**14.4 (12.8)**	**10.0 (8.5)**	**2.9^a [-3.2^a]**
LVC Sale	40.6 (37.6)	40.0 (32.2)	0.1 [-0.6]

Note: Significance levels are denoted by a) for 1%, b) for 5%, and c) for 10%. Source: own calculations.

Although the equity shares of venture capitalists decrease considerably in the wake of IPOs, the aggregate post-issue equity holdings of both, the venture capital syndicate and the lead financier, are still large. By retaining their capital stake after the offering, venture capitalists can provide assurance of continued monitoring and signal their belief in the prospects of a firm.[522] In contrast to the prior results, the comparative analysis of the selling intensity[523] does not reveal any significant differences between different types of venture capitalists. Nonetheless, it is worth mentioning that the documented selling intensity is generally higher than that of US venture capitalists. While in IPOs of the present sample about 40 percent of the shares are sold on average, Barry *et al.* (1990) report a mean selling intensity of only 6.6 percent for venture capitalists in the US.[524] This difference suggests that, by selling a higher portion of shares in IPOs, the average European venture capitalist cares less about reputation than its US equivalent.

All in all, this section provides a number of crucial insights with regard to the hypothesized differences in the experience and involvement of different types of venture capitalists (*H7*). First, it shows that international venture capitalists participated with different frequencies in the considered IPO markets. They were most active in the British IPO sample and the least participating in the German *Neuer Markt*. Moreover, it turns out that most international investors are British. This outcome supports the view that European venture capital is most developed in the UK, as shown in chapter 3.

---

522 See Barry *et al.* (1990), p. 461.

523 Following Barry *et al.* (1990), the selling intensity of venture capitalists at the IPO is defined as the number of shares sold divided by the pre-IPO holdings.

524 Similarly low selling figures of venture investors in the US are reported by Gompers and Lerner (1999b).

Second, there are substantial variations in the sophistication and business experience of different types of venture capitalists. On average, international venture capitalists are older than national investors, back a larger number of IPOs in the sample, hold board representations more often, and invest with more syndication partners. This result suggests that international venture capitalists are more experienced and sophisticated than national ones in backing portfolio firms.

Finally, as far as the involvement in IPOs is concerned, both types of venture capitalists can be considered as active investors, holding large equity positions in firms they bring public. However, international venture capitalists hold on average larger pre- and post-issue equity shares than national investors. Overall, the findings support the hypothesis that international venture capitalists are more experienced and involved in ventures than their nationally operating equivalents (*H7*). The following section analyzes whether or not venture capitalists have an effect on the choice of underwriters firms going public.

5.1.4 Underwriter Quality

Previous studies argue that venture capitalists are likely to influence the choice of underwriters given their active participation in the operations of firms they bring public. Moreover, Megginson and Weiss (1991) provide evidence that venture-backed firms are underwritten by higher quality underwriters than non venture-backed ones. Therefore, different reputation proxies for the lead underwriters of venture-backed IPOs, as described in section 4.4.4, are considered and compared to those of non venture-backed issues.

Table 27: Lead Underwriter Reputation of Venture- and Non Venture-Backed IPOs

Values for venture- and non venture-backed firms are expressed as means (medians in parentheses). To detect significant differences between the means [medians] of the two firm groups, Student's *t*-tests [non-parametric Mann-Whitney tests] are conducted. Bold values indicate statistically significant differences at the 1% or 5% confidence levels. *Mkt. Share* denotes the relative market share of proceeds brought to the IPO markets of France, Germany, and the UK by the lead underwriter between 1995 and 1999. *URank* is the underwriter reputation rank on a scale of 1 to 3, where 1 is the most and 3 the least prestigious underwriter. The statistics consider only IPO firms backed by domestic lead venture capitalists.

	France			*Germany*			*UK*		
Variable	VC-backed	non VC-backed	*t*-test [z-score]	VC-backed	non VC-backed	*t*-test [z-score]	VC-backed	non VC-backed	*t*-test [z-score]
Mkt. Share	6.3	6.5	-0.1	6.6	5.3	0.9	**4.6**	**1.7**	**2.1[b]**
	(1.3)	(0.6)	[-0.2]	(2.9)	(2.9)	[-0.2]	**(1.3)**	**(0.4)**	**[-2.7[a]]**
URank	1.9	2.1	-1.4	1.9	2.0	-0.5	**1.9**	**2.6**	**-3.0[a]**
	(2.0)	(3.0)	[-1.4]	(2.0)	(2.0)	[-0.6]	**(2.0)**	**(3.0)**	**[-2.8[a]]**

Note: Significance levels are denoted by [a)] for 1%, [b)] for 5%, and [c)] for 10%. Source: own calculations.

As shown in Table 27, neither in the French nor in the German sample appear any significant differences between the qualities of lead underwriters being involved in venture- and non venture-backed IPOs. There are at least two possible explanations for this finding. First, venture capitalists might be indifferent to the reputation of underwriters because they do not need them to certify financial prospects and the IPO quality.[525] Non venture-backed firms, by contrast, might rely to a larger extent on the certification of investment banks for a successful going public. Second, as the earlier findings revealed that a considerable number of IPOs in France and Germany was backed by national venture capitalists, it might be that these typically younger and less experienced investors have not yet built up relationships with top-tier investment banks. Therefore, venture-backed issuers might use underwriters, which are more or less as prestigious as those employed by non venture-backed firms going public.

In the British sample, however, where international venture capitalists are more frequently involved in IPOs than national venture investors, both the average market share and reputation rank of lead underwriters bringing venture-backed firms to the market are significantly higher than those of non venture-backed issues. These findings are in line with the evidence of prior studies, suggesting that the average venture-backed IPO is of higher quality as compared to non venture-backed issues.[526] The cross-country differences in the results might be due to the fact that venture capital is more developed in the UK and has a longer tradition than in the two continental European countries, allowing venture capitalists to establish relationships with prestigious underwriters.

Table 28: Lead Underwriter Reputation of IPOs Backed by International VCs

Values for venture- and non venture-backed firms are expressed as means (medians in parentheses). To detect significant differences between the means [medians] of the two firm groups, Student's *t*-tests [non-parametric Mann-Whitney tests] are conducted. Bold values indicate statistically significant differences at the 1% or 5% confidence levels. *Mkt. Share* denotes the relative market share of proceeds brought to the IPO markets of France, Germany, and the UK by the lead underwriter between 1995 and 1999. *URank* is the underwriter reputation rank on a scale of 1 to 3, where 1 is the most and 3 the least prestigious underwriter. The statistics consider only IPO firms backed by domestic lead venture capitalists.

	France			***Germany***			***UK***		
Variable	Int_VC-backed	Nat_VC-backed	*t*-test [z-score]	Int_VC-backed	Nat_VC-backed	*t*-test [z-score]	Int_VC-backed	Nat_VC-backed	*t*-test [z-score]
Mkt. Share	9.0	5.1	1.3	7.1	6.5	0.2	**6.6**	**1.1**	**2.8[a]**
	(1.3)	(1.3)	[-0.8]	(2.9)	(2.9)	[-1.0]	**(6.1)**	**(0.6)**	**[-2.2[b]]**
URank	1.6	2.0	-1.7[c]	1.6	2.0	-1.7[c]	**1.6**	**2.5**	**-3.2[a]**
	(1.0)	(2.0)	[-1.7[c]]	(1.5)	(2.0)	[-1.7[c]]	**(1.0)**	**(3.0)**	**[-2.8[a]]**

Note: Significance levels are denoted by [a)] for 1%, [b)] for 5%, and [c)] for 10%. Source: own calculations.

In order to check whether the mixed results regarding the reputation of underwriters are due to the documented heterogeneity of venture capital, Table 28 provides a refinement of the analysis. It compares the average lead underwriter reputation of IPOs backed by international and national venture capitalists. The results indicate that international venture capitalists in all countries attracted more prestigious lead underwriters than their national equivalents in terms of the underwriter rank. Yet, in the French and German sample, the differences in the ranking scores are only weakly significant. Overall, only the findings in the British sample suggest that firms backed by international venture capitalists use more prestigious underwriters than issues backed by national venture investors.

[525] See e.g. Doukas and Goncenc (2001).

[526] See e.g. Megginson and Weiss (1991), Wang *et al.* (2003).

Concluding the most important results of the univariate analysis, there is evidence that IPOs in all three countries underperform seasoned stocks over the 3-years post-issue period. Moreover, the German IPOs perform significantly worse than issues in the two other countries. Aside from significant differences in the selling behavior of old shareholders, the findings do not reveal systematic disparities between the firm and offering characteristics of venture- and non venture-backed IPOs. However, the refined analysis considering different types of venture capitalists shows that issues backed by international venture capitalists are larger in terms of the offering proceeds and achieve higher 3-year wealth relatives than those backed by national venture investors. Also, international venture capitalists are on average older than their national equivalents, more often represented on boards, back more IPOs in the sample, syndicate with more investment partners, and hold larger pre- and post-issue equity shares in firms they bring public. Finally, at least in the UK sample, there is evidence that IPOs backed by international venture capitalists use more prestigious underwriters than those backed by national investors.

With respect to the derived hypotheses, the findings hitherto support the claim that IPOs underperform seasoned stocks in the long run (*H1*) and that there are differences in the experience and involvement between national and international venture capitalists (*H7*). With regard to the overall research question, it appears therefore essential to include the insights related to the heterogeneity of venture capitalists in the multivariate analysis when testing the remaining hypotheses. In the following sections, the impact of venture capitalists on IPO performance is examined in the cross section by using a regression analysis.

5.2 Multivariate Analysis

5.2.1 Full Sample Results

Performance differences between firms with and without venture-backing might be due to other factors than the participation of venture capitalists. To explore to what extent the observed operating and market performance figures are related to the presence of venture capitalists, cross-sectional regression analyses are conducted including various explanatory and control variables, as described in the sections 4.4.1 to 4.4.4. First, the linear regressions are estimated piecewise for each country, using intercept and interaction dummy variables, to detect constant and slope effects of venture-backing in different country environments. Second, regressions based on the full sample are estimated, including country dummies and variables for different types of venture capitalists, to discover cross-country differences and venture-backing effects of distinct venture investors which apply for all three considered countries.

5.2.1.1 Accounting Performance

In order to investigate the impact of venture capitalists on the post-issue operating performance, as hypothesized in section 4.2.2 (*H2a*), Table 29 reports the cross-sectional estimates of the relationship between the average operating cash flow return on assets of the two fiscal years following the IPO and various explanatory variables. In column I of each country, the involvement of venture capitalists is considered by a simple dummy variable to account for constant effects of venture-backing. Inspired by the regression setup of Mikkelson *et al.* (1997), the natural logarithm of the 2-year average post-issue assets, the natural logarithm of firm age (in years), and the fraction of secondary shares sold in the IPO, measured by the participation ratio are employed as independent variables. Additionally, the ranking of the underwriter reputation is included as regressor.

Table 29: Country-Specific Regressions on Post-IPO Operating Performance

The dependent variable is the average operating cash flow return on assets of the two fiscal years subsequent to the IPO. The independent variables include a dummy variable that equals 1 if a firm is venture-backed (*VC*), the natural logarithm of the firm's 2-year average post-IPO assets (*Size*), the natural logarithm of the firm age in years (*Age*), the participation ratio of old shareholders in the IPO (*Part*), defined as number of old shares sold divided by the number of shares outstanding before flotation, and the underwriter reputation rank on a scale of 1 to 3 (*URank*), where 1 is the most and 3 the least prestigious underwriter. All regressions employ an ordinary least squares specification. The estimations include industry, calendar year and accounting standard dummy variables, which are not reported. *t*-statistics are in parentheses. Bold values indicate statistically significant differences at the 1% or 5% confidence levels.

	France		*Germany*		*UK*	
Variables	*I*	*II*	*I*	*II*	*I*	*II*
VC	**-0.22[b]**	-0.61	-0.10	-0.09	0.08	1.41
	(-2.2)	(-1.0)	(-1.3)	(-0.2)	(0.6)	(1.6)
Size	**0.24[b]**	0.27	**0.36[a]**	**0.42[a]**	**0.68[a]**	**0.93[a]**
	(2.4)	(1.5)	**(4.4)**	**(3.8)**	**(3.9)**	**(4.1)**
Size*VC		-0.02		-0.20		-0.86
		(-0.0)		(-0.5)		(-1.5)
Age	0.10	0.07	0.13[c]	0.04	0.13	0.08
	(1.0)	(0.4)	(1.8)	(0.4)	(1.1)	(0.5)
Age*VC		0.11		0.31		0.07
		(0.3)		(1.3)		(0.2)
Part	**0.33[a]**	**0.47[b]**	**0.15[b]**	0.02	0.28[c]	0.54[c]
	(3.1)	**(2.5)**	**(2.1)**	(0.1)	(2.0)	(1.8)
Part*VC		-0.20		0.19		-0.29
		(-0.9)		(1.2)		(-0.9)
Urank	**0.26[b]**	0.06	0.09	0.15	0.23	0.53
	(2.5)	(0.4)	(1.2)	(1.4)	(1.4)	(1.7)
URank*VC		0.46[c]		-0.20		-0.56
		(1.8)		(-0.8)		(-1.2)
Constant	**-0.34[a]**	-0.26	**-0.53[a]**	**-0.53[a]**	**-1.20[a]**	**-1.78[a]**
	(-2.7)	(-1.3)	**(-3.8)**	**(-2.9)**	**(-3.3)**	**(-2.9)**
Adj. R^2	0.28	0.28	0.38	0.38	0.29	0.28
F-test	**3.4[a]**	**2.9[a]**	**7.6[a]**	**6.1[a]**	**2.9[a]**	**2.4[a]**
Chow-test	1.2	1.2	1.1	1.1	2.0	2.0
Sample size	90	90	151	151	62	62

Note: Significance levels are denoted by [a)] for 1%, [b)] for 5%, and [c)] for 10%. Source: own calculations.

Rather surprisingly, in the French sample venture capital has a significant negative effect on operating performance. Yet, the significance does not hold when slope effects of venture-backing are considered (column II). In the two other countries, venture capital involvement does not indicate any significant impact. In line with Mikkelson *et al.* (1997), in all three country samples firm size and participation ratio are positively related to operating performance, indicating that large firms and those undertaking an IPO with high secondary sales are more profitable than small firms and those with low secondary sales. The positive sign of the underwriter rank is only significant in the French sample. This finding is consistent with the view that issuing firms with less profitable post-issue earnings use more prestigious underwriters to go public.

Table 30: Full Sample Regressions on Post-IPO Operating Performance

The dependent variable is the average operating cash flow return on assets of the two fiscal years subsequent to the IPO. The independent variables include a dummy variable that equals 1 if a firm is venture-backed (*VC*), the natural logarithm of the firm's 2-year average post-IPO assets (*Size*), the natural logarithm of the firm age in years (*Age*), the participation ratio of old shareholders in the IPO (*Part*), defined as number of old shares sold divided by the number of shares outstanding before flotation, and the underwriter reputation rank on a scale of 1 to 3 (*URank*), where 1 is the most and 3 the least prestigious underwriter. *France* and *UK* are country dummy variables, which equal 1 if a firm went public in the respective country. *Int_VC* is a dummy variable equal to 1 if an IPO was backed by a venture capitalist that backed firms in at least two countries of the sample. All regressions employ an ordinary least squares specification. The estimations include industry, calendar year and accounting standard dummy variables, which are not reported. *t*-statistics are in parentheses. Bold values indicate statistically significant differences at the 1% or 5% confidence levels.

Variables	*I*	*II*	*III*	*IV*	*V*
VC	-0.05 (-1.0)	-0.03 (-0.1)	-0.07 (-1.1)	-0.50 (-1.2)	**-0.89[b] (-2.2)**
Int_VC			0.03 (0.5)	**0.78[b] (2.0)**	**0.90[b] (2.3)**
Size	**0.48[a] (7.2)**	**0.54[a] (6.3)**	**0.47[a] (7.0)**	**0.54[a] (6.2)**	**0.40[a] (4.6)**
Size*VC		-0.30 (-1.3)		-0.07 (-0.2)	0.24 (0.8)
Size*Int_VC				-0.43[c] (-1.8)	-0.52[c] (-1.7)
Age	**0.11[b] (2.1)**	0.05 (0.7)	**0.11[b] (2.2)**	0.05 (0.7)	0.09 (1.1)
Age*VC		0.16 (0.9)		0.26 (1.2)	0.27 (1.3)
Age*Int_VC				-0.14 (-0.7)	-0.18 (-0.9)
Part	**0.27[a] (4.8)**	**0.23[b] (2.3)**	**0.26[a] (4.7)**	**0.23[b] (2.3)**	**0.28[a] (2.7)**
Part*VC		0.05 (0.4)		0.01 (0.0)	-0.03 (-0.2)
Part*Int_VC				0.04 (0.3)	0.04 (0.3)
URank	**0.18[a] (3.2)**	0.13 (1.5)	**0.19[a] (3.2)**	0.13 (1.5)	0.10 (1.2)
URank*VC		0.11 (0.7)		0.28 (1.5)	0.33 (1.8)
URank*Int_VC				-0.22 (-1.4)	-0.21 (-1.3)
France	0.12 (1.0)	0.11 (0.9)	0.11 (0.9)	0.08 (0.7)	0.10 (0.8)
UK	**-0.29[b] (-2.4)**	**-0.29[b] (-2.4)**	**-0.30[a] (-2.5)**	**-0.31[b] (-2.5)**	-0.21 (-1.7)
Constant	**-0.58[a] (-4.9)**	**-0.55[a] (-3.7)**	**-0.58[a] (-4.8)**	**-0.53[a] (-3.5)**	**-0.38[a] (-2.9)**
Adj. R^2	0.28	0.28	0.28	0.28	0.29
F-test	**7.5[a]**	**6.4[a]**	**7.1[a]**	**5.3[a]**	**5.6[a]**
Chow-test	1.4	1.4	**2.6[b]**	**2.6[b]**	**2.6[b]**
Sample size	303	303	303	303	302

Note: Significance levels are denoted by [a)] for 1%, [b)] for 5%, and [c)] for 10%. Source: own calculations.

The correlation matrix in Appendix A 17 shows that the participation ratio in all three countries is positively correlated with the venture capital dummy, suggesting that venture-backed IPOs are characterized by higher secondary sales than non venture-backed issues, as discussed in the univariate analysis. Moreover, larger firms are underwritten by investment banks with higher reputation, and venture-backed firms in the UK tend to use underwriters with higher reputation than non venture-backed ones, as shown in section 5.1.4. To explore the influence of multicollinearity[527] on the results, separate regressions are estimated without the participation ratio and the underwriter rank. Yet, the overall results do not lead to qualitatively different outcomes, and are therefore not stated.

Since venture-backing might also have slope effects, additional regressions in column II are performed with interactive dummies. The results confirm the insignificant impact of venture capital involvement on the operating performance of firms in Germany and the UK. Likewise, none of the variables accounting for interaction is significant at conventional levels. In France, where the previously significant coefficient of the venture capital intercept dummy loses its negative significance, the weakly significant coefficient of the interaction term *URank*VC* indicates that venture-backed firms with lower post-IPO operating performance employ underwriters with higher reputation. The finding suggests that venture capitalists in the French sample use the certification ability of prestigious investment banks to bring less profitable firms to the market. Finally, in line with the estimates accounting for interaction, the reported score of the *Chow*-test does not point out any significant structural differences between the regressions of venture- and non venture-backed issues.

Regressions based on the full sample, including intercept dummies for the different countries, are outlined in Table 30. Although only the dummy variables of France and the UK are listed by name, the constant term can be interpreted as the intercept of the German sample. In columns I and II, the constant and slope effects of venture capital are explored, respectively. Conform to the previous analysis the results do not indicate any significant impact of venture involvement on post-issue profitability.

To account for the heterogeneity of venture capitalists, in columns III and IV the regression setup of columns I and II is extended by the inclusion of an additional variable for the involvement of international venture capitalists (*Int_VC*). While the use of the dummy as an intercept (column III) does not lead to significant changes with respect to the effect of venture-backing, the consideration of slope effects (column IV) provides an interesting insight: IPOs backed by international venture capitalists exhibit significantly higher post-issue performance figures than non venture-backed issues and firms backed by national venture investors. Furthermore, the results indicate slope effects related to the presence of international venture investors, i.e. the post-issue operating performance of firms backed by international venture capitalists is less dependent on firm size. The significant *Chow*-test statistic supports the indicated differences between the individual regressions of the firm groups.

To make sure that the inference from the regression analysis is not driven by outliers, the British non venture-backed IPO firm *Manpower Software* with the extremely low average post-issue operating cash flow return on assets of −219 percent is taken out of the estimated equation in column V.[528] The results based on the smaller sample display a better regression

---

[527] The term "multicollinearity" is defined as a linear functional relationship between two or more variables that is so strong that it can significantly affect the estimation of the coefficients of the variables. See Studenmund (1997), pp. 263-328.

[528] According to Jobson (1991), a common measure for the influence of individual outliers on the regression results is *Cook's D*. In the regression of column IV, Manpower Software displays the highest *Cook's D* of all IPO with a value of 0.44.

fit, as shown by the higher values for the adjusted R^2 and the F-statistic, and confirm the significant positive impact of international venture capitalists. Moreover, they reveal a negative impact of national venture capitalists on the post-issue operating performance.

In order to investigate whether or not venture-backed firms display lower accounting profitability figures than non venture-backed firms when they go public, as conjectured by the certification hypothesis (*H4b*), regressions similar to those in the previous analysis are run, based on the pre-issue operating performance. However, only in the country-specific regressions of the French sample the venture capital intercept initially displays a negative sign at the 5 percent significance level which does not hold after considering interaction effects. Likewise, the distinction between different types of venture capitalists in the complete sample regressions, analogous to the setups of the columns III and IV in Table 30, does not provide further insights. Hence, there is no support for the notion that venture capitalists are able to bring relatively less profitable firms public.[529]

Summing up, the results do not provide support for the hypothesis that venture-backed firms generally outperform non venture-backed ones in terms of post-issue operating performance (*H2a*). Merely for firms backed by international venture capitalists, there is evidence for the conjecture of *H2a*. As far as the hypothesized certification role of venture capitalists is concerned (*H4b*), the analysis of the pre-issue firm profitability does not support the idea that venture capitalists are able to bring firms with lower operating performance to the market than non venture-backed firms, as suggested by prior studies.[530] In order to elucidate certification effects of venture capitalists on the valuation of IPOs, the following section explores the relationship between venture-backing and Tobin's Q.

5.2.1.2 Tobin's Q

To examine the influence of the certification role of venture capitalists on IPO valuation, as hypothesized in section 4.2.2 (*H4a*), this section focuses on the determinants of the approximated value of Tobin's Q at the first trading day. The country-specific estimations in column I of Table 31 employ a venture capital dummy and a number of explanatory variables, such as firm size, measured by the natural logarithm of the firm's market value of equity at the first day of trading, and the equity ratio, defined as the book value of equity divided by the book value of total assets. In compliance with previous studies on corporate value, the ratio of sales over assets is included in the equations to control for the importance of the firms' soft capital.[531] Finally, to consider the impact of the ownership of old shareholders immediately after the IPO, the variable *Alpha* considering the post-issue percentage of shares held by old shareholders is comprised in the regression estimates.[532]

As reported in column I of each country, venture-backing does not reveal any significant impact on market valuation measured by Tobin's Q. Among the remaining explanatory variables, only the ownership variable *Alpha* is significant. The positive sign of *Alpha* suggests

---

529 As the findings do not indicate any sustainable significant impact of venture capital on the pre-IPO operating performance, they are not reported.

530 See e.g. Barry *et al.* (1990), Jain and Kini (1995), Wang *et al.* (2003).

531 See e.g. Himmelberg *et al.* (1999).

532 *Alpha* has been applied in earlier studies, such as Jain and Kini (1994, 1995).

that the market appraises old shareholders to hold a higher portion of shares in the post-issue firm. Since the *Alpha* correlates significantly with the venture capital dummy in France and Germany, as shown in the correlation matrix in Appendix A 18, separate regressions are estimated without *Alpha*. However, the estimates do not qualitatively differ from the reported results, and are therefore not stated.

Table 31: Country-Specific Regressions on Tobin's *Q* at the First Trading Day

The dependent variable is the approximated value of Tobin's *Q* at the first trading day, defined as the market value of equity and book value of debt divided by the book value of total assets. The independent variables are a dummy variable that equals 1 if a firm is venture-backed (*VC*), the natural logarithm of the firm's market value of equity at the first trading day (*Size*), the equity ratio (*EqR*), defined as the book value of equity divided by the book value of assets at the fiscal year prior to the IPO, the inverse of capital intensity (*S_A*), defined as the ratio of sales over assets at the fiscal year prior to the IPO, and the fraction of the equity retained by old shareholders after the IPO (*Alpha*). All regressions employ an ordinary least squares specification. The estimations include industry, calendar year and accounting standard dummy variables, which are not reported. *t*-statistics are in parentheses. Bold values indicate statistically significant differences at the 1% or 5% confidence levels.

	France		*Germany*		*UK*	
Variables	*I*	*II*	*I*	*II*	*I*	*II*
VC	-0.10	1.09	0.08	1.07	-0.06	0.23
	(-1.0)	(1.1)	(1.0)	(1.6)	(-0.4)	(0.2)
Size	**0.50**[a]	**0.54**[a]	**0.40**[a]	**0.44**[a]	0.19	0.23
	(5.2)	**(3.7)**	**(5.3)**	**(4.3)**	(1.2)	(1.1)
Size*VC		-0.21		-0.34		-0.06
		(-0.5)		(-0.7)		(-0.1)
EqR	0.02	-0.07	0.02	0.02	-0.22[c]	-0.21
	(0.2)	(-0.5)	(0.2)	(0.27)	(-1.7)	(-1.5)
EqR*VC		0.15		0.03		-0.09
		(0.7)		(0.4)		(-0.5)
S_A	**0.22**[b]	**0.31**[b]	0.02	-0.03	0.01	0.09
	(2.5)	**(2.5)**	(0.3)	(-0.3)	(0.0)	(0.4)
S_A*VC		-0.22		0.11		-0.24
		(-1.1)		(0.9)		(-0.7)
Alpha	**0.19**[b]	0.33[c]	**0.21**[a]	**0.41**[a]	**0.36**[a]	0.37
	(2.0)	(1.9)	**(2.7)**	**(2.9)**	**(2.9)**	(1.3)
Alpha*VC		-0.86		-0.76[c]		-0.06
		(-1.0)		(-1.8)		(-0.1)
Constant	-1.07	-2.29[c]	**-6.23**[a]	**-9.47**[a]	-0.76	-1.31
	(-1.3)	(-1.8)	**(-3.2)**	**(-3.5)**	(-0.4)	(-0.4)
Adj. R^2	0.43	0.42	0.27	0.27	0.24	0.19
F-test	**5.8**[a]	**4.6**[a]	**4.9**[a]	**4.1**[a]	**2.5**[b]	1.8[c]
Chow-test	1.7	1.7	1.8	1.8	1.5	1.5
Sample size	90	90	151	151	62	62

Note: Significance levels are denoted by [a)] for 1%, [b)] for 5%, and [c)] for 10%. Source: own calculations.

Similar to the analysis of the preceding section, the regressions in column II of Table 31 control for slope effects of venture-backing by using interaction variables. Yet, again the obtained results do not reveal any systematic impact of venture-backing. In the German sample, the analysis of venture-backing-related slope effects suggests that the positive relationship between *Alpha* and Tobin's *Q* does not apply for venture-backed firms. Instead, the weakly significant interaction term *Alpha*VC* in conjunction with the coefficient of *Alpha* points out a negative relationship between the *Q* ratio and the post-issue percentage of shares held by old shareholders of venture-backed issues.

Table 32: Full Sample Regressions on Tobin's *Q* at the First Trading Day

The dependent variable is the approximated value of Tobin's *Q* at the first day of trading, defined as the market value of equity and book value of debt divided by the book value of total assets. The independent variables are a dummy variable that equals 1 if a firm is venture-backed (*VC*), the natural logarithm of the firm's market value of equity at the first days of trading (*Size*), the equity ratio (*EqR*), defined as the book value of equity divided by the book value of assets at the fiscal year prior to the IPO, the inverse of capital intensity (*S_A*), defined as the ratio of sales over assets at the fiscal year prior to the IPO, and the fraction of the equity retained by old shareholders after the IPO (*Alpha*). *France* and *UK* are country dummy variables, which equal 1 if a firm went public in the respective country. *Int_VC* is a dummy variable equal to 1 if an IPO was backed by a venture capitalist that backed firms in at least two countries of the sample. All regressions employ an ordinary least squares specification. The estimations include industry, calendar year and accounting standard dummy variables, which are not reported. *t*-statistics are in parentheses. Bold values indicate statistically significant differences at the 1% or 5% confidence levels.

Variables	*I*	*II*	*III*	*IV*
VC	0.01 (0.2)	0.50 (1.2)	0.00 (0.1)	0.41 (0.9)
Int_VC			0.01 (0.2)	0.25 (0.6)
Size	**0.42ᵃ (6.3)**	**0.43ᵃ (5.2)**	**0.42ᵃ (6.1)**	**0.43ᵃ (5.1)**
Size*VC		-0.05 (-0.2)		0.00 (0.0)
Size*Int_VC				-0.11 (-0.5)
EqR	-0.02 (-0.3)	-0.02 (-0.3)	-0.02 (-0.3)	-0.02 (-0.3)
EqR*VC		0.03 (0.5)		0.05 (0.6)
EqR*Int_VC				-0.07 (-0.8)
S_A	0.08 (1.5)	0.07 (1.1)	0.08 (1.5)	0.07 (1.1)
S_A*VC		0.02 (0.2)		0.05 (0.5)
S_A*Int_VC				-0.12 (-1.0)
Alpha	**0.23ᵃ (4.3)**	**0.35ᵃ (3.7)**	**0.23ᵃ (4.3)**	**0.35ᵃ (3.7)**
Alpha*VC		**-0.47ᵃ (-1.6)**		-0.46 (-1.4)
Alpha*Int_VC				-0.02 (-0.1)
France	-0.22ᶜ (-2.0)	**-0.24ᵇ (-2.1)**	-0.23ᶜ (-2.0)	**-0.24ᵇ (-2.0)**
UK	-0.08 (-0.7)	-0.08 (-0.7)	-0.08 (-0.7)	-0.09 (-0.8)
Constant	**-3.21ᵃ (-2.8)**	**-4.75ᵃ (-3.0)**	**-3.20ᵃ (-2.8)**	**-4.71ᵃ (-2.9)**
Adj. R^2	0.35	0.35	0.35	0.34
F-test	**10.2ᵃ**	**8.4ᵃ**	**9.6ᵃ**	**6.8ᵃ**
Chow-test	0.8	0.8	1.8ᶜ	1.8ᶜ
Sample size	303	303	303	303

Note: Significance levels are denoted by [a)] for 1%, [b)] for 5%, and [c)] for 10%. Source: own calculations.

In Table 32 regressions based on the full sample are estimated to account for significant cross-country differences in market valuations and the effects of venture-backing on Tobin's *Q*. The equations in the columns I and II employ setups similar to those in Table 31 to control for constant and slope effects of venture capital involvement, respectively. To account for country-related effects, dummy variables for France and the UK are implemented in the model specification. The findings suggest that larger firms experience higher market valuations than smaller ones at the first trading day. However, once more the outcome does not show any significant influence of venture-backing on Tobin's *Q*. In agreement with the univariate analysis in section 5.1.1, the regressions in columns I and II indicate that French IPOs undergo significantly lower market valuations in terms of Tobin's *Q* than issuing firms in the two other country samples.

In columns III and IV of Table 32, the setup of columns I and II is modified by including an additional variable for the involvement of international venture capitalists (*Int_VC*). Yet, the regressions accounting for constant and slope effects of international venture-backing do not yield any new insights concerning the impact of venture-backing on firm valuation. Accordingly, the reported *Chow*-test statistic is only weakly significant. Hence, there is no strong support for the existence of structural differences between the individual firm groups.

Separate estimations controlling for interactions between venture capital in each country and the different independent variables are conducted without leading to qualitatively different results. For brevity, the relatively comprehensive but inconclusive regressions are not reported. For robustness, alternative model estimations are conducted, comprising intercept and slope dummy variables for different nationalities of lead venture investors, but do not generate materially different outcomes. Finally, given the relatively high market valuations towards the end of the sample period, regressions based on a reduced sample without the IPOs of 1999 are estimated without yielding any significant findings with regard to the influence of venture-backing on Tobin's *Q*.

Summing up, the reported evidence does not indicate any significant effects of venture capitalist participation on the valuation of IPOs in terms of Tobin's *Q*. Even the refinement of the analysis considering the impact of different types of venture capitalists does not change this result. Thus, there is no support for the conjecture that the market appraises firms with venture involvement by higher market valuations than non venture-backed issues (*H4a*). Accordingly, the hypothesized certification role of venture capitalists with respect to the initial firm valuation of IPOs has to be rejected. The following section considers the influence of venture capitalist participation on the long-run market performance of IPOs.

5.2.1.3 Wealth Relatives

In this section, the long-term market performance of the sample firms is investigated by means of 3-year wealth relatives in order to test the hypotheses *H2b* and *H5*. Table 33 reports the cross-sectional estimates of the determinants of wealth relatives by country, using the respective country's growth market index as benchmark.[533] Following the regression setup employed by Brav and Gompers (1997), the dependent variable is the natural logarithm of the 3-year wealth relatives. The independent variables in column I of each country include a dummy variable indicating venture capital involvement, the approximated value of Tobin's *Q*, and the natural logarithm of the firm's market value of equity at the end of the first trading

---

[533] Alternatively, wealth relatives have been estimated by using the *Nasdaq Composite* index as benchmark. However, the overall results do not differ in terms of the coefficient signs and significance levels.

day. Moreover, inspired by the regression setup of Carter *et al.* (1998), the reputation rank of the lead underwriter bringing an IPO to the market is included as an explanatory variable in the regressions.

Table 33: Country-Specific Regressions on 3-Year Wealth Relatives

The dependent variable is the natural logarithm of the 3-year wealth relative using the respective country's growth market index as benchmark. The independent variables are a dummy variable that equals 1 if a firm is venture-backed (*VC*), the approximated value of Tobin's *Q* at the first day of trading, defined as the market value of equity and book value of debt divided by the book value of total assets, the natural logarithm of the IPO firm's market capitalization at the end of the first trading day (*Size*), and the underwriter reputation rank on a scale of 1 to 3 (*URank*), where 1 is the most and 3 the least prestigious underwriter. All regressions employ an ordinary least squares specification. The estimations include industry, calendar year and accounting standard dummy variables, which are not reported. *t*-statistics are in parentheses. Bold values indicate statistically significant differences at the 1% or 5% confidence levels.

	France		*Germany*		*UK*	
Variables	***I***	***II***	***I***	***II***	***I***	***II***
VC	0.05	0.42	**-0.23[a]**	0.25	-0.08	0.11
	(0.4)	(0.6)	**(-2.9)**	(0.3)	(-0.6)	(0.1)
Q	-0.13	0.08	-0.10	-0.09	0.01	0.03
	(-1.0)	(0.4)	(-1.0)	(-0.8)	(0.0)	(0.1)
Q*VC		-0.68		-0.02		-0.03
		(-1.6)		(-0.1)		(-0.1)
Size	-0.01	-0.12	-0.12	-0.04	**0.46[b]**	0.46[c]
	(-0.1)	(-0.5)	(-1.2)	(-0.3)	**(2.2)**	(1.7)
Size*VC		0.39		-0.60		-0.56
		(0.6)		(-0.9)		(-0.8)
URank	-0.01	0.05	-0.11	-0.16	0.06	-0.19
	(-0.1)	(0.3)	(-1.2)	(-1.3)	(0.3)	(-0.6)
URank*VC		-0.1		0.15		0.31
		(-0.3)		(0.5)		(0.6)
Constant	-0.39	-0.78	0.55	0.21	-2.44	-1.37
	(-0.4)	(-0.6)	(0.6)	(0.1)	(-1.5)	(-0.5)
Adj. R^2	0.07	0.06	0.22	0.21	0.27	0.26
F-test	1.5	1.4	**4.2[a]**	**3.5[a]**	**2.8[a]**	**2.5[a]**
Chow-test	0.6	0.6	1.8	1.8	0.7	0.7
Sample size	90	90	151	151	62	62

Note: Significance levels are denoted by [a)] for 1%, [b)] for 5%, and [c)] for 10%. Source: own calculations.

While venture-backing does not appear to have any significant effect on the market performance of firms in France and the UK, the results in column I of the German sample suggest that venture-backed IPOs underperform non venture-backed issues in the long run, as suggested by the hypothesis *H5*. Although this finding is consistent with previous studies on the long-term performance of German IPOs, it does not hold when slope effects are taken into account (column II).[534] Contrary to the conjectured relationship between firm size and market performance, bigger firms in the UK sample tend to perform better than smaller firms over the 3-year post-IPO period. Yet, in the two other country samples, firm size does not show any

[534] See e.g. Klein and Zoeller (2003), Neuhaus and Schremper (2003).

significant effect. Likewise, neither Tobin's *Q* nor the underwriter reputation reveal any significant impact on the long-run performance of the IPOs.

As shown by the correlation matrix in Appendix A 19, the reputation rank of the lead underwriter is correlated with firm size in all countries, suggesting that bigger firms are underwritten by more prestigious underwriters. Moreover, larger firms tend to have a higher Tobin's *Q* at the first trading day, and there is evidence that venture-backed IPOs in the UK are underwritten by investment banks with higher underwriter ranks. To make sure that the obtained results in column I of Table 33 are not biased by multicollinearity problems, Tobin's *Q* and the underwriter variable are taken out of the regressions. Yet, the results regarding the constant effects of venture-backing do not qualitatively differ from the findings in column I. To explore the slope effects of venture-backing on market performance, interaction dummy variables are added to the equations in column II. However, no particular impact of venture capital involvement on the wealth relatives can be detected. This finding is confirmed by the *Chow*-test, which rejects the existence of structural differences between the regressions of venture- and non venture-backed firms.

In a modified setup, as shown in Table 34, regressions based on the complete sample are estimated to detect significant cross-country differences in the effect of venture-backing. The effects of venture-backing in different country environments are investigated by replacing the simple venture capital variable with the interaction dummies *VC*Germany*, *VC*France*, and *VC*UK*. Conform to the previous results, German venture-backed issues perform significantly worse than the remainder of the sample, when accounting for constant effects of venture capital participation (column I and II). For robustness purposes, the regressions are also performed without the IPOs of 1999, but do not lead to materially different outcomes.

Nonetheless, as in the previous country-specific regressions, the full sample estimates considering interaction effects (column III) do not confirm the significant negative performance of German venture-backed IPOs. In order to check for the effects of multicollinearity, column IV shows the results of a reduced model specification omitting the underwriter rank as explanatory variable. The findings provide weak support for the positive impact of venture capital involvement in the UK on the market performance of firms. Furthermore, the estimates suggest that for venture-backed firms in the UK size is a significantly less important determinant of success than for non venture-backed issues. The results are consistent with venture capitalists in the UK being more qualified in bringing high quality small firms to the market than the market in general.

Analogous to the analysis in the previous sections, regressions based on the complete sample are run in Table 35, including variables for different countries and types of venture capitalists. The estimates in columns I and II control for constant and slope effects of venture-backing, respectively. While the results do not indicate any particular effect of venture capital participation on wealth relatives, the model specification in column II points out a negative size-related slope effect of venture-backing. The score of the *Chow*-test, though, does not indicate any significant structural differences between venture- and non venture-backed IPOs. To check for the influence of multicollinearity, the equations of columns I and II are replicated in additional regressions, omitting Tobin's *Q* and the underwriter rank as explanatory variables. Here also the results do not provide qualitatively different results with regard to venture-related effects on market performance.

Table 34: Full-Sample Regressions on 3-Year Wealth Relatives

The dependent variable is the natural logarithm of the 3-year wealth relative using the respective country's growth market index as benchmark. The independent variables are a dummy variable that equals 1 if a firm is venture-backed (*VC*), the approximated value of Tobin's *Q* (*Q*) at the first day of trading, defined as the market value of equity and book value of debt divided by the book value of total assets, the natural logarithm of the IPO firm's market capitalization at the end of the first day of trading (*Size*), and the underwriter reputation rank on a scale of 1 to 3 (*URank*), where 1 is the most and 3 the least prestigious underwriter. *France* and *UK* are country dummy variables, which equal 1 if a firm went public in the respective country. *France*, *Germany*, and *UK* are country dummy variables, which equal 1 if a firm went public in the respective country. All regressions employ an ordinary least squares specification. The estimations include industry, calendar year, and accounting standard dummy variables, which are not reported. *t*-statistics are in parentheses. Bold values indicate statistically significant differences at the 1% or 5% confidence levels.

Variables	*I*	*II*	*III*	*IV*
VC*France	0.06 (0.7)	0.06 (0.8)	0.25 (0.5)	0.14 (0.4)
VC*UK	-0.02 (-0.3)	-0.02 (-0.3)	0.07 (0.1)	0.59[c] (1.7)
VC*Germany	**-0.17[b] (-2.4)**	**-0.17[b] (-2.5)**	0.21 (0.3)	0.39 (0.9)
Q*France	-0.16 (-0.9)	-0.16 (-0.9)	0.14 (0.6)	0.14 (0.6)
Q*UK	-0.04 (-0.3)	-0.03 (-0.2)	-0.13 (-0.7)	-0.18 (-1.0)
Q*Germany	-0.19[c] (-1.8)	-0.13 (-1.2)	-0.14 (-1.1)	-0.20[c] (-1.7)
Q*VC*France			-0.46 (-1.6)	-0.46[c] (-1.7)
Q*VC*UK			0.19 (1.0)	0.23 (1.3)
Q*VC*Germany			0.01 (0.1)	0.01 (0.1)
Size*France	0.25 (1.0)	0.28 (1.1)	0.02 (0.1)	-0.03 (-0.1)
Size*UK	0.12 (0.5)	0.10 (0.3)	0.27 (0.6)	**0.61[b] (2.0)**
Size*Germany	-0.28 (-0.9)	-0.51 (-1.5)	-0.23 (-0.5)	0.05 (0.1)
Size*VC* France			0.34 (0.8)	0.40 (1.0)
Size*VC*UK			-0.64 (-1.3)	**-0.91[a] (-2.7)**
Size*VC*Germany			-0.52 (-0.9)	-0.57 (-1.2)
URank*France		0.05 (0.4)	0.08 (0.4)	
URank*UK		-0.01 (-0.07)	-0.39 (-1.1)	
URank*Germany		-0.20 (-1.4)	-0.31 (-1.5)	
URank*VC*France			-0.06 (-0.3)	
URank*VC*UK			0.28 (1.0)	
URank*VC*Germany			0.13 (0.6)	
France	-0.36 (-1.0)	-0.78[c] (-1.7)	-0.71 (-1.1)	-0.12 (-0.2)
UK	-0.38 (-1.0)	-0.65 (-1.2)	-0.16 (-0.2)	-0.42 (-1.0)
Constant	-0.15 (-0.2)	0.92 (0.8)	0.47 (0.3)	-1.01 (-0.9)
Adj. R^2	0.09	0.09	0.11	0.11
F-test	**2.4[a]**	**2.1[a]**	**2.1[a]**	**2.4[a]**
Sample size	303	303	303	303

Note: Significance levels are denoted by [a)] for 1%, [b)] for 5%, and [c)] for 10%. Source: own calculations.

Consistent with the discussed relationship between the initial firm valuation and the post-issue stock price performance, Tobin's *Q* in Table 35 turns out to be negatively related to the market performance. This result suggests that IPOs with higher valuations in terms of Tobin's *Q* at the first trading day perform worse in the long run than firms with a lower Tobin's *Q*. However, the relationship between Tobin's *Q* and the stock price performance in the after-market of venture-backed IPOs does not systematically differ from non venture-backed issues, as the interaction term *Q*VC* remains insignificant.[535]

Table 35: Full Sample Regressions on 3-Year Wealth Relatives

The dependent variable is the natural logarithm of the 3-year wealth relative using the respective country's growth market index as benchmark. The independent variables are a dummy variable that equals 1 if a firm is venture-backed (*VC*), the approximated value of Tobin's *Q* (*Q*) at the first day of trading, defined as the market value of equity and book value of debt divided by the book value of total assets, the natural logarithm of the IPO firm's market capitalization at the end of the first day of trading (*Size*), and the underwriter reputation rank on a scale of 1 to 3 (*URank*), where 1 is the most and 3 the least prestigious underwriter. *France* and *UK* are country dummy variables, which equal 1 if a firm went public in the respective country. *Int_VC* is a dummy variable equal to 1 if an IPO was backed by a venture capitalist that backed firms in at least two countries of the sample. All regressions employ an ordinary least squares specification. The estimations include industry, calendar year, and accounting standard dummy variables, which are not reported. *t*-statistics are in parentheses. Bold values indicate statistically significant differences at the 1% or 5% confidence levels.

Variables	*I*	*II*	*III*	*IV*	*V*
VC	-0.08 (-1.3)	0.39 (1.1)	**-0.18^a (-2.7)**	0.09 (0.2)	0.27 (1.0)
Int_VC			**0.22^a (3.1)**	0.41 (1.5)	**0.63^b (2.4)**
Q	**-0.15^b (-2.1)**	-0.15^c (-1.8)	**-0.15^b (-2.1)**	-0.15^c (-1.8)	**-0.18^b (-2.2)**
Q*VC		0.02 (0.1)		0.00 (0.0)	0.02 (0.1)
Q*Int_VC				0.02 (0.1)	0.07 (0.4)
Size	-0.01 (-0.1)	0.13 (1.1)	-0.04 (-0.4)	0.09 (0.8)	0.15 (1.5)
Size*VC		**-0.64^b (-2.1)**		-0.41 (-1.1)	-0.49 (-1.4)
Size*Int_VC				-0.36 (-0.9)	-0.49 (-1.5)
URank	-0.05 (-0.8)	-0.11 (-1.2)	-0.03 (-0.4)	-0.12 (-1.2)	
URank*VC		0.13 (0.7)		0.13 (0.6)	
URank*Int_VC				0.14 (0.8)	
France	0.21 (1.6)	0.18 (1.3)	0.15 (1.1)	0.12 (0.9)	0.13 (1.0)
UK	0.14 (1.1)	0.14 (1.1)	0.07 (0.6)	0.10 (0.8)	0.10 (0.7)
Constant	-0.93 (-1.1)	-1.38 (-1.4)	-0.74 (-0.9)	-1.03 (-1.1)	**-1.76^b (-2.5)**
Adj. R^2	0.09	0.11	0.12	0.14	0.14
F-test	**2.7^a**	**2.8^a**	**3.2^a**	**3.0^a**	**3.4^a**
Chow-test	1.2	1.2	**3.0^a**	**3.0^a**	**3.1^a**
Sample size	303	303	303	303	303

Note: Significance levels are denoted by ^a) for 1%, ^b) for 5%, and ^c) for 10%. Source: own calculations.

---

535 See also Figure 7 in section 4.4.2.2, p. 103.

In order to account for the heterogeneity of venture capital, in columns III and IV, an additional variable for the involvement of international venture capitalists is introduced. The refinement of the model specification in column III shows that IPOs backed by international venture investors perform significantly better than the average venture-backed issue. Firms backed by national venture capitalists, by contrast, tend to perform worse than the remainder of the sample in terms of wealth relatives.

Although the highly significant score of the *Chow*-test indicates structural differences between the individual regressions of the different firm groups, the variable for international venture capitalist involvement becomes insignificant in the full model specification accounting for slope effects of venture-backing (column IV). A possible explanation for this observation is the multicollinearity related to the similar use of dummy and interaction variables, on the one hand, and partly correlated variables, such as venture capital involvement, size, and the underwriter rank, on the other.[536] Even though the underlying simple correlations as reported in Appendix A 19 are not very large, it is possible that a strong linear relationship exists among several explanatory variables.[537] Thus, when all variables are included at once, the standard errors of the individual variables tend to be large and *t*-values of coefficients can become insignificant.[538]

According to Jobson (1991), multicollinearity can be a result of too few observations for too many variables. Thus, additional data are often the best way of reducing the undesired effects. Since it is not possible in the present analysis to generate additional data for the observed period, an alternative is to eliminate variables on a theoretical basis to reduce collinearity but avoid underspecification at the same time.[539] As prior studies on market performance consider merely firm size and the book-to-market ratio, the latter being per definition highly correlated with the inverse of Tobin's Q,[540] column V reports the regression estimates of 3-year wealth relatives without the underwriter rank as explanatory variable. Consistent with the findings of column III and the *Chow*-test statistic, the results of the reduced model specification confirm the positive effect of international venture capitalists on the observed wealth relatives.[541] Finally, the values of the adjusted R^2 and the F-statistic of the reduced setup in column V indicate that the omission of the underwriter rank as explanatory variable does not negatively affect the overall regression fit.

Summarizing, the main findings of this section do not provide general support for the conjecture that venture-backed IPOs outperform issues without venture capital participation in the aftermarket (*H2b*). However, similar to the earlier results, the refined examination distinguishing between the backing of international and national venture capitalists shows that firms supported by international venture investors outperform the average venture-backed issue. Although the model specification by country suggests that German venture-backed

---

[536] The analysis by means of the variance inflation factor (VIF), which represents a widespread method to detect multicollinearity, shows scores of higher than five for the coefficients of the underwriter rank and firm size. Some authors, such as Studenmund (1997), argue that a VIF score of higher than five can indicate multicollinearity problems. See Studenmund (1997), pp. 274-277.

[537] See e.g. Jobson (1991), pp. 320-321.

[538] See Gujarati (1988), p. 292.

[539] See Jobson (1991), p. 285.

[540] See e.g. Fama and French (1992), Brav and Gompers (1997).

[541] The same result is obtained if firm size is omitted as an explanatory variable in the regression setup.

IPOs perform significantly worse than non venture-backed issues, the estimations controlling for slope effects do not support this result. Given the limited support for the conjecture that venture capitalists in Germany suffer from particularly severe adverse selection due to the bank-based financial system, hypothesis *H5* is not accepted.

Since the employed dummy variables for venture capital involvement might be too coarse to reflect differences in the quality and effectiveness of venture-backing adequately, the next section performs a further refinement of the analysis. In order to explore the remaining hypotheses, it takes into account various characteristics of venture capitalists and their involvement in firms they brought public. Also, IPOs backed by various organizational types of venture capitalists are investigated about performance differences, as conjectured in section 4.2.4.

5.2.2 Venture-Backed IPOs

5.2.2.1 Certification and Monitoring Quality

To assess the certification and monitoring quality of venture capitalists, and its impact on the operating and market performance of IPO, as hypothesized in section 4.2.2 (*H3a*, *H3b*), the present section focuses on the subsample of venture-backed firms. Based on the approach followed by Barry *et al.* (1990), several variables reflecting the reputation and involvement of venture capitalists in IPOs are examined with respect to potential effects on the different performance measures. The variables include the number of venture capitalists having equity positions in an IPO (*NVC*), the representation of venture investors on the board of directors (*Board*), the age of the lead venture capitalist at the time of the going public (*VC_Age*), the pre- and post-issue equity shares held by all venture investors (*AVC PrEq*, *AVC PosEq*), the percentage of venture shareholdings sold (*AVC Sale*), and the number of IPOs in the sample backed by venture investors (*No_IPOs*). Since the rationales for the use of the individual variables have been addressed in section 4.4.3, they are not further discussed in the present section.

As documented in the correlation matrix of Table 36, most of the proxies for the certification and monitoring quality of venture capitalists are correlated with each other. Thus, the number of syndicating venture capitalists correlates positively with the representation of venture investors on the board of portfolio firms, the age of the lead financier, the pre- and post-issue venture shareholdings as well as the number of IPOs backed by a venture capitalist. Most of the reported positive correlations are in line with the findings of section 5.1.3 regarding the experience and involvement of international versus national venture capitalists. Yet, Table 36 also shows that the selling intensity of venture capitalists in the IPO process correlates significantly with the monitoring proxies and the number of sample IPOs backed by a venture capitalist. This outcome is rather surprising and suggests that experienced venture capitalists being more involved in portfolio firms sell a higher portion of their shares in IPOs.

First, to explore the soundness of hypothesis *H3a*, regressions on the post-issue operating performance of the IPO firms are estimated by country and based on the full sample. The regressions are run against each of the monitoring and certification proxy variables as well as the explanatory variables size, age, participation ratio, and underwriter rank. Due to the partial multicollinearity between the individual certification and monitoring proxies, the variables are included one by one in the regressions.[542] Nevertheless, the estimates on the operating performance are not conclusive, and none of the monitoring and certification proxies turns out to be significant. To account for differences between the certification and monitoring qualities of venture capitalists belonging to different countries, regressions are run with dummy variables for the nationalities of lead venture capitalists. Yet, the results do not provide any significant insights regarding the influence of certification and monitoring on operating performance. Likewise, regressions based on the reduced sample of firms backed by international venture capitalists do not yield any significant results.

Table 36: Correlation Matrix of Proxies for Quality of Venture Capital Monitoring

NVC denotes the number of venture capitalists backing a firm at the time of the IPO. *Board* is a dummy variable equal to 1 if a venture capitalist is on the board of a portfolio firm at the time of the IPO. *VC_Age* is the natural logarithm of the lead venture capitalist's age in years at the time of the IPO, whereby the lead venture capitalist is defined as the one with the largest equity position in the firm going public. *AVC PrEq* and *AVC PosEq* is the percentage of equity owned by all venture capitalists before and immediately after the IPO, respectively. *AVC Sale* denotes the percentage of shareholdings sold by all venture capitalists in the IPO, and *No_IPOs* stands for the number of IPOs in the sample backed by a venture capitalist.

Variables	(1)	(2)	(3)	(4)	(5)	(6)
NVC (1)	1.00					
Board (2)	**0.60**[a]	1.00				
VC_Age (3)	**0.27**[a]	0.09	1.00			
AVC PrEq (4)	**0.73**[a]	**0.66**[a]	**0.26**[a]	1.00		
AVC PosEq (5)	**0.75**[a]	**0.62**[a]	**0.27**[a]	**0.90**[a]	1.00	
AVC Sale (6)	**0.53**[a]	**0.58**[a]	0.03	**0.62**[a]	**0.39**[a]	1.00
No_IPOs (7)	**0.45**[a]	**0.39**[a]	**0.49**[a]	**0.53**[a]	**0.50**[a]	**0.47**[a]

Note: [a)] significant at the 1% level. Source: own calculations.

Second, to detect whether or not the conjectured certification of venture capitalists is related to the monitoring and value-adding hypothesis, Tobin's *Q* of the venture-backed IPOs at the first trading day is investigated in the cross-section against the different proxies to identify effects of the monitoring intensity and certification on the initial valuation in the capital markets. In analogy to the full sample analysis, firm size, equity ratio, sales over assets, and the ownership variable *Alpha* are used as explanatory variables in the equations. Yet, the regressions do not yield any significant findings with regard to the relationship between venture capital certification and monitoring and Tobin's *Q*.

Finally, to explore the validity of *H3b*, regressions of 3-year wealth relatives are estimated against each of the monitoring and certification proxies as well as the explanatory variables Tobin's *Q*, firm size, and the underwriter rank. The regression setup, which is partly similar to the model specification employed by Gompers and Lerner (1999a), is reported in Table 37.[543] However, the overall findings are rather inconclusive, and merely the fraction of sharehold-

542 A similar approach is followed by Barry *et al.* (1990)

543 Instead of using the post-issue shareholdings as a proxy for venture capital monitoring, in an alternative model specification the pre-issue shareholdings have been included as explanatory variable, without leading to qualitatively different results.

ings sold by venture capitalists at the IPO turns out to have a weakly significant positive influence on market performance (column V). The result suggests that the more stakes venture investors sell, the better the company performs in terms of wealth relatives. Although this outcome is only significant at the 10 percent level, it could be explained with reputational concerns of venture capitalists combined with the need to return capital to investors. While venture capitalists have to return funds to investors in order to raise new funds, they typically avoid selling a lot of equity in IPOs if a firm might decline in value. Given the negative reputational impact of "cashing out" to the detriment of new shareholders, venture capitalists might only sell if they are certain that future news regarding the firm will be positive.[544]

Table 37: Impact of Certification and Monitoring Variables on Wealth Relatives

The dependent variable is the natural logarithm of the 3-year wealth relative using the respective country's growth market index as benchmark. The monitoring variables comprise the number of venture capitalists backing an IPO (*NVC*), a dummy variable equal to 1 if a venture capitalist is on the board at the time of the IPO (*Board*), the natural logarithm of the lead venture capitalist's age in years at the IPO (*VC_Age*), whereby the lead venture capitalist is defined as the one with the largest equity position in the firm going public, the percentage of equity owned by all venture capitalists immediately after the IPO (*AVC PosEq*), the percentage of shareholdings sold by all venture capitalists in the IPO (*AVC Sale*), and the number of IPOs in the sample backed by a venture capitalist. The remaining independent variables include Tobin's *Q* at the first trading day, defined as the market value of equity and book value of debt divided by the book value of total assets, the natural logarithm of the market capitalization at the end of the first trading day (*Size*), and the underwriter reputation rank on a scale of 1 to 3 (*URank*), where 1 is the most and 3 the least prestigious underwriter. All regressions employ an ordinary least squares specification. The estimations include industry, calendar year and accounting standard dummy variables, which are not reported. *t*-statistics are in parentheses. Bold values indicate statistically significant differences at the 1% or 5% confidence levels.

	I	*II*	*III*	*IV*	*V*	*VI*
NVC	-0.01 (-0.1)					
Board		0.06 (0.7)				
VC_Age			0.04 (0.5)			
AVC PosEq				0.05 (0.6)		
AVC Sale					0.15[c] (1.7)	
No_IPOs						0.05 (0.6)
Q	-0.10 (-0.9)	-0.09 (-0.8)	-0.10 (-0.9)	-0.09 (-0.8)	-0.09 (-0.9)	-0.10 (-1.0)
Size	-0.15 (-1.0)	-0.17 (-1.1)	-0.17 (-1.1)	-0.16 (-1.1)	-0.13 (-0.9)	-0.13 (-0.9)
URank	0.00 (0.0)	0.01 (0.1)	0.01 (0.1)	0.01 (0.1)	0.02 (0.2)	-0.02 (-0.2)
France	0.32[c] (1.8)	0.33[c] (1.8)	0.31[c] (1.7)	0.32[c] (1.8)	0.35[c] (2.0)	0.30[c] (1.7)
UK	0.21 (1.2)	0.21 (1.2)	0.20 (1.1)	0.21 (1.2)	0.20 (1.1)	0.20 (1.1)
Constant	-0.83 (-0.7)	-0.90 (-0.8)	-0.95 (-0.8)	-0.91 (-0.8)	-1.50 (-1.3)	-0.91 (-0.8)
Adj. R^2	0.16	0.16	0.16	0.16	0.18	0.15
F-test	**2.7[a]**	**2.7[a]**	**2.7[a]**	**2.7[a]**	**2.9[a]**	**2.5[a]**
Sample size	154	154	154	154	154	154

Note: significance levels are denoted by [a)] for 1%, [b)] for 5%, and [c)] for 10%. Source: own calculations.

[544] This line of arguments is also followed by Gompers and Lerner (1999a), p. 18.

Concluding the main findings, the use of different monitoring and certification proxies provides only limited additional insights concerning the observed performance figures. Going back to the hypotheses derived in section 4.2, the insignificant results with regard to the influence of different monitoring proxies do not support the conjecture that venture-backed firms outperform non venture-backed issues due to the monitoring and value-adding ability of venture capitalists (*H3a*, *H3b*). Furthermore, there is no evidence for the claim that the certification role of venture capitalists is related to their degree of involvement in firms. In order to detect whether different types of venture capitalists have an influence on the performance of firms going public, the following section presents a further refinement of the analysis.

5.2.2.2 Backing of Different Types of Venture Capitalists

To explore the effects of different types of venture capitalists on the performance of firms going public, as stated by the hypotheses *H6a, H6b*, and *H8*, this section carries out an additional refinement of the analysis, using dummy variables for IPOs backed by independent and public lead venture capitalists. Moreover, using the insights of the earlier analysis, a distinction is made between firms backed by nationally and internationally operating venture capital investors. Finally, performance differences between firms financed by bank-affiliated and non bank-affiliated venture capitalists are examined in more detail.

First, as outlined in column I of Table 38, the market performance of firms backed by independent and public lead venture capitalists is analyzed in the cross-section. In line with the conjectured relationship of *H8*, the results indicate that firms backed by public lead venture capitalists underperform the remaining venture-backed IPOs of the sample in terms of 3-year wealth relatives. However, against the expectations of hypothesis *H6b*, issues backed by independent venture capitalists do not perform significantly different compared to the average venture-backed firm. The same results are obtained when a dummy variable for the involvement of international venture capitalists (*Int_VC*) is included in the estimates (column II), which leads to an overall improvement of the goodness of fit, as shown by the higher values of R^2 and the *F*-statistic. In line with earlier findings, issues backed by internationally operating venture investors perform significantly better than the average.

In order to test hypothesis *H6a*, in column III of Table 38, the regression setup is extended by including variables for the involvement of bank-affiliated venture capitalists. In order to account for potential country-related differences between the backing of bank-affiliated venture capitalists in France and Germany, the interaction variables *Bank*France* and *Bank*Germany* are considered in the estimates. Even though the results in column IV reveal a significant positive impact of bank-affiliated venture-backing on firm performance in France, this finding is only weakly significant when the dummy variable for international venture-backing is included (column III). Moreover, as the coefficient of the variable *Bank*Germany* remains insignificant in both model specifications, there is no uniform evidence indicating systematic performance differences between issues backed by bank-affiliated venture capitalists and the remaining firms of the sample. Accordingly, *H6a* has to be rejected.

The previous analysis exposes that the dichotomous categorical variable for the international involvement of venture capitalists is a good indicator for their experience in backing IPOs. Therefore, a logistic regression is used to identify the determinants of the underlying but unobservable experience of venture investors. The logistic approach estimates the probability that a venture capitalist is backing IPOs in at least two countries of the sample, by using internationality (*Int_VC*) as dependent variable and different characteristics of venture capital investors, such as their age, board representation, the number of syndication partners and backed IPOs, as well as the pre-issue shareholdings in firms, and the selling intensity, as explanatory variables.[545]

Table 38: Wealth Relatives of IPOs Backed by Different Types of Venture Capitalists

The dependent variable is the natural logarithm of the 3-year wealth relative using the respective country's growth market index as benchmark. The independent variables are the approximated value of Tobin's Q at the first trading day, defined as the market value of equity and book value of debt divided by the book value of total assets, the natural logarithm of the firm's market capitalization at the end of the first day of trading (*Size*), and the underwriter reputation rank on a scale of 1 to 3 (*URank*), where 1 is the most and 3 the least prestigious underwriter. *Indep* and *Public* are dummy variables equal to 1 if a venture capitalist is independent or owned by a public shareholder, respectively. *Int_VC* is a dummy variable equal to 1 if an IPO was backed by a venture capitalist that backed firms in at least two countries of the sample. *Bank* is a dummy variable equal to 1 if an IPO is backed by a bank-affiliated venture capitalist. All regressions employ an ordinary least squares specification. The estimations include industry, calendar year and accounting standard dummy variables, which are not reported. *t*-statistics are in parentheses. Bold values indicate statistically significant differences at the 1% or 5% confidence levels.

	I	*II*	*III*	*IV*
Indep	-0.04 (-0.4)	-0.18^c (-1.7)	0.01 (0.0)	
Public	**-0.20^b (-2.1)**	**-0.21^b (-2.4)**	**-0.20^b (-2.3)**	
Int_VC		**0.32^a (3.7)**	**0.35^a (4.0)**	**0.35^a (4.0)**
Bank*France			0.28^c (1.8)	**0.23^b (2.1)**
Bank*Germany			0.08 (0.5)	0.01 (0.1)
Q	-0.09 (-0.8)	-0.09 (-0.9)	-0.08 (-0.8)	-0.09 (-0.9)
Size	-0.19 (-1.3)	-0.23 (-1.6)	-0.24^c (-1.8)	-0.18 (-1.3)
URank	-0.04 (-0.4)	0.03 (0.3)	0.02 (0.3)	0.03 (0.4)
France	0.29 (1.6)	0.14 (0.8)	0.05 (0.3)	0.05 (0.3)
UK	0.19 (1.1)	0.10 (0.6)	0.09 (0.5)	0.10 (0.6)
Constant	-0.49 (-0.4)	-0.40 (-0.4)	-0.90 (-0.8)	-1.04 (-0.9)
Adjusted R^2	0.17	0.24	0.25	0.23
F-test	**2.7^a**	**3.5^a**	**3.4^a**	**3.4^a**
Sample size	154	154	154	154

Note: Significance levels are denoted by $^{a)}$ for 1%, $^{b)}$ for 5%, and $^{c)}$ for 10%. Source: own calculations.

Table 39 outlines the logistic estimates of the distinct variables as predictors for the internationality of venture capitalists. The *Wald*-test statistic of the coefficients suggests that the age and the number of IPOs backed by a venture capitalist are significant determinants of international involvement. In order to interpret the effects of changes in the logistic coefficients on the dependent variable, the estimates are transformed back into odds ratios in the last column of Table 39. The transformation reveals that each additional year of the age increases the odds of a venture capitalist being international by the factor of 8.52, holding all other vari-

545 See e.g. Long (1997), pp. 34-84, for details on regression models for categorical dependent variables.

ables constant. Likewise, for each additional IPO backed by venture investors the odds for the involvement in international activities are increased by the factor 1.67. The overall robustness of the estimates is underlined by the statistically significant likelihood ratio (*LR*) of the χ^2-test statistics.

Table 39: The Determinants of the Backing by International Venture Capitalists

The table estimates the determinants of backing by international venture capitalists using a logit model. The dependent variable is the dummy variable *Int_VC*, indicating whether a venture capitalist is involved in IPOs in more than one country of the sample. The independent variables include the age (in years) of the venture capitalist at the time of the IPO (*VC_Age*), a dummy variable for board representation (*Board*), the total number of venture capitalists having invested in the same portfolio firm (*NVC*), the number of backed IPOs in the sample (*No_IPOs*), the percentage of equity owned by all venture capitalists before the IPO (*AVC PrEq*), and the percentage of shares sold by all venture capitalists in the IPO (*AVC Sale*). The Wald-test that equals the square of the ratio of the coefficient to its standard error is reported to test the significance of the individual model coefficients. Bold values indicate statistically significant differences at the 1% confidence level.

Variable	*Coefficient (B)*	*S.E.*	*Wald*	*p-value*	*Exp(B)*
VC_Age	**2.14**[a]	**0.57**	**14.31**	**0.00**	**8.52**
Board	0.47	0.59	0.66	0.42	1.61
NVC	0.24	0.21	1.30	0.25	1.27
No_IPOs	**0.52**[a]	**0.13**	**16.37**	**0.00**	**1.67**
AVC PrEq	0.03	0.02	2.10	0.15	1.03
AVC Sale	-0.02	0.01	1.19	0.27	0.99
Constant	**-9.95**[a]	**1.99**	**24.90**	**0.00**	
-2 ln L	89.40				
LR χ^2	**114.73**[a]				
Cox & Snell's R^2	0.54				
Nagelkerke R^2	0.72				

Note: S.E. – standard error. Significance levels are denoted by [a)] for 1%. Source: own calculations.

Going back to the conjecture that the categorical variable of internationality represents an adequate indicator for the unobservable experience of venture capitalists, the results of the logistic regression suggest that the age and number of backed IPOs should equally serve as proxies for the experience of venture investors. Yet, in the previous analysis on firm performance both variables turned out to be insignificant. There are at least two possible explanations for the observed insignificance. First, it could be that the distinct variables *VC_Age* and *No_IPOs* do not sufficiently reflect the experience of venture capitalists, which is captured by the dummy variable for international involvement. Second, as the logistic regression is based on a non-linear estimation technique, it might be that a non-linear relationship exists between firm performance and the explanatory variables of venture capital experience, which is not effectively considered by using linear OLS regressions.

To explore the potential non-linear relationship between firm performance and the proxies for the experience of venture capitalists, such as their age and the number of backed IPOs, separate regressions are run, using the fitted values of the logit estimation in Table 39 for the internationality of venture capitalists instead of the dummy variable. However, as the influence of the predicted internationality remains insignificant, the regressions are not reported. Alternatively, regressions employing the probability values of the logit estimates for international involvement of venture capitalists are run. Yet, as shown in Table 40, merely in the reduced model specification (column III), there is weak support for the positive influence of the probability values of venture capital internationality on the firms' market performance. Furthermore, the results slightly indicate that the previously documented negative relationship between Tobin's Q and market performance loses in importance, the higher the probability that a venture capitalist operates internationally. Hence, the market performance of firms backed by venture capitalists which are more probable to be international reveals a tendency to follow a random walk. Finally, the negative sign of the weakly significant interaction variable *Size*Pr(Int_VC)* points that the stocks of smaller firms are more profitable in the aftermarket, the higher the probability that they are backed by an international venture capitalist.

Table 40: Regressions Using Probabilities for International VC Involvement

The dependent variable is the natural logarithm of the 3-year wealth relative using the respective country's growth market index as benchmark. *Pr(Int_VC)* is the probability that an IPO is backed by an international venture capitalist, using the estimation approach of the logistic regression in Table 39. The remaining independent variables include the approximated value of Tobin's Q at the first day of trading, defined as the market value of equity and book value of debt divided by the book value of total assets, the natural logarithm of the IPO firm's market capitalization at the end of the first day of trading (*Size*), and the underwriter reputation rank on a scale of 1 to 3 (*URank*), where 1 is the most and 3 the least prestigious underwriter. *France* and *UK* are country dummy variables, which equal 1 if a firm went public in the respective country. All regressions employ an ordinary least squares specification. The estimations include industry, calendar year and accounting standard dummy variables, which are not reported. *t*-statistics are in parentheses. Bold values indicate statistically significant differences at the 1% or 5% confidence levels.

Variables	*I*	*II*	*III*
Pr(Int_VC)	0.06 (0.7)	0.48 (0.9)	0.62^{c} (1.7)
Q	-0.10 (-0.9)	-0.23 (-1.5)	-0.24 (-1.7)
Q*Pr(Int_VC)		0.30 (1.3)	0.32 (1.5)
Size	-0.17 (-1.1)	0.03 (0.2)	0.06 (0.3)
Size*Pr(Int_VC)		-0.82 (-1.5)	-0.91^{c} (-2.0)
URank	0.01 (0.1)	-0.03 (-0.2)	
URank*Pr(Int_VC)		0.08 (0.3)	
France	0.30^{c} (1.7)	0.32^{c} (1.8)	0.32^{c} (1.8)
UK	0.19 (1.1)	0.27 (1.5)	0.27 (1.5)
Constant	-0.86 (-0.8)	-1.53 (-1.1)	-1.69 (-1.5)
Adj. R^2	0.16	0.17	0.18
F-test	$\mathbf{2.7^{a}}$	$\mathbf{2.5^{a}}$	$\mathbf{2.8^{a}}$
Sample size	154	154	154

Note: Significance levels are denoted by a) for 1%, b) for 5%, and c) for 10%. Source: own calculations.

Summarizing, the findings of this section support the hypothesis that firms backed by public lead venture investors underperform firms backed by private venture capitalists (*H8*). However, the evidence does not confirm the conjecture that IPOs backed by independent venture capitalists outperform issues backed by captive venture investors (*H6b*). Likewise, the claim that firms supported by bank-affiliated venture capitalists in France and Germany outperform issues backed by non bank-affiliated venture investors has to be rejected (*H6a*). Finally, assuming that international venture-backing is a good indicator for the experience of venture capitalists, the findings suggest that the age and backing frequency of venture investors in IPOs of the sample are good predictors for their experience. In the following section, several checks are documented, which were performed to underline the overall robustness of the empirical findings.

5.2.3 Robustness Checks

In addition to the documented robustness checks, a number of alternative model specifications have been applied to test the overall robustness of the results. Thus, instead of using 2-year averages of operating performance measures, the regressions are run with the individual values of each year without leading to qualitatively different results. Also, different cash flow definitions are used to compute the operating profitability measures, and pure accounting-based measures such as the operating return on assets are employed in place of cash flow-related profitability measures as dependent variables in the regression analysis. As an alternative for the computation of wealth relatives, the *Nasdaq Composite* index is used as a benchmark instead of the indices of the respective countries' growth markets. Nevertheless, the obtained results generally confirm the previously reported findings.

In case the data contain outliers, the least squares fit might be affected. A measure for the influence of outliers on the regression results is *Cook's D*. The *Cook's D* statistic measures the change in the parameter estimates caused by deleting each observation. Jobson (1991) argues that a useful criterion for *Cook's D* is to conclude that the influence is relatively large if the measure exceeds the value of one. Although the value of *Cook's D* is not bigger than this critical value in any of the equations, the outlying observations with the highest *Cook's D* values are taken out of the individual regressions to assess the robustness of the results. Yet, the findings of the modified regressions confirm the earlier coefficient signs and significance levels, suggesting no influence of outliers on the inferences from the regression analysis.

Finally, to detect the potential presence of heteroscedasticity, which might cause a bias in the OLS estimates, scatter plots of the variance of disturbance terms are scrutinized, without indicating any violations regarding the assumed homoscedasticity. Given the overall robustness of the results, the subsequent section provides a discussion of the main findings of the empirical analysis. Moreover, it points out the limitations of the study and addresses alternative explanations for the obtained results.

5.3 Discussion, Limitations, and Alternative Explanations

The most important results of the empirical analysis are recapitulated in Table 41. As shown in the table, the overall evidence provides support for the hypotheses *H1*, *H7*, and *H8*. As far as the hypotheses *H2a* and *H2b* are concerned, the assumed performance effects are limited to a subsample of firms backed by international venture capitalists. Finally, the hypotheses *H3a*, *H3b*, *H4a*, *H4b*, *H6a*, and *H6b* are rejected on the basis of the empirical findings.

With respect to the overall research question, the empirical results suggest that only a subgroup of internationally investing venture capitalists is able to spur the performance of firms going public in the considered European stock markets. Relating this outcome to the theory of the multinational enterprise it suggests that venture capitalists decide to become international if they possess a countervailing advantage over national venture capitalists, such as more investment experience and/or better skills in the assessment of firm profitability, which is necessary to balance their limited knowledge of foreign markets.

Firms backed by nationally operating venture capitalists, by contrast, do not outperform non venture-backed IPOs. This finding is inconsistent with the results of prior US studies, suggesting that venture-backed IPO generally outperform non venture-backed issues. An explanation for the variations in the performance figures of firms backed by different types of venture capitalists might be related to the lack of experience of many players in the relatively young and immature European venture capital market.[546]

Still, there are several limitations to the approach of the present study, requiring a cautious interpretation of the empirical results. First, it is possible that the investigation does not take into account the quality and intensity of venture backing sufficiently. Although various types of venture capital firms and monitoring proxies are considered, the intensity of venture investors' support is typically related to the length of their involvement in the portfolio firms. Yet, the employed data set does not allow controlling for this issue. The analysis does neither regard the specialization of venture capitalists in specific investment stages or industries, even though this aspect might be a determinant of their advisory capacities.

Second, the sample only covers the 3-year period after the flotation of venture-backed firms and may not allow detecting whether or not venture-backed IPOs outperform their non venture-backed counterparts in the long term. Previous studies, such as Brav and Gompers (1997), focus on a 5-year post-issue time horizon to investigate the long-run performance of venture-backed firms. Thus, it seems useful to extend the analyzed post-issue period beyond the employed time interval to check the robustness of the results. Unfortunately, due to the relatively recent opening of the considered European stock market segments, such a study can only be performed for a relatively small number of firms presently.

Third, the findings related to the market performance are assumed to be strongly influenced by the "Internet bubble" period of 1999-2000 with many excessive and unsustainable high stock market valuations.[547] The unselective high valuations of firms listed at European stock markets at the end of the 1990s may have distorted the true firm values and the market-based performance measures of the considered IPOs. Although for robustness reasons the regressions were also estimated without the IPOs of 1999, an "overshooting" phenomenon in the venture capital markets during the observed period may have led to a situation, where some venture firms ended up funding too many opportunities.[548] Lerner (2002) argues that "during boom periods, the prevalence of over-funding of particular sectors can lead to sharp decline in the effectiveness of venture funds".[549] Hence, the "overshooting" argument may account in part for the observed performance of venture-backed IPOs in general.

---

546 See e.g. Schwienbacher (2002), Hege *et al.* (2003).

547 See Ljungqvist and Wilhelm (2003) for an analysis of the pricing of IPOs in the dot-com bubble.

548 Gompers and Lerner (2001b) argue that supply of venture capital may "overshoot" demand when there is a long lag between a shortage and an attempt to fill that shortage. See Gompers and Lerner (2001b), p. 117 and p. 133.

549 Lerner (2002), p. 1.

In order to conclude, the following chapter summarizes the main findings of the overall analysis and provides an outlook with respect to future research.

Table 41: Summary of the Main Findings

	Hypothesis	*Finding*
H1	Lon-run IPO performance	In all three countries, the average IPO underperforms relative to seasoned stocks over the 3-year post-issue period, irrespective of the employed benchmark. IPOs at the German *Neuer Markt* tend to perform significantly worse than issues in the French and the British sample.
H2a	VC selection and operating performance	Venture-backed firms do not generally outperform non venture-backed firms in terms of post-issue operating performance. Merely firms backed by international venture capitalists display significantly superior post-issue operating performance figures.
H2b	VC selection and market performance	Venture-backed firms do not generally outperform non venture-backed firms in terms of market performance. Only firms backed by international venture capitalists outperform the average firm over the 3-year post-issue period.
H3a	VC monitoring and operating performance	The analysis of different monitoring proxies does not indicate any significant effect on the operating performance of venture-backed firms.
H3b	VC monitoring and market performance	The analysis of different monitoring proxies does not indicate any significant effect on the performance of venture-backed firms in the aftermarket.
H4a	VC certification and Tobin's Q	Inconsistent with the certification hypothesis, there is no evidence that venture-backed firms undergo higher valuations in terms of Tobin's Q than non venture-backed firms at the first trading day.
H4b	VC certification and firm profitability	Inconsistent with the certification hypothesis, the profitability of venture-backed firms at the time of the IPO does not differ from non venture-backed firms.
H5	VC and adverse selection in bank-based financial systems	Venture-backed IPOs in the German sample appear to underperform non venture-backed issues in the aftermarket, but this regularity does not hold after controlling for slope effects of venture-backing.
H6a	Bank-affiliated VCs in France and Germany	Firms backed by French and German bank-affiliated venture capitalists do not outperform firms backed by non bank-affiliated venture investors.
H6b	Independent vs. captive VCs	Firms backed by independent venture capitalists do not outperform firms backed by captive venture capitalists in the aftermarket.
H7	Experience and involvement of international vs. national VCs	International venture capitalists are more experienced and more involved in portfolio firms than their national counterparts, i.e. they are older, syndicate with a larger number of investment partners, back more IPOs in the sample, are more often represented on the board, and hold higher pre- and post-IPO shareholdings.
H8	Public vs. private VCs	Firms backed by public lead venture capitalists underperform firms backed by private venture capitalists in the aftermarket.

> *"Venture capital firms in Europe are more deal makers and less active monitors; they seem to be lagging still in their capacity to add value to innovative firms."*[550]

6 Conclusion

This dissertation provides new stylized facts on the role of venture capitalists and their impact on the performance of firms going public. The main findings suggest that European venture-backed issues do not generally outperform non venture-backed ones. Instead, only internationally operating venture capitalists are able to spur the performance of portfolio firms. Given the recent rise in the importance of IPOs in continental European stock markets and the relatively young phenomenon of venture capital in many European countries, the study represents a first comparative empirical assessment of the role of venture capitalists in the going public process and their impact on the long-term performance of IPOs in France, Germany, and the UK. A further contribution is related to the international dimension of the analysis that allows cross-country comparisons and a robustness test of theoretical frameworks in venture capital finance in different financial systems outside the US, and thus outside the environment in which they were uncovered.

The following sections compare the main findings of the analysis (section 6.1) and provide a summarizing discussion including an outlook with respect to future research (section 6.2).

6.1 Comparison of the Main Results

The literature review in *chapter 2* recapitulates the most important findings of previous studies on venture capital. It highlights that many prior contributions concentrate on the role of venture capitalists played in the US, which seems to be related to the advanced development of the US market for venture capital. Accordingly, the synopsis of the chapter points out the general lack of international studies and empirical assessments of European venture capital. In comparison to prior work investigating the performance of venture-backed IPOs, an important contribution of the present analysis is the focus on the relatively young European venture capital market taking into account the heterogeneity of venture capital firms in the considered countries.

In order to understand the histories and structures of venture capital in the countries concerned, *chapter 3* considers the institutional contexts in France, Germany, and the UK, and portrays the different paths of development of their national markets for venture capital. The analysis shows that the different national approaches and roles of public institutions with respect to the promotion of venture capital and the provision of funding for small innovative firms have a number of features in common. Yet, the comparison indicates that funds are channeled through different ways and institutions to their recipients. Thus, each of the three considered countries relies to some extent on the idiosyncrasy of its financial system. While in France the government plays a crucial role for the development of the venture capital market, the first but unsuccessful attempt to promote venture capital in Germany was strongly influenced by banks. In the UK, the early existence of liquid IPO markets favored the development

---

[550] Hege *et al.* (2003), p. 4.

of venture capital. Therefore, the government restricted its role widely to establishing an adequate business environment for venture capital investors. Overall, venture capital developed earlier in the UK than in France and Germany. Conform to the claim of Black and Gilson (1998), the market for venture capital only increased after the creation of stock market segments suited to the exit requisites of venture investors.

Chapter 4 outlines the research design of the empirical analysis. Based on the findings of previous theoretical and empirical contributions, hypotheses were derived with respect to the impact of venture capitalists on the performance of IPOs and expected variations between different types of venture capitalists. Moreover, the chapter describes details on the collection and sources of the employed data set, sample characteristics, and selection issues. In the methodological section, different accounting and market performance measures as well as proxies for the quality of venture-backing are discussed with respect to their suitability to test the individual hypotheses. Finally, the empirical model specifications and test procedures are sketched out.

The main findings of the empirical study are presented in *Chapter 5*. In accord with previous studies, the univariate analysis suggests that both venture- and non venture-backed IPOs in the three considered countries underperform the market over the 3-year post-issue period. Yet, the cross-country comparison reveals that IPOs at the German *Neuer Markt* perform significantly worse than issues in the two other countries. The univariate comparison of venture- and non venture-backed IPOs did not yield systematic differences in the offering and firm characteristics, notwithstanding that venture-backed issues show higher secondary sales than their non venture-backed counterparts. However, a refinement of the analysis suggests that firms backed by international venture capitalists are able to raise more capital in the IPO market and perform better in terms of 3-year wealth relatives than those backed by national venture capitalists. Consistent with this finding, there is evidence that international venture capitalists are more experienced and involved in portfolio firms than national investors. International venture capitalists are on average older than national ones, back more IPOs in the sample, are represented more often on the board, invest with more syndication partners, and hold larger equity positions in portfolio firms. In line with the documented heterogeneity among venture capitalists, the multivariate analysis indicates that venture-backed firms do not generally outperform non venture-backed issues, irrespective of the employed performance measure. Instead, only the subgroup of internationally operating venture capitalists reveals positive effects on the operating and market performance of IPOs.

6.2 Discussion and Outlook

Given the overall results, the empirical work conducted above can now be translated into a coherent picture. Starting with the insights of the analysis comparing the evolution of the venture capital markets in the considered countries, it is shown that the industry developed first in the UK due to the more favorable institutional environment. The existence of liquid IPO markets matched the needs of venture capitalists concerning an exit channel for investments. In France and Germany, however, venture capital rose only recently, stimulated by the launch of the *Nouveau Marché* and the *Neuer Markt*. As the empirical study reveals, the flourishing IPO markets at the end of the 1990s induced many young and experienced venture capitalists to enter the European market. Thus, it seems plausible that the booming environment during the observed period is related to the heterogeneity of European venture capitalists.

In accordance with the heterogeneity of European venture capital, the empirical analysis shows that venture-backed IPOs do not generally surpass non venture-backed issues in terms of operating and market performance, as suggested by previous US studies. Instead, using the internationality of investors as a proxy for business experience, merely firms backed by international venture capitalists outperform the average IPO of the sample. Since the internationality of venture capitalists is related to their age and the number of backed IPOs, the findings suggest that experience is a critical factor for the success of venture-backing. Given that nationally operating venture capitalists are generally younger and back a smaller number of IPOs than international ones, they appear to lack the necessary experience and skills to assess the quality of portfolio firms.

The result that IPOs backed by international venture capitalists are more successful than those backed by national venture capitalists is consistent with theoretical arguments about the multinational enterprise. According to Hymer (1970, 1976), multinational enterprises have to possess a countervailing advantage over locally operating firms in order to make international activities viable. Relating this line of reasoning to the venture capital industry, the reported evidence supports the view that venture capitalists become international if they possess an advantage over nationally operating venture capital firms, i.e. in terms of more investment experience and/or better skills in the assessment of firm profitability, which allows compensating their restricted knowledge of foreign markets.

The main finding that venture-backed issues do not commonly outperform non venture-backed ones has an important implication for the research on venture capital finance. It indicates that the findings of previous studies on the role of venture capitalists in the US and their influence on the operating and long-run market performance of IPO firms cannot be generally transferred to European countries. This outcome is consistent with the results of related empirical studies, documenting considerable differences in the impact of venture capital on portfolio firms between the US and Europe, as discussed in the literature review in chapter 2. The institutional diversity of the countries from which the international data set employed in the present study derives thereby underlines the robustness of the inferences.

The refinement of the analysis, considering different types of venture capital investors and their involvement in the firms going public, represents a further contribution of this dissertation. It reveals that the heterogeneity in the business experience of venture capitalists operating in the European market is an important explanatory factor for the overall results. Using the international composition of the data set, the refined investigation suggests that the cross-country involvement of venture capitalists in IPOs is an adequate proxy for their business experience which affects their ability to spur firm performance. Since the refined examination turned out to be fruitful to understand the origins of the overall observed performance patterns of European venture-backed IPOs, this approach might represent a promising direction for future empirical research in the area of venture capital finance.

Interpreting the results in a more general context, the documented differences in the effects of venture-backing seem to be related to the lacking degree of maturity and critical size of the relatively young European venture capital industry as a whole. The current downturn and consolidation process corroborates that many European venture capitalists are relatively inexperienced and cannot guide their portfolio firms over the difficult market situation. The findings with respect to differences in the quality of venture-backing explain why earlier studies on

European venture capital not distinguishing between different types of venture investors did not show systematic performance variations between venture- and non venture-backed issues.

The analysis reveals a number of issues to further inquiry. First, future research should encounter the discussed limitations of the present dissertation and employ a more comprehensive data set. It should refine the analysis by considering more variables capturing the support of venture capitalists, such as the length of their involvement in firms going public as well as their degrees of specialization on specific investment stages or industries. Second, upcoming contributions should extend the time span of the observed post-issue period and widen the scope of the study, by considering a larger number of countries as well as of IPOs per country to check the robustness of the findings. Finally, future work should include IPOs of the post "Internet bubble" period to provide more insights concerning the "overshooting" argument of Gompers and Lerner (2001b). Such an approach may also help to detect whether or not the ongoing downturn in the market has reduced the heterogeneity of European venture capital. Overall, further empirical evidence is clearly required to shed additional light on the role performed by venture capitalists in Europe.

Appendices

A 1: Stages of Venture Capital Investments

Seed Investments

- Small amounts of capital are provided to an inventor or entrepreneur to determine whether an idea deserves further consideration.

Start-Up

- Companies are less than one year old.
- The company uses the money for product development, prototype testing, and test making.

First Stage – Early Development

- Investment continues through the first stage only if the prototypes look good enough for further technical risk to be considered minimal.
- First-stage companies are unlikely to be profitable.

Second Stage – Expansion

- A company in the second stage has shipped enough products to enough customers to have real feedback from the market.
- The firm is probably still unprofitable.
- The firm probably needs more capital for equipment purchases, inventory and receivables financing.

Third Stage – Profitable but Cash Poor

- Sales growth is probably fast.
- New venture capital may be used for further expansion of manufacturing facilities, expanded marketing, or product enhancements.

Fourth Stage – Rapid Growth toward Liquidity

- A company may still need outside cash to sustain growth.
- The risk to outside investors is much reduced, and the cash-out point and method are undetermined.

Bridge Stage – Mezzanine Investment

- Despite potentially knowing the approximate timing and form of exit of the venture capital from the company, the company still needs capital to continue growth.

Liquidity Stage – Cash-Out or Exit

- Investors can gain liquidity of a substantial portion of their holdings in a company.

Source: Sahlman (1990), p. 479.

A 2: Evolution of Venture Capital in France, Germany, and the UK

	France	*Germany*	*UK*
	Fundamentals		
Legal system	French civil law	German civil law	English common law
Financial system	Formerly bank-based; current trend toward market-based	Bank-based	Market-based
Primary locus of industrial innovation	Research universities, public research institutions, companies	Research universities, companies	Research universities, companies
	History		
First public effort to foster enterprise creation	1955: SDRs (Regional Development Companies)	1965: KBGs (Capital Investment Companies)	1945: ICFC (Industrial and Commercial Finance Corporation, since 1983 known as 3i)
First venture capital organization involving non-profit institutions	1968: ANVAR (national agency for the promotion of research)	1975: WFG (semi-public venture capital firm)	1945: ICFC/3i
First private venture capital firm	1972: Sofinnova	1979: Hannover Finanz	1974: Finance for Industry (later renamed in 3i)
Dates of important legal changes	1972: SFIs (financial companies for innovation) 1975: IRPs (regional equity investment institutes) 1983: FCPRs (similar to US limited partnerships) 1985: SCRs (venture capital companies)	1987: UBGG (law for investment companies): elimination of certain tax disadvantages 1989: BJTU Program: government loans for technology-oriented start-ups 1999: modified bankruptcy law, extended form of receivership	1983: BES (Business Expansion Scheme) 1994: EIS (Enterprise Investment Scheme) 1995: VCT (Venture Capital Trust) 1999: EMI (Enterprise Management Incentives)
Date of creation of first public equity market dedicated to high-growth companies	1983: Second Marché 1996: Nouveau Marché	1987: Geregelter Markt 1997: Neuer Markt	1980: Unlisted Securities Market 1995: Alternative Investment Market (AIM)

Sources: Bank of England (1996, 2001), Clark (1987), Dubocage and Rivaud-Danset (2002), Kuemmerle (2001), La Porta et al. (1998).

A 3: Listing Requirements at the Nouveau Marché, Neuer Markt, and TechMARK

	Nouveau Marché	*Neuer Markt*	*TechMARK*[1]
Size	Equity book value ≥ 1.5 million EUR	Equity book value ≥ 1.5 million EUR	Equity book value ≥ 0.7 million GBP
Age	No minimum age required	≥ 3 years	Financial record of at least 3 years
Accounting Standards	No specific listing requirements	IAS or US GAAP Quarterly reports required Annual Financial Analysts Conference	IAS, US or UK GAAP (Unaudited) half-yearly figures required
Prospectus	Required	Required; must also be published in English	Required
IPO Volume	≥ 5 million EUR ≥ 100,000 shares to be issued	≥ 5 million EUR ≥ 100,000 shares to be issued	No specific listing requirements
Free Float	≥ 20% of capital ≥ 50% of free float must come from capital increase	≥ 20% of capital ≥ 25% recommended not to be closely held ≥ 50% of free float must come from capital increase	≥ 25% of capital
Type of Shares	No restrictions to free negotiability	Only ordinary share (for the first issue) No restrictions to free negotiability	No specific listing requirements
Sponsor	At least one	At least two	At least one
Lock-up Period	To be chosen by the company, ≥ 12 months (insiders must lock at least 80 percent of their shares) Sales by managing shareholders are made public	To be chosen by the company, ≥ 6 months	No specific listing requirements
Takeover Rules	No specific requirements	Must comply with German Takeover Code	Compliance with corporate governance requirements as set out in the Combined Code is voluntary
Foreign Companies	Admissible	Admissible; home rule and regulations apply	Admissible; regulations of the UKLA apply

Source: Deutsche Börse (1997, 2000), Bourse de Paris (1999), London Stock Exchange (2001). Note: [1)] Since techMARK is a segment of the main board, the listing requirements of the UK Listing Authority (UKLA) apply, and firms are admitted according to their business activity.

A 4: Comparison of the Financial Systems in France, Germany, and the UK

	France	*Germany*	*UK*
		Banking System	
Commercial banks	Since 1945, there have been waves of nationalization and privatization. Until 1981, 3 banks were nationalized: Société Générale, Crédit Lyonnais, and Banque Nationale de Paris (BNP). After the election of a socialist government in 1981, many banks were nationalized, but five years later this policy was reversed and many banks were privatized.	Includes the 3 major banks: Deutsche Bank, Dresdner Bank, and Commerzbank. These are universal banks that engage in a full range of activities, including taking deposits, granting loans and mortgages, providing life insurance, underwriting security issues, and investing directly in securities, including equities.	Traditionally dominated by the four clearing banks: Barclays, National Westminster, Midland, and Lloyds.
Saving banks	These have the important privileges that interest is tax free up to a certain level of savings. Loans can be granted only to individuals or bodies whose principal activities are neither industrial nor commercial.	These are not profit-maximizing entities but are operated in the public interest. There are 3 tiers: local savings banks, state savings banks, and the central savings banks.	
Mutual and cooperative banks	There are various kinds of mutual banks with different specific purposes. For example, the aim of Crédit Agricole Mutuel is to provide loans to farmers at reasonable rates. There is typically a pyramid structure with local, regional, and national levels.	The depositors are the shareholders. As with savings banks, there is a three-tier structure with local, state, and central levels.	
		Financial Markets	
Stock markets	The most important exchange is Paris Bourse, but there are 6 other regional exchanges. The Paris market grew significantly in the 1980s and 1990s and is now one of the major markets in Europe.	7 regional exchanges, with Frankfurt being the most important. Relatively few companies are listed. These markets have traditionally been an insignificant source of funds for firms.	Important for domestic and overseas companies. Large number of listed companies. Source of primary funds through IPOs.
Debt markets	There is a significant market for government debt as well as for banks. There is also an important commercial paper market.	Important source of funds for all levels of government as well as for banks. Not important for non-financial firms.	Important for government debt and bank debt. The domestic bond market has been replaced by the less regulated Eurobond market based in London.

Sources: Allen and Gale (2000), tables 3.9, 3.11, and 3.12 on the pages 60, 69, and 72, respectively.

A 5: Sample Selection

The IPOs sample employed in the empirical analysis is limited to offerings of non-financial firms at the *Nouveau Marché*, *Neuer Markt*, and the *techMARK* completed between 1996 and 1999. Lists of the firms going public at the stock exchanges were obtained from Deutsche Börse, Euronext/Bourse de Paris, and the London Stock Exchange (LSE). The lists were checked in order to exclude relistings, transfers from other stock markets or market tiers, as well as mergers and demergers of previously listed firms. Moreover, listings not accompanied by an equity offer remain unconsidered.

The final sample only contains IPOs of non-financial companies for which consistently reported data are available. 21 firms changed their level of consolidation during the period of observation or presented financial statements on periods with varying time-spans, and were therefore excluded from the sample. 19 observations remain unconsidered because of liquidations, mergers, and acquisitions by other companies, as well as missing data. The final sample only includes firms with available data for the whole period of interest, i.e. the end of the fiscal year prior to the IPO as well as the two fiscal years succeeding the offering. The following panels A to C report details of the applied selection procedures which led to the final composition of each country's IPO sample. The panels D to F outline the individual IPO firms which constitute the final sample of each country.

A 5: Sample Selection (Continued)

Panel A: Selection Procedure of the French IPO Sample (Nouveau Marché)		
104	**Total number of IPOs**	
3	IPOs of financial institutions	Astra, Bourse Direct, GPE Envergure/Joliez-Regol
9	IPO firms with missing data	Alphamedia, Appligene Oncor, ATN, Cross System Company, IT Link, Kalisto Entertainment, Regina Rubens, SERP, Union Technology
2	IPO firms having merged or been acquired by another firm	Fabmaster, Omnicom
90	**Total number of IPOs considered in the sample**	
36	non venture-backed IPOs	
54	venture-backed IPOs	

Panel B: Selection Procedure of the German IPO Sample (Neuer Markt)		
181	**Total number of IPOs**	
4	IPOs of financial institutions	Consors, Foris, Entrium Direct Bankers, DAB
4	IPO firms with missing data	Gigabell, Infomatec, Kabel New Media, Refugium
21	IPO firms without comparable data over time	Aixtron, BinTec Communication, Brokat, Computec Media, Edel Music, Euromicron, ID Media, Heyde, Jack White Productions, Jetter, Kinowelt, mb Software, Micrologica, MSH International Service, Mobilcom, musicmusicmusic, Pixelpark, PSB, Sinner-Schrader, Systematics, Trintech Group
1	IPO firm having merged or been acquired by another firm	SVC Schmidt
151	**Total number of IPOs considered in the sample**	
86	non venture-backed IPOs	
65	venture-backed IPOs	

Panel C: Selection Procedure of the British IPO Sample (techMARK)		
64	**Total number of IPOs**	
0	IPOs of financial institutions	–
2	IPO firms whose prospectus was not available	Alphyra Group (formerly ITC Group), Emblaze Systems (formerly GEO Interactive Media Group)
62	**Total number of IPOs considered in the sample**	
27	non venture-backed IPOs	
35	venture-backed IPOs	

A 5: Sample Selection (Continued)

Firm Name	IPO Date	Market
Panel D: Final Composition of the French IPO Sample (Nouveau Marché)		
Infosources / ex Infonie	20-Mar-96	Nouveau Marché
High Co	17-Apr-96	Nouveau Marché
Jean Claude Aubry / ex Proxidis	02-May-96	Nouveau Marché
LaCie /ex Electronique D2	29-May-96	Nouveau Marché
Picogiga	05-Jun-96	Nouveau Marché
Genset	06-Jun-96	Nouveau Marché
Coil	26-Jun-96	Nouveau Marché
Mille Amis	03-Jul-96	Nouveau Marché
R2I Santé	23-Oct-96	Nouveau Marché
Naturex	30-Oct-96	Nouveau Marché
BVRP Software	05-Dec-96	Nouveau Marché
Thermatec Ingénérie	06-Dec-96	Nouveau Marché
Olitec	17-Dec-96	Nouveau Marché
Belvédère	21-Jan-97	Nouveau Marché
IGE+XAO	06-Mar-97	Nouveau Marché
La Tête Dans Les Nues	10-Mar-97	Nouveau Marché
Lexibook Linguistic Electronic System	25-Mar-97	Nouveau Marché
Duran Duboi	22-May-97	Nouveau Marché
IDP - Interactif Delta Production	23-May-97	Nouveau Marché
Esker	02-Jul-97	Nouveau Marché
Aldeta	10-Jul-97	Nouveau Marché
Quantel	30-Sep-97	Nouveau Marché
HF Company	15-Oct-97	Nouveau Marché
Saveurs de France	17-Oct-97	Nouveau Marché
Eurofins Scientific	24-Oct-97	Nouveau Marché
AB Soft	03-Dec-97	Nouveau Marché
Hologram Industries	19-Jan-98	Nouveau Marché
CEREP	18-Feb-98	Nouveau Marché
Transgène	26-Mar-98	Nouveau Marché
Alpha Mos	02-Apr-98	Nouveau Marché
Barbara Bui	03-Apr-98	Nouveau Marché
Mondial Pêche	28-Apr-98	Nouveau Marché
Phone Systems & Network	30-Apr-98	Nouveau Marché
Durand Allize	07-May-98	Nouveau Marché
Adlpartner	12-May-98	Nouveau Marché
Recif	14-May-98	Nouveau Marché
DMS - Dinostic Medical Systems	19-May-98	Nouveau Marché
Synelec	02-Jun-98	Nouveau Marché
Desk	09-Jun-98	Nouveau Marché
Cyrano / ex-Imes Multi Medias	10-Jun-98	Nouveau Marché
Réponse	11-Jun-98	Nouveau Marché
Radoux International / Diosos	12-Jun-98	Nouveau Marché
Silicomp	16-Jun-98	Nouveau Marché
Biodome / Baxter International, Inc.	22-Jun-98	Nouveau Marché
Floreane Medical Implants	30-Jun-98	Nouveau Marché
Prologue Software	02-Jul-98	Nouveau Marché
Western Telecom	03-Jul-98	Nouveau Marché
CAC Systèmes	09-Jul-98	Nouveau Marché

A 5: Sample Selection (Continued)

Panel D: Final Composition of the French IPO Sample (Nouveau Marché)		
Firm Name	**IPO Date**	**Market**
Imecom Group	09-Jul-98	Nouveau Marché
Generix	15-Jul-98	Nouveau Marché
Medidep	17-Jul-98	Nouveau Marché
Genesys	01-Oct-98	Nouveau Marché
FI System	07-Oct-98	Nouveau Marché
Perfect Technologies	16-Oct-98	Nouveau Marché
Cyber Press Publishing	23-Oct-98	Nouveau Marché
Prosodie / ex ABFI	17-Nov-98	Nouveau Marché
Staci	19-Nov-98	Nouveau Marché
Avenir Telecom	20-Nov-98	Nouveau Marché
V CON Telecommunications	25-Nov-98	Nouveau Marché
Guillemot Corporation	27-Nov-98	Nouveau Marché
Cryo Interactive Entertainment	08-Dec-98	Nouveau Marché
European Cargo Services	10-Dec-98	Nouveau Marché
Europstat / Umanis	11-Dec-98	Nouveau Marché
Groupe D! / D Interactive	20-Jan-99	Nouveau Marché
Infotel	21-Jan-99	Nouveau Marché
TR Services	22-Jan-99	Nouveau Marché
Effik	01-Feb-99	Nouveau Marché
Soitec	09-Feb-99	Nouveau Marché
GL Trade	16-Feb-99	Nouveau Marché
Visiodent	19-Feb-99	Nouveau Marché
Algoriel	23-Feb-99	Nouveau Marché
Rigiflex International / Prismaflex	11-Mar-99	Nouveau Marché
Intercall	26-Mar-99	Nouveau Marché
A Novo	12-Apr-99	Nouveau Marché
Valtech	12-Apr-99	Nouveau Marché
CMT Medical Technologies	17-May-99	Nouveau Marché
Cast	26-May-99	Nouveau Marché
Integra	02-Jun-99	Nouveau Marché
Egide	11-Jun-99	Nouveau Marché
Wavecom	14-Jun-99	Nouveau Marché
Coheris Atix	30-Jun-99	Nouveau Marché
Ipsos	01-Jul-99	Nouveau Marché
Métrologic Group	30-Sep-99	Nouveau Marché
Consodata	07-Oct-99	Nouveau Marché
Brime Technologies	22-Oct-99	Nouveau Marché
Devoteam	27-Oct-99	Nouveau Marché
Nicox	02-Nov-99	Nouveau Marché
Alti	12-Nov-99	Nouveau Marché
Automa-Tech	12-Nov-99	Nouveau Marché
Access Commerce	16-Nov-99	Nouveau Marché

A 5: Sample Selection (Continued)

Panel E: Final Composition of the German IPO Sample (Neuer Markt)		
Firm Name	**IPO Date**	**Market**
Beta Systems Software	30-Jun-97	Neuer Markt
Saltus Technology	14-Jul-97	Neuer Markt
SER Systeme	14-Jul-97	Neuer Markt
Mensch und Maschine Software	21-Jul-97	Neuer Markt
Sachsenring Automobiltechnik	02-Oct-97	Neuer Markt
SCM Microsystems Inc.	09-Oct-97	Neuer Markt
EM.TV & Merchandising	30-Oct-97	Neuer Markt
Singulus Technologies	25-Nov-97	Neuer Markt
technotrans	10-Mar-98	Neuer Markt
United Internet / ex 1&1	23-Mar-98	Neuer Markt
Hunzinger	30-Mar-98	Neuer Markt
transtec	03-Apr-98	Neuer Markt
Drillisch	22-Apr-98	Neuer Markt
Ceyoniq / CE Computer Equipment	27-Apr-98	Neuer Markt
W.E.T. Automotive Systems	28-Apr-98	Neuer Markt
Augusta Technology	05-May-98	Neuer Markt
Cenit	06-May-98	Neuer Markt
Graphisoft N.V.	08-Jun-98	Neuer Markt
ELSA	15-Jun-98	Neuer Markt
ce Consumer Electronic	23-Jun-98	Neuer Markt
TDS Informationstechnologie	26-Jun-98	Neuer Markt
Teles	30-Jun-98	Neuer Markt
TelDaFax	01-Jul-98	Neuer Markt
Mühlbauer Holding	10-Jul-98	Neuer Markt
Intershop Communications	16-Jul-98	Neuer Markt
Höft & Wessel	20-Jul-98	Neuer Markt
SoftM Software und Beratung	21-Jul-98	Neuer Markt
plenum	03-Aug-98	Neuer Markt
PSI	31-Aug-98	Neuer Markt
Lintec Computer	07-Sep-98	Neuer Markt
Deutsche Entertainment	14-Sep-98	Neuer Markt
iXOS Software	07-Oct-98	Neuer Markt
Articon Information Systems	28-Oct-98	Neuer Markt
CeoTronics	09-Nov-98	Neuer Markt
DataDesign	09-Nov-98	Neuer Markt
Teleplan International N.V.	23-Nov-98	Neuer Markt
AC Service	30-Nov-98	Neuer Markt
LPKF Laser & Electronics	30-Nov-98	Neuer Markt
Centrotec	08-Dec-98	Neuer Markt
Plambeck Neue Energien	15-Dec-98	Neuer Markt
hancke & peter IT Service / Arxes Information Design	25-Jan-99	Neuer Markt
Senator Film	29-Jan-99	Neuer Markt
CineMedia Film	03-Feb-99	Neuer Markt
Intertainment	08-Feb-99	Neuer Markt
Utimaco Safeware	16-Feb-99	Neuer Markt
PrimaCom	22-Feb-99	Neuer Markt
Medion	26-Feb-99	Neuer Markt
i:FAO	01-Mar-99	Neuer Markt

A 5: Sample Selection (Continued)

Panel E: Final Composition of the German IPO Sample (Neuer Markt)		
Firm Name	**IPO Date**	**Market**
MorphoSys	09-Mar-99	Neuer Markt
Brain International	10-Mar-99	Neuer Markt
Endemann!! Internet / Abacho	10-Mar-99	Neuer Markt
Nemetschek	10-Mar-99	Neuer Markt
FortuneCity.Com Inc.	19-Mar-99	Neuer Markt
Basler	23-Mar-99	Neuer Markt
JUMPtec Industrielle Computertechnik	26-Mar-99	Neuer Markt
ADVA Services & Solutions	29-Mar-99	Neuer Markt
Wizcom Technologies Inc.	29-Mar-99	Neuer Markt
Odeon Film	12-Apr-99	Neuer Markt
CPU Software	19-Apr-99	Neuer Markt
Easy Software	19-Apr-99	Neuer Markt
NSE Software	20-Apr-99	Neuer Markt
Rhein Biotech N.V.	21-Apr-99	Neuer Markt
telegate	22-Apr-99	Neuer Markt
realTech	26-Apr-99	Neuer Markt
Prout	27-Apr-99	Neuer Markt
D.Logistics	28-Apr-99	Neuer Markt
MWG Biotech	07-May-99	Neuer Markt
aap Implantate	10-May-99	Neuer Markt
Tria Software	10-May-99	Neuer Markt
Highlight Communications	11-May-99	Neuer Markt
infor business solutions	11-May-99	Neuer Markt
IDS Scheer	12-May-99	Neuer Markt
Sanochemia Pharmazeutika	12-May-99	Neuer Markt
STE HamaTech	12-May-99	Neuer Markt
artnet.com	17-May-99	Neuer Markt
Süss MicroTec	18-May-99	Neuer Markt
SHS Informationssysteme	19-May-99	Neuer Markt
Eckert & Ziegler	25-May-99	Neuer Markt
Netlife	01-Jun-99	Neuer Markt
Softmatic	01-Jun-99	Neuer Markt
Kleindienst Datentechnik	02-Jun-99	Neuer Markt
SZ Testsysteme	02-Jun-99	Neuer Markt
Prodacta	07-Jun-99	Neuer Markt
RTV Family Entertainment	08-Jun-99	Neuer Markt
Maxdata	09-Jun-99	Neuer Markt
Brain Force Software	10-Jun-99	Neuer Markt
b.i.s. börsen-informations-systeme	14-Jun-99	Neuer Markt
BEKO Holding	14-Jun-99	Neuer Markt
Euromed	16-Jun-99	Neuer Markt
Parsytec	16-Jun-99	Neuer Markt
Vectron Systems	16-Jun-99	Neuer Markt
TePla	21-Jun-99	Neuer Markt
Management Data Software Engineering	22-Jun-99	Neuer Markt
GFT Technologies	28-Jun-99	Neuer Markt
ACG	01-Jul-99	Neuer Markt
Mosaic Software	01-Jul-99	Neuer Markt

A 5: Sample Selection (Continued)

Panel E: Final Composition of the German IPO Sample (Neuer Markt)		
Firm Name	**IPO Date**	**Market**
mediantis / ex bücher.de	05-Jul-99	Neuer Markt
Computer Links	07-Jul-99	Neuer Markt
Met@box	07-Jul-99	Neuer Markt
P&I Personal & Informatik	07-Jul-99	Neuer Markt
TTL Information Technology	12-Jul-99	Neuer Markt
Haitec	14-Jul-99	Neuer Markt
teamwork information management	14-Jul-99	Neuer Markt
Silicon Sensor International	15-Jul-99	Neuer Markt
WWL Internet	15-Jul-99	Neuer Markt
AT&S Austria Technologie & Systemtechnik	16-Jul-99	Neuer Markt
EMPRISE Management Consulting	16-Jul-99	Neuer Markt
ricardo.de	21-Jul-99	Neuer Markt
MANIA Technologie	26-Jul-99	Neuer Markt
Adcon Telemetry	28-Jul-99	Neuer Markt
UB / ex InternetMediaHouse.com	30-Jul-99	Neuer Markt
Advanced Medien	06-Aug-99	Neuer Markt
Das Werk	25-Aug-99	Neuer Markt
PC-Spezialist Franchise	25-Aug-99	Neuer Markt
OTI On Track Innovations Ltd.	31-Aug-99	Neuer Markt
Constantin Film	13-Sep-99	Neuer Markt
WaveLight Laser Technologie	15-Sep-99	Neuer Markt
Cancom IT Systems	16-Sep-99	Neuer Markt
splendid medien	24-Sep-99	Neuer Markt
Sunburst Merchandising	24-Sep-99	Neuer Markt
GEDYS Internet Products	27-Sep-99	Neuer Markt
Fantastic Corp.	28-Sep-99	Neuer Markt
fluxx.com	28-Sep-99	Neuer Markt
Fabasoft	01-Oct-99	Neuer Markt
NorCom Information Technology	01-Oct-99	Neuer Markt
Helkon Media	07-Oct-99	Neuer Markt
ELMOS Semiconductor	11-Oct-99	Neuer Markt
Dino Entertainment	12-Oct-99	Neuer Markt
tiscon Infosystems	14-Oct-99	Neuer Markt
LIPRO Holding	15-Oct-99	Neuer Markt
Gauss Interprise	28-Oct-99	Neuer Markt
digital advertising	29-Oct-99	Neuer Markt
Pandatel	02-Nov-99	Neuer Markt
buch.de internetstores	08-Nov-99	Neuer Markt
RT-SET Ltd. / VI[Z]RT Ltd.	08-Nov-99	Neuer Markt
PLAUT	09-Nov-99	Neuer Markt
Secunet Security Networks	09-Nov-99	Neuer Markt
EVOTEC BioSystems	10-Nov-99	Neuer Markt
Novasoft	15-Nov-99	Neuer Markt
Orad Hi-Tec Systems Ltd.	16-Nov-99	Neuer Markt
POET Holdings, Inc.	16-Nov-99	Neuer Markt
ADS Argyrakis Dein System	17-Nov-99	Neuer Markt
Phenomedia	22-Nov-99	Neuer Markt
Balda	23-Nov-99	Neuer Markt

A 5: Sample Selection (Continued)

Panel E: Final Composition of the German IPO Sample (Neuer Markt)

Firm Name	IPO Date	Market
Brainpool TV	23-Nov-99	Neuer Markt
3U Telekommunikation	25-Nov-99	Neuer Markt
CyBio	25-Nov-99	Neuer Markt
ComROAD	26-Nov-99	Neuer Markt
Tomorrow Internet	30-Nov-99	Neuer Markt
Bäurer	02-Dec-99	Neuer Markt
freenet.de	03-Dec-99	Neuer Markt

Panel F: Final Composition of the British IPO Sample (techMARK)

Firm Name	IPO Date	Market
Shire Pharmaceuticals Group	15-Feb-96	techMARK
Triad Group	21-Mar-96	techMARK
Easynet Group	22-Mar-96	techMARK
Dicom Group	03-Apr-96	techMARK
Xansa / FI Group	10-Apr-96	techMARK
Phytopharm	25-Apr-96	techMARK
Vernalis / Vanguard Media Group	09-May-96	techMARK
PPL Therapeutics	11-Jun-96	techMARK
Fibernet Group	18-Jun-96	techMARK
Staffware	20-Jun-96	techMARK
Vocalis Group	02-Jul-96	techMARK
BATM Advanced Communications	15-Jul-96	techMARK
Alizyme	18-Jul-96	techMARK
Plasmon	26-Jul-96	techMARK
Ultra Electronics Holdings	03-Oct-96	techMARK
Advanced Power Components	28-Nov-96	techMARK
Aveva Group / Cadcentre Group	05-Dec-96	techMARK
Colt Telecom Group	17-Dec-96	techMARK
IMS Interactive Media Services Group	06-Feb-97	techMARK
Izodia / Infobank	18-Feb-97	techMARK
Cenes Pharmaceuticals / Core	03-Mar-97	techMARK
KBC Advanced Technologies	20-Mar-97	techMARK
London Bridge Software Holdings	20-Mar-97	techMARK
Diagonal	24-Mar-97	techMARK
CAT - Cambridge Antibody Technology Group	25-Mar-97	techMARK
Comino Group	17-Apr-97	techMARK
Gladstone / Versalite Group	14-May-97	techMARK
Royalblue Group	09-Jun-97	techMARK
PowderJect Pharmaceuticals	17-Jun-97	techMARK
AIT Group	27-Jun-97	techMARK
Galen Holdings	10-Jul-97	techMARK
I S Solutions	09-Sep-97	techMARK
NSB Retails Systems	07-Oct-97	techMARK
Xaar	17-Oct-97	techMARK
Gyrus Group	13-Nov-97	techMARK

A 5: Sample Selection (Continued)

Panel F: Final Composition of the British IPO Sample (techMARK)		
Firm Name	**IPO Date**	**Market**
Energis	16-Dec-97	techMARK
Guardian IT	16-Mar-98	techMARK
Oxford GlycoSciences	16-Apr-98	techMARK
ARM Holdings	24-Apr-98	techMARK
Convergent Communications	05-May-98	techMARK
ICM Computer Group	21-May-98	techMARK
Computacenter	29-May-98	techMARK
ITNET	12-Jun-98	techMARK
Surfcontrol /JSB Software Technologies	15-Jun-98	techMARK
Ecsoft Group	01-Jul-98	techMARK
Manpower Software	05-Nov-98	techMARK
Financial Objects	11-Dec-98	techMARK
Bioglan Pharma	17-Dec-98	techMARK
Synstar	05-Mar-99	techMARK
Axon Group	11-Mar-99	techMARK
Morse Holdings plc	30-Mar-99	techMARK
Affinity Internet Holdings	20-Apr-99	techMARK
Netbenefit	04-Jun-99	techMARK
Terence Chapman Group	07-Jul-99	techMARK
Kingston Communications	19-Jul-99	techMARK
Exchange FS Group	13-Aug-99	techMARK
QXL Ricardo / QXL.com	14-Oct-99	techMARK
Easyscreen	21-Oct-99	techMARK
Redstone	25-Oct-99	techMARK
Eckoh Technologies/ 365 Corporation	02-Dec-99	techMARK
Horizon Technology Group	06-Dec-99	techMARK
SDL Software	07-Dec-99	techMARK

A 6: Data Preparation

This appendix explains how the individual variables used throughout the analysis are defined and extracted from the various sources.

Firm Specific Information

The firm age in years at the time of the IPO is employed as an independent variable in the empirical analysis. The dates of a firm's creation and incorporation are typically provided in the firm history section of the IPO prospectus. Since some firms only incorporated when they decided to go public, the age is preferably measured as the number of years since the creation of the company. In addition to the age, the number of employees by the end of the fiscal year prior to the IPO as well as the two fiscal years succeeding the IPO was collected. For most of the firms, the number of employees is documented in the listing prospectus, the annual reports, or on the corporate website.

Accounting Items

Accounting items, such as data from balance sheets, profit and loss accounts, and cash flow statements, are obtained from the listing prospectuses and annual reports of firms. In general, the analysis considers the financial statements featuring the broadest level of consolidation. If company accounts are stated both in domestic and international accounting standards, i.e. US GAAP or IAS, the figures based on international standards are recorded. To compute the operating performance ratios, as outlined in the methodology part of the text, the following items are extracted from the financial records: sales, operating income, net income, depreciation and amortization, provisions, shareholder's equity, and total assets. All items quoted in foreign currencies are converted into Euro by using the exchange rates at the end of the respective fiscal year. In order to compute cash flow items which are related to the firms' operations, the operating income is adjusted for non cash-related items, such as depreciation and provisions. Based on the operating income, the operating cash flow is calculated as the operating income plus depreciation, amortization, and provisions. For robustness checks, different cash flow definitions are constructed end employed in the empirical analysis, such as the cash flow from operating activities, as stated in the company's cash flow statement, and the operating income plus depreciation and amortization.

Underwriter Reputation

To account for the reputation of underwriters, the number of IPOs, in which they acted as lead underwriters, is considered as well as their relative market share of proceeds in the IPO markets of France, Germany, and the UK between 1995 and 1999. Based on the market share, a ranking on a scale of 1 to 3 is applied. Investment banks with a lead management market share of higher than 5 percent are assigned the most prestigious rank of 1, those with a market share between 1 and 5 percent the rank 2, and those with a market share of lower than 1 percent the least prestigious rank 3. To account for the reputation of international high quality underwriters, which are less active in the European IPO markets, investment banks classified as prestigious underwriters by Loughran and Ritter (2003) are assigned the value 1, irrespective of their market share in the European IPO markets. An overview of the ranks and participation of the top underwriters by country is provided in Appendix A 9.

A 6: Data Preparation (Continued)

Share Prices, Benchmark Indices, and Market Capitalization Data

Share prices of the firms going public and different benchmark indices are obtained from *Thomson Financial Datastream* in form of performance indices, which are calculated on the basis of monthly closing prices in relation to the first day of trading closing price. The performance indices are adjusted for dividend payments, stock splits, share issues, and changes in nominal values. Monthly returns in the aftermarket are calculated as follows:

$$r_{i,t} = \frac{P_{i,t} + D_{i,t} - P_{i,t-1}}{P_{i,t-1}},$$

where $P_{i,t}$ is the stock price for firm *i* at time *t*, and $D_{i,t}$ is the dividend at time *t*.

Market capitalizations of individual firms at the end of the first day of trading are taken from *Thomson Financial Datastream*. Firm values in foreign currencies are converted into Euro by using the exchange rates at the first day of trading.

Venture Capital Involvement

British venture-backed IPOs are initially identified by means of a list of "UK Venture Backed Flotations" provided by the British Venture Capital Association (BVCA). The classification of venture-backed IPOs at the *Neuer Markt* and *Nouveau Marché* follows a different approach since neither the French Venture Capital Association (AFIC) nor its German equivalent (BVK) keep an official record of national venture-backed IPOs. The definition of venture-backed IPOs is based on a search for names of venture capitalists as shareholders in the IPO prospectuses. As in related studies, venture capitalists are considered, which have a full member status of AFIC, BVCA, BVK, or EVCA.[551] In addition, members of other countries' associations, such as the Belgian, Dutch, Israeli, and US national venture capital associations are classified as venture capital investors.

Based on the information provided by the venture capital associations and venture capital firms' websites, venture capitalists are classified as independent, captive, bank-affiliated, and public organizations. In accordance with the definition of the EVCA, funds are classified as independent if less than 20 percent of their financing sources derive from one source. Venture capital funds are categorized as public, if more than 50 percent of their financing sources originate from governmental organizations. Board representation and details of the equity ownership of venture capitalists in the pre- and post-IPO firms were identified by means of the listing prospectuses. However, in several cases, there is not enough information disclosed to gauge the proportion of venture capitalists' pre-IPO shareholdings. Sometimes, only the aggregate figure of shares sold by the original shareholders is provided.

---

[551] See Espenlaub *et al.* (1999), Engel (2002).

A 6: Data Preparation (Continued)

Industry Classifications

As in previous studies on IPOs at European stock markets, six industry categories derived from the *Financial Times* classification are employed in the analysis.[552] The categories are based on the following aggregations:

- *Biomedical* (incl. chemicals, health, pharmaceuticals, and personal care),
- *IT, software,* and *Internet* (incl. automotive services and retailers),
- *Media and entertainment* (incl. leisure and media),
- *Technology* (incl. construction, engineering, and electronics),
- *Telecom,*
- *Traditional* (incl. food, forestry, steel, and distribution).

Tobin's *Q*

An approximation of Tobin's Q that is sometimes referred to "simple Q" is calculated in the following way:

$$Q = \frac{BV(Assets) - BV(Equity) + MV(Equity)}{BV(Assets)},$$

where BV and MV denote the book and the market value, respectively. In the present analysis, an approximated value of the "simple Q" at the first day of trading is calculated, as follows:

$$Q_{IPO} = \frac{BV(Assets_{old}) - BV(Equity_{old}) + MV(Equity_{old}) + MV(Equity_{new})}{BV(Assets_{old}) + BV(Assets_{new})},$$

where the sum of $MV(Equity_{old})$ and $MV(Equity_{new})$ equals an IPO's market capitalization at the end of the first day of trading, and $BV(Assets_{new})$ is identical with $BV(Equity_{new})$, i.e. the number of new shares offered multiplied with the offer price per share in the IPO.

[552] A similar industry attribution is used by Bottazzi and Da Rin (2002a, 2002b).

A 7: Definition of Variables

Panel A: Dependent Variables	
The panel describes the different dependent variables which are employed throughout the empirical analysis.	
Operating / Accounting Performance Variables	
ROA	operating return on assets, defined as the operating income divided by total assets
CFROA	operating cash flow return on assets, defined as the operating income plus depreciation, amortization and provisions divided by total assets
ROS	operating return on sales, defined as the operating income divided by total sales
CFROS	operating cash flow return on sales, defined as the operating income plus depreciation, amortization and provisions divided by sales
Market Performance Variables	
Q	Tobin's *Q*, i.e. the approximated value of simple *Q* at the first day of trading, defined as the market value of equity and book value of debt divided by the book value of total assets, where the market value of equity equals the share price times the number of shares outstanding
BHR	buy and hold return of an IPO firm, calculated by compounding monthly returns for a given period after the first day of trading (adjusted for dividends, stock splits, and share issues)
BHER	buy-and-hold excess return for a given period, defined as the difference between the buy and hold return of an IPO firm and the buy and hold return of a benchmark portfolio for the same period
WR	wealth relative, defined as the ratio of one plus the IPO portfolio return over one plus the return on the chosen benchmark over the same period

A 7: Definition of Variables (Continued)

Panel B: Independent Variables	
Firm Variables	
Age	number of years since the firm was founded at the time of the IPO
Alpha	fraction of the firm's equity retained by old shareholders immediately after the IPO
Assets	total assets in million EUR
Empl	number of employees of the firm (at the end of the specified fiscal year)
EqR	equity ratio, defined as the book value of equity divided by the book value of total assets
IPO_Size	gross proceeds of the IPO in million EUR, incl. exercised overallotments ("greenshoe")
Mkt_Cap	market capitalization of equity at the end of the first trading day in million EUR
Part	participation ratio of old shareholders in the IPO, defined as number of old shares sold divided by the number of shares outstanding before flotation
S_A	inverse of capital intensity, or sales turnover, defined as sales over assets
Sales	total sales in million EUR
URank	IPO underwriter reputation rank on a scale of 1 to 3, where 1 is the most and 3 the least prestigious underwriter
Industry Dummy Variables	
Biomed	equal to 1 if the IPO firm belongs to the biomedical sector
ITSINT	equal to 1 if the IPO firm belongs to the IT, software and internet sector
Media	equal to 1 if the IPO firm belongs to the media and entertainment sector
Techno	equal to 1 if the IPO firm belongs to the technology sector
Telecom	equal to 1 if the IPO firm belongs to the telecom sector
Accounting Standard Dummy Variables	
IAS	equal to 1 if the IPO firm's statements are reported under IAS
USGAAP	equal to 1 if the IPO firm's statements are reported under US GAAP
Country Dummy Variables	
France	equal to 1 if a firm went public in France
Germany	equal to 1 if a firm went public in Germany
UK	equal to 1 if a firm went public in the UK
VC Variables	
VC	dummy variable equal to 1 if an IPO is venture-backed
Int_VC	dummy variable equal to 1 if an IPO is backed by an international venture capitalist that backed firms in at least two countries of the sample
VC_Age	number of years since the lead venture capitalists' foundation at the time of the IPO
Bank	dummy variable equal to 1 if a venture capitalist is a bank-affiliated organization
Indep	dummy variable equal to 1 if a venture capitalist is independent
Public	dummy variable equal to 1 if a venture capitalist is owned by a public shareholder
Board	dummy variable equal to 1 if a venture capitalist is on the board of an IPO firm
No_IPOs	number of IPOs in the sample backed by a venture capitalist
NVC	number of venture capitalists backing a firm at the time of the IPO
AVC (LVC) PrEq	percentage of equity owned by all venture capitalists (the lead venture capitalist) at the time of the IPO
AVC (LVC) PosEq	percentage of equity owned by all venture capitalists (the lead venture capitalist) immediately after the IPO
AVC (LVC) Sale	percentage of shares sold by all venture capitalists (the lead venture capitalist) at the IPO

A 8: Participation of Venture Capitalists in the IPO Market

Panel A: France (Nouveau Marché)	*Origin*	*1996*	*1997*	*1998*	*1999*	*Total*
BNP Paribas Développement	France	1	0	3	6	10
Galileo Partners	France	0	2	5	1	8
Financière Natexis Banques Populaires	France	0	0	4	2	6
Sofinnova Partners	France/Int.	2	0	2	2	6
3i Group / 3i Gestion	UK/Int.	2	1	1	1	5
Banexi Ventures Partners	France	1	2	1	1	5
CDC Participations	France/Int.	0	1	2	2	5
Innovacom (France Telecom)	France/Int.	1	-	1	3	5
ABN Amro Venture France	NL/Int.	0	0	2	2	4
Apax	UK/Int.	0	0	2	2	4

Panel B: Germany (Neuer Markt)	*Origin*	*1996*	*1997*	*1998*	*1999*	*Total*
3i Group / 3i Germany	UK/Int.	0	1	3	4	8
Goldzack	Germany	0	1	5	2	8
Apax	UK/Int.	0	0	3	2	5
Atlas Venture Germany	UK/Int.	0	0	0	4	4
Commerzbank Unternehmensbeteiligung	Germany	0	3	1	0	4
TVM Techno Venture Management	Germany	0	1	0	3	4
Deutsche Effekten- und Wechselbeteiligungsgesellschaft (DEWG)	Germany	0	0	0	3	3
Innovacom (France Telecom)	France/Int.	0	0	1	2	3
Schroders	UK/Int.	0	1	0	2	3
TFG Venture Capital	Germany	0	0	0	3	3

Panel C: UK (techMARK)	*Origin*	*1996*	*1997*	*1998*	*1999*	*Total*
3i Group	UK/Int.	5	7	2	1	15
Schroders	UK/Int.	2	1	1	1	5
Apax	UK/Int.	1	0	1	2	4
Mercury	UK	3	1	0	0	4
NatWest	UK	1	1	1	0	3
Prelude	UK	1	2	0	0	3
Rothschild	France/Int.	3	0	0	0	3
Advent	UK/Int.	1	1	0	0	2
Alta Berkeley	US/Int.	0	2	0	0	2
Baronsmead	UK	1	0	0	1	2
Charterhouse	UK	1	1	0	0	2
JAFCO	Japan/Int.	1	1	0	0	2
Quester	UK	0	1	1	0	2
Sofinnova	France/Int.	1	0	1	0	2
Thompson Clive & Partners	UK/Int.	0	0	1	1	2

Note: Int. *denotes international venture capitalists being involved in IPOs of at least two countries of the sample.*
Source: own calculations.

A 9: Participation of the Top Underwriters in the IPO Market

Panel A: France (Nouveau Marché)	*Proceeds[1,2]*	*No. of Issues[1]*	*Market Share[1]*	*Ranking*	*CM / LR[3]*
BNP Paribas	10,882.2	26	37.8	1	9.0 / 8.1
Crédit Lyonnais	5,086.8	25	17.9	1	9.0 / 7.1
Société Générale	3,758.2	40	12.7	1	9.0 / 7.1
Credit Suisse First Boston	2,332.7	6	7.9	1	9.0 / 9.1
ABN AMRO[4]	717.8	4	2.4	1	9.0 / 8.1
Crédit Agricole Indosuez	518.0	25	1.8	2	9.0 / 7.1
Lehman Brothers	499.3	3	1.7	1	9.0 / 9.1
Natexis Banques Populaires	373.2	55	1.3	2	– / –
Groupe Oddo Pinatton	162.0	21	0.6	3	– / –

Notes: [1)] between 1995 and 1999; [2)] in million USD; [3)] CM: Carter et al. (1998) ranking; LR: Loughran and Ritter (2003) ranking; [4)] incl. Rothschild and Banque de Neuflize Schlumberger Mallett. Source: IFR Thomson Financial.

Panel B: Germany (Neuer Markt)	*Proceeds[1,2]*	*No. of Issues[1]*	*Market Share[1]*	*Ranking*	*LR / CM[3]*
Dresdner Kleinwort Wasserstein	12,860.0	44	32.3	1	9.0 / 7.1
Deutsche Bank	5,027.3	40	12.6	1	9.0 / 9.1
Goldman Sachs & Co.	3,886.6	13	9.8	1	9.0 / 9.1
UBS Warburg	3,878.3	12	9.8	1	9.0 / 9.1
Commerzbank	1,339.7	18	3.4	2	9.0 / 7.1
HypoVereinsbank[4]	1,256.8	25	3.2	2	– / –
Credit Suisse First Boston	1,227.0	12	3.1	1	9.0 / 9.1
DZ-Bank[5]	1,162.3	32	2.9	2	– / –
ING BHF-Bank	890.8	9	2.2	2	9.0 / 7.1
HSBC Trinkaus	520.1	9	2.2	1	9.0 / 8.1

Notes: [1)] between 1995 and 1999; [2)] in million USD; [3)] CM: Carter et al. (1998) ranking; LR: Loughran and Ritter (2003) ranking; [4)] incl. Bayerische Vereinsbank and Bayerische Hypobank before the merger to Bayerische HypoVereinsbank; [5)] Former DG-Bank. Source: IFR Thomson Financial.

Panel C: UK (techMARK)	*Proceeds[1,2]*	*No. of Issues[1]*	*Market Share[1]*	*Ranking*	*LR / CM[3]*
UBS Warburg	10,874.5	32	17.5	1	9.0 / 8.1
Cazenove & Co.	10,751.9	38	17.3	1	9.0 / 8.1
Merrill Lynch & Co.	9,658.0	27	15.5	1	9.0 / 9.1
Goldman Sachs & Co.	3,961.1	10	6.4	1	9.0 / 9.1
Dresdner Kleinwort Wasserstein	3,818.7	17	6.1	1	9.0 / 7.1
Morgan Stanley	3,238.6	17	5.2	1	9.0 / 9.1
Robert Fleming / JP Morgan	2,698.7	12	4.3	2	9.0 / 7.1
ING[4]	897.3	15	1.4	2	9.0 / 7.1
Deutsche Bank	874.7	12	1.4	1	9.0 / 9.1
Credit Suisse First Boston	780.8	9	1.3	1	9.0 / 9.1

Notes: [1)] between 1995 and 1999; [2)] in million USD; [3)] CM: Carter et al. (1998) ranking; LR: Loughran and Ritter (2003) ranking; [4)] incl. Barings and Williams de Broe. Source: IFR Thomson Financial.

A 10: Comparison of Market Indices and Sample-Based Indices over Time

Panel A: France (Nouveau Marché Index versus Sample-Based Index)

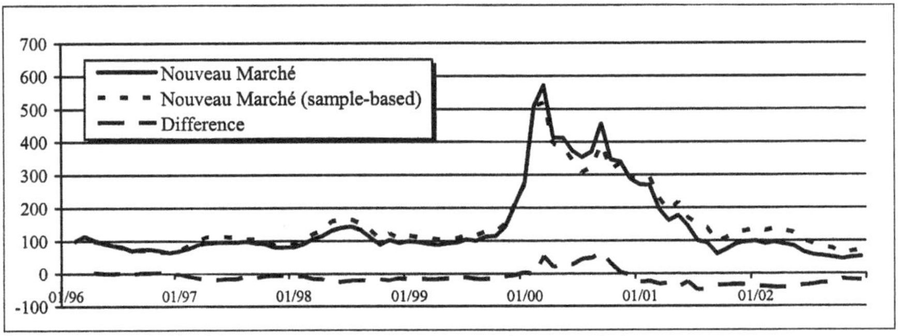

Source: own calculations.

Panel B: Germany (Nemax All Share Index versus Sample-Based Index)

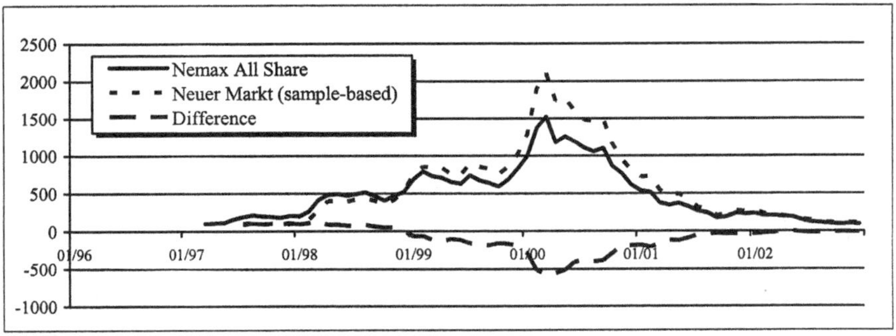

Source: own calculations.

Panel C: UK (techMARK All Share Index versus Sample-Based Index)

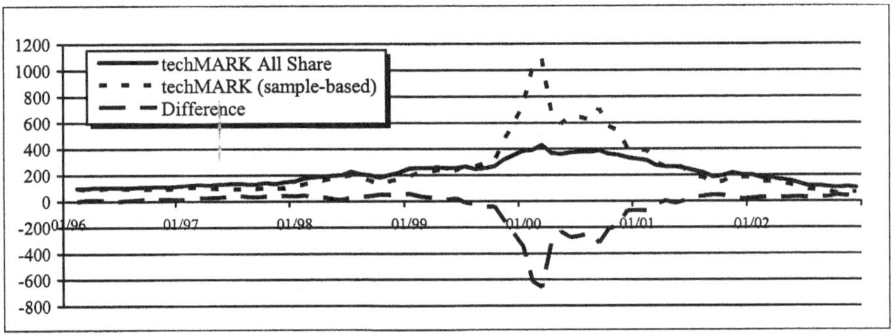

Source: own calculations.

A 11: Offering and Firm Characteristics at the Time of the IPO

The values for firms backed by international and national venture capitalists are expressed as means (medians in parentheses). To detect significant differences between the means [medians] of the two firm groups, Student's *t*-tests [non-parametric Mann-Whitney tests] are conducted. Bold values indicate statistically significant differences at the 1% or 5% confidence levels. The market capitalization of equity (*Mkt_Cap*) is expressed in million EUR and related to the first day of trading. *IPO_Size* represents the gross proceeds of the IPO incl. exercised overallotments. *Part* is the participation ratio (in percent), defined as number of old shares sold divided by the number of shares outstanding before flotation. *Alpha* is the percentage of the firm's equity retained by old shareholders after the IPO. Tobin's *Q* is the approximated value at the end of the first trading day, defined as the sum of the market value of equity and book value of debt, divided by the book value of total assets. Total sales (*Sales*), assets (*Assets*), and net income *(Income)* are expressed in million EUR and related to the end of the fiscal year prior to the IPO. Likewise, the number of employees *(Empl)* is related to the end of the fiscal year prior to the IPO. The equity ratio (*EqR*) is expressed in percent, defined as the book value of equity divided by the book value of total assets at the end of the fiscal year prior to the IPO. The firm age (*Age*) is expressed in years at the time of the IPO. All accounting profitability ratios are expressed in percent and based on accounting figures at the end of the fiscal year prior to the IPO. The operating return (cash flow return) on assets is defined as the operating income (operating cash flow) divided by total assets and the operating return (cash flow return) on sales is defined as the operating income (operating cash flow) divided by sales.

	France			***Germany***			***UK***		
Variable	Int_VC-backed	Nat_VC-backed	*t*-test [*z*-score]	Int_VC-backed	Nat_VC-backed	*t*-test [*z*-score]	Int_VC-backed	Nat_VC-backed	*t*-test [*z*-score]
Mkt_Cap	61.2	44.7	1.0	**324.9**	**177.9**	**3.0^{a}**	278.1	113.5	1.4
	(44.0)	(28.1)	[-1.4]	**(263.1)**	**(133.8)**	**[-2.3^{b}]**	(158.5)	(60.5)	[-1.7^{c}]
IPO_Size	18.9	14.2	0.9	**72.4**	**35.3**	**3.5^{a}**	91.2	36.0	1.4
	(12.0)	**(7.6)**	**[-2.0^{b}]**	**(55.1)**	**(27.9)**	**[-3.7^{a}]**	**(48.1)**	**(20.4)**	**[-3.2^{a}]**
Part	11.3	8.6	1.0	19.7	13.5	1.6	20.9	19.8	0.2
	(10.1)	(7.1)	[-1.5]	(17.8)	(10.1)	[-1.6]	(18.6)	(16.5)	[-1.9^{c}]
Alpha	71.7	74.3	-1.2	59.4	63.4	-0.9	61.1	64.1	-0.5
	(71.5)	(75.0)	[-0.9]	(65.4)	(70.4)	[-1.6^{c}]	(65.1)	(68.8)	[-1.7^{c}]
Q	2.4	2.3	0.5	5.1	4.5	0.9	3.7	4.1	-0.6
	(2.3)	(2.3)	[-0.6]	(3.6)	(4.0)	[-0.2]	(3.5)	(3.4)	[-1.1]
Sales	16.4	21.5	-0.6	28.8	23.3	0.7	107.3	37.6	0.6
	(10.5)	(10.4)	[-0.2]	(17.3)	(8.1)	[-1.4]	(16.6)	(10.6)	[-1.2]
Assets	16.3	16.4	0.0	51.7	18.9	1.6	44.5	27.3	0.5
	(11.0)	(10.5)	[-0.7]	**(16.1)**	**(8.5)**	**[-2.3^{b}]**	**(18.8)**	**(5.6)**	**[-3.2^{a}]**
Income	0.0	-0.1	0.2	-0.8	-0.1	-1.0	2.0	0.4	0.5
	(0.4)	(0.2)	[-0.5]	(-0.6)	(0.2)	[-1.1]	(1.6)	(0.5)	[-0.2]
EqR	**35.7**	**24.5**	**2.2^{b}**	**13.7**	**37.6**	**-2.3^{b}**	13.6	30.9	-0.7
	(36.2)	**(25.4)**	**[-2.2^{b}]**	(23.0)	(36.2)	[-1.8^{c}]	(33.6)	(31.6)	[-0.1]
Empl	128	169	-0.7	194	148	0.9	350	346	0.0
	(85)	(86)	[-0.4]	**(122)**	**(68)**	**[-2.0^{b}]**	(113)	(100)	[-2.0^{c}]
Age	8.1	10.8	-0.6	12.5	13.2	-0.2	10.1	9.0	-0.5
	(7.0)	(8.0)	[-0.4]	(12.0)	(9.0)	[-0.3]	(8.0)	(8.0)	[-1.1]
ROA	5.1	3.5	0.3	-5.5	-3.7	-0.2	-14.4	-19.3	0.2
	(7.5)	(7.1)	[-0.3]	(-6.4)	(1.7)	[-1.0]	(8.4)	(7.2)	[-0.4]
CFROA	11.6	9.1	0.5	5.3	5.3	0.0	-9.0	-11.5	0.1
	(14.7)	(15.0)	[-0.2]	(4.3)	(8.9)	[-0.6]	(11.6)	(15.3)	[-0.2]
ROS	-4.4	-248.0	1.0	-17.4	-24.5	0.4	-4,056.1	-38.2	-0.6
	(7.6)	(5.6)	[-0.7]	(-10.3)	(2.3)	[-0.9]	(5.4)	(3.6)	[-0.3]
CFROS	3.7	-208.8	1.0	-1.7	-12.0	0.6	-3,902.1	-31.2	-0.6
	(13.5)	(9.1)	[-0.5]	(4.7)	(6.5)	[-0.4]	(7.4)	(9.0)	[-0.2]

Note: Significance levels are denoted by $^{a)}$ for 1%, $^{b)}$ for 5%, and $^{c)}$ for 10%. Source: own calculations.

A 12: Tobin's *Q* by Yearly Cohorts

Values for venture- and non venture-backed firms are expressed as means (medians in parentheses). To detect significant differences between the means [medians] of the two firm groups, Student's *t*-tests [non-parametric Mann-Whitney tests] were conducted. Bold values indicate statistically significant differences at the 1% or 5% confidence levels. Tobin's *Q* is the approximated value at the end of the first trading day, defined as the sum of the market value of equity and book value of debt, divided by the book value of total assets.

	France			*Germany*			*UK*		
Year	VC-backed	non VC-backed	*t*-test [*z*-score]	VC-backed	non VC-backed	*t*-test [*z*-score]	VC-backed	non VC-backed	*t*-test [*z*-score]
1996	2.8	2.6	0.5				3.4	3.3	0.3
	(3.0)	(2.3)	[-0.4]				(3.4)	(3.3)	[-0.0]
1997	2.2	2.4	-0.3	2.4	2.8	-0.6	3.2	4.3	-1.7
	(2.3)	(2.3)	[-0.0]	(2.2)	(2.4)	[-1.0]	(2.9)	(4.3)	[-1.5]
1998	2.2	2.3	-0.4	6.4	5.3	1.0	4.3	3.6	0.5
	(2.2)	(2.3)	[-0.3]	(5.9)	(4.3)	[-1.6]	(3.7)	(2.1)	[-1.0]
1999	**2.3**	**3.5**	**-3.2[a]**	4.3	4.9	1.2	4.5	5.7	-0.8
	(2.1)	**(3.1)**	**[-2.6[a]]**	(3.7)	(4.0)	[-1.1]	(4.1)	(5.7)	[-1.3]

Note: Significance levels are denoted by [a)] for 1%, [b)] for 5%, and [c)] for 10%. Source: own calculations.

A 13: International Comparison of Tobin's *Q* by Yearly Cohorts

Values for venture- and non venture-backed firms are Student *t*-tests statistics (non-parametric Mann-Whitney *z*-scores in parentheses) to detect significant differences between the means (medians) of firm groups in different countries. Bold values indicate statistically significant differences at the 1% or 5% confidence levels. Tobin's *Q* is the approximated value at the end of the first trading day, defined as the sum of the market value of equity and book value of debt, divided by the book value of total assets.

Comparison	***France vs. UK***		***Germany vs. UK***		***Germany vs. France***	
Variable	VC-backed	Non VC-backed	VC-backed	Non VC-backed	VC backed	Non VC-backed
1996	-1.0	-1.4				
	(-0.7)	(-1.3)				
1997	**-2.1[b]**	**-2.4[b]**	-1.6	-1.4	0.3	0.5
	(-1.8[c])	**(-2.1[b])**	(-1.5)	(-1.6)	(-0.2)	(-1.0)
1998	**-4.4[a]**	-1.8[c]	**2.2[b]**	0.8	**7.2[a]**	**3.6[a]**
	(-3.8[a])	(-0.2)	**(-2.2[b])**	(-1.1)	**(-4.7[a])**	**(-3.7[a])**
1999	**-3.2[a]**	**-2.3[b]**	-0.3	-0.7	**4.4[a]**	1.3
	(-2.5[b])	**(-2.0[b])**	(-0.1)	(-1.5)	**(-4.7[a])**	(-1.6)

Note: Significance levels are denoted by [a)] for 1%, [b)] for 5%, and [c)] for 10%. Source: own calculations.

A 14: Post-IPO Accounting Profitability Ratios (2-Year Averages)

The values for firms backed by international and national venture capitalists are expressed as means (medians in parentheses). To detect significant differences between the means [medians] of the two firm groups, Student's *t*-tests [non-parametric Mann-Whitney tests] are conducted. Bold values indicate statistically significant differences at the 1% or 5% confidence levels. All accounting profitability ratios are expressed in percent and based on accounting figures at the end of the fiscal year prior to the IPO. The operating return (cash flow return) on assets is defined as the operating income (operating cash flow) divided by total assets and the operating return (cash flow return) on sales is defined as the operating income (operating cash flow) divided by sales.

	France			*Germany*			*UK*		
Variable	Int_VC-backed	Nat_VC-backed	*t*-test [*z*-score]	Int_VC-backed	Nat_VC-backed	*t*-test [*z*-score]	Int_VC-backed	Nat_VC-backed	*t*-test [*z*-score]
ROA	-0.6	-11.1	1.5	-5.9	-20.1	1.1	-9.9	0.3	-0.8
	(2.8)	(1.1)	[-0.7]	(1.4)	(-2.4)	[-0.7]	(7.7)	(6.2)	[-0.4]
CFROA	4.1	-5.0	1.5	2.9	-3.4	1.0	-0.9	6.4	-0.9
	(8.7)	(6.1)	[-0.9]	(6.0)	(4.7)	[-0.3]	(7.7)	(12.7)	[-0.6]
ROS	-50.0	-51.0	0.0	-37.2	-46.9	0.4	-1,726.2	-15.9	-0.7
	(1.8)	(0.8)	[-0.1]	(-1.5)	(-2.7)	[-0.2]	(5.2)	(6.0)	[-0.7]
CFROS	-41.4	-36.0	-0.1	-7.9	-17.4	0.6	-1,605.1	-5.5	-0.7
	(9.5)	(5.5)	[-0.5]	(6.3)	(5.7)	[-0.1]	(7.4)	(10.5)	[-0.8]

Note: Significance levels are denoted by [a)] for 1%, [b)] for 5%, and [c)] for 10%. Source: own calculations.

A 15: Long-Run Performance of IPOs in the Aftermarket

Panel A: All IPOs in France (Nouveau Marché)

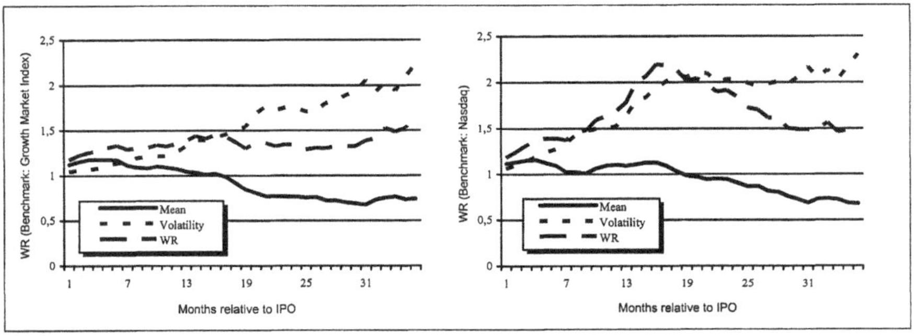

Note: WR - wealth relatives. Source: own calculations.

Panel B: Venture-Backed IPOs in France (Nouveau Marché)

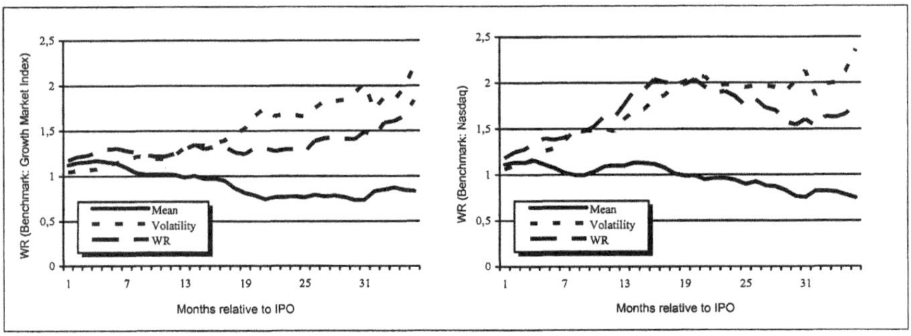

Note: WR - wealth relatives. Source: own calculations.

Panel C: Non Venture-Backed IPOs in France (Nouveau Marché)

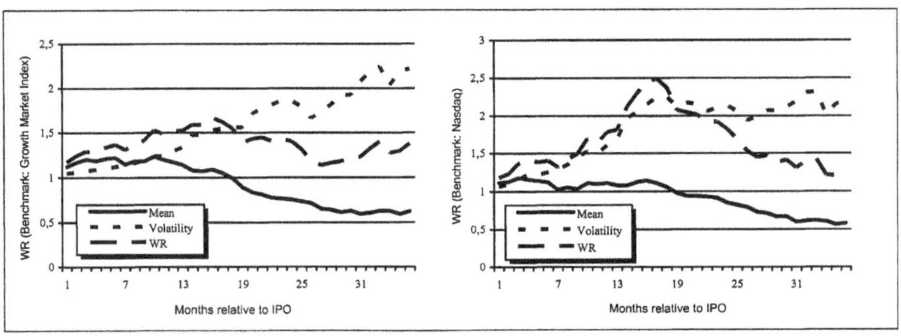

Note: WR - wealth relatives. Source: own calculations.

A 15: Long-Run Performance of IPOs in the Aftermarket (continued)

Panel D: All IPOs in Germany (Neuer Markt)

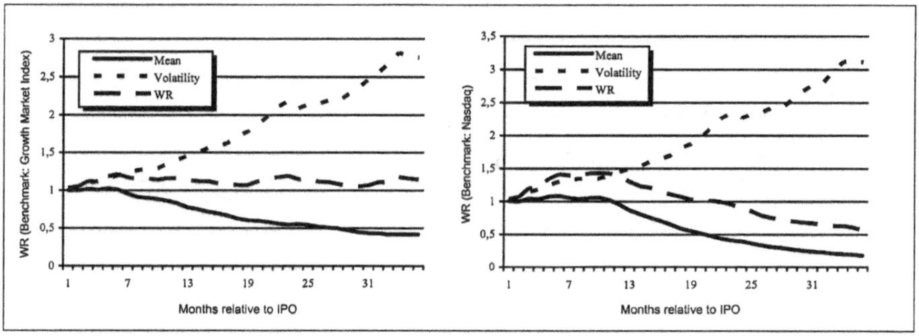

Note: WR - wealth relatives. Source: own calculations.

Panel E: Venture-Backed IPOs in Germany (Neuer Markt)

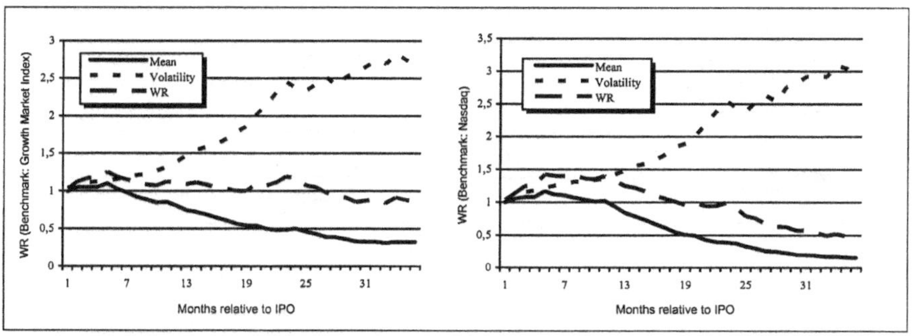

Note: WR - wealth relatives. Source: own calculations.

Panel F: Non Venture-Backed IPOs in Germany (Neuer Markt)

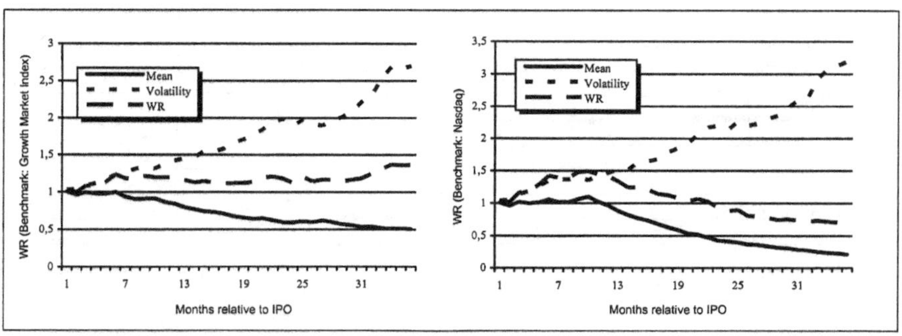

Note: WR - wealth relatives. Source: own calculations.

A 15: Long-Run Performance of IPOs in the Aftermarket (continued)

Panel G: All IPOs in the UK (techMARK)

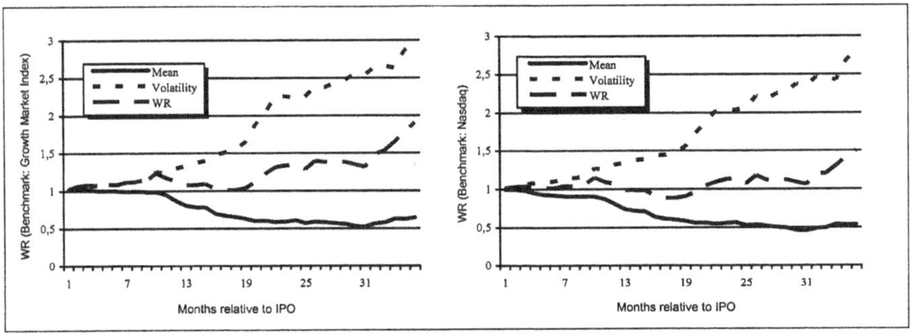

Note: WR - wealth relatives. Source: own calculations.

Panel H: Venture-Backed IPOs in the UK (techMARK)

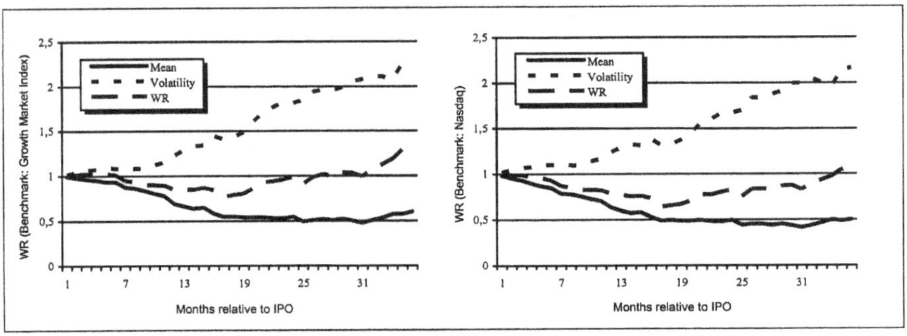

Note: WR - wealth relatives. Source: own calculations.

Panel I: Non Venture-Backed IPOs in the UK (techMARK)

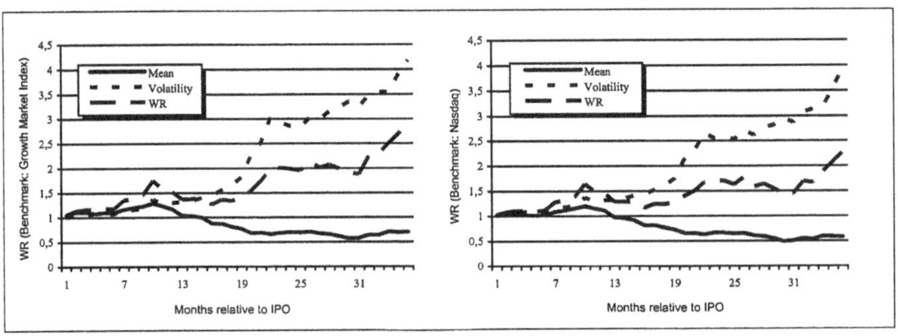

Note: WR - wealth relatives. Source: own calculations.

A 16: 3-Year Wealth Relatives of IPOs by Yearly Cohorts against Different Benchmarks

3-year equal weighted returns on IPOs are compared with alternative benchmarks (medians in parentheses). The buy and hold returns (*BHR*) and benchmark returns are calculated by compounding monthly returns for 36 months after the IPO date. Wealth relatives are computed by taking the ratio of 1 plus the IPO portfolio return over 1 plus the return on the chosen benchmark over the same period. *GM* denotes the index of the country's growth market, i.e. *Nouveau Marché*, *Nemax All* Share, and *techMARK All Share* index for *Nouveau Marché*, *Neuer Markt*, and *techMARK*, respectively. *Nasdaq* denotes the *Nasdaq Composite* index.

	Venture-Backed IPOs					*Non Venture-Backed IPOs*				
Year	BHR (%)	Benchmark Return (%)		Wealth Relative		BHR (%)	Benchmark Return (%)		Wealth Relative	
		GM	Nasdaq	GM	Nasdaq		GM	Nasdaq	GM	Nasdaq
					France (Nouveau Marché)					
1996	38.0 (-4.9)	33.7 (9.5)	126.4 (121.4)	0.87 (0.78)	0.57 (0.40)	-1.4 (4.8)	92.6 (56.8)	153.1 (135.4)	0.48 (0.35)	0.38 (0.35)
1997	84.3 (82.5)	303.5 (243.1)	164.3 (154.5)	0.48 (0.39)	0.73 (0.56)	323.4 (-43.6)	276.3 (232.0)	162.0 (129.5)	1.24 (0.13)	1.94 (0.20)
1998	124.8 (-7.2)	3.6 (-12.9)	14.3 (13.4)	2.78 (1.07)	2.04 (0.85)	28.1 (-33.1)	-7.9 (-11.0)	7.3 (5.0)	1.59 (0.69)	1.20 (0.67)
1999	52.0 (2.0)	-32.9 (-25.1)	-36.4 (-34.2)	2.29 (1.37)	2.35 (1.41)	4.1 (-66.4)	-39.1 (-24.8)	-40.2 (-32.5)	1.70 (0.94)	1.67 (0.71)
					Germany (Neuer Markt)					
1997	111.7 (53.1)	438.0 (420.3)	126.8 (111.1)	0.50 (0.24)	1.07 (0.57)	3,216.9 (373.0)	437.6 (465.3)	155.4 (180.7)	6.82 (0.84)	15.97 (1.69)
1998	-67.6 (-77.9)	-36.1 (-35.8)	8.7 (10.4)	0.50 (0.30)	0.30 (0.21)	-13.9 (-25.2)	-42.8 (-50.7)	6.1 (2.4)	1.65 (1.08)	0.83 (0.65)
1999	-78.5 (-93.7)	-79.0 (-81.6)	-42.7 (-47.8)	0.91 (0.39)	0.35 (0.14)	-80.3 (-91.0)	-79.0 (-81.2)	-43.7 (-47.4)	0.95 (0.43)	0.34 (0.14)
					UK (techMARK)					
1996	92.5 (46.7)	152.4 (146.1)	237.5 (235.8)	0.76 (0.45)	0.55 (0.39)	656.4 (94.8)	162.7 (151.3)	260.0 (247.0)	2.57 (0.65)	1.92 (0.52)
1997	631.3 (393.4)	209.5 (211.1)	321.1 (315.7)	2.33 (1.47)	1.73 (1.15)	760.9 (502.0)	190.9 (185.3)	287.2 (278.6)	2.98 (2.12)	2.37 (1.58)
1998	20.1 (-41.0)	23.5 (26.6)	36.8 (39.6)	0.91 (0.44)	0.86 (0.41)	180.8 (-78.1)	27.3 (39.9)	32.6 (41.0)	2.02 (0.12)	2.00 (0.12)
1999	-75.3 (-88.4)	-52.3 (-52.3)	-52.2 (-60.7)	0.42 (0.31)	0.42 (0.28)	-63.7 (-89.1)	-49.8 (-55.2)	-49.6 (-59.9)	0.59 (0.28)	0.58 (0.28)

Source: own calculations.

A 17: Correlation Matrix for the Independent Variables of Operating Performance

VC is a dummy variable that equals 1 if a firm is venture-backed, *Size* is the natural logarithm of the firm's 2-year average post-IPO assets, *Age* is the natural logarithm of the firm age in years, *Part* is the participation ratio of old shareholders in the IPO, defined as number of old shares sold divided by the number of shares outstanding before flotation, and *URank* is the underwriter reputation rank on a scale of 1 to 3, where 1 is the most and 3 the least prestigious underwriter.

	France				*Germany*				*UK*			
Variables	(1)	(2)	(3)	(4)	(1)	(2)	(3)	(4)	(1)	(2)	(3)	(4)
VC (1)	1.00				1.00				1.00			
Size (2)	0.04	1.00			-0.13	1.00			0.18	1.00		
Age (3)	-0.06	0.04	1.00		-0.08	0.16	1.00		0.07	0.11	1.00	
Part (4)	**0.23**[b]	-0.03	**0.25**[b]	1.00	**0.26**[a]	-0.04	0.09	1.00	**0.30**[b]	0.19	**0.28**[b]	1.00
URank (5)	-0.15	**-0.24**[b]	0.07	0.02	-0.04	**-0.45**[a]	0.06	0.00	**-0.36**[a]	**-0.64**[a]	-0.02	-0.15

Note: a) significant at the 1% level; b) significant at the 5% level. Source: own calculations.

A 18: Correlation Matrix for the Independent Variables of Tobin's *Q*

VC is a dummy variable that equals 1 if a firm is venture-backed, *Size* is the natural logarithm of the firm's market value of equity at the first days of trading, *EqR* is the equity ratio, defined as the book value of equity divided by the book value of assets at the fiscal year prior to the IPO, *S_A* is the inverse of capital intensity, defined as the ratio of sales over assets at the fiscal year prior to the IPO, and *Alpha* is the fraction of the equity retained by old shareholders after the IPO.

	France				*Germany*				*UK*			
Variables	(1)	(2)	(3)	(4)	(1)	(2)	(3)	(4)	(1)	(2)	(3)	(4)
VC (1)	1.00				1.00				1.00			
Size (2)	0.07	1.00			-0.07	1.00			0.24	1.00		
EqR (3)	0.10	0.08	1.00		0.14	-0.04	1.00		0.19	0.13	1.00	
S_A (4)	-0.20	-0.10	-0.12	1.00	-0.14	0.03	-0.07	1.00	-0.11	-0.19	0.19	1.00
Alpha (5)	**-0.34**[a]	0.08	-0.09	0.17	**-0.26**[a]	0.11	-0.06	0.00	-0.24	-0.12	-0.02	0.02

Note: a) significant at the 1% level. Source: own calculations.

A 19: Correlation Matrix for the Independent Variables of Wealth Relatives

VC is a dummy variable that equals 1 if a firm is venture-backed, *Q* the approximated value of Tobin's *Q* at the first day of trading, *Size* the natural logarithm of the IPO firm's market capitalization at the end of the first day of trading, and *URank* the underwriter reputation rank on a scale of 1 to 3, where 1 is the most and 3 the least prestigious underwriter.

	France			*Germany*			*UK*		
Variables	(1)	(2)	(3)	(1)	(2)	(3)	(1)	(2)	(3)
VC (1)	1.00			1.00			1.00		
Q (2)	-0.15	1.00		-0.04	1.00		-0.11	1.00	
Size (3)	0.07	**0.53**[a]	1.00	-0.07	**0.42**[a]	1.00	0.24	0.13	1.00
URank (4)	-0.15	-0.16	**-0.38**[a]	-0.04	0.14	**-0.36**[a]	**-0.36**[a]	0.18	**-0.67**[a]

Note: a) significant at the 1% level. Source: own calculations.

References

3i, 2002, Annual Report, London.

Admati, A.R. and P. Pfleiderer, 1994, Robust Financial Contracting and the Role of Venture Capitalists, *Journal of Finance 49*, 371-402.

Admati, A.R., P. Pfleiderer, and J. Zechner, 1992, Large Shareholder Activism, Risk Sharing, and Financial Market Equilibrium, *Journal of Political Economy 102*, 1097-1130.

AFIC, 1996, Rapport sur l'Activité du Capital Investissement en France, Paris.

Aghion, P. and P. Bolton, 1992, An Incomplete Contracts Approach to Financial Contracting, *Review of Economic Studies 59*, 473-493.

Akerlof, G.A., 1970, The Market for Lemons: Qualitative Uncertainty and the Market Mechanism, *Quarterly Journal of Economics 84*, 488-500.

Albach, H., D. Hunsdiek, and L. Kokalj, 1986, Finanzierung mit Risikokapital, Stuttgart.

Allen, F., and D. Gale, 1995, A Welfare Comparison of Intermediaries and Financial Markets in Germany and the U.S., *European Economic Review 39*, 179-209.

Allen, F., and D. Gale, 2000, Comparing Financial Systems, Cambridge, MA.

Allen, F. and A.M. Santomero, 1996, The Theory of Financial Intermediation, *Journal of Banking and Finance 21*, 1461-1485.

Amit, R, J. Brander, and C. Zott, 1998, Why Do Venture Capital Firms Exist?, Theory and Canadian Evidence, *Journal of Business Venturing 13*, 441-466.

Amit, R, L. Glosten, and E. Muller, 1990, Entrepreneurial Ability, Venture Investment, and Risk Sharing, *Management Science 36*, 1232-1245.

Amit, R. and P. Schoemaker, 1993, Strategic assets and organizational rent, *Strategic Management Journal 13*, 33-46.

Aoki, M., 1994, The Contingent Governance of Teams: An Analysis of Institutional Complementarity, *International Economic Review 35*, 657-676.

Aoki, M., 2000, Information and Governance in the Silicon Valley Model, in: Vives, X. (ed.), Corporate Governance. Theoretical and Empirical Perspectives, Cambridge, 169-195.

Arrow, K.J. and G. Debreu, 1954, Existence of an Equilibrium for a Competitive Economy, *Econometrica 22*, 265-290.

Audretsch, D. and E. Lehmann, 2003, The Role of Venture Capital Financing in the New Economy in Germany, Working Paper, Universities of Indiana and Konstanz.

Baker, M. and P.A. Gompers, 1999, Executive Ownership and Control in Newly Public Firms: The Role of Venture Capitalists, Working Paper, Harvard Business School.

Baker, M. and P.A. Gompers, 2003, The Determinants of Board Structure at Initial Public Offering, *Journal of Law and Economics (forthcoming)*.

Balzer, K., 2000, Die Bedeutung des Venture Capital für innovative Unternehmen, Aachen.

Bank of England, 1996, The Financing of Technology-Based Small Firms, London.

Bank of England, 2001, The Financing of Technology-Based Small Firms, London.

Barber, B.M. and J.D. Lyon, 1997a, Detecting Long-Run Abnormal Stock Returns: The Empirical Power and Specification of Test Statistics, *Journal of Financial Economics 43*, 341-372.

Barber, B.M. and J.D. Lyon, 1997b, Firm Size, Book-to-Market Ratio, and Security Returns: A Holdout Samples of Financial Firms, *Journal of Finance 52*, 875-884.

Barney, J.B., 1991, Firm resources and sustained competitive advantage, *Journal of Management 17*, 99-120.

Barry, C.B., 1989, Initial Public Offering Underpricing: The Issuer's View – A Comment, *Journal of Finance 27*, 447-471.

Barry, C.B., 1994, New Directions in Research on Venture Capital Finance, *Financial Management 23*, 3-15.

Barry, C.B., C. Muscarella, J. Peavy III., and M. Vetsuypens, 1990, The Role of Venture Capital in the Creation of Public Companies: Evidence from the Going-Public Process, *Journal of Financial Economics 27*, 447-471.

Bascha, A., 2000, Why Do Venture Capitalists Hold Different Types of Equity Securities?, Working Paper, University of Tübingen, 447-471.

Bascha, A. and U. Walz, 2001a, Convertible Securities and Optimal Exit Decisions in Venture Capital Finance, *Journal of Corporate Finance 7*, 285-306.

Bascha, A. and U. Walz, 2001b, Financing Practice in the German Venture Capital Industry: An Empirical Assessment, Working Paper, University of Tübingen.

Battini, P., 1999, Le capital risque: mode d'emploi, Paris.

Beatty, R and J.R. Ritter, 1986, Investment Banking and Reputation and the Underpricing of Initial Public Offerings, *Journal of Financial Economics 15*, 213-232.

Becker, R. and T. Hellmann, 2003, The Genesis of Venture Capital: Lessons from the German Experience, in Keuschnigg, C. and V. Kanniainen (eds.), Venture Capital, Entrepreneurship, and Public Policy (forthcoming).

Bender, C., 2003, Changing Firm Boundaries in a New Information and Communication Environment: Evidence from the Manufacturing and Music Industry, Dissertation, University of Münster.

Bender, C., R. Harms, and G. Rindermann, 2002, Do Clusters Matter? Evidence from Germany's Neuer Markt, Working Paper, University of Münster, presented at the European International Business Academy (EIBA) Conference in Athens.

Benston, G.J. and C.W. Smith, 1976, A Transaction Cost Approach to the Theory of Financial Intermediation, *Journal of Finance 31*, 215-231.

Bergemann, D. and U. Hege, 1998, Venture Capital Financing, Moral Hazard and Learning, *Journal of Banking and Finance 22*, 703-735.

Berger, A.N. and G.F. Udell, 1998, The Economics of Small Business Finance: The Roles of Private Equity and Debt Markets in the Financial Growth Cycle, *Journal of Banking and Finance 22*, 613-673.

Berglöf, E., 1994, A Control Theory of Venture Capital Finance, *Journal of Law, Economics and Organization 10*, 247-267.

Berglöf, E., 2002, Comment on Bottazzi, L. and M. Da Rin: Venture Capital in Europe and the Financing of Innovative Companies, *Economic Policy 34*, 266-267.

Berglöf, E. and E. Perotti, 1994, The Governance Structure of Japanese Financial Keiretsu, *Journal of Financial Economics 36*, 259-284.

Bertero, E., 1994, The Banking System, Financial Markets, and Capital Structure: New Evidence from France, *Oxford Review of Economic Policy 10*, 42-54.

Bester, H., 1985, Screening versus Rationing in Credit Markets with Imperfect Information, *American Economic Review 75*, 850-855.

Betsch, O., A.P. Groh, and K. Schmidt, 2000, Gründungs- und Wachstumsfinanzierung innovativer Unternehmen, Munich.

Bhattacharya, S. and A.V. Thakor, 1993, Contemporary Banking Theory, *Journal of Financial Intermediation 3*, 2-50.

Binks, M. and C.T. Ennew, 1996, Growing Firms and Credit Constraints, *Journal of Small Business Finance 8*, 17-25.

Bishop, B., 1996, Venture Capital in the United Kingdom, in: OECD, Venture Capital and Innovation, 80-82.

Black, B.S. and R.J. Gilson, 1998, Venture Capital and the Structure of Capital Markets, *Journal of Financial Economics 47*, 243-377.

Blass, A. and Y. Yafeh, 2001, Vagabond Shoes Longing to Stray: Why Foreign Firms List in the US, *Journal of Banking and Finance 25*, 555-572.

de Boisseu, C., 1993, The French Banking Sector in the Light of European Financial Integration, in: Dermine, J. (ed.), European Banking in the 1990s, 2nd edition, Oxford.

Bottazzi, L. and M. Da Rin, 2002a, Venture Capital in Europe and the Financing of European Innovative Firms, *Economic Policy 34*, 229-262.

Bottazzi, L. and M. Da Rin, 2002b, Europe's New Stock Markets, Discussion Paper No. 3521, Center for Economic Policy Research.

Bottazzi, L. and M. Da Rin, 2002c, Financing Entrepreneurial firm in Europe: Facts, Issues, and Research Agenda, Working Paper, presented at the CESifo Conference on Venture Capital, Entrepreneurship, and Public Policy in Munich.

Bourse de Paris, 1999, Brochure Nouveau Marché, Paris.

Brav, A., 2000, Inference in Long-Horizon Event Studies: A Bayesian Approach with Application to Initial Public Offerings, *Journal of Finance 55*, 1979-2016.

Brav, A. and P. Gompers, 1997, Myth or Reality? The Long-Run Underperformance of Initial Public Offerings: Evidence from Venture- and Non-Venture-Capital-Backed Companies, *Journal of Finance 52*, 1791-1822.

Brody, P. and D. Ehrlich, 1998, Can Big Companies Become Successful Venture Capitalists? *McKinsey Quarterly 2*, 50-63.

Buckley, P.J. and M.C. Casson, 1976, The Future of Multinational Enterprise, London.

Büschgen, H.G., 1985, Banken und Venture-Capital-Finanzierung, *Die Bank 6*, 284-292.

Bundesministerium für Wirtschaft und Technologie, 2000, ERP-Programme 2000 – Wirtschaftsförderung für den Mittelstand – Förderprogramme und Richtlinien, Berlin.

Burghof, H.-P. and T. Kraus, 2003, Post IPO Performance and the Exit of Venture Capitalists, Working Paper, University of Munich.

BVCA, 1999, The Economic Impact of Venture Capital in the UK, London.

BVCA, 2000, Sources of Business Angel Capital, London.

BVK, 1999, Annual Report, Berlin.

BVK, 2000, Annual Report, Berlin.

Bygrave, W. and J. Timmons, 1992, Venture Capital at the Crossroads, Harvard University School Press.

Calvet, A.L., 1981, A Synthesis of Foreign Direct Investment Theories and Theories of the Multinational Firm, *Journal of International Business 12*, 43-59.

Campbell, T. and W. Kracaw, 1980, Information Production and, Market Signaling, and the Theory of Intermediation, *Journal of Finance 35*, 863-882.

Carter, R. and S. Manaster, 1990, Initial Public Offerings and Underwriter Reputation, *Journal of Finance 45*, 1045-1067.

Carter, R., S. Manaster, and A. Singh, 1998, Underwriter Reputation, Initial Returns, and the Long-Run Performance of IPO Stocks, *Journal of Finance 53*, 285-312.

Cassamata, C., 1999, Financing and Advising: Optimal Financial Contracts with Venture Capitalists, Working Paper, University of Toulouse.

Casson, M.C., 1992, Internationalization Theory and Beyond, in: Buckley, P., New Directions in International Business, Aldershot, 4-27.

Chabbal, R., 1995, Le système financier français face à l'investissement innovation, Rapport au ministre des entreprises, La Documentation Française, Paris.

Chan, Y., 1983, On the Positive Role of Financial Intermediation in Allocation of Venture Capital in a Market with Imperfect Information, *Journal of Finance 38*, 1543-1568.

Chan, Y., D. Siegel, and A. Thakor, 1990, Learning, Corporate Control and Performance Requirements in Venture Capital Contracts, *International Economic Review 31*, 365-381.

Chan, L., N. Jegadeesh, and J. Lakonishok, 1995, Evaluating the Performance of Value versus Glamour Stock: The Impact of Selection Bias, *Journal of Financial Economics 38*, 269-296.

Chapman, J., 1986, Venture Capital in the UK and its Impact on the Business on the Small Business Sector, National Economic Development Office, London.

Charkham, C., 1994, Keeping Good Company: A Study of Corporate Governance in Five Countries, Oxford.

Chesbrough, H.W., 1999, The Organizational Impact of Technological Change: A Comparative Theory of National Institutional Factors, *Industrial and Corporate Change 8*, 447-485.

Cho, M.-H., 1998, Ownership Structure, Investment, and the Corporate Value: An Empirical Analysis, *Journal of Financial Economics 47*, 103-121.

Chow, G., 1960, Tests of Equality between Sets of Coefficients in Two Linear Regressions, *Econometrica 28*, 591-605.

Chung, K. H. and S. Pruitt, 1994, A Simple Approximation of Tobin's Q, *Financial Management 23*, 70-74.

Clark, R., 1987, Venture Capital in Britain, America, and Japan, London & Sydney.

Clarkson, P.M. and R. Thompson, 1990, Empirical Estimates of Beta When Investors Face Estimation Risk, *Journal of Finance 45*, 431-453.

Coase, R., 1937, The Theory of the Firm, *Economica 4*, 386-405.

Cochrane, J.H., 2001, Risk and Return of Venture Capital, Working Paper, University of Chicago.

Cornelli, F. and O. Yosha, 2003, Stage Financing and the Role of Convertible Securities, *Review of Economic Studies 70*, 1-32.

Crane, D.B., K.A. Froot, S.P. Mason, Z. Bodie, and A.F. Perold, 1995, The Global Financial System: A Functional Perspective, Boston.

Cumming, D.J., 2001, The Convertible Preferred Equity Puzzle in Canadian Venture Capital Finance, Working Paper, University of Alberta.

Cumming, D.J., 2002, Contracts and Exits in Venture Capital Finance, Working Paper, University of Alberta.

Cumming, D.J. and J.G. MacIntosh, 2001, Venture Capital Investment Duration in Canada and the United States, *Journal of Multinational Financial Management 11*, 445-463.

Cumming, D.J. and J.G. MacIntosh, 2002, Crowding Out Private Equity: Canadian Evidence, Working Paper, presented at the CESifo Conference on Venture Capital, Entrepreneurship, and Public Policy in Munich.

Cumming, D.J. and J.G. MacIntosh, 2003a, A Cross-Country Comparison of Full and Partial Venture Capital Exits, *Journal of Banking and Finance 27*, 511-548.

Cumming, D.J. and J.G. MacIntosh, 2003b, Venture Capital Exits in Canada and the United States, *University of Toronto Law Journal 53*, 101-200.

Davis, J., 1994, The Cross-Section of Realized Stock Returns: The Pre-Compustat Evidence, *Journal of Finance 49*, 1579-1593.

De, D. and S. Wimmers, 1994, Mittelstand und Mittelstandspolitik in den Mitgliedstaaten der Europäischen Union: Belgien, Frankreich, Großbritannien, Irland. Schriften zur Mittelstandsforschung No. 57, Institut für Mittelstandsforschung, Bonn.

DeAngelo, L., 1981, Auditor Independence, 'Low Balling', and Disclosure Regulation, *Journal of Accounting and Economics 3*, 113-127.

DeBondt, W. and R. Thaler, 1985, Does the Stock Market Overreact?, *Journal of Finance 40*, 793-808.

Degeorge, F. and R. Zeckhauser, 1993, The Reverse LBO Decision and Firm Performance, *Journal of Finance 48*, 1323-1348.

Denis, A., 1999, Vingt ans de soutien aux PME, l'Anvar, levier d'innovations, *Les Echos*, 20.10.1999, 67-77.

Densing, C., 1997, Venture Capital für mittelständische Unternehmen – dargestellt am Beispiel Deutschland, Frankreich und Großbritannien, Wirtschaftswissenschaftliche Schriftenreihe: Handels- und Industriepolitik, Bd. 2, Bochum.

Department of Trade and Industry, 1994, An Evaluation of the Small Firms Merit Award for Research and Technology (SMART), DTI Assessment Paper No. 27, London.

Department of Trade and Industry, 1999, White Paper on Competitiveness: Our Competitive Future – Building the Knowledge Driven Economy, HMSO, London.

Department of Trade and Industry, 2001, White Paper on Enterprise, Skills, and Innovation, Opportunity for All in a World of Change, HMSO, London.

Deutsche Ausgleichsbank, 2001, Annual Report, Bonn.

Deutsche Börse, 1997, Rules and Regulations – Neuer Markt, Frankfurt.

Deutsche Börse, 2000, Information Folder – Neuer Markt, Frankfurt.

Deutsche Börse, 2001, Fact Book, Frankfurt.

Dewatripont, M. and J. Tirole, 1994, A Theory of Debt and Equity: Diversity of Securities and Manager-Shareholder Congruence, *Quarterly Journal of Economics 109*, 1027-1054.

Diamond, D.W., 1984, Financial Intermediation as Delegated Monitoring, *Review of Economic Studies 51*, 393-414.

Diamond, D.W., 1989, Reputation Acquisition in Debt Markets, *Journal of Political Economy 97*, 828-862.

Diamond, D.W., 1991, Monitoring and Reputation: The Choice between Bank Loans and Directly Placed Debt, *Journal of Political Economy 99*, 689-721.

Diamond, D.W. and P.H. Dybvig, 1983, Bank Runs, Depot Insurance, and Liquidity, *Journal of Political Economy 91*, 401-419.

Dixit, A.K. and R.S. Pindyck, 1994, Investment under Uncertainty, Princeton.

Doukas, J. and H. Gonenc, 2001, Long-Term Performance of New Equity Issues, Venture Capital, and Reputation of Investment Bankers, Working Paper, New York University and Hacettepe University, Ankara.

Dubocage, E. and D. Rivaud-Danset, 2002, Government Policy on Venture Capital Support in France, *Venture Capital 4*, 25-43.

Ducourtieux, C., 2001, Le capital risque en France, Le Monde Interactif, 18.06.2001.

Ducourtieux, C., 2003, La frilosité du capital-risque pèse sure les sociétés innovantes, Le Monde, 13.03.2003.

Dunning, J.H., 1993, The Determinants of MNE Activity, Wokingham.

Edwards, J. and K. Fischer, 1994, Banks, Finance and Investment in Germany, Cambridge.

Elsas, R. and J.P. Krahnen, 1998, Is Relationship Lending Special? Evidence from Credit-File Data in Germany, *Journal of Banking and Finance 22*, 1283-1316.

Engel, D., 2002, The Impact of Venture Capital on Firm Growth: An Empirical Investigation, ZEW Discussion Paper No. 02-02, Center for European Economic Research.

Espenlaub, S., I. Garrett, and W.P. Mun, 1999, Conflicts of Interest and the Performance of Venture-Capital-Backed IPOs: A Preliminary Look at the UK, *Venture Capital 1*, 325-349.

Euronext, 2001, Fact Book, Paris.

European Commission, 1994, Growth, Competitiveness, Employment: The Challenges and the Ways Forward into the 21st Century (White Paper), Brussels.

European Commission, 1995, Green Paper on Innovation, Brussels.

European Commission, 1996, First Action Plan for Innovation in Europe, Brussels.

European Commission, 1998, Risk Capital: A Key to Job Creation in the European Union, Brussels.

European Commission, 2000a, The European Observatory for SME, 6th Report, Brussels.

European Commission, 2000b, Funding of New Technology-Based Firms by Commercial Banks in Europe, Brussels.

European Commission, 2000c, Innovation in a Knowledge-Driven Economy, Brussels.

European Commission, 2000d, Progress Report on the Risk Capital Action Plan, Brussels.

EVCA, 1997, Yearbook, Zaventem.

EVCA, 2000, Yearbook, Zaventem.

EVCA, 2002, Survey of the Economic and Social Impact of Venture Capital in Europe, Research Paper, Zaventem.

Fama, E.F., 1970, Efficient capital markets: a review of theory and empirical work, *Journal of Finance 25*, 383-417.

Fama, E.F., 1998, Market Efficiency. Long-Term Returns, and Behavioral Finance, *Journal of Financial Economics 49*, 283-306.

Fama, E. and K. French, 1992, The Cross-Section of Expected Stock Returns, *Journal of Finance 47*, 427-465.

Fama, E. and K. French, 1993, Common risk Factors in the Returns of Stocks and Bonds, *Journal of Financial Economics 33*, 3-55.

Fama, E. and K. French, 1995, Size and Book-To-Market Factors in Earnings and Returns, *Journal of Finance 50*, 131-156.

Faugère, J.-P. and C. Voisin, 1994, Le système financier français, Paris.

Fenn, G. and N. Liang, 1998, New Resources and New Ideas: Private Equity for Small Businesses, *Journal of Banking and Finance 22*, 1077-1084.

Fenn, G., N. Liang, and S. Prowse, 1997, The Private Equity Market: An Overview, *Financial Markets, Institutions and Instruments 6*, 1-106.

Fessler, B, G. Hegmann, and C. Lemaitre, 2000, Die Behandlung vom Börseneinführungskosten unter Beachtung internationaler Rechnungslegungsnormen bei Unternehmen des Neuen Markes, Deutsches Steuerrecht, 1069-1072.

Fiedler, M.O. and T. Hellmann, 2001, Against All Odds: The Late but Rapid Development of the Germany Venture Capital Industry, *Journal of Private Equity 4*, 31-45.

Fischer, C., 2000, Why Do Companies Go Public? Empirical Evidence from Gemany's Neuer Markt, Working Paper, University of Munich.

Francis, B. and I. Hasan, 2001, The Underpricing of Venture and Nonventure Capital IPOs: An Empirical Investigation, *Journal of Financial Services Research 19*, 99-113.

Franzke, S., 2001, Underpricing of Venture-Backed and Non Venture-Backed IPOs: Germany's Neuer Markt, CFS Working Paper No. 2001/01, Center for Financial Studies, Frankfurt.

Friderichs, H., B. Paranque, and A. Sauvé, 1999, Structures of Corporate Finance in Germany and France: A Comparative Analysis for West German and French Incorporated Enterprises with Special References to Institutional Factors, in: Sauvé, A. and M. Scheuer (eds.), Corporate Finance in Germany and France, Deutsche Bundesbank, 73-137.

Fried, V.H. and R.D. Hisrich, 1994, Toward a Model of Venture Capital Decision Making, *Financial Management 23*, 28-37.

Galesne, A., 1999, Le financement de l'entreprise, Rennes.

Garnsey, E., 1995, High Technology Renewal and the UK Investment Problem, *Journal of General Management 20*, 1-22.

Gebhardt, G. and K.M. Schmidt, 2001, Der Markt für Venture Capital: Anreizprobleme, Governance Strukturen und staatliche Intervention, Working Paper, Munich.

Gerke, W., 1996, Venture Capital in Germany, in: OECD, Venture Capital and Innovation, 90-93.

Gerschenkron, A., 1962, Economic Backwardness in Historical Perspective. A Book of Essays, Cambridge, MA.

Gilson, R. and M. Roe, 1993, Understanding Japanese Keiretsu: Overlaps between Corporate Governance and Industrial Organization, *Yale Law Review 102*, 871-906.

Gilson, R. and D. Schizer, 2002, Understanding Venture Capital Structure: A Tax Explanation for Convertible Preferred Stock, Working Paper, Columbia University.

Giudici, G. and S. Paleari, 2001, 'New Stock Markets': Risky Assets for the European Economy, Working Paper, University of Bergamo.

Giudici, G. and P. Roosenboom, 2002, Pricing Initial Public Offerings on 'New' Stock European Stock Markets, Working Paper, Universities of Bergamo and Tilburg.

Goergen, M., 1999, Corporate Governance and Financial Performance, A Study of German and UK Initial Public Offerings, Cheltenham.

Gompers, P.A., 1993, The Theory, Structure and Performance of Venture Capital, Unpublished D.Phil. Thesis, Harvard University, Boston.

Gompers, P.A., 1994, The Rise and Fall of Venture Capital, *Business and Economic History 23*, 1-24.

Gompers, P.A., 1995, Optimal Investment, Monitoring and the Staging of Venture Capital, *Journal of Finance 50*, 1461-1489.

Gompers, P.A., 1996, Grandstanding in the Venture Capital Industry, *Journal of Financial Economics 43*, 133-156.

Gompers, P.A., 1999, Ownership and Control in Entrepreneurial Firms: An Examination of Convertible Securities in Venture Capital Investment, Working Paper, Harvard Business School.

Gompers, P.A. and J. Lerner, 1996, The Use of Covenants: An Empirical Analysis of Venture Partnership Agreements, *Journal of Law and Economics 39*, 463-498.

Gompers, P.A. and J. Lerner, 1997, Venture Capital and the Creation of Public Companies: Do Venture Capitalists Really Bring more than Money?, *Journal of Private Equity 1*, 15-32.

Gompers, P.A. and J. Lerner, 1998, Venture Capital Distributions: Short-Run and Long-Run Reactions, *Journal of Finance 53*, 2161-2183.

Gompers, P.A. and J. Lerner, 1999a, Conflict of Interest in the Issuance of Public Securities: Evidence from Venture Capital, *Journal of Law and Economic 42*, 1-28.

Gompers, P.A. and J. Lerner, 1999b, An Analysis of Compensation in the US Venture Capital Partnership, *Journal of Financial Economic 51*, 3-44.

Gompers, P.A. and J. Lerner, 1999c, The Venture Capital Cycle, Cambridge, MA.

Gompers, P.A. and J. Lerner, 2001a, The Venture Capital Revolution, *Journal of Economic Perspectives 15*, 145-168.

Gompers, P.A. and J. Lerner, 2001b, The Money of Invention: How Venture Capital Creates New Wealth, Boston.

Gorman, M. and W. Sahlman, 1989, What Do Venture Capitalists Do?, *Journal of Business Venturing 4*, 231-248.

Greenbaum, S. and A. Thakor, 1995, Contemporary Financial Intermediation, 2nd edition, Fort Worth *et al.*

Güllmann, P., 2000, Die Rolle des Staates bei der Mobilisierung von (privatem) Risikokapital im Seed- und Start-up-Segement: Am Beispiel von Deutschland, Großbritannien und den Niederlanden, Cologne.

Gueslain, A., 1992, L'Etat et le système financier, in: Histoire économique de la France au XXe siècle, *Les Cahiers Français 255*, 48-49.

Gujarati, N.G., 1988, Basic Econometrics, 2nd edition, New York.

Gupta, A.K. and H.J. Sapienza, 1992, Determinants of Capital Firms' Preferences Regarding the Industry Diversity and Geographic Scope of their Investments, *Journal of Business Venturing 7*, 347-362.

Gurley, J.G. and E.S. Shaw, 1955, Financial Aspects of Economic Development, *American Economic Review 45*, 515-538.

Gurley, J.G. and E.S. Shaw, 1960, Money in a Theory of Finance, The Brookings Institution, Washington D.C.

Habib, M.A. and A.P. Ljungqvist, 2001, Underpricing and Entrepreneurial Wealth Losses in IPOs: Theory and Evidence, *Review of Financial Studies 14*, 433-458.

Hackethal, A. and R.H. Schmidt, 2000, Finanzsystem und Komplementarität, Working Paper Series: Finance & Accounting, University of Frankfurt.

Hamao, Y., F. Packer, and J.R. Ritter, 2000, Institutional Affiliation and the Role of Venture Capital: Evidence from Initial Public Offerings in Japan, *Pacific-Basin Finance Journal 8*, 529-558.

Hampden-Turner, C., and A. Trompenaars, 1993, The Seven Cultures of Capitalism, New York.

Hanley, K., 1993, The Underpricing of Initial Public Offerings and the Partial Adjustment Phenomenon, *Journal of Financial Economics 34*, 231-250.

Harding, R., 2002, Plugging the Knowledge Gap: An International Comparison of the Role for Policy in the Venture Capital Market, *Venture Capital 4*, 59-76.

Hardymon, G.F., M.J. DeNino, and M.S. Salter, 1983, When Corporate Venture Capital Does Not Work, *Harvard Business Review 61*, 114-121.

Harhoff, D. and T. Körting, 1999, Lending Relationships in Germany - Empirical Evidence from Survey Data, *Journal of Banking and Finance 22*, 1317-1353

Harm, C., 1992, The Financing of Small Firms in Germany, *Policy, Planning, and Research Working Papers*, WPS 899, May.

Harm, C., 1998, Corporate Governance Policy and Relationship Banking, Working Paper, Copenhagen Business School.

Harm, C., 2000, The Limits of Corporate Governance – An Examination of the Manager-Shareholder Conflict, Working Paper, University of Münster.

Harrison, E.M., 1990, The West German Venture Capital Market: An Analysis of its Market Structure and Economic Performance, Frankfurt.

Harrison, R. and C. Mason, 1996, Informal Venture Capital – Evaluating the Impact of Business Introduction Services, London.

Harrison, R. and C. Mason, 1999, Public Policy and the Development of the Informal Venture Capital Market, in: Cowling, K. (ed.), Industrial Policy in Europe, London.

Hart, O., 1995, Firms, Contracts, and Financial Structure, Oxford.

Hart, O. and J. Moore, 1988, Incomplete Contracts and Renegotiation, *Econometrica 56*, 755-786.

Hartmann-Wendels, T., 1987, Venture Capital aus finanzierungstheoretischer sich, *Zeitschrift für betriebswirtschaftliche Forschung 39*, 16-30.

Hege, U., F. Palomino, and A. Schwienbacher, 2003, Determinants of Venture Capital Performance: Europe and the United States, Working Paper, HEC School of Management, Tilburg University, and University of Namur.

Hellmann, T., 1993, Financial Structure and Control in Venture Capital, Unpublished Manuscript, Stanford University.

Hellmann, T., 1997, Venture Capital: A Challenge for Commercial Banks, *Journal of Private Equity (Fall)*, 49-55.

Hellmann, T., 1998a, The Allocation of Control Rights in Venture Capital Contracts, *Rand Journal of Economics 29*, 57-76.

Hellmann, T., 1998b, Teaching Note for Symantec Corporation (S-SM-27), Stanford Business School. Joint Venture, *1988 Index of Silicon Valley*, Palo Alto, CA.

Hellmann, T., 1998c, A Theory of Corporate Venture Investing, Stanford Business School, Working Paper, Stanford University.

Hellmann, T., 2000, Venture Capitalists: The Coaches of Silicon Valley, in: Miller, W., C.M. Lee, M. Gong Hanock and H. Rowen (eds.), The Silicon Valley Edge: A Habitat for Innovation and Entrepreneurship, Stanford University Press.

Hellmann, T., 2001, Developing a Venture Capital Industry, in: Kochendörfer-Lucius, G. and B. Pleskovic (eds.), The Institutional Foundations of a Market Economy, Villa Borsig Workshop Series 2000, German Foundation for International Development (DSE).

Hellmann, T., L. Lindsey, and M. Puri, 1999, Banks in the Venture Capital Market: An Empirical Analysis, Working Paper, Stanford University.

Hellmann, T. and M. Puri, 2000, The Interaction between Product Market and Financing Strategy: The Role of Venture Capital, *Review of Financial Studies 13*, 959-984.

Hellmann, T. and M. Puri, 2002, Venture Capital and the Professionalization of Start-Up Firms: Empirical Evidence, *Journal of Finance 57*, 169-197.

Hellwig, M., 1990, Banking, Financial Intermediation and Corporate Finance, in: Giovannini, A and C. Mayer (eds.), European Financial Integration, New York, 35-63.

Hellwig, M., 1994, Banking and Finance at the End of the Twentieth Century, WWZ Discussion Paper 26, Basel.

Hellwig, M., 1998, Banks, Markets and the Allocation of Risk in the Economy, *Journal of Institutional and Theoretical Economics 154*, 328-345.

Himmelberg, C., G. Hubbard, and D. Palia, 1999, Understanding the Determinants of Managerial Ownership and the Link between Ownership and Performance, *Journal of Financial Economics 53*, 353-384.

Hofstede, G., 1984, Culture's Consequences: International Differences in Work-Related Values, Beverly Hills, CA.

Holmström, B., 1979, Moral Hazard and Observability, *Bell Journal of Economics 10*, 74-91.

Holthausen R. and D. Larcker, 1996, The Financial Performance of Reverse Leveraged Buyouts, *Journal of Financial Economics 42*, 293-332.

Hughes, A., 1997, Finance for SMEs: A U.K. Perspective, *Small Business Economics 9*, 151-166.

Hughes, P.J. and A.V. Thakor, 1992, Litigation Risk, Intermediation, and the Underpricing of Initial Public Offerings, *Review of Financial Studies 5*, 709-742.

Huntsman, B. and J.P. Hoban, Jr., 1980, Investment in New Enterprise: Some Empirical Observations on Risk, Return, and Market Structure, *Financial Management 9*, 44-51.

Hurry, D., A.T. Miller, and E.H. Bowman, 1992, Calls on High-Technology: Japanese Exploration of Venture Capital Investments in the United States, *Strategic Management 13*, 85-101.

Hymer, S.H., 1970, The Efficiency (Contradictions) of Multinational Corporations, *American Economic Review 60*, 441-448.

Hymer, S.H., 1976, The International Operations of National Firms: A Study of Foreign Direct Investment, Cambridge, MA.

Hyytinen, A. and M. Pajarinen, 2002, Financial Systems and Venture Capital in Nordic Countries: A Comparative Study, Discussion Paper No. 774, The Research Institute of the Finnish Economy.

IMF, 2001, International Financial Statistics, Washington D.C.

Jain, B.A. and O. Kini, 1994, The Post-Issue Operating Performance of IPO Firms, *Journal of Finance 49*, 1699-1726.

Jain, B.A. and O. Kini, 1995, Venture Capitalist Participation and the Post-Issue Operating Performance of IPO Firms, *Managerial and Decision Economics 6*, 593-606.

Jakobsen, J. and O. Sørensen, 2001, Decomposing and Testing Long-Run Returns: An Application on Danish IPOs, *European Financial Management 7*, 393-417.

Jakobsen, J. and T. Voetmann, 2003, Post-Acquisition Performance in the Short and Long Run, *European Journal of Finance (forthcoming)*.

Jeng, L.A. and P.C. Wells, 2000, The Determinants of Venture Capital Funding: Evidence Across Countries, *Journal of Corporate Finance 6*, 241-289.

Jenkinson, T. and A. Ljungqvist, 2001, Going Public: The Theory and Evidence on How Companies Raise Equity Finance, 2nd edition, Oxford.

Jensen, M.C. and W.H. Meckling, 1976, Theory of the Firm: Managerial Behavior, Agency Costs, and Ownership Structure, *Journal of Financial Economics 3*, 305-360.

Jobson, J.D., 1991, Applied Multivariate Data Analysis, Vol. I: Regression and Experimental Design, New York.

John, K. and L.W. Senbet, 1998, Corporate Governance and Board Effectiveness, *Journal of Banking and Finance 22*, 371-403.

Johnson, J. and R. Miller, 1988, Investment Banking Prestige and the Underpricing of Initial Public Offerings, *Financial Management 17*, 19-29.

Kane, E.J., 1988, Interaction of Financial and Regulatory Innovation, *American Economic Review 78*, 328-334.

Kaplan, S. and P. Strömberg, 2002, Characteristics, Contracts and Actions: Evidence from Venture Capitalist Analyses, Discussion Paper No. 3243, Centre for Economic Policy Research, London.

Kaplan, S. and P. Strömberg, 2003, Financial Contracting Theory Meets the Real World: An Empirical Analysis of Venture Capital Contracts, *Review of Economic Studies 70*, 281-315.

Kaplan, S., F. Martel, and P. Strömberg, 2002, Venture Capital Contracts around the World, Working Paper, University of Chicago.

Kaserer, C. and V. Kempf, 1995, Das Underpricing-Phänomen am deutschen Kapitalmarkt, *Zeitschrift für Bankrecht und Bankwirtschaft 7*, 45-68.

Katzenstein, P., 1978, Between Power and Plenty: Foreign Economic Policies of Advanced Industrial Countries, Madison, WI.

Kindleberger, C.P., 1969, American Business Abroad: Six Lectures on Direct Investment, New Haven.

Kirilenko, A., 2001, Valuation and Control in Venture Capital Finance, *Journal of Finance 56*, 565-587.

Kiwitz, T, and G. Melzer, 2001, Die Kosten der Börseneinführung eines mittelständischen Unternehmens, Deutsches Steuerrecht, 42-56.

Klein, P. and K. Zoeller, 2003, Universal-Bank Underwriting and Conflict of Interest: Evidence from German Initial Public Offerings, Working Paper, University of Missouri, presented at the Annual Meeting of the Midwestern Finance Association in St. Louis.

Kluth, M.F. and J.B. Andersen, 1999, Globalisation and Financial Diversity: The Making of Venture Capital Markets in France, Germany, and the U.K., in: Archibugi, D., J. Howells, and J. Michie, Innovation Policy in a Global Economy, New York, 120-138.

Kogut, B. and U. Zander, 1992, Knowledge of the Firm, Combinative Capabilities, and the Replication of Technology, *Organization Science 3*, 383-397.

Kortum, S. and J. Lerner, 2000, Assessing the Contribution of Venture Capital to Innovation, *Rand Journal of Economics 31*, 674-692.

Kothari, S.P. and J.B. Warner, 1997, Measuring Long-Horizon Security Price Performance, *Journal of Financial Economics 43*, 301-340.

Kraus, T., 2002, Underpricing of IPOs and the Certification Role of Venture Capitalists: Evidence from Germany's Neuer Markt, Working Paper, Universities of Warwick and Munich.

Kreditanstalt für Wiederaufbau, 2002, Power for the Economy – Progress for All, Introducing Ourselves and Our Services, Frankfurt.

Kuemmerle, W., 2001, Comparing Catalysts of Change: Evolution and Institutional Differences in the Venture Capital Industries in the United States, Japan and Germany, in: Chesbrough, H. and R.A. Burgelman (eds.), Research on Technological Innovation, Management and Policy, Greenwich, CT.

Küting, K., 2000, Bilanzierung und Bilanzanalyse am Neuer Markt, Stuttgart.

Kulicke, M. and U. Wupperfeld, 1996, Beratung junger Technologieunternehmen, in: Koschatzky, K. (ed.), Technologieunternehmen im Innovationsprozess: Management, Finanzierung und regionale Netzwerke, Fraunhofer-Institut für Systemtechnik und Innovationsforschung (ISI), Heidelberg.

Kutsuna, K., M. Cowling, and P. Westhead, 2000, The Short-Run Performance of JASDAQ Companies and Venture Capital Involvement before and after IPO, *Venture Capital 2*, 1-25.

La Porta, R., F. Lopez-de-Silanes, A. Shleifer, and R. Vishny, 1998, Law and Finance, *Journal of Political Economy 106*, 305-360.

Lang, L. and R. Stulz, 1994, Tobin's *q*, Corporate Diversification, and Firm Performance, *Journal of Political Economy 102*, 1248-1280.

Lang, L., R. Stulz, and R. Walking, 1991, Managerial Performance, Tobin's *q* and the Gains From Successful Tender Offers, *Journal of Financial Economics 24*, 137-154.

Lee, P.M. and S. Wahal, 2002, Grandstanding, Certification and the Underpricing of Venture Capital Backed IPOs, Working Paper, Emory University, Atlanta.

Leland, H.E. and D. Pyle, 1977, Informational Asymmetries, Financial Structure, and Financial Intermediation, *Journal of Finance 32*, 371-387.

Leleux, B. and B. Surlemont, 2001, Public versus Private Venture Capital: Seeding or Crowding Out? A Pan-European Analysis, *Journal of Business Venturing 17, 1-21*.

Leopold, G. and H. Frommann, 1998, Eigenkapital für den Mittelstand. Venture Capital im In- und Ausland, Munich.

Lerner, J., 1994a, The Syndication of Venture Capital Investments, *Financial Management 23*, 16-27.

Lerner, J., 1994b, Venture Capitalists and the Decision to Go Public, *Journal of Financial Economics 35*, 293-316.

Lerner, J., 1995, Venture Capital and the Oversight of Private Firms, *Journal of Finance 50*, 301-317.

Lerner, J., 1998, "Angel" Financing and Public Policy: An Overview, *Journal of Banking and Finance 22*, 773-783.

Lerner, J., 1999, The Government as Venture Capitalist: The Long-Run Impact of the SBIC Program, *Journal of Business 72*, 285-318.

Lerner, J., 2002, Boom and Bust in the Venture Capital Industry and the Impact on Innovation, Working Paper, Harvard University.

Lessat, V. et al., 1999, Beteiligungskapital und technologieorientierte Unternehmensgründung: Markt – Finanzierung – Rahmenbedingungen, Wiesbaden.

Lewellen, W. and S.G. Badrinath, 1997, On the Measurement of Tobin's *q*, *Journal of Financial Economics 44*, 77-122.

Lin, T.H., 1996, The Certification Role of Large Block Shareholders in Initial Public Offerings: The Case of Venture Capitalists, *Quarterly Journal of Business and Economics 35*, 55-65.

Lin, T.H. and R.L. Smith, 1998, The Unwinding of Venture Capital Investments: Insider Selling During Equity IPOs, *Journal of Corporate Finance 4*, 241-263.

Lindenberg, E. and S. Ross, 1981, Tobin's *q* Ratio and Industrial Organization, *Journal of Business 54*, 1-33.

Ljungqvist, A.P., 1995, The Timing, Pricing and Long Term Performance of Initial Public Offerings, unpublished D.Phil. thesis, Nuffield College, Oxford University.

Ljungqvist, A.P., 1997, Pricing initial public offerings: further evidence from Germany, *European Economic Review 41*, 1309-1320.

Ljungqvist, A.P., 1999, IPO Underpricing, Wealth Loss and the Curious Role of Venture Capitalist in the Creation of Public Companies, Working Paper, Oxford University and CEPR.

Ljungqvist, A.P. and W. Wilhelm Jr., 2003, IPO Pricing in the Dot-Com Bubble, *Journal of Finance 58*, 723-752.

Locket, A., G. Murray, and M. Wright, 2002, Do UK venture capitalists still have a bias against investments in new technology firms?, *Research Policy 31*, 1009-1030.

Loderer, C. and K. Martin, 1997, Executive Stock Ownership and Performance, *Journal of Financial Economics 45*, 223-255.

London Stock Exchange, 2001, techMARK – London's International Technology Market, London.

Long, J. S., 1997, Regression Models for Categorical and Limited Dependent Variables, Advanced Quantitative Techniques in the Social Sciences Series 7, Thousand Oaks.

Loughran, T. and J.R. Ritter, 1995, The New Issues Puzzle, *Journal of Finance 50*, 23-52.

Loughran, T. and J.R. Ritter, 2003, Why Has IPO Underpricing Changed Over Time?, Working Paper, Universities of Notre Dame and Florida.

Loughran, T., J.R. Ritter, and K. Rydqvist, 1994, Initial Public Offerings: International Insights, *Pacific Basin Journal of Finance 2*, 165-199.

Lyon, J.D., B.M. Barber, and C.-L. Tsai, 1999, Improved Methods for Tests of Long-Run Abnormal Stock Returns, *Journal of Finance 54*, 165-201.

MacMillan, I., L. Zemann and P.N. Subbanarasimha, 1987, Criteria Distinguishing Successful from Unsuccessful Ventures in the Venture Screening Process, *Journal of Business Venturing 2*, 127-137.

MacMillan, I., D. Kulow, and R. Khoylian, 1988, Venture Capitalists Involvement in their Investments: Extent and Performance, *Journal of Business Venturing 3*, 27-47.

Manigart, S., 1994, The Founding Rate of Venture Capital Firms in Three European Countries (1970-1990), *Journal of Business Venturing 9*, 525-541.

Manigart, S. and W.V. Hyfte, 1999, Post-Investment Evolution of Belgian Venture-Capital Backed Companies: An Empirical Study, in: Reynolds, P. et al. (eds.), Frontiers of Entrepreneurship Research. Wellesley, MA.

Manigart, S., K. De Waele, M. Wright, K. Robbie, P. Desbrières, H. Sapienza, and A. Beckham, 2000, Venture Capitalists, Investment Appraisals, and Accounting Information: A Comparative Study of the USA, UK, France, Belgium, and Holland, *European Financial Management 6*, 389-403.

Manigart, S., K. De Waele, M. Wright, K. Robbie, P. Desbrières, H. Sapienza, and A. Beckham, 2002a, Determinants of Required Return in Venture Capital Investments: A Five Country Study, *Journal of Business Venturing 17*, 291-312.

Manigart, S., A. Locket, M. Meuleman, H. Landström, P. Desbrières, and H. Brujning, 2002b, The Syndication of Venture Capital Investments in Europe: Evidence from Five European Countries, Working Paper presented at the CMI Conference in Toronto on "What Next for Private Equity and Venture Capital?".

Markowitz, H., 1952, Portfolio Selection, *Journal of Finance 7*, 77-91.

Markowitz, H., 1959, Portfolio Selection, Efficient Diversification of Investments, New York.

Martin, J.D. and J.W. Petty, 1983, An Analysis of the Performance of Publicly Traded Venture Capitalists Companies, *Journal of Financial and Quantitative Analysis 18*, 401-410.

Marx, L., 1998, Efficient Venture Capital Financing Combining Debt and Equity, *Review of Economic Design 3*, 371-387.

Mayer, C., K. Schoors, and Y. Yafeh, 2001, Sources of Funds and Investment Strategies of Venture Capital Funds: Evidence from Germany, Israel, Japan and the UK, CEPR Working Paper.

Mayer, M., 2001, Venture Capital Backing als Qualitätsindikator beim IPO am Neuen Markt?, *Zeitschrift für Betriebswirtschaft 9*, 1043-1063.

Mayer, M. and R. Müller, 1991, Die Deutsche Wagnisfinanzierungs-Gesellschaft mbH (WFG) - Erfahrungen und Ergebnisse eines Modellvorhabens Fraunhofer-Institut für Systemtechnik und Innovationsforschung, ISI-B-20-91.

McGlue, D., 2002, The Funding of Venture Capital in Europe: Issues for Public Policy, *Venture Capital 4*, 45-58.

Megginson, W.L. and K.A. Weiss, 1991, Venture Capitalists Certification in Initial Public Offering, *Journal of Finance 46*, 879-903.

Michaely, R. and W.H. Chaw, 1994, The Pricing of Initial Public Offering: Tests of Adverse-Selection and Signaling Theories, *Review of Financial Studies 7*, 279-319.

Mikkelson, W., M. Partch, and K. Shah, 1997, Ownership and Operating Performance of Companies that Go Public, *Journal of Financial Economics 44*, 281-307.

Milgrom, P. and J. Roberts, 1990, The Economics of Modern Manufacturing: Technology, Strategy, and Organization, *American Economic Review 80*, 511-528.

Milgrom, P. and J. Roberts, 1992, Economics, Organization and Management, Englewood Cliffs.

Milgrom, P. and J. Roberts, 1994, Complementarities and Systems: Understanding Japanese Economic Organization, *Estudios Economicos 9*, 3-42.

Milgrom, P. and J. Roberts, 1995, Complementarities and Fit: Strategy, Structure, and Organizational Change in Manufacturing, *Journal of Accounting and Economics 19*, 179-208.

Milhaupt, C., 1997, The Market for Innovation and in the United States and Japan: Venture Capital and the Comparative Governance Debate, *Northwestern University Law Review 91*, 865-898.

Miller, E.M., 1977, Risk, Uncertainty, and Divergence of Opinion, *Journal of Finance 32*, 1155-1168.

Mitchell, M. and E. Stafford, 2000, Managerial Decisions and Long-Term Stock Price Performance, *Journal of Business 73*, 287-329.

Mohnen, P. and L.H. Röller, 2000, Complementarities in Innovation Policy, Discussion Paper FS IV 00-18, Wissenschaftszentrum Berlin.

Moore, B., 1994, Financial Constraints to the Growth and Development of Small High Technology Firms, in: Hughes, A. and D.J. Storey (eds.), Finance and the Small Firm, London.

Moore, I. and E. Garnsey, 1993, Funding for Innovation in Small Firms: The Role of Government, *Research Policy 22*, 507-519.

Mørck, R., A. Shleifer, and R. Vishny, 1988, Management Ownership and Market Valuation, *Journal of Financial Economics 20*, 293-315.

Murray, G., 1991, The Changing Nature of Competition in the UK Venture Capital Industry, *National Westminster Quarterly Review (November)*, 65-80.

Murray, G., 1994, The European Union's Support for New Technology-Based Firms: An Assessment of the First 3 Years of the European Seed Capital Fund, *European Planning Studies 2*, 435-461.

Murray, G., 1995, Evolution and Change: An Analysis of the First Decade of the UK Venture Capital Industry, *Journal of Business Finance and Accounting 22*, 1077-1106.

Murray, G. and J. Lott, 1995, Have UK Venture Capitalists a Bias against Investment in New Technology?, *Research Policy 24*, 283-299.

Murray, G. and R. Mariott, 1998, Why has the Investment Performance of Technology-Specialist European Venture Capital Funds been so Poor?, *Research Policy 27*, 947-976.

Mustar, P. and P. Larédo, 2002, Innovation and Research Policy in France (1980-2000) or the Disappearance of the Colbertist State, *Research Policy 31*, 55-72.

Myers, S.C. and N. Majluf, 1984, Financing Decisions When Firms Have Information That Investors Do Not Have, *Journal of Financial Economics 13*, 187-221.

Nelson, R., 1993, National Innovation Systems: A Comparative Study, Oxford.

Neuhaus, S. and R. Schremper, 2003, Langfristige Performance von Initial Public Offerings am deutschen Kapitalmarkt, *Zeitschrift für Betriebswirtschaft 73*, 445-472.

Norton, E., 1993, Venture Capital Finance: Review and Synthesis, in: Lee, C.F. (ed.), Advances in Quantitative Analysis of Finance and Accounting, Vol. 2, Greenwich, CT.

Norton, E., 1995, Venture capital as an alternative means to allocate control, an agency-theoretic approach, *Entrepreneurship Theory and Practice (Winter)*, 19-29.

Norton, E. and B.H. Tenenbaum, 1993, Specialization versus Diversification as a Venture Capital Investment Strategy, *Journal of Business Venturing 8*, 431-442.

NVCA, 2002, Yearbook, Arlington, VA.

OECD, 1993, National Systems for Financing Innovation, Paris.

OECD, 1995, Best Practice Policies – for Small and Medium Sized Enterprises, Paris.

OECD, 1996a, Government Programmes for Venture Capital, Paris.

OECD, 1996b, Main Science and Technology Indicators, No. 2, Paris.

OECD, 1997, Government Venture Capital for Technology-Based Firms, Paris.

Ooghe, H., S. Manigart, and Y. Fassin, 1991, Growth Patterns in the European Venture Capital Industry, *Journal of Business Venturing 6*, 381-404.

Ordelheide, D. and KPMG, 1995, Transnational Accounting, Vol. 1, London.

Osano, H., 1996, Intercorporate Shareholdings and Corporate Control in the Japanese Firm, *Journal of Banking and Finance 20*, 1047-1068.

O'Shea, M., 1996, Government Programs for Venture Capital, OECD Working Group on Innovation and Technology Policy, Paris.

Petersen, M.A. and R.G. Rajan, 1994, The Benefits of Lending Relationships: Evidence from Small Business Data, *Journal of Finance 49*, 3-37.

Petty, W., W. Bygrave, and J. Shulman, 1994, Harvesting the Entrepreneurial Venture: A Time for Creating Value, *Journal of Applied Corporate Finance (Spring)*, 48-58.

Pfirrmann, O., U. Wupperfeld, and J. Lerner, 1996, Venture Capital and New Technology Based Firms. An US-German Comparison, Heidelberg.

Prowse, S., 1992, The Structure of Corporate Ownership in Japan, *Journal of Finance 57*, 1121-1145.

Prowse, S., 1996, Corporate Finance in International Perspective: Legal and Regulatory Influences on Financial System Development, Economic Review – Federal Reserve Bank of Dallas, Third Quarter.

Prowse, S., 1998, Angel Investor and the Market for Angel Investments, *Journal of Banking and Finance 22*, 785-792.

Rajan, R.G., 1992, Insiders and Outsiders: The Choice between Informed and Arm's-Length Debt, *Journal of Finance 47*, 1367-1400.

Rajan, R.G. and H. Servaes, 1994, The Effect of Market Conditions on Initial Public Offerings, Working Paper, University of Chicago.

Rajan, R.G. and L. Zingales, 1995, What Do We Know About Capital Structure? Some Evidence from International Data, *Journal of Finance 50*, 1421-1460.

Rehkugler, H. and A. Schenek, 2001, Underpricing oder Overpricing neuer Aktien? IPOs am deutschen Kapitalmarkt, in: Wirtz, B.W. and E. Salzer (eds.), IPO Management, Wiesbaden, 278-308.

Rindermann, G., 1996, Privatisierungen in Frankreich und ihr Einfluss auf das französische Finanzsystem, Unpublished Diploma Thesis, University of Frankfurt.

Ritter, J.R., 1984, The 'Hot Issue' Market of 1980, *Journal of Business 57*, 215-240.

Ritter, J.R., 1991, The Long-Run Performance of Initial Public Offerings, *Journal of Finance 42*, 365-394.

Roberts, E.B., 1991, High Stakes for High-Tech Entrepreneurs: Understanding Venture Capital Decision Making, *Sloan Management Review 9*, 9-20.

Rock, E.B., 2001, Greenhorns, Yankees and Cosmopolitans: Venture Capital, IPOs, Foreign Firms & U.S. Markets, Research Paper No. 01-07, Institute for Law ands Economics, University of Pennsylvania.

Röder, K., 1999, Kurswirkungen von Meldungen deutscher Aktiengesellschaften, in: Bomsdorf, E., W. Kösters, and W. Mathes (eds.), Quantitative Ökonomie 98, Lohmar/Cologne.

Rugman, A.M., 1980, Internationalization as a General Theory of FDI, *Weltwirtschaftliches Archiv 116*, 365-379.

Rybczynski, T, 1984, Industrial Financial Systems in Europe, US, and Japan, *Journal of Economical Behavior and Organization 5*, 275-286.

Sagari, S.B. and G. Guidotti, 1992, Venture Capital – Lessons from the Developed World for the Developing Markets, Discussion Paper 13, International Finance Corporation, The World Bank, Washington D.C.

Sah, R.K. and J.E. Stiglitz, 1986, The Architecture of Economic Systems: Hierarchies and Polyarchies, *American Economic Review 76*, 716-727.

Sahlman, W., 1988, Aspects of Financial Contracting in Venture Capital, *The Continental Bank Journal of Applied Corporate Finance 1*, 23-36.

Sahlman, W., 1989, Notes on Financial Contracting, Harvard Business School 9-288-014.

Sahlman, W., 1990, The Structure and Governance of Venture Capital Organizations, *Journal of Financial Economics 27*, 473-521.

Santhanakrishnan, M., 2002, The Influence of Complementarity on the Performance of Entrepreneurial Companies, Working Paper, Arizona State University, presented at the Northern Finance Association Conference in Banff, Canada.

Sapienza, H., 1992, When Do Venture Capitalists Add Value?, *Journal of Business Venturing 7*, 9-27.

Sapienza, H., S. Manigart, and W. Vermeir, 1996, Venture Capitalist Governance and Value Added in Four Countries, *Journal of Business Venturing 11*, 439-469.

Saxenian, A., 1994, Regional Advantage: Culture and Competition in Silicon Valley and Route 128, Cambridge, Mass.

Schäfer, D., 2002, How Important is the Venture Capital Industry? An Assessment Based on a Comparison of the U.S. and the German Financial Systems, Working Paper, American Institute for Contemporary Studies, Johns Hopkins University.

Schefczyk, M., 2000, Erfolgsstrategien deutscher Venture Capital Gesellschaften, Stuttgart.

Schertler, A., 2002, Comparative Advantages of Public Loan and Public Equity Schemes in Venture Capital Markets, Kiel Working Paper No. 1118, Institute of World Economics, Kiel.

Schertler, A. and M. Stolpe, 2000, Venture Mania in Europe: Its Causes and Consequences, Kiel Discussion Papers 358, Institute of World Economics, Kiel.

Scherzinger, A., 2002, die Technologiepolitik der Länder in der Bundesrepublik Deutschland – ein Überblick, DIW Discussion Paper No. 164, Berlin.

Schmeisser, W. and D. Krimphove, 2001, Vom Gründungsmanagement zum Neuen Markt: Strategien für technologieorientierte kleine und mittlere Unternehmen, Wiesbaden.

Schmidt, K., 2003, Convertible Securities and Venture Capital Finance, *Journal of Finance 58*, 1139-1166.

Schmidt, R.H., 1985, Venture-Capital aus Sicht der Finanzierungstheorie, *Betriebswirtschaftliche Forschung und Praxis 5*, 421-437.

Schmidt, R.H., 1988, Venture Capital in Deutschland – ein Problem der Qualität?, *Die Bank 4*, 184-188.

Schmidt, R.H., 2001, The Future of Banking in Europe, Working Paper Series: Finance & Accounting, No. 72, J.W. Goethe-University, Frankfurt.

Schmidt, R.H., A. Hackethal, and M. Tyrell, 1999, Disintermediation and the Role of Banks in Europe: An International Comparison, *Journal of Financial Intermediation 8*, 36-67.

Schmidt, R.H., A. Hackethal, and M. Tyrell, 2002, The Convergence of Financial Systems in Europe, in: German Financial Markets and Institutions: Selected Studies, *Schmalenbach Business Review Special Issue 1-02*, 7-53.

Schreyer, P., 2000, High-Growth Firms and Unemployment, OECD STI Working Paper 2000/03.

Schween, K., 1996, Corporate Venture Capital, Wiesbaden.

Schwienbacher, A., 2001, Innovation and Venture Capital Exits, Working Paper, University of Namur.

Schwienbacher, A., 2002, An Empirical Analysis of Venture Capital Exits in Europe, Working Paper, University of Namur.

Sharpe, W.F., 1964, Capital Asset Prices: A Theory of Market Equilibrium under Conditions of Risk, *Journal of Finance 19*, 425-442.

Sharpe, W.F., 1970, Portfolio Theory and Capital Markets, New York.

Shleifer, A. and R. Vishny, 1986, Large Shareholders and Corporate Control, *Journal of Political Economy 94*, 461-488.

Shleifer, A. and R. Vishny, 1997, A Survey of Corporate Governance, *Journal of Finance 52*, 737-783.

Shonfield, A., 1965, Modern Capitalism, London.

Sieber, C., 2002, Venture Capital in Großbritannien: Bestandsaufnahme und Vergleich mit den USA, Unpublished Diploma Thesis, University of Münster.

Siegel, R., E. Siegel, and I.C. MacMillan, 1988, Corporate Venture Capitalists: Autonomy, Obstacles, and Performance, *Journal of Business Venturing 3*, 233-247.

Simon, H.A., 1979, Rational Decision Making in Business Organizations, *American Economic Review 69*, 493-513.

Simon, H.A., 1982, Behavioral Economics and Business Organization, in: Models of Bounded Rationality, Vol. 2, Cambridge, Mass.

Simon, H.A., 1992, Bounded Rationality, in: Eatwell, J., M. Milgate and P. Newman (eds.), The New Palgrave Dictionary of Money and Finance, Vol. 1, New York, 226-227.

Sohl, J.E., 1998, The Early-Stage Equity Market in the USA, *Venture Capital 1*, 101-120.

Spence, M.A., 1973, Job Market Signaling, *Quarterly Journal of Economics 87*, 355-374.

Stedler, H., 1987, Venture Capital und geregelter Freiverkehr: Eine empirische Studie, Frankfurt/Main.

Stehle, R. and O. Erhardt, 1999, Renditen bei Börseneinführungen am deutschen Kapitalmarkt, *Zeitschrift für Betriebswirtschaft 69*, 1395-1422.

Stein, T., 2000, VCs Go Global, *Red Herring (July)*, 175-182.

Stephany, E., 2001, L'évolution des pratiques du capital-risque en France, *Revue Française de Gestion 135*, 63-75.

Stiglitz, J.E. and A. Weiss, 1981, Credit Rationing in Markets with Imperfect Information, *American Economic Review 71*, 393-409.

Storey, D.J. and B. Tether, 1996, Review of Empirical Knowledge and an Assessment of the Statistical Data on the Economic Importance of New Technology-Based Firms in Europe, Warwick.

Studenmund, A.H., 1997, Using Econometrics. A Practical Guide, 3rd edition, Reading, Mass.

Swary, I. and B. Topf, 1992, Global Financial Deregulation: Commercial Banking at the Cross Roads, Cambridge, Mass.

Tabourin, F., 1989, Le capital risqué en France, principes et bilan, *Cahier de recherche 8901*, Université Paris IX Dauphine.

TBG, 2001, Durch Beteiligung zum Markterfolg: Beteiligungskapital für kleine Technologieunternehmen (BTU-Programm), Bonn.

Teece, D., 1986, Transaction Cost Economics and the Multinational Enterprise, *Journal of Economic Behavior and Organization 7*, 21-45.

Thakor, A.V., 1996, The Design of Financial Systems: An Overview, *Journal of Banking and Finance 20*, 917-948.

Titman, S. and B. Trueman, 1986, Information Quality and the Valuation of New Issues, *Journal of Accounting and Economics 8*, 159-172.

Tobin, J., 1969, A General Equilibrium Approach to Monetary Theory, *Journal of Money, Credit, and Banking 1*, 15-29.

Tobin, J. and W. Brainard, 1968, Pitfalls in Financial Model Building, *American Economic Review 59*, 99-102.

Trester, J.J., 1998, Venture Capital Contracting Under Asymmetric Information, *Journal of Banking and Finance 22*, 675-699.

Trigeorgis, L., 1996, Real Options, Cambridge.

Tyebjee, T. and A. Bruno, 1984, A Model of Venture Capitalist Investment Activity, *Management Science (September)*, 1051-1066.

Tyebjee, T. and L. Vickery, 1988, Venture Capitalist in Western Europe, *Journal of Business Venturing 3*, 123-136.

Tykvová, T., 2000, Venture Capitalist in Germany and its Impact on Innovation, Working Paper, Center for European Economic Research, Mannheim.

Ueda, M., 2000, Bank versus Venture Capital, Working Paper, Universitat Pompeu Fabra.

Uhlir, H., 1989, Der Gang an die Börse und das Underpricing-Phänomen, *Zeitschrift für Bankrecht und Bankwirtschaft 1*, 2-16.

van Osnabrugge, M., 1998, The Financing of Entrepreneurial Firms in the UK: A Comparison of Venture Capital and Business Angels Investments Procedures, unpublished doctoral thesis, Hertford College [recited in: Mayer *et al.* (2001)].

van Osnabrugge, M. and R.J. Robinson, 2001, The Influence of Venture Capitalist's Sources of Funds, *Venture Capital 3*, 25-39.

Venture Economics, 1999, Venture Capital Association Yearbook, Newark, NJ.

w.a., 2002, Regierung bündelt Mittelstandsförderung, *Frankfurter Allgemeine Zeitung 288*, 11.12.2002, 11.

Wang, K., C. Wang, and Q. Lu, 2002, Differences in Performance of Independent and Finance-Affiliated Venture Capital Firms, *The Journal of Financial Research 25*, 59-80.

Wang, K., C. Wang, and Q. Lu, 2003, Effects of Venture Capitalists' Participation in listed Companies, *Journal of Banking and Finance 27*, 2015-2034.

Wernerfelt, B., 1984, A resource-based view of the firm, *Strategic Management Journal 5*, 171-180.

Williamson, O.E., 1975, Markets and Hierarchies: Analysis and Antitrust Implications, New York.

Williamson, O.E., 1983, Credible Commitments: Using Hostage to Support Exchange, *American Economic Review 73*, 519-540.

Williamson, O.E., 1985, Institutions and Capitalism: Firms, Markets, Relational Contracting, New York.

Williamson, O.E., 1991, Strategizing, Economizing, and Economic Organization, *Strategic Management Journal 12*, 75-94.

Wolken, J.D., 1998, "New" Data Sources for Research on Small Business Finance, *Journal of Banking and Finance 22*, 1067-1076.

World Federation of Exchanges, 2001, Annual Report, Paris.

Wright, M. and K. Robbie, 1996, Venture Capital to the Next Millenium, British Venture Capital Association, London.

Yermack, D., 1995, Do corporations award CEO stock options effectively?, *Journal of Financial Economics 39*, 237-269.

Zemke, I., 1995, Die Unternehmensverfassung von Beteiligungskapital-Gesellschaften, Wiesbaden.

Zerah, D., 1993, Le système financier français: Dix ans de mutations, Paris.

Zysman, J., 1983, Governments, Markets and Growth, Ithaca, NY.

Peter Lang · Europäischer Verlag der Wissenschaften

Christian M. Pfeil

Capital Structure, Managerial Incentives and Corporate Governance

Frankfurt am Main, Berlin, Bern, Bruxelles, New York, Oxford, Wien, 2002.
203 pp.
Development and Finance.
Edited by Manfred Nitch, Reinhard H. Schmidt and Claus-Peter Zeitinger. Vol. 9
ISBN 3-631-38574-9 / US-ISBN 0-8204-5475-3 · pb. € 39.00*

What factors determine a firm's financing decision? Informational economics and contract theory have contributed a great deal to answer this question. This book contains three essays that further contribute to this strand of literature with the focus on theories that view capital structure as a disciplining instrument for a self-interested management. Some of the existing theories abstract from other disciplining devices such as ordinary incentive wages to justify debt as a mean to mitigate a moral hazard problem between managers and owners of a firm. Two of the models presented here turn to the quetion of whether debt can play a role as an incentive device when other incentive mechanisms are available as well. A third model revisits the signaling literature on capital structure in the light of new empirical evidence. All models are embedded into a corporate governance framework that allows to set the conclusions into a broader perspective.

Contents: Capital structure · Firm financing · Mangerial incentives · Corporate governance · Financial systems · Asymmetric information · Contract theory · Financial contracting

Frankfurt am Main · Berlin · Bern · Bruxelles · New York · Oxford · Wien
Distribution: Verlag Peter Lang AG
Moosstr. 1, CH-2542 Pieterlen
Telefax 00 41 (0) 32 / 376 17 27

*The €-price includes German tax rate
Prices are subject to change without notice

Homepage http://www.peterlang.de

FSC
www.fsc.org
MIX
Papier | Fördert
gute Waldnutzung
FSC® C083411

Kolibri 360 GmbH
Ferdinand-Jühlke-Straße 7
99095 Erfurt, Deutschland
produktsicherheit@kolibri360.de

Druck:
CPI Druckdienstleistungen GmbH
im Auftrag der
Kolibri 360 GmbH
Ferdinand-Jühlke-Str. 7
99095 Erfurt